Edexcel
GCSE MATHEMATICS

HIGHER COURSE

Edexcel
Success through qualifications

WHITMORE HIGH SCHOOL
HARROW

Heinemann
Inspiring generations

About this book

This book is designed to provide you with the best possible preparation for your Edexcel GCSE Mathematics Examination. The authors are examiners and coursework moderators themselves and have a good understanding of Edexcel's requirements.

Finding your way around

To help you find your way around when you are studying and revising use the:

- **edge marks** (shown on the front pages) – these help you get to the right unit quickly;
- **contents list** – this lists the headings that identify key syllabus ideas covered in the book so you can turn straight to them. (Codes are included to show which part of the programmes of study and left-hand column of the syllabus these relate to. For example **S4d** means the content relates to Shape, space and measures section **4** understanding and using measures subsection **d**).

Remembering key ideas

We have provided clear explanations of the key ideas and techniques you need throughout the book. **Key points** you need to remember are listed in a **summary** at the end of each unit and marked like this where they appear in the units themselves:

■ **Each digit in a number has a face value and a place value.**

Exercises and examination questions

In this book questions are carefully graded so they increase in difficulty and gradually bring you up to examination standard.

- **past examination questions** are marked with an [E];
- **worked examples** and **worked examination questions** show you how to answer questions;
- **examination practice papers** – these are included to help you prepare for the examination itself;
- **answers** are included at the end of the book.

Investigations and information technology

Three units focus on particular skills required for your course and examination:

- **using and applying mathematics** (unit 11A) – shows how investigative work is assessed, and the skills required to carry out such work;
- **data handling** (unit 11B) – shows you how to approach your data handling coursework project, and how the project will be assessed;
- **calculators and computers** (unit 29) – shows you how to use a variety of methods of solving problems using calculators and computers.

17
18
19
20
21
22
23
24
25
26
27
28
29

Contents

4 Collecting and presenting data

5 Using basic number skills

10 Brackets in algebra

11A Using and applying mathematics

15 Averages and measures of spread

16 Measure and mensuration

17 Proportion

18 Graphs and higher order equations

19 Advanced mensuration

20 Simplifying algebraic expressions

24 Applying transformations to sketch graphs

25 Vectors

26 Circle theorems

Heinemann Educational Publishers
Halley Court, Jordan Hill, Oxford OX2 8EJ
a division of Reed Educational & Professional Publishing Ltd
Heinemann is a registered trademark of Reed Educational & Professional Publishing Ltd

OXFORD   MELBOURNE   AUCKLAND
JOHANNESBURG   BLANTYRE   GABORONE
IBADAN   PORTSMOUTH NH (USA)   CHICAGO

© Heinemann Educational Publishers

First published 2001

ISBN: 0435 53271 5

06   05   04
10 9 8

Designed and typeset by Tech-Set Ltd, Gateshead, Tyne and Wear
Original edition produced by Gecko Limited, Bicester, Oxon
Cover design by Macdesign Ltd
Printed and bound by Mateu, Spain

Acknowledgements

The publisher's and authors' thanks are due to Edexcel for permission to reproduce questions from past examination papers. These are marked with an [E]. The answers have been provided by the authors and are not the responsibility of Edexcel.

The publishers and authors would like to thank the following for permission to use photographs: p43 and p71 Ander McIntyre; p56 Mirror Syndication International; p98 Fiat; p106 Milepost; p115 and p119 SPL/ESA; p117, p124 and p361 SPL; p126: SPL/Colin Cuthbert; p134 SPL/Tim Davis; p162 The Anthony Blake Photo Library; p170 Britstock-IFA/ Weststock T Collicott; p177 Eye Ubiquitous/Tim Page; p228 and p565 PA News Photo Library; p270 John Walmsley; p302 SPL/Dave Roberts; p337 top and bottom Photodisc; p349 The Environmental Picture Library/John Morrison.

Publishing team	Design	Author team	
Editorial	Phil Richards	John Casson	David Kent
Sue Bennett	Colette Jacquelin	Tony Clough	Andrew Killick
Philip Ellaway	Mags Robertson	Gareth Cole	Christine Medlow
Maggie Rumble		Ray Fraser	Graham Newman
Nick Sample	**Production**	Barry Grantham	Keith Pledger
Harry Smith	Jason Wyatt	John Hackney	Sally Russell
Juliet Smith		Karen Hughes	Rob Summerson
		Trevor Johnson	John Sylvester
		Peter Jolly	Roy Woodward

Tel: 01865 888058   www.heinemann.co.uk

1 Exploring numbers

You will need to use and apply numbers in a wide variety of situations throughout your GCSE course. You also need to know many terms that describe numbers and their properties. This unit will help you remember them.

1.1 Multiples, factors and primes

■ **The factors of a number are whole numbers that divide exactly into the number. The factors include 1 and the number itself.**

■ **Multiples of a number are the results of multiplying the number by a positive whole number.**

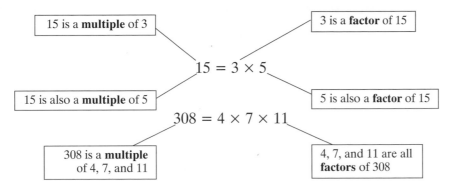

■ **A prime number is a number greater than 1 which has only two factors: itself and 1.**

The only factors of 7 are 1 and 7. 7 is a **prime number**.

Exercise 1A

1 Find all the factors of:
   (a) 24   (b) 36
   (c) 308   (d) 1001
   (e) 1400   (f) 53

2 Find all the prime numbers between 1 and 100.

3 Can the sum of two prime numbers be a prime number?
   Explain your answer.

1.2 Prime factor form

You can write any number as the product of its prime factors. These factors are prime numbers:

$$720 = 2 \times 360$$
$$= 2 \times 2 \times 180$$
$$= 2 \times 2 \times 2 \times 90$$
$$= 2 \times 2 \times 2 \times 2 \times 45$$
$$= 2 \times 2 \times 2 \times 2 \times 3 \times 15$$
$$= 2 \times 2 \times 2 \times 2 \times 3 \times 3 \times 5$$

This can be written as: $720 = 2^4 \times 3^2 \times 5$

Here 720 is in **prime factor form** / \ 2, 3, and 5 are **prime factors** of 720

■ **A number written as the product of prime numbers is written in prime factor form.**

1.3 Finding the Highest Common Factor (HCF)

Sometimes you will need to find the largest factor two numbers have in common, called the **Highest Common Factor** or **HCF**. For example, to find the highest common factor of 720 and 84:

First write each number in prime factor form:

$$720 = 2 \times 2 \times 2 \times 2 \times 3 \times 3 \times 5$$
$$84 = 2 \times 2 \times 3 \times 7$$

Then pick out the common factors: those that appear in *both* numbers. These are:

$$2 \times 2 \times 3$$

The Highest Common Factor of 720 and 84 is $2 \times 2 \times 3 = 12$

■ **The Highest Common Factor (HCF) of two numbers is the highest factor common to both of them.**

■ **Two numbers which have a Highest Common Factor of 1 are called co-prime numbers.**

Exercise 1B

1   Find the Highest Common Factor of:
   (a) 36 and 48   (b) 720 and 252   (c) 19 600 and 756

2   Find out whether these pairs of numbers are co-prime. Give a reason for each answer:
   (a) 6 and 15   (b) 81 and 111   (c) 35 and 36

Euclid's algorithm for the HCF

You may find this method of finding the HCF of two numbers useful. It is called Euclid's algorithm. It *will not be tested* in your GCSE examination.

> An **algorithm** is a method or procedure of computation. It often involves several steps.

Example 1

Find the HCF of 64 and 24.

First find the largest multiple of 24 which is less than 64:

The largest multiple is $2 \times 24 = 48$, to make 64 remainder 16.

Now find the largest multiple of the remainder 16 which is less than 24:

The largest multiple is $1 \times 16 = 16$, to make 24 remainder 8.

Find the largest multiple of the remainder 8 which is less than or equal to 16:

The largest multiple is $2 \times 8 = 16$, remainder 0.

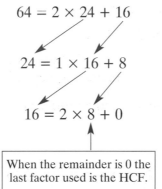

$$64 = 2 \times 24 + 16$$
$$24 = 1 \times 16 + 8$$
$$16 = 2 \times 8 + 0$$

When the remainder is 0 the last factor used is the HCF.

The HCF of 64 and 24 is 8.

Exercise 1C

1 Use Euclid's algorithm to find the HCF of:
  (a) 6 and 15   (b) 24 and 130   (c) 36 and 1960

2 Write down at least five pairs of numbers of your own choice. For each pair find the HCF by:
  ● using the prime factor form
  ● using Euclid's algorithm
  Check that both methods give the same result.

The Sigma Function: an investigation

In questions 3 to 6 $f(n)$ means the sum of the factors of n.
So $f(36) = 1 + 2 + 3 + 4 + 6 + 9 + 12 + 18 + 36 = 91$

3 Find $f(n)$ where n is:
  (a) 24   (b) 48   (c) 100
  (d) 256   (e) 32

4 Write down an expression for $f(p)$ where p is any prime number.

5 Show that $f(2^n) = 1 + 2 + 2^2 + 2^3 + \ldots + 2^n$

6 Find an expression for $f(p^n)$ where p is any prime number.

1.4 Finding the lowest common multiple (LCM)

The multiples of 6 are: 6 12 18 24 30 36 42 48 54 60 66 72 …
The multiples of 8 are: 8 16 24 32 40 48 56 64 72 80 88 96 …

6 and 8 have some
common multiples.

The Lowest Common Multiple (**LCM**) of 6 and 8 is 24.

■ **The Lowest Common Multiple of two numbers is the lowest number that is a multiple of them both.**

You can work out an LCM by looking at the prime factor forms of the two numbers.

Example 2

Find the LCM of 6 and 8.

First write the numbers in prime factor form:
$$8 = 2 \times 2 \times 2 \quad \text{and} \quad 6 = 2 \times 3$$

The LCM must contain all the prime factors of 8 and all the prime factors of 6:

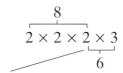

8 and 6 have a
common prime factor of 2
which is only counted once.

The LCM of 6 and 8 is $2 \times 2 \times 2 \times 3 = 24$.

Exercise 1D

1   Find the Lowest Common Multiple (LCM) of:
    (a) 12 and 15   **(b)** 36 and 16   **(c)** 50 and 85

2   A ship is at anchor between two lighthouses L and H.
    The light from L shines on the ship every 30 seconds.
    The light from H shines on the ship every 40 seconds.
    How often do both lights shine on the ship at once?

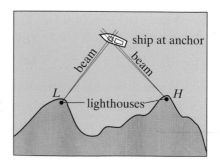

3   Write down an expression for the LCM of:
    (a) two prime numbers p and q.
    (b) two co-prime numbers n and m.

1.5 Triangular, square and cube numbers

You need to be able to recognize the following types of numbers.

Triangular numbers can be shown as a triangular pattern of dots.
Here are the first four triangular numbers:

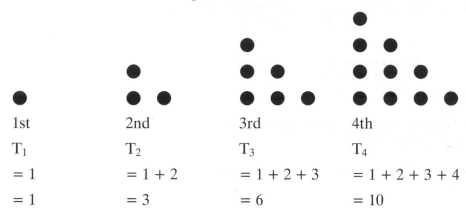

1st	2nd	3rd	4th
T_1	T_2	T_3	T_4
$= 1$	$= 1 + 2$	$= 1 + 2 + 3$	$= 1 + 2 + 3 + 4$
$= 1$	$= 3$	$= 6$	$= 10$

Square numbers can be shown as a square pattern of dots:

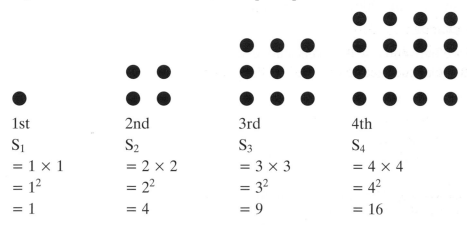

1st	2nd	3rd	4th
S_1	S_2	S_3	S_4
$= 1 \times 1$	$= 2 \times 2$	$= 3 \times 3$	$= 4 \times 4$
$= 1^2$	$= 2^2$	$= 3^2$	$= 4^2$
$= 1$	$= 4$	$= 9$	$= 16$

Cube numbers can be arranged in 3-D as a cubic pattern of dots:

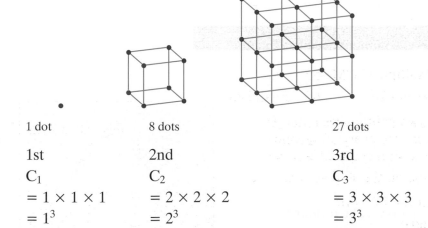

1 dot	8 dots	27 dots
1st	2nd	3rd
C_1	C_2	C_3
$= 1 \times 1 \times 1$	$= 2 \times 2 \times 2$	$= 3 \times 3 \times 3$
$= 1^3$	$= 2^3$	$= 3^3$
$= 1$	$= 8$	$= 27$

Exercise 1E

1   (a) Show that the 5th triangular number $T_5 = \frac{1}{2} \times 5 \times 6$.

    (b) Find an expression in n for the nth triangular number.

2   (a) Use the diagrams of the 3rd and 4th triangular numbers, T_3 and T_4 to show that their sum is the 4th square number, that is $S_4 = T_3 + T_4$.

    (b) What is the connection between the nth triangular number, the $(n + 1)$th triangular number and the nth square number?

3   (a) Write down a table of square numbers from the first to the tenth.

    (b) Find two square numbers which add to give a square number.

    (c) Repeat part (b) for at least two other pairs of square numbers.

4   Explain whether:

    (a) 441 is a square number   (b) 2001 is a square number

    (c) 1007 is a square number.

5   Show that the difference between any two consecutive square numbers is an odd number.

6   Show that the difference between:

    (a) the 7th square number and the 4th square number is a multiple of 3

    (b) the 8th square number and the 5th square number is a multiple of 3

    (c) the 11th square number and the 7th square number is a multiple of 4.

    (d) generalise the statement implied in parts (a), (b) and (c).

    (e) 64 is equal to the 8th square number $S_8 = 8^2$
         64 is equal to the 4th cube number $C_4 = 4^3$

         Find other cube numbers which are also square numbers.

         If you can, make a general comment about such cube numbers.

1.6 Index numbers

Section 1.5 shows that

$$3 \times 3 = 3^2 = 9 \text{ is a } \textbf{square number}$$

and

$$4 \times 4 \times 4 = 4^3 = 64 \text{ is a } \textbf{cube number}$$

because they can be shown as square and cubic patterns of dots.

We cannot draw dot patterns which go beyond three dimensions but we can extend the notation for multiplying numbers like this:

$$3 \times 3 \times 3 \times 3 = 3^4 = 81$$

we say that 81 is **3 raised to the power 4**, or 81 is **3 to the fourth**.

Also:

$$2 \times 2 \times 2 \times 2 \times 2 = 2^5 = 32$$

we say that 32 is **2 raised to the power 5**, or 32 is **2 to the fifth**.

These are examples of **index numbers**.

32 written in **index form** is 2^5.

■ **The power a number is raised to is called the index** (plural **indices**).

Example 3

(a) Calculate 4^7

(b) Write 243 in index form.

For part **(a)**:

$$4^7 = 4 \times 4 \times 4 \times 4 \times 4 \times 4 \times 4$$
$$= 16\,384$$

For part **(b)**:

$$243 = 3 \times 81$$
$$= 3 \times 9 \times 9$$
$$= 3 \times 3 \times 3 \times 3 \times 3$$
$$= 3^5$$

Exercise 1F

1 Calculate:
   (a) 2^6 (b) 10^5
   (c) 5^4 (d) 7^3
   (e) 12^4 (f) $(0.9)^5$

2 Write in index form:
   (a) 1000   (b) 125
   (c) 512   (d) 2401
   (e) 625   (f) 19 683

3 Show that 64 can be written as either 2^6 or 4^3.
   Explain why this is the case.

4 Which is the greater, and by how much:
   2^5 or 5^2

General form for index numbers

We can write about index numbers in a general way. For any number a three lots of a multiplied together can be written:

$$a \times a \times a = a^3$$

and n lots of a multiplied together can be written:

$$a \times a \times a \times \ldots \times a = a^n$$
$$\longleftarrow n \text{ lots of } a \longrightarrow$$

■ **A number written in the form a^n is written in index form.**

1.7 The rules for indices

We shall now look at the rules for manipulating numbers written in index form.

Multiplication

$3^2 \times 3^3$ is:

$$(3 \times 3) \times (3 \times 3 \times 3) = 3 \times 3 \times 3 \times 3 \times 3$$
$$= 3^5$$

so:
$$3^2 \times 3^3 = 3^5 = 3^{2+3}$$

So to multiply two powers of the same number you add the indices. The multiplication rule is:

■ $a^n \times a^m = a^{n+m}$

Division

$2^5 \div 2^3$ is:

$$\frac{2 \times 2 \times 2 \times 2 \times 2}{2 \times 2 \times 2}$$

which simplifies to:

$$2 \times 2 = 2^2$$

so:
$$2^5 \div 2^3 = 2^2 = 2^{5-3}$$

So to divide two powers of the same number subtract the indices. The division rule is:

■ $a^n \div a^m = a^{n-m}$ or $\dfrac{a^n}{a^m} = a^{n-m}$

Power of zero

Notice that:

$$1 = \frac{8}{8} = \frac{2^3}{2^3} = 2^{3-3} = 2^0$$

The 2 could be replaced by any other non-zero number a giving the same result:

■ $a^0 = 1$ (where $a \neq 0$)

Negative powers

Sometimes numbers are raised to a negative power. For example:

$$2^2 \div 2^5 = 2^{2-5} = 2^{-3} \quad \text{and}$$

$$2^2 \div 2^5 = \frac{2 \times 2}{2 \times 2 \times 2 \times 2 \times 2} = \frac{1}{2 \times 2 \times 2} = \frac{1}{2^3}$$

So:

$$2^{-3} = \frac{1}{2^3} = \frac{1}{2 \times 2 \times 2} = \frac{1}{8}$$

In general:

■ $a^{-n} = \dfrac{1}{a^n}$ (where $a \neq 0$)

Powers raised to powers

$$(2^3)^2 = (2 \times 2 \times 2)^2$$
$$= (2 \times 2 \times 2) \times (2 \times 2 \times 2)$$
$$= 2 \times 2 \times 2 \times 2 \times 2 \times 2 = 2^6$$

So:

$$(2^3)^2 = 2^{3 \times 2} = 2^6$$

To raise a power to a power multiply the indices together.
The rule is:

■ $(a^n)^m = a^{n \times m}$

Example 4

Calculate each of the following using the multiplication, division and power rules for indices. Give your answers both in index form and without using indices where possible.

(a) $2^5 \times 2^3$ (b) $3^7 \div 3^5$

(c) $(5^3)^2$ (d) 2^{-3}

(e) 117^0 (f) $2^2 + 2^3$

Answer

(a) $2^5 \times 2^3 = 2^{5+3} = 2^8 = 256$

(b) $3^7 \div 3^5 = 3^{7-5} = 3^2 = 9$

(c) $(5^3)^2 = 5^{3 \times 2} = 5^6 = 15\ 625$

(d) $2^{-3} = \dfrac{1}{2^3} = \dfrac{1}{2 \times 2 \times 2} = \dfrac{1}{8} = 0.125$

(e) $117^0 = 1$

(f) $2^2 + 2^3 = 4 + 8 = 12$

Exercise 1G

1 Work out each of the following, giving your answer where possible both in index form and without using indices.

(a) $5^2 \times 5^4$ (b) $5^{10} \div 5^2$ (c) $10^3 \div 10$

(d) $12^2 \times 12$ (e) $3^3 \times 3^4$ (f) 6^0

(g) $(1.3)^0$ (h) $\dfrac{1}{25}$ (i) $6^{10} \div 6^9$

(j) $5^3 + 5^2$ (k) $2^1 + 2^2 + 2^3$ (l) $(4^3)^2$

(m) $(5^3)^3$ (n) $(2^5)^2$ (o) $\dfrac{3^{-2}}{3^2}$

(p) $\dfrac{4^3}{4^{-2}}$ (q) $\dfrac{6}{6^2}$ (r) $5^2 \div 5^{-3}$

(s) $\dfrac{2^3 \times 2^5}{(2^2)^3}$ (t) $\dfrac{(4^2)^3}{4^2 \times 4^3}$ (u) $(5^{-2})^3$

(v) $(3^2)^{-2}$ (w) $\dfrac{2^3 \times 2^{-3}}{(2^2)^2}$ (x) $\dfrac{(3^{-2})^3}{3^{-2} \times 3^{-6}}$

2 Investigate to find possible solutions of $n^m = m^n$.

1.8 Fractional indices

You should know that:
$$\sqrt{9} = 3$$
so:
$$\sqrt{9} \times \sqrt{9} = 3 \times 3 = 9 = 9^1$$

We can use this to evaluate $9^{\frac{1}{2}}$

Using the multiplication rule:
$$9^{\frac{1}{2}} \times 9^{\frac{1}{2}} = 9^{\frac{1}{2} + \frac{1}{2}} = 9^1$$

But $\sqrt{9} \times \sqrt{9} = 9^1$ so $9^{\frac{1}{2}} = \sqrt{9}$

or $9^{\frac{1}{2}} = 3$

Similarly:
$$8^{\frac{1}{3}} \times 8^{\frac{1}{3}} \times 8^{\frac{1}{3}} = 8^{\frac{1}{3} + \frac{1}{3} + \frac{1}{3}} = 8^1$$

But $\sqrt[3]{8} \times \sqrt[3]{8} \times \sqrt[3]{8} = 8$ so $8^{\frac{1}{3}} = \sqrt[3]{8}$

or $8^{\frac{1}{3}} = 2$

In general:

■ $a^{\frac{1}{n}} = \sqrt[n]{a}$

cube roots
The cube root of 8 is 2 because
$$2 \times 2 \times 2 = 8$$
The cube root is written:
$$\sqrt[3]{8}$$

Example 5

Work out:

(a) $8^{\frac{1}{3}}$ (b) $8^{\frac{2}{3}}$ (c) $8^{-\frac{1}{3}}$

Answer

(a) $8^{\frac{1}{3}} = \sqrt[3]{8} = 2$ because $8 = 2 \times 2 \times 2$

(b) $8^{\frac{2}{3}} = 8^{\frac{1}{3}} \times 8^{\frac{1}{3}} = 2 \times 2 = 4$

(c) $8^{-\frac{1}{3}} = \dfrac{1}{8^{\frac{1}{3}}} = \dfrac{1}{2}$

Example 6

Work out:

(a) $64^{\frac{1}{3}}$ (b) $64^{-\frac{2}{3}}$

Answer

(a) $64^{\frac{1}{3}} = \sqrt[3]{64} = 4$ because $4 \times 4 \times 4 = 64$

(b) $64^{-\frac{2}{3}} = 64^{-\frac{1}{3}} \times 64^{-\frac{1}{3}} = \dfrac{1}{64^{\frac{1}{3}}} \times \dfrac{1}{64^{\frac{1}{3}}}$

$= \dfrac{1}{4} \times \dfrac{1}{4} = \dfrac{1}{16}$

so $64^{-\frac{2}{3}} = \dfrac{1}{64^{\frac{2}{3}}} = \dfrac{1}{16}$

Exercise 1H

1   Work out:

    (a) $4^{\frac{1}{2}}$ (b) $4^{-\frac{1}{2}}$ (c) $4^{\frac{3}{2}}$

    (d) $125^{-\frac{1}{3}}$ (e) $16^{\frac{1}{4}}$ (f) $16^{-\frac{1}{4}}$

    (g) $16^{-\frac{3}{4}}$ (h) $25^{-\frac{1}{2}}$ (i) $100^{-\frac{1}{2}}$

    (j) $1000^{\frac{1}{3}}$ (k) $25^{\frac{3}{2}}$ (l) $8^{-\frac{2}{3}}$

2   Work out the following, writing your answers where possible in both index form and without using indices:

    (a) $2^3 \times 2^7$ (b) $3^4 \div 3^2$

    (c) 12^0 (d) $\dfrac{(5^2)^3}{5^6}$

    (e) $3^0 + 3 + 3^2 + 3^3$

    (f) $(4^2)^{-3}$ (g) $\dfrac{10^3}{10^5}$ (h) $(5^{-2})^{-2}$

3   Work out:

    (a) $27^{\frac{1}{3}}$ (b) $25^{\frac{1}{2}}$ (c) $49^{-\frac{1}{2}}$

    (d) $(8^{-\frac{1}{3}})^{-2}$ (e) $16^{\frac{1}{4}} \times 32^{\frac{2}{5}}$ (f) $1000^{-\frac{2}{3}}$

1.9 Powers of 2 and 10

You should be able to recall very quickly the powers of 2 and the powers of 10.

Remember 2^3 means $2 \times 2 \times 2$ and this equals 8

whilst $2^{-3} = \dfrac{1}{2^3} = \dfrac{1}{8}$

So the table for the powers of 2 from 2^{-5} to 2^5 is:

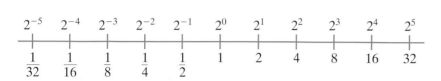

In a similar manner,

$$10^3 = 10 \times 10 \times 10 = 1000 \text{ and}$$

$$10^{-3} = \frac{1}{10^3} = \frac{1}{1000} = 0.001$$

So the table of values for the powers of 10 from 10^{-5} to 10^5 is:

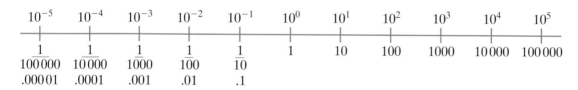

Example 7

Work out the values of:

(a) 2^6 (b) 2^{10} (c) 2^{-10}

(d) 10^7 (e) 10^{-7} (f) $2^2 \times 10^3$

Answer

(a) $2^6 = 2 \times 2 \times 2 \times 2 \times 2 \times 2 = 64$

(b) $2^{10} = 2 \times 2 \times 2 \times 2 \times 2 \times 2 \times 2 \times 2 \times 2 \times 2 = 1024$

(c) $2^{-10} = \dfrac{1}{2^{10}} = \dfrac{1}{1024}$

(d) $10^7 = 10 \times 10 \times 10 \times 10 \times 10 \times 10 \times 10 = 10\,000\,000$

(e) $10^{-7} = \dfrac{1}{10^7} = 0.000\,0001$

(f) $2^2 \times 10^3 = (2 \times 2) \times (10 \times 10 \times 10)$
$$= 4 \times 1000$$
$$= 4000$$

Example 8

Solve the equation

$$2^{3n-1} = 64$$

First recognise that $64 = 2^6$ so:

$$2^{3n-1} = 2^6$$

Hence, equating the indices:

$$3n - 1 = 6$$
$$3n = 7$$
$$n = \tfrac{7}{3} \quad \text{or} \quad n = 2\tfrac{1}{3}$$

Example 9 (Summing a series of powers)

Find an expression for the sum of the series

$$1 + 10 + 10^2 + 10^3 + \ldots + 10^n$$

The starting point of this will generalise for other similar series.

Write

$$S = 1 + 10 + 10^2 + 10^3 + \ldots + 10^n$$

then multiply by 10 so

$$10S = 10 + 10^2 + 10^3 + \ldots + 10^{n+1}$$
$$\text{and} \quad S = 1 + 10 + 10^2 + \ldots + 10^n$$

subtract

$$10S = 10 + 10^2 + 10^3 + \ldots + 10^{n+1}$$
$$-S = -1 - 10 - 10^2 - \ldots - 10^n$$

$$9S = 10^{n+1} - 1$$
$$S = \frac{10^{n+1} - 1}{9}$$

Exercise 1I

1   Work out the values of
    (a) 2^3 (b) 2^{-2} (c) 10^{-4} (d) 10^{10} (e) $2^3 \times 10^5$

2   Round 2^{10} to the nearest 100

3   For values of x from -3 to 5, draw the graphs of
    (a) $y = 2^x$ (b) $y = 10^x$

4   Solve the equations
    (a) $10^{5n-2} = 100$ (b) $2^{5-2n} = 8$ (c) $10^{2n+3} = 0.00001$

5   Find an expression for the sum of the series
    $$1 + 2 + 2^2 + 2^3 + \ldots + 2^n.$$

Exercise 1 J (Mixed questions)

1 For any positive whole number n its **Tau Function** $\tau(n)$ is defined as the number of positive whole number factors of n.

7 is a prime number. It has two factors, 1 and 7, so $\tau(7) = 2$

(a) Show that if p is *any* prime number then $\tau(p) = 2$

(b) For any prime number p and any positive value of n find an expression for $\tau(p^n)$.

(c) *Investigation*

$6 \times 7 = 42$

The factors of 42 are 1, 2, 3, 6, 7, 14, 21 and 42.

So 42 has 8 factors and $\tau(42) = 8$

$6 \times 7 = 42$ and $\tau(42) = 4 \times 2 = \tau(6) \times \tau(7)$

So $\tau(6 \times 7) = \tau(6) \times \tau(7)$

Investigate to see whether or not $\tau(n \times m) = \tau(n) \times \tau(m)$ for various values of n and m.

Remember $\tau(6) = 4$ and $\tau(7) = 2$ from above.

2 One way of making the number 5 by adding ones and threes is $5 = 3 + 1 + 1$

Another different way is $5 = 1 + 3 + 1$

Investigate the number of different ways of making any number by adding ones and threes.

3 $1^3 + 2^3 + 3^3 = 1 + 8 + 27 = 36 = 6^2 = (1 + 2 + 3)^2$

$1^3 + 2^3 + 3^3 + 4^3 = 1 + 8 + 27 + 64 = 100 = 10^2$
$$= (1 + 2 + 3 + 4)^2$$

Investigate this situation further.

4 The Last Digit of 146 is 6. This is written as LD (146) = 6

What comments can you make about:

(a) LD $(n \times m)$ (b) LD (any square number)

(c) Show that $10n + 7$ can never be a square number for any positive whole number value of n.

5 Solve each of these equations

(a) $2^{5n - 2} = 1024$

(b) $2^{3n + 1} = 0.125$

(c) $10^{4n + 3} = 0.000\,000\,1$

6 Find an expression for the sum of the series

$1 + 5 + 25 + 125 + 625 + \ldots + 5^n$

1.10 Mixed fractions: the four rules

A mixed fraction is made up of a whole number and a proper fraction.

Example 10 (addition)

$$2\tfrac{2}{3} + 1\tfrac{1}{5}$$

add the whole number $\quad 2 + 1 = 3$

now add $\qquad\qquad \tfrac{2}{3} + \tfrac{1}{5} = \tfrac{10}{15} + \tfrac{3}{15} = \tfrac{13}{15}$

so $\qquad\qquad\quad 2\tfrac{2}{3} + 1\tfrac{1}{5} = 3\tfrac{13}{15}$

Example 11 (subtraction)

$$3\tfrac{1}{12} - 1\tfrac{3}{8}$$

For a subtraction, it is usually best to express both mixed numbers as a 'top heavy' fraction.

$$3\tfrac{1}{12} = \tfrac{37}{12} \qquad 1\tfrac{3}{8} = \tfrac{11}{8}$$

Now express $\tfrac{37}{12}$ and $\tfrac{11}{8}$ in terms of fractions with the same numerator. In the case of 12 and 8 the 'common' denominator is the LCM of 12 and 8, that is 24

$$\tfrac{37}{12} = \tfrac{74}{24} \quad \text{and} \quad \tfrac{11}{8} = \tfrac{33}{24}$$

so $\qquad 3\tfrac{1}{12} - 1\tfrac{3}{8} = \tfrac{37}{12} - \tfrac{11}{8} = \tfrac{74}{24} - \tfrac{33}{24} = \tfrac{41}{24} = 1\tfrac{17}{24}$

Example 12 (multiplication)

$$1\tfrac{1}{3} \times 2\tfrac{3}{5}$$

$$1\tfrac{1}{3} = \tfrac{4}{3} \qquad 2\tfrac{3}{5} = \tfrac{13}{5}$$

so $\qquad 1\tfrac{1}{3} \times 2\tfrac{3}{5} = \tfrac{4}{3} \times \tfrac{13}{5}$

$$= \tfrac{52}{15}$$

$$= 3\tfrac{7}{15}$$

Example 13 (division)

$$2\tfrac{1}{4} \div 1\tfrac{3}{5}$$

$$2\tfrac{1}{4} = \tfrac{9}{4} \qquad 1\tfrac{3}{5} = \tfrac{8}{5}$$

$$2\tfrac{1}{4} \div 1\tfrac{3}{5} = \tfrac{9}{4} \div \tfrac{8}{5}$$

$$= \tfrac{9}{4} \times \tfrac{5}{8}$$

$$= \tfrac{45}{32}$$

$$= 1\tfrac{13}{32}$$

Example 14

Work out
$$\frac{1\frac{3}{5} - \frac{1}{4}}{2\frac{1}{2} + 1\frac{1}{5}}$$

$$1\frac{3}{5} - \frac{1}{4} = \frac{8}{5} - \frac{1}{4}$$
$$= \frac{32}{20} - \frac{5}{20}$$
$$= \frac{27}{20}$$

$$2\frac{1}{2} + 1\frac{1}{5} = 3 + \left(\frac{1}{2} + \frac{1}{5}\right)$$
$$= 3 + \frac{5}{10} + \frac{2}{10}$$
$$= 3\frac{7}{10}$$
$$= \frac{37}{10}$$

So
$$\frac{1\frac{3}{5} - \frac{1}{4}}{2\frac{1}{2} + 1\frac{1}{5}} = \frac{\frac{27}{10}}{\frac{37}{10}} \text{ or } \frac{27}{20} \div \frac{37}{10}$$
$$= \frac{27}{20} \times \frac{10}{37} \qquad \text{cancel through}$$
$$= \frac{27}{20_2} \times \frac{10^{1}}{37} \qquad \text{by 10}$$
$$= \frac{27 \times 1}{2 \times 37}$$
$$= \frac{27}{74}$$

Exercise 1 K

1 Work out

(a) $1\frac{1}{4} + 3\frac{2}{5}$ **(b)** $1\frac{1}{4} \times 3\frac{2}{5}$ **(c)** $3\frac{1}{4} - 1\frac{1}{2}$

(d) $\dfrac{\frac{4}{7} + \frac{1}{3}}{\frac{1}{4} + \frac{1}{5}}$ **(e)** $3\frac{2}{5} \div 1\frac{1}{4}$ **(f)** $\dfrac{1\frac{1}{2} - \frac{1}{3}}{\frac{3}{4}}$

(g) $\dfrac{3\frac{1}{2} + 2\frac{1}{5}}{\frac{1}{3} - \frac{1}{4}}$ **(h)** $\dfrac{4\frac{2}{5} - 3\frac{1}{4}}{3\frac{1}{2} \div \frac{1}{4}}$ **(i)** $\dfrac{\frac{1}{3}\left(\frac{2}{5} - \frac{1}{4}\right)}{\frac{3}{7}}$

(i) $\dfrac{4\frac{1}{2} + 2\frac{1}{3}}{5\frac{1}{7} - 3\frac{1}{2}}$

2 A common error made in adding fractions is to add the tops and add the bottoms

that is $\frac{2}{3} + \frac{1}{5} \rightarrow \frac{2+1}{3+5} \rightarrow \frac{4}{8}$ [This is wrong]

Investigate whether this error can ever lead to a correct result.

Summary of key points

1 A **factor** is a whole number which divides into another whole number exactly. The factors include 1 and the number itself.

2 A number is a **multiple** of any number that divides into it exactly.

3 A **prime number** is a number greater than 1 which has only two factors: itself and 1.

4 When a number is written as the product of its prime number factors it is in **prime factor form**.

5 The **Highest Common Factor (HCF)** of two numbers is the highest factor common to them both.

6 Two numbers which have a Highest Common Factor of 1 are called **co-prime numbers**.

7 The **Lowest Common Multiple** of two numbers is the lowest number that is a multiple of them both.

8 The following types of numbers can be shown visually using dot patterns:

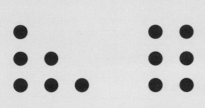

triangular numbers   square numbers   cube numbers

9 The power a number is raised to is called the index (plural indices).

10 A number written in the form a^n is written in index form, where n is the index.

11 $a^n \times a^m = a^{n+m}$

12 $a^n \div a^m = a^{n-m}$ or $\dfrac{a^n}{a^m} = a^{n-m}$

13 $a^0 = 1$ (where $a \neq 0$)

14 $a^{-n} = \dfrac{1}{a^n}$ (where $a \neq 0$)

15 $(a^n)^m = a^{n \times m}$

16 $a^{\frac{1}{n}} = \sqrt[n]{a}$

17 A **mixed fraction** is made up of a whole number and a proper fraction.

2 Solving equations and inequalities

2.1 Solving simple equations

Sometimes you will need to solve equations such as $4x + 1 = 7$. Solving means finding the value or values of x that make the equation true.

To solve an equation like this rearrange it so that the x term is on its own on the left hand side of the equation. The x term in $4x + 1 = 7$ is $4x$.

■ **To rearrange an equation you can:**
- **add the same quantity to both sides**
- **subtract the same quantity from both sides**
- **multiply both sides by the same quantity**
- **divide both sides by the same quantity.**

■ **Whatever you do to one side of an equation you must do to the other side.**

Here are two different ways of solving the equation $4x + 1 = 7$

Method 1

$$4x + 1 = 7$$

subtract 1 from both sides: $\qquad 4x = 7 - 1$

$$4x = 6$$

divide both sides by 4: $\qquad x = 1\tfrac{1}{2}$

Check by substituting $1\tfrac{1}{2}$ in place of x in $4x + 1 = 7$:

$$(4 \times 1\tfrac{1}{2}) + 1 = 6 + 1 = 7$$

so $1\tfrac{1}{2}$ is a solution.

Another way of thinking about this is to imagine $+1$ moving from the left side to the right side and becoming -1.

■ **You can also use a flow diagram to solve an equation.**

Method 2: Use a flow diagram to build the expression $4x + 1$

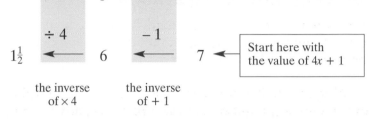

As this is equal to 7 you can use a reverse flow diagram to find the value of x.
Start at the right hand side and reverse the steps in the arrows:

$$1\tfrac{1}{2} \xleftarrow{\div 4} 6 \xleftarrow{-1} 7 \longleftarrow \boxed{\begin{array}{l}\text{Start here with}\\ \text{the value of } 4x + 1\end{array}}$$

the inverse   the inverse
of × 4   of + 1

so $x = 1\tfrac{1}{2}$

Exercise 2A

Solve these equations and check your answers.

1	$3x + 2 = 17$	**2**	$5x - 2 = 18$	**3**	$2x + 5 = 10$
4	$5x - 1 = 12$	**5**	$7x = 3$	**6**	$4x + 3 = 3$
7	$5 + 6x = 10$	**8**	$9x - 4 = 3$	**9**	$2x + 7 = 1$
10	$3a + 8 = 2$	**11**	$4b + 17 = 7$	**12**	$5c - 1 = -13$
13	$8d + 7 = 2$	**14**	$8 + 10e = 3$	**15**	$10f + 7 = 3$

When the 'unknown' x is on both sides

$$5x - 4 = 3x - 10$$

In this equation there is an x term on both sides. To solve this
rearrange the equation with all the x terms on one side:

Example 1

Solve $5x - 4 = 3x - 10$

subtract $3x$ from both sides: $\quad 5x - 3x - 4 = -10$

$2x - 4 = -10$

add 4 to both sides: $\quad\quad\quad 2x = -10 + 4$

$2x = -6$

divide both sides by 2: $\quad\quad\quad x = -3$

Another way of thinking
about this is to imagine:

$3x$ moving from right side to
left and becoming $-3x$

-4 moving from left side to
right and becoming $+4$

Check by substituting -3 in place of x in $5x - 4 = 3x - 10$

left side: $\quad (5 \times -3) - 4 \ = -15 - 4 \ = \ -19$

right side: $\quad (3 \times -3) - 10 \ = -9 - 10 \ = \ -19$

Exercise 2B

Solve these equations and check your answers.

1   $2a + 5 = a + 9$ 2 $6b - 5 = 4b + 1$

3   $6c + 1 = 4c + 2$ 4 $2d + 3 = 7d - 1$

5   $5e + 1 = 2e - 5$ 6 $3f - 2 = 8f + 3$

7   $4g - 5 = 8g - 11$ 8 $7h - 3 = 4h + 8$

9   $2j - 9 = 8j - 1$ 10 $9k + 8 = 4k - 3$

When the x term is negative

$$5 - 3x = 7$$

Here the number multiplying the unknown x is negative. This number is called a **coefficient.**

Here are three ways of solving the equation:

Method 1	**Method 2**
Solve $5 - 3x = 7$	Solve $5 - 3x = 7$
add $3x$ to both sides: $5 = 3x + 7$	subtract 5 from both sides: $-3x = 2$
subtract 7 from both sides: $-2 = 3x$	divide both sides by -3: $x = -\frac{2}{3}$
divide both sides by 3 $x = -\frac{2}{3}$	Check the solution by substituting $-\frac{2}{3}$ in place of x in $5 - 3x = 7$
Check the solution by substituting $-\frac{2}{3}$ in place of x in $5 - 3x = 7$	$$5 - (3 \times -\tfrac{2}{3}) = 5 + 2 = 7$$
$$5 - (3 \times -\tfrac{2}{3}) = 5 + 2 = 7$$	

Method 1 is very useful in the algebra work you will do elsewhere in this book.

Method 3: Use a flow diagram to build the expression $5 - 3x$

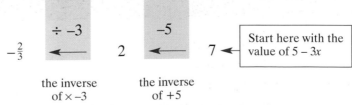

As this is equal to 7 you can use a reverse flow diagram to find the value of x. Start at the right hand side and reverse the steps in the arrows:

So $x = -\frac{2}{3}$

Exercise 2C

Solve these equations. Check your answers by substituting them back in the equations.

1	$6 - x = 5$	**2**	$8 - x = 10$
3	$7 - 2y = 3$	**4**	$4 - 3y = 16$
5	$-2a = 12$	**6**	$-5b = 13$
7	$5 - 2c = 2$	**8**	$2 - 6d = 1$
9	$5 - 8e = 7$	**10**	$3 - 5f = 12$

When the x term includes a fraction

$$\tfrac{1}{4}x - 5 = 2$$

Here the coefficient of x is a fraction. Here are two ways of solving the equation:

Method 1

Solve $\tfrac{1}{4}x - 5 = 2$

add 5 to both sides: $\qquad\qquad \tfrac{1}{4}x = 2 + 5$

$\qquad\qquad\qquad\qquad\qquad \tfrac{1}{4}x = 7$

multiply both sides by 4: $\qquad x = 28$

-5 moves from left side to right side and becomes $+5$

Method 2

Solve $\tfrac{1}{4}x - 5 = 2$

multiply both sides by 4: $\quad x - 20 = 8$

add 20 to both sides: $\qquad\quad x = 8 + 20$

$\qquad\qquad\qquad\qquad\qquad\; x = 28$

-20 moves from left side to right side and becomes $+20$

Check by substituting 28 for x in $\tfrac{1}{4}x - 5 = 2$:

$$\tfrac{1}{4}(28) - 5 = 7 - 5 = 2$$

So x is 28.

Exercise 2D

Solve these equations. Check your answers.

1	$\tfrac{1}{3}x - 4 = 2$	**2**	$\tfrac{1}{2}x + 7 = 4$	**3**	$\tfrac{1}{4}x - 5 = 1$
4	$\tfrac{3}{4}x + 2 = 9$	**5**	$\tfrac{2}{3}x - 4 = -1$	**6**	$9 - \tfrac{1}{2}x = 2$
7	$7 - \tfrac{1}{3}x = 9$	**8**	$6 - \tfrac{1}{4}x = 9$	**9**	$5 - \tfrac{2}{3}x = 1$
10	$3 - \tfrac{3}{4}x = 8$				

Unknown on both sides and negative coefficients

$$5x + 4 = 2 - 3x$$

Here the 'unknown' x is on both sides and one of the x terms is negative: $-3x$

To solve this type of equation rearrange it so that the unknown appears only on one side and is positive.

Example 2

Solve $5x + 4 = 2 - 3x$

add $3x$ to both sides:

$$5x + 3x + 4 = 2$$
$$8x + 4 = 2$$

subtract 4 from both sides:

$$8x = 2 - 4$$
$$8x = -2$$

divide both sides by 8:

$$x = -\frac{2}{8} = -\frac{1}{4}$$

$-3x$ moves from right side to left and becomes $+3x$

$+4$ moves from left to right and becomes -4

Example 3

Solve $1 - 5x = 10 - 2x$ Here both x terms are negative.

add $5x$ to both sides:

$$1 = 10 - 2x + 5x$$
$$1 = 10 + 3x$$

subtract 10 from both sides:

$$1 - 10 = 3x$$
$$-9 = 3x$$

divide both sides by 3:

$$-3 = x$$

So $x = -3$

Exercise 2E

Solve these equations. Check your answers.

1 $4x + 3 = 8 - x$
2 $3x + 4 = 7 - 3x$
3 $2 - 5x = 3x + 14$
4 $9 - 2x = 3 - 4x$
5 $7 - 8a = 4 - 3a$
6 $5 - 6b = 8 - 2b$
7 $9 - 7c = 3 - 2c$
8 $5d - 7 = 2 - 4d$
9 $9 - 4e = 9 + 3e$
10 $8 - 3f = 1 - 6f$

Exercise 2F   (Mixed questions)

Solve these equations. Check your answers.

1 $7x - 2 = 12$ **2** $5x + 9 = 5$

3 $4x - 3 = -4$ **4** $8x + 7 = 6x + 10$

5 $3x - 8 = 7x - 12$ **6** $9a - 5 = 7a + 11$

7 $9 - 4b = 6$ **8** $-6c = 9$

9 $-3d = -2$ **10** $8 - 3e = 8$

11 $\frac{1}{4}f + 3 = 8$ **12** $9 - \frac{1}{3}g = 13$

13 $2 - 7h = 3h$ **14** $1 - 4j = -2j$

15 $7 - 4k = -2$ **16** $6 - 5m = m - 9$

17 $4 - 3n = 2n - 7$ **18** $8 - 7p = 12 - 5p$

19 $9 - 6q = 12 - 2q$ **20** $4 - 8r = 2 - 5r$

2.2 Using equations to help solve problems

Example 4

The sizes of the angles of a triangle are $x + 20°$, $x + 50°$ and $x - 10°$. Find the size of the smallest angle of the triangle.

To set up the equation, you use the fact that the sum of the angles of a triangle is $180°$.

$$x + 20° + x + 50° + x - 10° = 180°$$
$$3x + 60° = 180°$$
$$3x = 120°$$
$$x = 40°$$

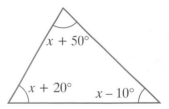

$x - 10°$ is the smallest angle.

The size of the smallest angle $= 40° - 10° = 30°$.

Example 5

I multiply a mystery number by 3 and add 11. The result is the same as when I multiply the number by 7 and subtract 9. Find the number.

What number am I thinking of?

You can represent this problem as an equation using x as the mystery number. Then:

$$3x + 11 = 7x - 9$$
$$11 = 4x - 9$$
$$20 = 4x$$
$$x = 5$$

The mystery number is 5.

Exercise 2G

1 The sum of three numbers $a + 3$, $2a + 5$ and $5a - 3$ is 37.
   Find the value of a.

2 In the diagram PQR is a straight line.
   Find the size of angle PQS.

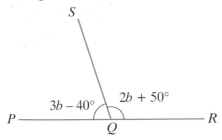

3 The diagram shows three angles at a point.
   Find the size of each of the angles.

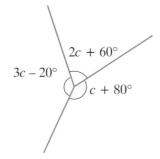

4 The lengths, in centimetres, of the sides of a triangle are
   $2d + 5$, $4d - 1$ and $d + 6$. The perimeter of the triangle is
   45 cm. Find the length of each of its sides.

5 The lengths, in centimetres, of the sides of a quadrilateral are
   $5e + 8$, $3e + 4$, $2e + 5$ and $4e - 1$. The perimeter of the
   quadrilateral is 44 cm. Find the value of e.

6 I multiply a number by 7 and add 5. The result is 89.
   Find the number. (Hint: $7x + 5 = 89$)

7 I multiply a number by 4 and add 9. The result is the same as
   when I multiply the number by 6 and subtract 13.
   Find the number.

8 I multiply a number by 6 and subtract 5. The result is the same
   as when I multiply the number by 3 and subtract it from 58.
   Find the number.

9 The sum of three consecutive integers is 72. Find the smallest
   of these integers.

10 Mrs Jones is 24 years older than her daughter. The sum of their
   ages is 70 years. How old is Mrs Jones?

11 Philip is three times as old as his daughter Louise. The sum of their ages is 72 years. How old is Philip?

12 Harry and Sally live 21 miles apart. One day they leave their houses at noon. Harry walks towards Sally's at 4 mph and Sally cycles towards Harry's at 10 mph. At what time do they meet?

13 Alan buys some 'Choc bars' which cost 24p each. He pays for them with three £1 coins. Baljit buys the same number of 'Super Choc bars' which cost 34p each. He pays for them with four £1 coins. They each receive the same amount of change. How many bars did each person buy?

14 One of the angles of an isosceles triangle is 96°. Find the sizes of the other two angles.

15 Fazal has £2.60 and Gareth has £5.30. Gareth gives Fazal some 5p coins. Each boy then has the same amount of money. How many 5p coins did Gareth give Fazal?

2.3 Rearranging formulae

Sometimes you will need to rearrange a formula so that a letter such as x is on one side of the equals sign with the rest of the formula on the other side.

This is called **changing the subject** of the formula. You can use the same methods you use to rearrange equations:

■ **x is the subject of the formula when it appears on its own on one side of the formula and does not appear on the other side.**

Example 6

Rearrange $ax + b = c$ to make x the subject of the formula.

$$ax + b = c$$

subtract b from both sides: $$ax = c - b$$

divide both sides by a: $$x = \frac{c - b}{a}$$

Example 7

Use a flow diagram to rearrange $ax + b = c$ to make x the subject of the formula.

First build the left side of the formula $ax + b$:

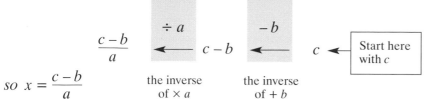

As this is equal to c you can use a reverse flow diagram to make c the subject:

$$so \quad x = \frac{c - b}{a}$$

the inverse   the inverse
of $\times a$ of $+ b$

Example 8

Make x the subject of the formula $p - 5x = q$

add $5x$ to both sides: $\qquad p = 5x + q$

subtract q from both sides: $\qquad p - q = 5x$

divide both sides by 5: $\qquad \dfrac{p - q}{5} = x$

$-5x$ goes from left side to right and becomes $+5x$

q goes from right to left and becomes $-q$

Example 9

Make y the subject of $2x + 3y = 6$

subtract $2x$ from both sides: $\qquad 3y = 6 - 2x$

divide both sides by 3: $\qquad y = \dfrac{6 - 2x}{3}$

or $\qquad y = 2 - \frac{2}{3}x$

Exercise 2H

In questions **1–14** make x the subject of the formula.

1 $x + p = q$

2 $x + 7 = t$

3 $x - a = b$

4 $5x = w$

5 $nx = p$

6 $\frac{1}{3}x = f$

7 $\dfrac{x}{m} = d$

8 $\dfrac{ax}{b} = c$

9 $nx - p = q$

10 $5 - ax = b$

11 $c - dx = h$

12 $p - 2x = 3x + q$

13 $a - 4x = b - 7x$

14 $\dfrac{x}{m} + n = p$

In questions **15–20** make y the subject.

15 $x + 2y = 6$ **16** $3x + 4y = 12$

17 $2x + 5y = 10$ **18** $x - 2y = 8$

19 $5x - 4y = 20$ **20** $3x - 8y = 24$

2.4 Showing inequalities on a number line

■ **You can use a number line like this to show an inequality like $x > 2$ which means x is greater than 2:**

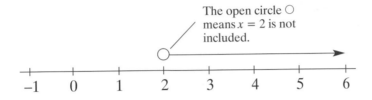

The open circle ○ means $x = 2$ is not included.

Here is $x \leqslant 3$ which means x is less than or equal to 3:

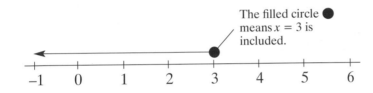

The filled circle ● means $x = 3$ is included.

And this shows $-2 \leqslant x < 1$ which means -2 is less than or equal to x which is less than 1:

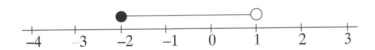

Exercise 2 I

Write down the inequalities shown in questions **1–4**.

1

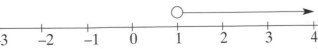

2

3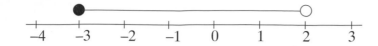

4

Draw number lines to show the inequalities in questions **5–10**.

5 $x < 4$
6 $x \geqslant -2$

7 $x \leqslant -1$
8 $1 < x < 5$

9 $-2 < x \leqslant 0$
10 $-3 \leqslant x < -1$

2.5 Solving inequalities

$3 > -6$ is an inequality. It is also a true statement, so $3 > -6$ is a true inequality.

But does the inequality remain true when you do the following things to it?

(a) add 4 to both sides:
$$3 + 4 > -6 + 4$$
$$7 > -2 \text{ is still true}$$

(b) subtract 2 from both sides:
$$3 - 2 > -6 - 2$$
$$1 > -8 \text{ is still true}$$

(c) multiply both sides by 2:
$$3 \times 2 > -6 \times 2$$
$$6 > -12 \text{ is still true}$$

(d) divide both sides by 3:
$$3 \div 3 > -6 \div 3$$
$$1 > -2 \text{ is still true}$$

(e) multiply both sides by –5:
$$3 \times -5 > -6 \times -5$$
$$-15 > 30 \text{ is } not \text{ true}$$

(f) divide both sides by –3:
$$3 \div -3 > -6 \div -3$$
$$-1 > 2 \text{ which is } not \text{ true}$$

 To solve an inequality you can:
- **add the same quantity to *both* sides**
- **subtract the same quantity from *both* sides**
- **multiply *both* sides by the same *positive* quantity**
- **divide *both* sides by the same *positive* quantity**

but you must *not*:
- **multiply both sides by a *negative* quantity**
- **divide both sides by a *negative* quantity.**

You can solve inequalities like equations *except* that you can't multiply or divide by a *negative* number.

Example 8

Solve $3x + 4 > 22$

subtract 4 from both sides: $\qquad$ $3x > 18$

divide both sides by 3: $\qquad$ $x > 6$

Example 9

Solve $6 - 5x \geqslant 3x + 2$

add $5x$ to both sides: $\qquad$ $6 \geqslant 3x + 5x + 2$

$\qquad\qquad\qquad\qquad$ $6 \geqslant 8x + 2$

subtract 2 from both sides: $\qquad$ $6 - 2 \geqslant 8x$

$\qquad\qquad\qquad\qquad$ $4 \geqslant 8x$

divide both sides by 8: $\qquad$ $\frac{1}{2} \geqslant x$

$-5x$ moves from left side to right and becomes $+5x$

$+2$ moves from right side to left and becomes -2

Another way of writing $\frac{1}{2} \geqslant x$ is $x \leqslant \frac{1}{2}$

Exercise 2J

Solve these inequalities:

1 $\quad 3x - 5 \leqslant 4$ $\qquad$ 2 $\quad 5x + 2 > 4$

3 $\quad 4x + 7 \geqslant 7x + 2$ $\qquad$ 4 $\quad 7x + 2 < 5x - 4$

5 $\quad 8 - 3a \leqslant 5$ $\qquad$ 6 $\quad 1 - 6b > 7$

7 $\quad -5c < 20$ $\qquad$ 8 $\quad -2d \geqslant -3$

9 $\quad \frac{1}{2}e + 3 \leqslant 1$ $\qquad$ 10 $\quad 5 - \frac{1}{4}f > 12$

11 $\quad 2 - \frac{1}{2}g \geqslant 1$ $\qquad$ 12 $\quad 4 - 5h \leqslant 3h$

13 $\quad 8 - 3j < 2j + 13$ $\qquad$ 14 $\quad 7 - 5k \leqslant 12 - 8k$

15 $\quad 8 - 2m \geqslant 1 - 4m$ $\qquad$ 16 $\quad 4 - 9n > 1 - 11n$

Example 10

Solve $7 \leqslant 3x - 2 < 9$

Treat this as two inequalities:

$$7 \leqslant 3x - 2 \qquad 3x - 2 < 9$$
$$9 \leqslant 3x \qquad 3x < 11$$
$$3 \leqslant x \qquad x < 3\frac{2}{3}$$

Write the answer as: $3 \leqslant x < 3\frac{2}{3}$

Exercise 2K

Solve these inequalities:

1 $7 < 4p + 3 \leqslant 27$ **2** $3 < 2q - 7 < 8$

3 $1 \leqslant 3r + 4 \leqslant 10$ **4** $-2 < \frac{1}{2}s + 1 < 3$

Giving integer solutions to inequalities

■ **Integers** are the positive and negative whole numbers and 0:

$$\ldots -3 \quad -2 \quad -1 \quad 0 \quad 1 \quad 2 \quad 3 \ldots$$

■ **Some questions will ask you for only the integer solutions of an inequality.**

Worked examination question 1

List all the possible integer values of n such that $-3 \leqslant n < 2$ [E]

$$n = -3 \quad -2 \quad -1 \quad 0 \quad 1$$

Worked examination question 2

Given that n is an integer find the greatest value for n for which
$4n + 3 < 18$ [E]

$$4n + 3 < 18$$
$$4n < 15$$
$$n < 3\tfrac{3}{4}$$

The greatest integer value of n is 3.

Exercise 2L

In questions **1–4** list the possible integer values of n.

1 $-2 \leqslant n \leqslant 1$ **2** $0 < n < 5$

3 $-3 \leqslant n < 0$ **4** $-1 < n \leqslant 3$

In questions **5–8** write down an inequality satisfied by the integers listed.

5 2, 3, 4, 5   **6** 0, 1, 2, 3

7 $-3, -2, -1$ **8** $-2, -1, 0, 1, 2$

In questions **9–12** find the greatest integer value of n.

9 $3n + 2 < 10$ **10** $8 - 5n \geqslant 2$

11 $4n + 7 < 5$ **12** $1 - 3n > 9$

In questions **13–16** find the least (smallest) integer value of n.

13 $4n + 5 \geqslant 1$ **14** $7 - 2n < 4$

15 $6n + 7 > 2$ **16** $5 - 4n < -2$

In questions **17–20** find all the possible integer values of n.

17 $5 \leqslant 2n + 1 < 9$ **18** $4 < 5n - 1 \leqslant 14$

19 $-4 < 3n + 5 < 12$ **20** $-2 \leqslant 1 - 2n < 3$

Exercise 2M   (Mixed questions)

Solve these equations:

1 $6a - 5 = 13$ **2** $5b + 2 = 9$

3 $8c + 10 = 2$ **4** $4d + 5 = 3d + 9$

5 $7e + 4 = 5e - 8$ **6** $6f - 1 = 3f + 1$

7 $8 - g = 10$ **8** $-4h = 10$

9 $7 - 5i = 4$ **10** $\frac{1}{2}j + 3 = 7$

11 $9 - \frac{1}{4}k = 11$ **12** $\frac{2}{3}m + 5 = 9$

13 $7n - 6 = 14 - 3n$ **14** $8 - 3p = 13 - 5p$

15 I multiply a number by 5 and add 3. The result is the same as when I multiply the number by 8 and subtract 18. Find the number.

16 Patterned tiles cost six times as much as plain tiles. The cost of 20 patterned tiles and 200 plain tiles is £64. Find the cost of a plain tile.

17 Rearrange these formulae to make x the subject:

   (a) $px - 4 = q$ **(b)** $n - 4x = 2x$

   (c) $5 - 3x = m - 5x$

18 Draw number lines to show these inequalities:

   (a) $x > 5$ **(b)** $x \leqslant -3$

   (c) $3 < x \leqslant 5$ **(d)** $-4 \leqslant x < 0$

19 Solve these inequalities:

   (a) $8x + 3 < 19$ **(b)** $6 - 5x \geqslant 16$

   (c) $3 - 2x < 7 - 5x$

20 Find **(a)** the greatest and **(b)** the least integer values of n for which these are true:

   (i) $4 \leqslant 3n - 2 < 10$ (ii) $1 < 4n + 5 < 20$

Summary of key points

1 To rearrange an equation you can:
   - add the same quantity to both sides
   - subtract the same quantity from both sides
   - multiply both sides by the same quantity
   - divide both sides by the same quantity.

2 Whatever you do to one side of an equation you must do to the other side.

3 You can use a **flow diagram** to solve an equation.

   > For example, build the expression $4x + 1 = 7$
   >
   >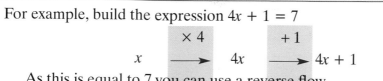
   >
   > As this is equal to 7 you can use a reverse flow diagram to find the value of x. Start at the right hand side and reverse the steps in the arrows:
   >
   > $$1\tfrac{1}{2} \xleftarrow{\ \div 4\ } 6 \xleftarrow{\ -1\ } 7$$
   >
   > so $x = 1\tfrac{1}{2}$ the inverse of $\times 4$ the inverse of $+ 1$

4 A letter such as x is the **subject of a formula** when it appears on its own on one side of the formula and does not appear on the other side.

5 Inequalities can be shown algebraically like this: $-2 < x \leqslant 1$ (which means -2 is less than x which is less than or equal to one), or on a **number line** like this:

   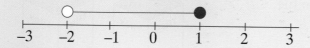

6 To solve an inequality you can:
   - add the same quantity to *both* sides
   - subtract the same quantity from *both* sides
   - multiply *both* sides by the same *positive* quantity
   - divide *both* sides by the same *positive* quantity
     but you must *not*:
   - multiply both sides by a *negative* quantity
   - divide both sides by a *negative* quantity.

7 Sometimes **integer solutions** only are given to an inequality such as: $-1 < x \leqslant 2$

   The integer solutions that make the inequality true are $x = 0, 1, 2$

3 Shapes

Euclid was a Greek mathematician who lived at Alexandria in Egypt about the year 300 BC. He wrote a book called *The Elements* explaining everything that was known about 2-D (plane) and 3-D (solid) shapes, like these:

This two-dimensional (2-D) shape is a parallelogram.

This three-dimensional (3-D) shape is a cone.

Euclid's book began with definitions of the mathematical meanings of everyday words such as **straight**, **line** and **segment**. This was followed by hundreds of theorems about 2-D and 3-D shapes. This unit will help you understand and use some of Euclid's theorems.

3.1 Angles

When two lines meet at a point the amount of turn or rotation in moving one line to the other is called an **angle**. Angles can be measured in units called degrees.

The turn can be made in a clockwise ⟲ or anticlockwise ⟳ direction. A complete turn or **revolution** is 360°.

■ **You need to know the names of these types of angles:**

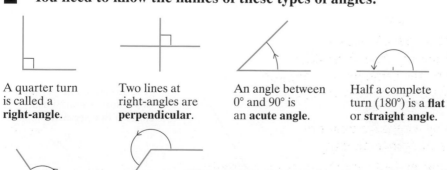

A quarter turn is called a **right-angle**.

Two lines at right-angles are **perpendicular**.

An angle between 0° and 90° is an **acute angle**.

Half a complete turn (180°) is a **flat** or **straight angle**.

An angle between 90° and 180° is called an **obtuse angle**.

An angle between 180° and 360° is called a **reflex angle**.

Here are three ways of describing the same angle:

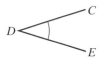 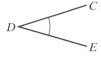

$\stackrel{\wedge}{CDE}$... with a 'hat' $<CDE$... with an angle sign d... with a letter

You can draw some angles accurately using a pencil, ruler and compasses. This is called a construction.
Here is how to construct a 45° angle.

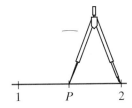 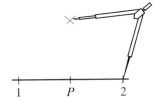 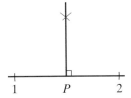

Mark point *P* on a straight line. Use compasses to mark points 1 and 2 on the line, each the same distance from *P*.

With the compass point on point 1, then point 2, mark two intersecting arcs above *P*.

Join *P* to the intersection. This line is at 90° to the original straight line.

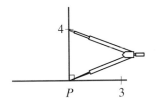 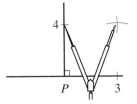 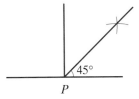

Use compasses to mark points 3 and 4 each the same distance from *P*.

With the compass point on point 3, then point 4, mark two intersecting arcs.

Join *P* to the intersection. This line is at 45° to the original straight line.

You can draw angles accurately using a protractor:

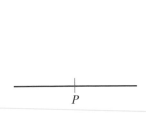 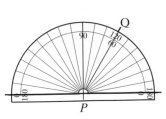 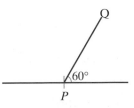

Mark point *P* on a straight line.

Put the 'cross' of the protractor exactly over *P*. Mark a point *Q* at 60°.

Join points *P* and *Q*. The line segment *PQ* is at 60° to the original line.

A section of a line between two points is sometimes called a **line segment**.

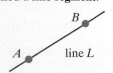

AB is a line segment of the line *L*.

Properties of angles

■ **You need to know these properties of angles:**

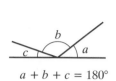

$$a + b + c = 180°$$

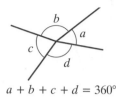

$$a + b + c + d = 360°$$

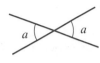

Angles which meet at a point on a straight line add up to 180°. They are called **supplementary angles**.

Angles which meet at a point add up to 360°.

Two straight lines which cross at a point form equal **vertically opposite angles**. Vertically comes from the word **vertex** meaning point.

■ **Angles are formed when a straight line crosses a pair of parallel lines:**

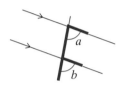

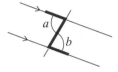

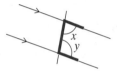

Corresponding angles are equal. So $a = b$. You can find them by looking for an F shape.

Alternate angles are equal. So $a = b$. Look for a Z shape.

Interior angles add up to 180°. So $x + y = 180°$. Look for a C shape.

Example 1

Find angles a and b giving reasons for your answers.

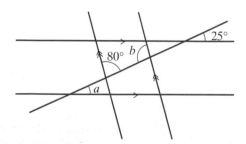

a and 25° are corresponding angles, so $a = 25°$

80° and b are interior angles, so:

$$80° + b = 180°$$
$$b = 180° - 80°$$
$$= 100°$$

Exercise 3A

1 Calculate the size of each lettered angle:

(a)

(b)

(c)

(d)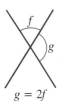

$g = 2f$

(e)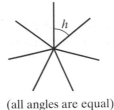

(all angles are equal)

2 Calculate each of the lettered angles.

(a)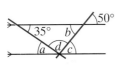

(b)

(c)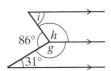

3 (a) Draw a line segment AB 8 cm long.

   (b) Using only pencil, ruler and compasses, construct a line segment XYZ which bisects AB at point Y and is at right-angles to it. This is called the perpendicular bisector of AB.

   (c) Construct the angle bisector of XYB.

   (d) Measure the size of the new angle.

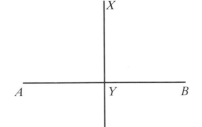

4 (a) Use only pencil, ruler and compasses to construct an equilateral triangle.

   (b) Use your diagram to construct an angle of 30°.

3.2 2-D (plane) shapes

A flat surface is called a **plane**. For example, a page of this book is plane when it is lying flat. A shape which lies entirely in a plane must be two dimensional (2-D). The photograph on page 56 is 2-D even though it shows a 3-D scene.

■ **Closed plane shapes like these are called polygons:**

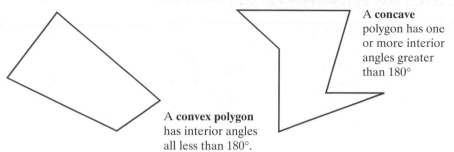

A **concave** polygon has one or more interior angles greater than 180°

A **convex polygon** has interior angles all less than 180°.

A polygon has the same number of sides as it has interior angles. A **regular polygon** has all its sides equal and all its interior angles equal.

The marks on the sides mean they are the same length.

3.3 Triangles

For a polygon to be closed it must have at least three sides. A three-sided polygon is a **triangle**.

■ **Some triangles have special properties and names:**

An **equilateral triangle** has: 3 sides equal 3 interior angles equal (60°).

An **isosceles triangle** has: 2 sides equal base angles equal.

A **right-angled triangle** has: 1 angle of 90°.

Triangles without any of these properties are called **scalene triangles**.

■ **The interior angles in a triangle add up to 180°.**

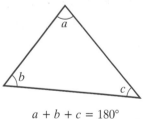

$a + b + c = 180°$

Example 2

Calculate the sizes of the lettered angles, giving reasons for your answers.

$34° + 92° + a = 180°$ (interior angles of a triangle)

$a = 180° - 126° = 54°$

$a + b = 180°$ (angles on a straight line)

$b = 180° - 54° = 126°$

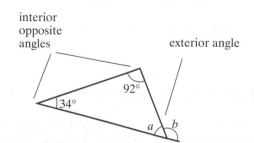

interior opposite angles

exterior angle

■ **The exterior angle of a triangle is equal to the sum of the interior opposite angles.**

Example 3

Using only a pencil, ruler and compasses, construct a triangle *ABC*
where *AB* = 7 cm, *BC* = 5 cm, *AC* = 4 cm.

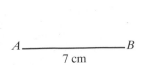 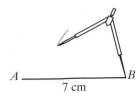 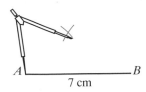 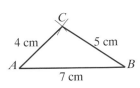

Draw any side. It is often easiest to draw the longest side horizontal.

With the compass point at *B* draw an arc of radius 5 cm.

With the point at *A* draw an arc of radius 4 cm.

Mark point *C* where the two arcs cross. Join *CB* and *CA*. Leave the arcs showing.

Exercise 3B

You need a ruler, protractor and compasses.

1   Calculate the sizes of the angles marked with letters. Write down what type each triangle is.

(a)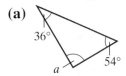

(b)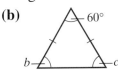

(c)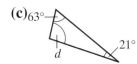

(d)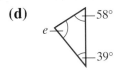

(e)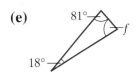

2   Calculate the sizes of the angles marked with letters:

(a)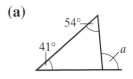

(b)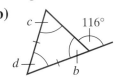

(c)

3   Construct these triangles:

(a) Triangle *ABC* in which *AB* = 8 cm, *BC* = 3 cm, *AC* = 6 cm.

(b) Equilateral triangle *PQR* with *PQ* = 7 cm.

(c) Right-angled triangle *DEF* where *DE* = 5 cm, *EDF* = 90°, *DF* = 6 cm. Then measure the length *EF*.

(d) Triangle *XYZ* in which *YZ* = 10 cm, $X\hat{Y}Z = 45°$, $X\hat{Z}Y = 60°$. Then measure *XZ*.

(e) Triangle *PQR* in which *QR* = 9 cm, $P\hat{Q}R = 48°$, $P\hat{R}Q = 35°$. Then measure *PR*.

4   How many triangles are there in this diagram?

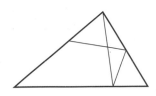

3.4 Quadrilaterals

A polygon with 4 sides is called a **quadrilateral**.

■ **Some quadrilaterals have special names and properties.**

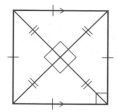

A **square** has:
all sides equal
opposite sides parallel
all interior angles 90°
diagonals that bisect at
　90°.

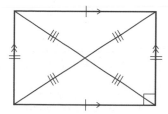

A **rectangle** has:
opposite sides equal
opposite sides parallel
all interior angles 90°
diagonals that bisect each
　other.

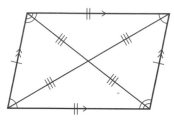

A **parallelogram** has:
opposite sides equal
opposite sides parallel
diagonally opposite
　angles equal
adjacent angles are
　supplementary angles
diagonals that bisect
　each other.

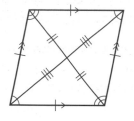

A **rhombus** has:
all sides equal
opposite sides parallel
opposite angles equal
adjacent angles are
　supplementary angles
diagonals that bisect at
　90°.

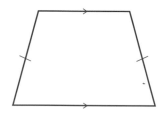

An **isosceles trapezium**
　has:
1 pair of parallel lines
1 line of symmetry.

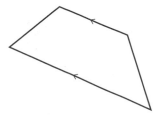

A **trapezium** has:
1 pair of parallel lines.

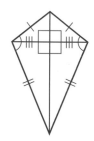

A **kite** has:
2 pairs of adjacent sides
　equal
1 pair of opposite angles
　equal
diagonals cut at 90°
1 line of symmetry.

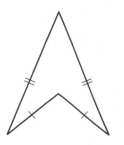

An **arrowhead** has:
2 pairs of adjacent sides
　equal
1 line of symmetry.

Each special name means the shape has
all the properties listed.

The interior angles in a quadrilateral add up
to 360°. You can see this by dividing it into
two triangles:

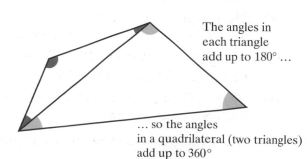

The angles in
each triangle
add up to 180° …

… so the angles
in a quadrilateral (two triangles)
add up to 360°

Example 4

ABCD is a rhombus. *AC* is a diagonal. $D\hat{A}C = 43°$

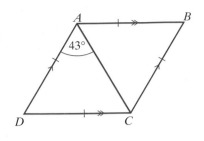

Calculate the sizes of these angles giving reasons for your answers:

(a) $A\hat{C}B$ **(b)** $A\hat{D}C$ **(c)** $A\hat{B}C$

Answer

(a) $A\hat{C}B = 43°$ (alternate angles)

(b) $A\hat{D}C = 180° - (2 \times 43°) = 94°$ (*AD, DC* equal)

(c) $A\hat{B}C = 94°$ (opposite angles)

Exercise 3C

You need a ruler, protractor and compasses.

1   *ABCD* is a quadrilateral. Write down all the possible names of the quadrilateral if:

   (a) all interior angles are 90° and both diagonals bisect at 90°

   (b) no pairs of sides are parallel and the diagonals cut at 90°

   (c) one pair of sides is parallel

   (d) diagonals and sides make 4 right-angled triangles

2   **(a)** *ABCD* is a rectangle   **(b)** *ABCD* is a kite

        BCEF is a square $A\hat{B}C = 82°$

        $D\hat{A}E = 38°$ Find: (i) $A\hat{B}X$ (ii) $B\hat{A}X$

        Find: (i) $A\hat{E}D$ (ii) $A\hat{E}F$

             (iii) $A\hat{E}B$

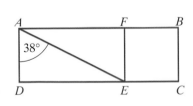
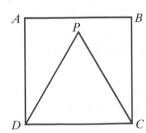

   (c) *MNOP* is a trapezium   **(d)** *ABCD* is a square

        MN = MP   Triangle *PDC* is equilateral

        $N\hat{P}O = 29°$ Calculate: (i) $P\hat{C}B$ (ii) $C\hat{P}B$

        Find: (i) $M\hat{N}P$ **(ii)** $P\hat{M}N$ (iii) $A\hat{P}D$ (iv) $A\hat{P}B$

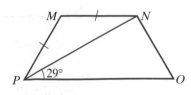

3 Construct kite *PQRS* like this:

 (a) draw *PR* = 7 cm

 (b) construct *PQR* with sides *PQ* = *QR* = 5 cm

 (c) construct *PRS* with sides *PS* = *SR* = 8 cm

 (d) measure *QS*

3.5 Angles in polygons

The sum of the interior angles of any polygon can be found by dividing the shape into triangles from one vertex:

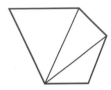

Pentagon
3 triangles
sum of interior angles:
3 × 180° = 540°

Hexagon
4 triangles
sum of interior angles:
4 × 180° = 720°

Octagon
6 triangles
sum of interior angles:
6 × 180° = 1080°

The number of triangles is two less than the number of sides of the polygon.

■ **Generalizing for a polygon with *n* sides:**

   sum of interior angles = (*n* – 2) × 180°

■ **If a polygon is regular each interior angle can be calculated from:**

$$\textbf{interior angle} = \frac{(\textbf{\textit{n}} - \textbf{2}) \times \textbf{180°}}{\textbf{\textit{n}}}$$

Sometimes it is simpler to calculate the exterior angle.

■ **The sum of the exterior angles of any polygon is 360°.**

So for a regular polygon with *n* sides:

 exterior angle $= \frac{360°}{n}$ and **interior angle = 180° – exterior angle**

LOGO can be used to draw polygons. The exterior angle is used in instructions like these:

```
to pentagon
repeat 5[ fd 100 rt 72]
end
```

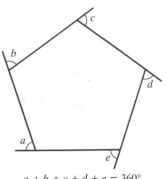

$a + b + c + d + e = 360°$

The exterior angle in a pentagon is 72°.

3.6 Circles

These names describe different parts of a circle:

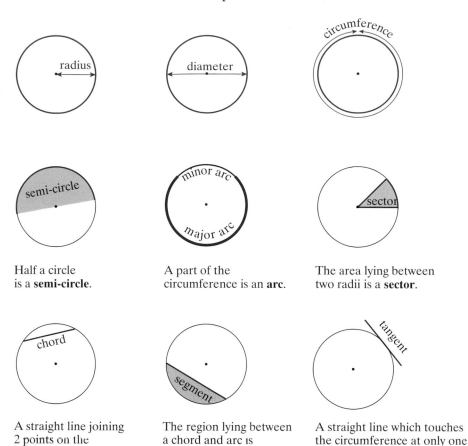

Half a circle
is a **semi-circle**.

A part of the
circumference is an **arc**.

The area lying between
two radii is a **sector**.

A straight line joining
2 points on the
circumference is a **chord**.

The region lying between
a chord and arc is
a **segment**.

A straight line which touches
the circumference at only one
point is a **tangent**.

3.7 3-D (solid) shapes

A solid shape with plane faces and straight edges is called a
polyhedron (plural: polyhedra).

Some polyhedra have special names:

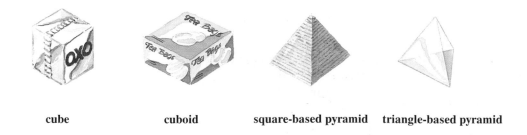

cube   **cuboid**   **square-based pyramid**   **triangle-based pyramid**

A **pyramid** takes its name from the shape of its base.

A **prism** is a polyhedron that has the same shape or cross-section wherever you slice it along its length. A prism takes its name from the shape of its cross-section.

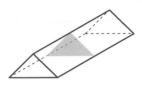

triangular prism

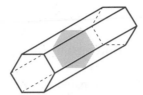

hexagonal prism

These solids have some circular edges so they are *not* polyhedra:

circular prism – usually called a **cylinder**

circle-based pyramid – usually called a **cone**

sphere

Drawing 3-D shapes in 2-D

You can use plans and elevations like these to draw accurate 2-D pictures of 3-D shapes:

The view from the front is a **front elevation.**

The view from the side is a **side elevation**.

The view from above is a **plan**.

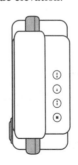

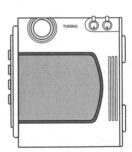

You can also draw accurate drawings of 3-D shapes on dotted paper:

A drawing on square dotted paper is an **oblique projection**.

A drawing on triangle dotted paper is an **isometric projection**.

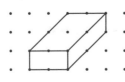

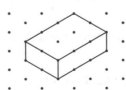

All these 2-D diagrams provide sufficient information to reproduce the object in 3-D if required. They do not show the effect of perspective (measurements can be made from them).

Nets: solid shapes unfolded

■ **The net of a 3-D shape is a 2-D shape that can be folded to make the 3-D shape.**
To fold up properly, edges which will touch must be the same length.

Either of these nets would make a cuboid with dimensions 1 cm by 2 cm by 6 cm:

Exercise 3D

1 Name these shapes:

(a)   **(b)**   **(c)**   **(d)**

2 Use this notation:
V = number of vertices   F = number of faces   E = number of edges

(a) Copy and complete this table:

Solid	V	F	E
Cube	8	6	12
Cuboid			
Triangular prism			
Hexagonal prism			
Octagonal prism			
Triangle-based pyramid			
Square-based pyramid			

(b) Write down a relationship between V, F and E.
This relationship is known as Euler's Theorem after the
Swiss mathematician who discovered it.

3 A regular polyhedron is a solid with faces which are regular
polygons. The same number of these polygons meet at each vertex.

tetrahedron   **cube**   **octahedron**   **dodecahedron**   **icosahedron**

These five regular solids were known to the Greek mathematician and philosopher
Plato, and are often called the **Platonic solids**. A tetrahedron is the same as a
triangle-based pyramid.

(a) Check that Euler's Theorem (from question 2) also works
for these polyhedra:

Solid	V	F	E
Tetrahedron			
Cube			
Octahedron			
Dodecahedron			
Icosahedron			

(b) Copy and complete this table:

Solid	Number of triangular faces meeting at a vertex …
Tetrahedron	
Octahedron	
Icosahedron	

(c) Explain why a regular polyhedron could not have 6
triangular faces meeting at a vertex.

(d) Explain why a regular polyhedron cannot be made with
regular hexagons as faces.

4 Sketch a plan, front elevation and side elevation for each of
these solids:

(a)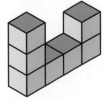   **(b)**

5   Draw six different nets which would make a cube.

6   Sketch a net to make a square-based pyramid.

7   Sketch the net to make this prism. Label the lengths for each side.

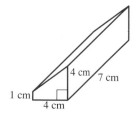

8   Make an isometric projection sketch of the solid shown by the plan and elevations:

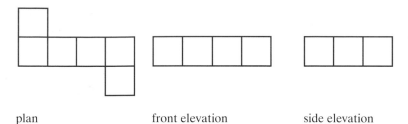

plan   front elevation   side elevation

9   Make an oblique projection sketch of the solid shown by the plan and elevations:

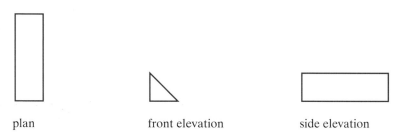

plan   front elevation   side elevation

3.8 Congruent shapes

■ **When 2-D shapes are exactly the same shape and size they are *congruent* to each other.**

One shape may have to be rotated or turned over to see the match. Corresponding sides and corresponding angles are identical.

3-D shapes are congruent if corresponding faces are identical and the numbers of faces meeting at corresponding vertices are the same.

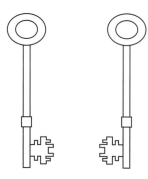

These keys are congruent.

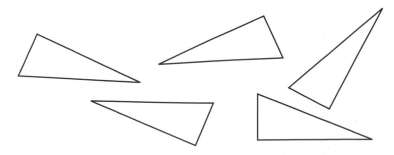

These triangles are congruent.

It is necessary to test triangles for congruency by cutting them out or knowing the size of each angle and the length of each side.

Triangles are congruent to each other if these facts are true:

Three pairs of sides are equal (remember this with the abbreviation **SSS** for three sides).

Two pairs of sides are equal and the angle between them (the included angle) is equal (**SAS**).

Two pairs of angles are equal and one pair of corresponding sides is equal (**AAS**).

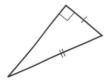 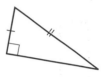

Both triangles have a right-angle, the hypotenuses are equal and one pair of corresponding sides is equal (**RHS**).

Notice that triangles are *not* congruent if corresponding angles only are equal. (There is more about this on page 48.)

Example 5

State whether these shapes are congruent. List the vertices in corresponding order and give reasons for congruency.

(a)

 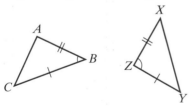

(b)

 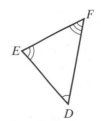 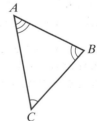

(c)

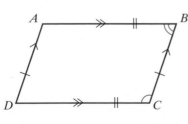 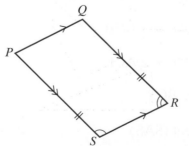

(a) Yes. *ABC* = *XZY* (SAS)

(b) No. Only angles are equal; corresponding sides may not be the same length.

(c) No. Parallelogram *ABCD* is not congruent to *QRSP*. It is not clear whether *AD* = *PQ* or *BC* = *RS*.

Example 6

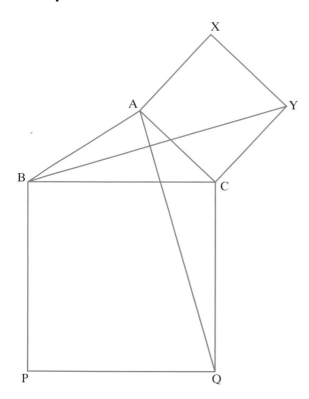

In the diagram, ACYX and BCQP are squares. Prove that triangles ACQ and BCY are congruent.

$$AC = CY \qquad \text{(sides of square ACYX)}$$

$$CQ = BC \qquad \text{(sides of square BCPQ)}$$

$$\angle ACQ = 90° + \angle ACB$$

$$\angle BCY = 90° + \angle ACB$$

$$\therefore \angle ACQ = \angle BCY$$

Triangles ACQ, YCB are congruent (SAS)

Exercise 3E

1 Which of the following drawings is not congruent to the other two:

(a) (i)   (ii)   (iii)

(b) (i)   (ii)   (iii)

(c) (i)   (ii)   (iii)

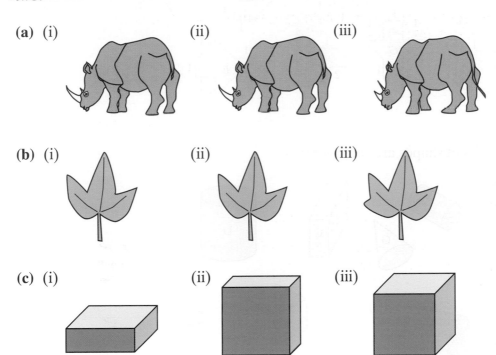

2 Which of these pairs of triangles are congruent? Give the vertices in corresponding order and the reason for congruency.

(a) (i)   (b) (i)   (c) (i)

(ii)   (ii)   (ii)

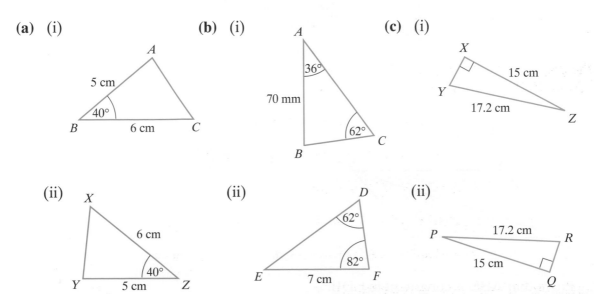

3 Which pairs of shapes are congruent:

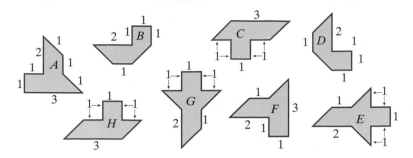

4 State which pairs of shapes are congruent:

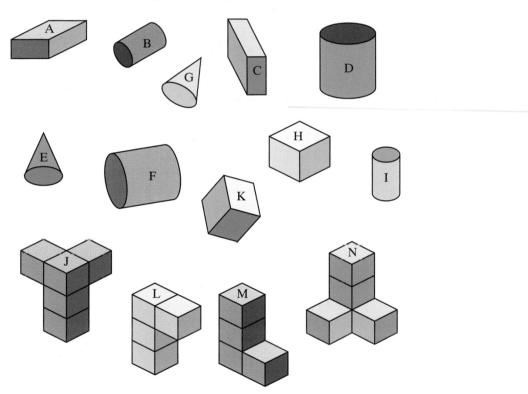

5 ABCD is a parallelogram.

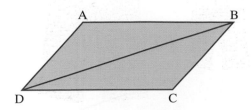

Prove that ABD is congruent to CDB.

6

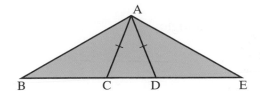

In the diagram AC = AD and BD = CE. Prove that triangles ABC and ADE are congruent.

7

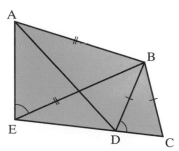

In the diagram,

$$AB = BE$$

$$BD = BC$$

$$\angle AEB = \angle BDC$$

Prove that triangles ABD and EBC are congruent.

6

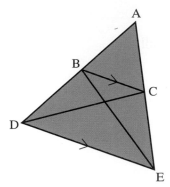

In the diagram, ABC is an isosceles triangle with AB = AC. Prove that triangles ACD and ABE are congruent.

3.9 Similar shapes

■ **Shapes are *similar* if one shape is an enlargement of the other.**

Polygons are similar if all corresponding angles are *equal* and the ratio of object length to image length is the same for all sides.

These pairs of trousers are similar. The pattern for one pair is an enlargement of the pattern for the other.

Example 7

These rectangles are similar. Find the length of the second rectangle.

7 cm

Using corresponding sides, the scale factor of the enlargement is $\frac{9}{4} = 2.25$

So the enlarged length is $7 \times 2.25 = 15.75$ cm

■ **The scale factor of the enlargement is the ratio:**

$$\frac{\textbf{length of a side on one shape}}{\textbf{length of corresponding side on other shape}}$$

To decide whether two triangles are similar, you need to check that all the corresponding angles are equal, or that the corresponding sides are in the same ratio. Triangles are similar if one of the following statements is true:

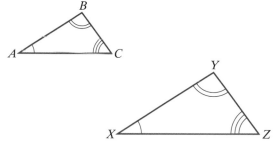

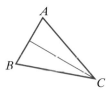

All corresponding sides are in the same ratio:

$$\frac{PQ}{AB} = \frac{QR}{BC} = \frac{PR}{AC} = \text{scale factor}$$

All corresponding angles are equal:

$$\hat{A} = \hat{X} \quad \hat{B} = \hat{Y} \quad \hat{C} = \hat{Z}$$

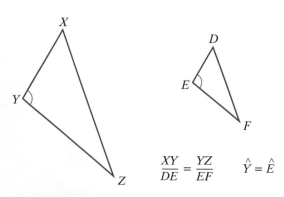

$$\frac{XY}{DE} = \frac{YZ}{EF} \quad \hat{Y} = \hat{E}$$

Two pairs of corresponding sides are in the same ratio and the included angles are equal:

Example 8

Find the length of the side marked x:

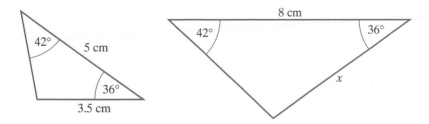

Two pairs of angles are equal, so the third pair must be equal and the triangles are similar. The scale factor of the enlargement is $\frac{8}{5} = 1.6$

So $x = 3.5 \times 1.6 = 5.6$ cm

Exercise 3F

1   Each pair of shapes is similar. Calculate the length of each side marked by a letter:

(a)

(b)

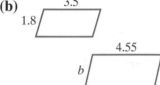

(c)

(d)

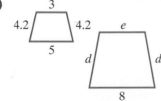

2 These shapes are similar. Calculate the lengths marked by letters:

(a)

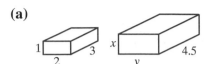

(b)

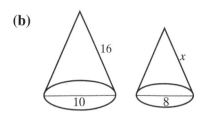

(c)

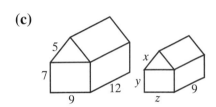

3 Which pairs of rectangles are similar:

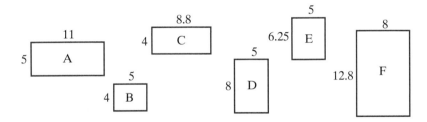

4 Which of these families of shapes have members which are always similar:

   (a) squares   **(b)** rectangles

   (c) parallelograms   **(d)** circles

   (e) ellipses   **(f)** equilateral triangles

   (g) isosceles triangles   **(h)** regular hexagons

   (i) trapeziums

5 Which of these families of solid shapes have members which are always similar:

   (a) cubes   **(b)** cuboids

   (c) spheres   **(d)** tetrahedrons

   (e) pyramids   **(f)** hexagonal prisms

   (g) octahedrons   **(h)** cones

6 Each group of three triangles has two similar and one 'different'. Which triangle is 'different':

(a) (i)

55°

(ii)

45°

(iii)

35°
55°

(b) (i)

30°

72°

(ii)

72° — 78°

(iii)

30°

68°

(c) (i)

7

50°   6

(ii)

7.8

50°

9.1

(iii)

6
40°

7

7 Write down why these pairs of triangles are similar and calculate the length of each side marked by a letter:

(a)

42°
6.24
2.8
x

42°
7.8
3.5
5.9

(b)

3.24   y
54°
36°

36°
2.7

1.5

(c)

15°
a
b
2.88
120°   45°
1.8   4.8
45°
6

8 (a) Explain why these two triangles are similar.

2.4
1.5   x
5.6   y
4.2

(b) Calculate the lengths of x and y.

9   (a) Name the similar triangles.
    (b) Explain why they are similar.
    (c) Calculate the length AB.
    (d) Calculate the length AY.

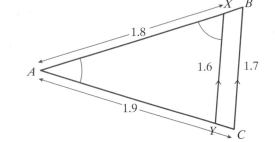

3.10 Symmetrical shapes

Lines of symmetry

The front of this building is symmetrical:
each part on the left of the dotted line has
a mirror image on the right.

- **When a straight line divides a shape
  into two identical (congruent) shapes
  the shape is symmetrical about the line.
  The line is called a line of symmetry.
  It is usually shown by a dotted line.**

Example 9

Draw all the lines of symmetry for a rectangle.

A rectangle has two lines of symmetry. The diagonal is *not* a line of
symmetry.

Example 10

Draw all the lines of symmetry for a regular hexagon.

A regular hexagon has six lines of symmetry.

- **A regular polygon with *n* sides has *n* lines of symmetry.**

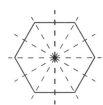

Planes of symmetry

- **Some 3-D shapes can be divided by a plane to produce two
  identical shapes. The plane is called a plane of symmetry.**

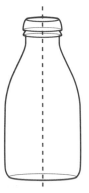

Example 11

Sketch the planes of symmetry of this cuboid:

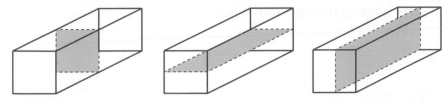

The cuboid has three planes of symmetry.

Example 12

Sketch the planes of symmetry in a cube:

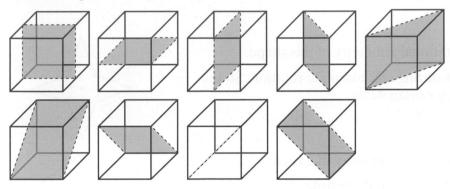

A cube has nine planes of symmetry.

Rotational symmetry

Shapes can be rotated (turned) about any point, called a **centre of rotation**.

This shape appears on the crest of the Isle of Man. Imagine tracing it, then rotating the tracing about a point at its centre.

■ **If the tracing fits exactly onto the shape two or more times during one complete rotation the shape has rotational symmetry.**

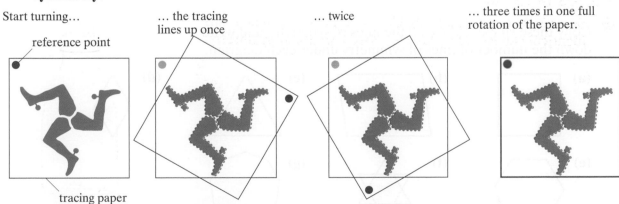

Start turning...   ... the tracing   ... twice   ... three times in one full
                        lines up once   rotation of the paper.

reference point

tracing paper

This shape does have rotational symmetry.

The number of times it fits onto itself is the **order of rotation**. It has rotational symmetry order 3.

When a shape looks the same after rotating through 180° it has **point symmetry**.

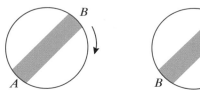

This road sign has point symmetry.

Example 13

Write down the order of rotational symmetry of this shape.

It looks the same four times during one complete rotation. So the shape has rotational symmetry order 4.

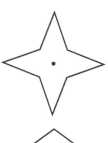

Example 14

Write down the order of rotational symmetry of a regular pentagon.

A regular pentagon has rotational symmetry order 5.

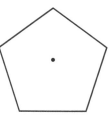

■ **A regular polygon with *n* sides has rotational symmetry order *n*. When *n* is even, the shape also has point symmetry.**

Exercise 3G

1 Copy these capital letters. Mark on all the lines of symmetry.

A   B   C   D   E

2 Copy these shapes. Mark all the lines of symmetry and write down the number of lines of symmetry under each diagram.

(a)   (b)   (c)   (d)

(e)   (f)   (g)

3 Write down how many planes of symmetry each of these solids has:

(a)

(b)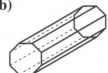

(c)

4 Write down the order of rotational symmetry of each of these shapes:

(a)

(b)

(c)

(d)

5 Write down the order of rotational symmetry of each of these shapes:

(a)

(b)

(c)

(d)

(e)

(f)

(g)

6 Write down which shapes in question 5 have point symmetry.

7 Draw an octagon with only 2 lines of symmetry and rotational symmetry order 2.

8 Copy and complete this diagram so that it has 4 lines of symmetry and rotational symmetry order 4.

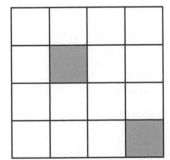

Summary of key points

1 **Angles** can be:

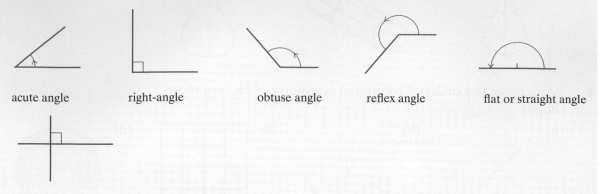

acute angle   right-angle   obtuse angle   reflex angle   flat or straight angle

two lines at right-angles are perpendicular

2 You need to know these properties of angles:

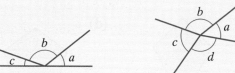 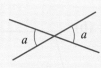

$a + b + c = 180°$
Angles which meet at a point on a straight line add up to 180°. They are called **supplementary angles**.

$a + b + c + d = 360°$
Angles which meet at a point add up to 360°.

Two straight lines which cross at a point form equal, **vertically opposite angles**. Vertically comes from the word **vertex**, meaning point.

3 A straight line crossing parallel lines creates:

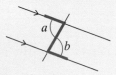 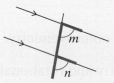 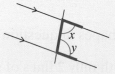

alternate angles,
$a = b$
Look for a Z shape.

corresponding angles, $m = n$
Look for an F shape.

interior angles,
$x + y = 180°$
Look for a C shape.

4 A closed plane figure is called a **polygon**.

5 **Triangles** can be:

equilateral   isosceles   right-angled   scalene

6 Interior angles of a triangle add up to 180°.

$a + b + c = 180°$

7 The exterior angle of a triangle is equal to the sum of the interior opposite angles.

8 A polygon with four sides is called a **quadrilateral**. A quadrilateral can be a:

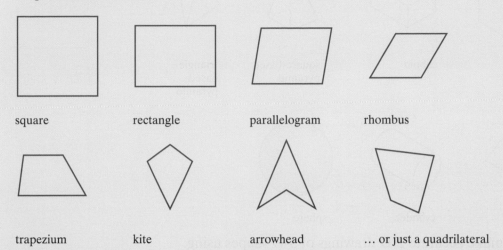

square   rectangle   parallelogram   rhombus

trapezium   kite   arrowhead   ... or just a quadrilateral

Interior angles in a quadrilateral add up to 360°.

9 The sum of the interior angles of an n-sided polygon is $(n - 2) \times 180°$.

10 Each interior angle of a regular, n-sided polygon $= \dfrac{(n - 2) \times 180°}{n}$

11 The sum of the exterior angles of a polygon is 360°.

12 Each exterior angle of an n-sided regular polygon is $\dfrac{360°}{n}$

13 You need to know these names for parts of a **circle**:

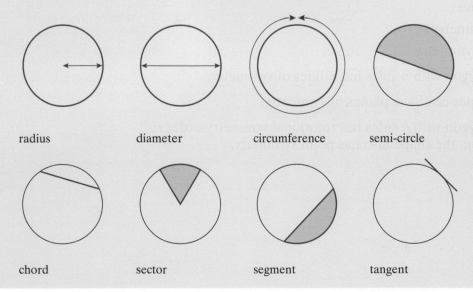

radius   diameter   circumference   semi-circle

chord   sector   segment   tangent

14 You should be able to recognize these **solid shapes**:

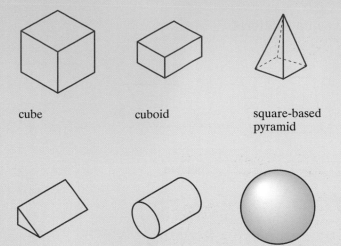

cube   cuboid   square-based   triangle-
                              pyramid   based
                                                   pyramid

prism   cylinder   sphere

15 You can make accurate 2-D drawings of 3-D shapes using:
 - plan and elevations   ● oblique projections
 - isometric projections.

16 A **net** is a 2-D shape which can be folded to make a 3-D (solid) shape.

17 **Congruent** shapes are identical to each other in shape and size.

18 Shapes are **similar** if one shape is an enlargement of the other.

19 The **scale factor** of enlargement is the ratio:

$$\frac{\text{length of a side on one shape}}{\text{length of corresponding side on the other shape}}$$

20 Shapes can have:
 - lines of symmetry or
 - rotational symmetry.

21 A regular polygon with n sides has n lines of symmetry.

22 Solid 3-D shapes can have planes of symmetry.

23 A regular polygon with n sides has rotational symmetry order n. When n is even, the shape also has point symmetry.

4 Collecting and presenting data

Every day you are bombarded with information:

This unit is about collecting and presenting information in ways that make it easier to work with it and to spot patterns.

We call the information we collect **data**. There are many different types of data.

4.1 Different types of data

■ **Qualitative data is described using words.**

Here is some data on the hair colour of students in year 11 at Lucea High School:

Hair colour	Brown	Black	Blonde	Red
Number of students	58	23	37	14

Hair colours, place names, and activities such as mountain biking are **qualitative data**.

■ **Quantitative data consists of numbers.**

Here is some GCSE data for the same students at Lucea High School:

Number of GCSEs taken	4	5	6	7	8	9
Number of students	5	10	18	48	23	28

The number of GCSEs taken and any other data that can be counted or measured such as the number of days in a holiday, temperatures and weights are **quantitative data**.

Quantitative data can be **discrete** or **continuous**.

■ **Discrete data can only take particular values.**

For example, you can buy shoes in these sizes:

$$6 \qquad 6\tfrac{1}{2} \qquad 7 \qquad 7\tfrac{1}{2} \qquad 8$$

These values are **discrete** (meaning separate). There are **no** values in between them.

Discrete data has an *exact* value.

■ **Continuous data can take any value.**

For example, your foot could be:

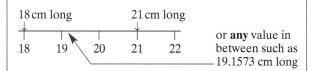

18 cm long   21 cm long

18   19   20   21   22   or **any** value in between such as 19.1573 cm long

Continuous data *cannot be measured exactly*. The accuracy of a measurement depends on the accuracy of the measuring device.

Exercise 4A

1   For each of these sets of data write down whether it is is qualitative or quantitative:
   (a) the ages of the students at Lucea High School
   (b) the students' favourite sports
   (c) the sizes of shoes sold in a shop
   (d) the colours of shoes sold in a shop
   (e) the subjects offered for A-level studies at a college

2   For each of these types of data write down whether it is discrete or continuous:
   (a) the ages of people in your town
   (b) dress sizes
   (c) the number of CDs that have been sold
   (d) the number of coins in circulation in the UK

(e) the length of a road

(f) the speed at which a car is travelling

(g) the area of a field

(h) the marks given by judges at an ice skating competition

4.2 Recording and presenting discrete data

It is easy to lose count when you are collecting data. One way of avoiding this is to use **tally marks** to collect the data, then arrange it in a **frequency distribution table** (or frequency table for short).

Example 1

Fred is collecting data on snails for a science project. He counts the number of snails in 30 different sections of woodland, each with an area of 1 m². Here is his data:

```
3  3  5  0  0  2  1  1  2  5  3  3  2  0  0
2  1  1  3  5  0  2  2  1  3  5  3  2  2  1
```

Record Fred's data in a tally chart and frequency table.

This data is discrete with a maximum value of 5 snails in each area. Here is the frequency table. It includes a tally column.

Number of snails in a section	Tally	Frequency
0	ЖЖ	5
1	ЖЖ I	6
2	ЖЖ III	8
3	ЖЖ II	7
4		0
5	IIII	4
	Total 30	

The frequency of a result is the number of times it occurred.

Notice that 5 tallies are recorded as ЖЖ

Fred could have collected his data directly in a blank frequency table. This is called a **data capture sheet**.

Presenting the data

The data from Example 1 can be presented visually in several ways ...

... in a **bar-line chart**

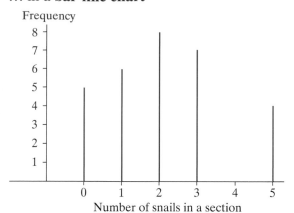

Number of snails in a section

... in a **bar chart**

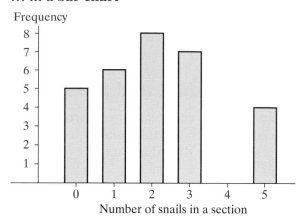

Number of snails in a section

You could also present this data in a **pie chart** to show what proportion of all the 1 m² patches contained 0 snails, 1 snail, 2 snails, and so on.

The angle of each sector represents the proportion of the total number of sections of woodland that contained a given number of snails.

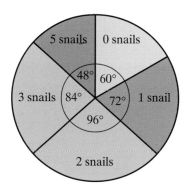

For example, the total number of sections in the frequency table is 30. Of these, 5 contained 0 (zero) snails. So the sector representing these sections has an angle of $\frac{5}{30}$ of the 360° in the circle.

$$\frac{5}{30} \times 360° = 60°$$

Stem and leaf diagrams

Another way of presenting data is to use a stem and leaf diagram. Here is an example.

The ages of the 30 members of an aerobics class are:

19	22	31	17	8	12
23	46	53	48	19	46
38	59	47	52	21	58
54	26	32	47	55	62
64	36	37	43	15	51

These are presented as a stem and leaf diagram by: using the first digit as stem, and the second digit as leaf

19 is written 1 | 9	that is	0	8	
22 is written 2 | 2		1	9, 7	
31 is written 3 | 1		2	2,	
		3	1,	
17 is then 1 | 9, 7		4		
		5		
and 8 is 0 | 8		6		

so the complete table becomes

```
0 | 8
1 | 9, 7, 2, 9, 5
2 | 2, 3, 1, 6
3 | 1, 8, 2, 6, 7
4 | 6, 8, 6, 7, 7, 3
5 | 3, 9, 2, 8, 4, 5, 1
6 | 2, 4
```

we now put the 'leaves' in order

```
0 | 8
1 | 2, 5, 7, 9, 9
2 | 1, 2, 3, 6
3 | 1, 2, 6, 7, 8
4 | 3, 6, 6, 7, 7, 8
5 | 1, 2, 3, 4, 5, 8, 9
6 | 2, 4
```

A stem and leaf diagram can be used to work out the median. In this case there are fifteen numbers up to and including 38 and fifteen numbers at 43 or above. So the median is

$$\frac{38 + 43}{2} = \frac{81}{2} = 40.5$$

4.3 Recording and presenting grouped data

When quantitative data has a wide range of values it makes sense to group sets of values together. This makes it easier to record the data and to spot any patterns or trends.

For example, scores out of 50 in a test might be grouped:

0–10,   11–20,   21–30,   31–40,   41–50

It often makes life easier if all the groups are the same size, but they do not have to be. In this case the first group has 11 members and the others have 10.

We call the groupings 0–10, 11–20 and so on **class intervals**.

■   **Class intervals are groupings of quantitative data.**

Here is a frequency table for a test taken by some pupils:

Score	Tally	Frequency
0–10	||	2
11–20	ЖЖ́ |||	8
21–30	ЖЖ́ ЖЖ́ ||||	14
31–40	ЖЖ́	5
41–50	|	1

The class intervals must not overlap. Each score can only be in one class interval.

The test scores are discrete data. You can score 31 or 32, but *not* 31.4.

You can also group continuous data. Here is some data from a traffic study. Kuljit has timed the interval in seconds between successive vehicles passing her school for 50 vehicles. The times are correct to the nearest 0.1 seconds:

```
 3.3  21.0  23.5   9.4  14.6  46.8  34.3  45.8  56.1  50.4
12.2   7.6  14.4  16.5  34.5  35.7   3.5   2.8   5.7  10.0
45.1  46.2  23.6  17.1   9.7  38.4   3.5   7.3  28.7  32.1
51.7  21.3   6.1  17.4  12.2  43.2  56.3  34.5  37.9  12.0
17.3   3.2  15.3   4.2  24.4  35.6  29.8   7.1   3.7  10.2
```

To record this data in a frequency table you need to group it into class intervals. If the class intervals are too narrow there will be too many groups to show up any pattern in the data. If the class intervals are too wide there will be too few groups to show up a pattern.

There is no golden rule for choosing the sizes of the class intervals or the number of groups, but it is usually best to have no fewer than 5 groups and no more than 20.

Here is a frequency table for the data grouped into class intervals of width 10 seconds:

Class interval (seconds)	Tally	Frequency
0 up to but not including 10	ЖЖ́ ЖЖ́ ||||	14
10 up to but not including 20	ЖЖ́ ЖЖ́ ||	12
20 up to but not including 30	ЖЖ́ ||	7
30 up to but not including 40	ЖЖ́ |||	8
40 up to but not including 50	ЖЖ́	5
50 up to but not including 60	||||	4

Presenting the data

■ **Data that is grouped and continuous can be displayed in a histogram**

Here is a histogram for the traffic data from the frequency table:

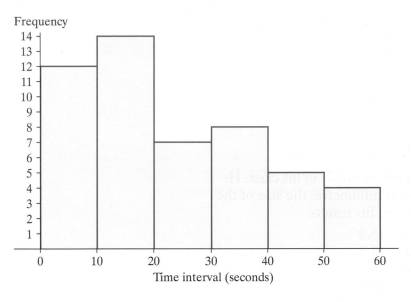

A histogram *looks* like a bar chart but:

● the data is continuous so there can be no gaps between the bars

● the data must be grouped into class intervals of equal width if you want to use the lengths of the bars to represent data values.

There is more about histograms in Unit 29.

Exercise 4B

1 Decide which of the following are qualitative or quantitative data. For quantitative data write down whether it is discrete or continuous.

    (a) age   **(b)** date of birth   **(c)** hair colour

    (d) height   **(e)** personality   **(f)** intelligence

    (g) favourite drink

2 Holly kept a record of the goals that her hockey team had scored:

    2 0 2 3 1

    5 3 2 1 2

    1 2 3 1 2

    5 0 0 0 1

    (a) Draw up a frequency table for this data.

    (b) Show the data as a bar graph.

3   As part of a science project Danielle recorded the number of dandelion plants in each square metre of a lawn:

0	1	3	1	3
0	1	2	3	1
1	1	1	0	4
0	1	2	3	2
1	0	3	5	2

Draw up a table to display her data.

4   Sam was studying the hand spans of pupils in his class. He measured, correct to the nearest millimetre, the size of the hand span for 30 pupils. Here are his results:

12.6   13.2   10.9   12.8   13.8   15.9
 9.8   12.1   15.4   13.7   14.0   11.8
10.5   11.9   12.5   9.9   15.0   14.3
11.1   11.1   13.4   14.2   15.3   15.1
12.2   12.7   13.0   14.8   14.4   13.7

(a) Draw up a table to display this data.

(b) Display the data as a histogram.

5   Peter took a random sample of 40 people and obtained their ages. The results were:

21   18   37   51   36   26   81   15   7   53
31   54   72   33   19   42   37   13   47   56
57   16   28   49   27   38   9   33   24   71
87   37   26   62   68   33   43   55   46   8

Present this information in the form of a stem and leaf diagram.

Use your stem and leaf diagram to work out the median of these forty ages.

4.4 Collecting data

You can collect data from many sources and for many purposes. For example:

From an experiment

You will have done experiments in your science work. For example, you could do an experiment to find out whether there is a relationship between the length of a pendulum and the time it takes to complete a swing.

To do this you would need to collect data about the time taken to complete, say 20 swings for pendulums of different lengths.

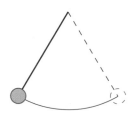

The time it takes a pendulum to complete a swing is related to its length.

From a survey

Companies and other organizations do market research to find out what customers like and dislike about their existing products and to see whether they would like new products. They may collect data by carrying out a survey, using interviews and a questionnaire.

To test a hypothesis

A **hypothesis** is a statement of an idea which you can test using an experiment or survey. For example:

'People prefer salt and vinegar crisps to plain crisps.'

'The taller the tree, the bigger its diameter.'

4.5 Questionnaires

One way of collecting data is to use a **questionnaire**. When you are writing questions for a questionnaire:

■ **be clear what you want to find out, and what data you need**

■ **ask short, simple questions**

Here are some good examples:

Are you:

Male ☐
Female ☐

This has a clear choice of two answers.

What age are you: **Which of these do you like:**

under 21 ☐   Soul ☐
21–40 ☐   Rock ☐
41–60 ☐   Pop ☐
over 60 ☐   Classical ☐

These both offer four choices.

■ **avoid questions which are too vague, too personal, or which may influence the answer.**

Do you go running:

sometimes ☐   occasionally ☐   often ☐

'Sometimes', 'occasionally' and 'often' may mean different things to different people.

Have you ever stolen anything from a shop:

Yes ☐   No ☐

Even a hardened criminal is unlikely to answer this question honestly!

> Do you agree that running is good for you:
>
> Yes ☐   No ☐

This is a leading question which suggests that the right answer is Yes. It is **biased**.

Test your questionnaire on a few people first to see if it works or needs to be improved. This is called a **pilot survey**.

Example 2

The Governors of Lucea High School want to find out whether parents think that girls attending the school should wear ties. Design a question to obtain parents' views on this which could be used in a questionnaire.

Several designs are possible. Here is one example:

> Please tick the box which most accurately reflects your view on this statement:
>
> 'Girls attending Lucea High School should wear ties.'
>
> ☐ strongly   ☐ agree   ☐ not sure   ☐ disagree   ☐ strongly
>   agree   disagree

You could use only the middle three boxes. Using more than the five boxes here would make it harder to find a pattern in the responses.

Avoiding bias

When you are collecting data you need to make sure that your survey or experiment is **fair** and avoids **bias**. A bias is anything that might make the data unrepresentative.

For example, if you want to find out which sport is most popular amongst 14–16 year olds in the UK, you will not be able to ask everyone in this age group. You will have to ask a **sample** of 14–16 year olds and treat their views as being representative of all people of their age.

If you only asked people at a football match your survey would be biased because you would be collecting the views of an unusually large number of male football enthusiasts.

The question 'Do you agree that running is good for you?' is biased too because it suggests that a particular answer is correct.

Exercise 4C

1   Write down, with reasons, whether the following questions would be suitable for use in a questionnaire. Suggest improved versions where necessary.

   (a) Do you watch films?

   (b) Do you agree that it is wrong not to prevent people from smoking when they are not indoors?

   (c) Don't you think that your dog deserves the best dog food around?

   (d) How often have you drunk too much?

   (e) Do you believe that all people should eat good food?

2   For each of the following identify a possible source of bias:

   (a) You want to estimate the amount of traffic past your school. You take a traffic survey during school lessons.

   (b) You want to estimate the number of people in cars passing your school. You count the number of people in cars during school lessons.

   (c) You want to find out how good peoples' short-term memories are. You ask people in your class.

   (d) You want to find out what people think of school dinners. You stand outside the school dinner hall and ask people as they go in.

3   For each of the tasks in question 2 suggest a less biased way of collecting data.

4   Joan is investigating how balls bounce on flat surfaces. Her hypothesis is that the height of the first bounce is proportional to the height from which the ball is dropped. Explain how she could test her hypothesis.

5   Fred is trying to find out which weekly magazines are read most by students in his school. He decides to ask 10 of his friends.

   (a) Explain why this is not a reliable method.

   (b) Suggest two steps that Fred could take to get more reliable results.

6   Design an experiment to decide whether there is any difference between boys' and girls' ability to estimate lengths.

7   Design a questionnaire to collect data on one of the following:

   (a) television viewing habits (b) time spent on sport

   (c) use of a local youth club (d) school uniform

   (e) memory

8   Do people with big hands have large feet? Collect data to find out whether this is true.

4.6 Sampling

To find out which is the most popular band amongst students at your school or college you could ask everyone the question 'Which is your favourite band?'

The group of students you are interested in is called the **population**. Asking the whole population may take too much time. Instead you can ask a sample of the population whose views are likely to be representative of the views of the whole population.

The sample size should be small enough to be manageable, but large enough that their views are representative of the whole. In most cases a random sample of around 5% to 10% should be sufficient.

Random sampling

If there are 600 students in your school a manageable and representative sample size would be 30 students.

It is important that your sample is taken at random. If you only ask Radiohead followers your sample's views will not be representative of those of the whole population.

■ **A random sample is one in which each member of the population is equally likely to be selected.**

To take your random sample of 30:

1 List all 600 pupils in any order and label their names from 000 to 599.

000	C.J. Adams
001	P.E. Ash
002	C.R. Ashbee

2 Choose a way of generating numbers from 000 to 599. Here are some possible methods:

Method 1

Use one 6-sided and two 10-sided spinners to generate the three digits.

Method 2
Use the **RND** function on a scientific calculator.

Stratified sampling

In a survey on bands it is likely that the popularity of a band will be influenced by the age of the student. Some bands have more appeal to younger pupils, some to older ones.

You might want to look at the popularity of the various bands according to the ages of students. To do this you can put the pupils in groups called **strata**, according to their ages.

■ **A stratified sample is one in which the population is divided into groups called strata and each strata is randomly sampled.**

A convenient way of doing this in a school would be to use the year groups of the students as the strata.

In a school of 600 students there might be the following number of students in each year group:

Year group	Number of students
7	150
8	150
9	100
10	100
11	100

Take a random sample from within each strata – in this case from each year group.

One of the important things to do is to ensure that the numbers in each part of the sample are in the same proportion as the proportion of each strata to the entire population.

Each sample should be the same proportion of its year group. For example you could take a 6% sample of each of years 7 to 11.

A 6% sample is in the range 5%–10% and gives whole numbers of students in each strata:

Year group	Number of students	Size of sample 6%
7	150	9
8	150	9
9	100	6
10	100	6
11	100	6

For each strata the sample you take must be random.

Selective sampling

Selective sampling is a technique often used by companies which make components or items such as light bulbs, batteries, computer chips and tinned food which need to be tested to ensure that they have been made to satisfactory standards.

■ **A selective sample is one in which every *n*th item is chosen.**

To take a 5% sample you select 1 in every 20 items as they are made and test them.

Similarly you could make a selective sample of the 600 students at your school. To do this you would list all the names. Randomly select any number from 1 to 20 such as 9. Then from your list choose the 9th, 29th, 49th, 69th and so on student names. This will give you a selective random sample.

It is, of course, always possible to take a selective random sample from within a strata. This method can also be applied to stratified sampling for each strata.

Example 3

There are 1200 pupils at Lucea High School. This table shows how pupils are distributed by year group and gender:

Year group	Number of boys	Number of girls
7	144	156
8	150	150
9	87	113
10	102	96
11	100	102

Jamilla is conducting a survey about the pupils' favourite school subject. She decides to use a stratified random sample of 100 pupils according to year group and gender:

How many Year 7 girls should there be in her sample?

The number of girls in Year 7 is 156.

As a proportion of the total of 1200 this is $\frac{156}{1200} = 0.13$

Jamilla's total sample is to be 100 so she should sample:

$$0.13 \times 100 = 13 \text{ girls from Year 7}$$

Exercise 4D

1   For the example on page 76, work out the number of pupils
    Jamilla should sample from:

    (a) Year 8   (b) boys in Year 10

    (c) Explain why the correct answer to (b) above will create a
         problem for Jamilla. How could Jamilla resolve this problem?

2   Just before a local election a market research company makes a
    survey of the voting intentions of the electorate. There are
    30 000 people who can vote at the election. They are categorized
    according to age and gender as in the table below:

Age range (years)	Number of males	Number of females
Under 30	5800	6200
30 or over	9500	8500

    The market research company will try to obtain the views of a
    2000 person sample.

    Work out the number in the sample who should be:

    (a) aged under 30   (b) male

    (c) female aged 30 or over

3   Explain how you could take a selective sample of 2% of the
    names in a telephone directory.

4   There are 1200 pupils at Russell High School.

    Barbara wishes to take a random sample of 60 of these pupils
    for her humanities project.

    Describe at least three different ways in which Barbara could
    take such a sample.

4.7 Frequency polygons

This table shows the frequency distribution of marks in a test:

Mark	0–9	10–19	20–29	30–39	40–49	50–60
Frequency	16	14	18	9	6	2

This is discrete data grouped into class intervals 0–9, 10–19 and so on.

One way of presenting this data is to plot a graph of frequency
against mark. Plot the midpoint of each class interval against the
frequency for that interval and join the points with straight lines:

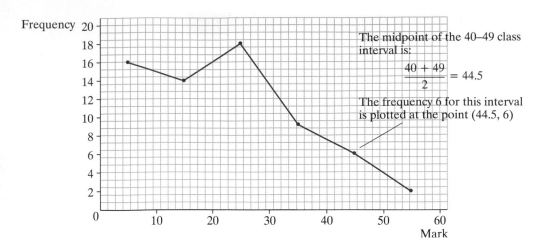

The resulting graph is called a **frequency polygon**. Here the midpoint of each class interval is treated as being representative of all the points in that interval.

1   James asked people to draw a line 5 cm long using a straight edge without any markings on it. Here are the lengths in centimetres of the lines drawn:

4.3   3.2   3.9   4.7   5.8   6.1
5.7   6.2   6.5   3.7   4.2   5.1
6.5   7.2   7.4   3.7   5.8   4.2
4.1   5.0   5.1   4.7   3.2   3.5
5.2   2.9   2.8   4.3   5.1   4.8

   (a) Draw up a grouped frequency table for the data. Use a class interval of 1 centimetre.

   (b) Draw a frequency polygon for the data.

2   This frequency polygon shows the times of 25 girls who each had to sew on two buttons with standard lengths of thread:

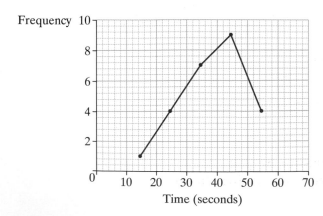

A group of 25 boys was challenged to do the same task. Here are their times, to the nearest second.

25	32	46	32	57
46	51	28	49	68
55	78	32	44	89
22	56	67	77	40
41	48	60	*	*

The stars represent two boys who gave up.

(a) Group the data and draw a frequency polygon.

(b) Compare the two frequency polygons. What conclusions can you draw from the two frequency polygons?

4.8 Cumulative frequency diagrams

■ **You can display both discrete and continuous data in a cumulative frequency diagram.**

Discrete data

Look at the test marks data given at the start of **Section 4.7** again. This is discrete data. You can show many people scored *up to a particular mark* by keeping a running total.

■ **The running total of the frequency at the end of each class interval is called the cumulative frequency.**

Here is a cumulative frequency table:

Mark	Cumulative frequency
0 up to but not including 10	16
0 up to but not including 20	30
0 up to but not including 30	48
0 up to but not including 40	57
0 up to but not including 50	63
0 up to 60	65

The cumulative frequency of marks up to but not including 20 is 16 + 14 = 30.

30 people scored marks up to but not including 20.

The same data can be presented in a **cumulative frequency diagram** by plotting the cumulative frequency against the end of each class interval.

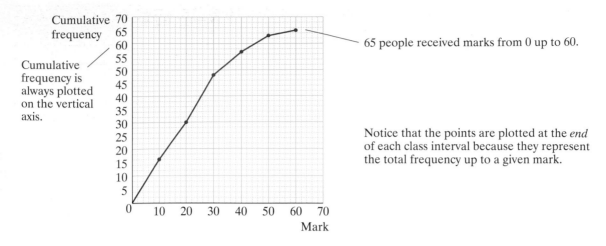

65 people received marks from 0 up to 60.

Notice that the points are plotted at the *end* of each class interval because they represent the total frequency up to a given mark.

Here the points are joined by straight lines giving a **cumulative frequency polygon**. They can also be joined by a continuous smooth curve giving a **cumulative frequency curve**.

Cumulative frequency diagrams can be used to solve a variety of problems. For example, if people gaining the top 20% of marks are to receive an A grade what is the lowest mark that will gain this grade?

If the top 20% of people gain an A grade then the 'bottom' 80% do not. There are 65 people altogether, so:

$$80\% \text{ of } 65 \quad \text{is} \quad 0.8 \times 65 = 52 \text{ people}$$

To find the highest mark scored by the people in this group:

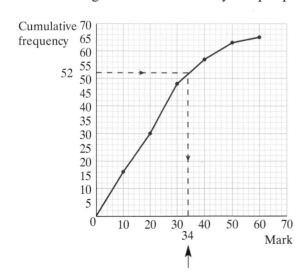

Draw a horizontal line from 52 on the cumulative frequency axis across to the graph…

… then draw a vertical line from the graph down to the mark axis.

The highest score in the bottom group was 34 marks.

From the cumulative frequency polygon, everyone scoring above a mark of 34 will receive an A grade. So the lowest mark that will gain an A grade is 35.

Worked examination question

The speeds in miles per hour (mph) of 200 cars travelling on the A320 road were measured. The results are shown in this table.

(a) On the grid draw a cumulative frequency graph to show these figures.

(b) Use your graph to find an estimate for the percentage of cars travelling at less than 48 miles per hour.

Speed (mph)	Cumulative frequency
not exceeding 20	1
not exceeding 25	5
not exceeding 30	14
not exceeding 35	28
not exceeding 40	66
not exceeding 45	113
not exceeding 50	164
not exceeding 55	196
not exceeding 60	200
Total	200

(a) Here is the cumulative frequency graph.

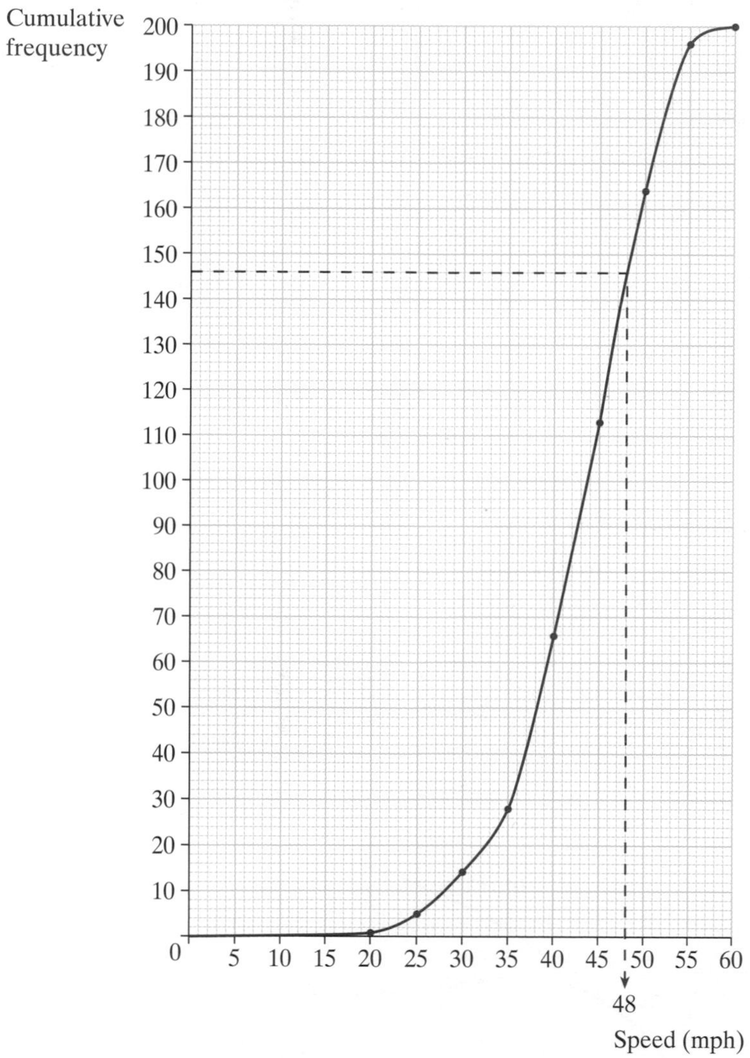

(b) The graph shows that 146 out of 200 cars are travelling at less than 48 mph.

$$\frac{146}{200} = \frac{73}{100} = 73\%$$

So 73% of the cars are travelling at less than 48 mph.

Continuous data

This table shows the time intervals between cars on the Lucea High Road:

The signs < (less than) and ⩽ (less than or equal to) show that a time of 10 seconds is in the first class interval.

Time t (seconds)	Frequency
$0 < t \leqslant 10$	13
$10 < t \leqslant 20$	13
$20 < t \leqslant 30$	7
$30 < t \leqslant 40$	8
$40 < t \leqslant 50$	5
$50 < t \leqslant 60$	4

The cumulative frequency table is:

Time (seconds)	Cumulative frequency
up to 10	13
up to 20	26
up to 30	33
up to 40	41
up to 50	46
up to 60	50

You plot the points
(10, 13)
(20, 26)
and so on,
and join them with a smooth curve.

Here is the **cumulative frequency curve**:

Remember to plot the point (0, 0) even though this is not a value in the table.

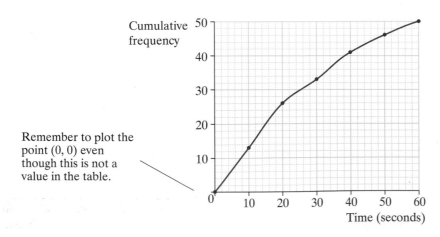

Here is another example. This table shows the time interval between cars on the Lucea Ring Road. The times in seconds are recorded to the nearest second:

Time interval (seconds)	Frequency
0–9	14
10–19	11
20–29	9
30–39	7
40–49	6
50–60	3

In this example the class interval 10–19 seconds is to the nearest second.

The least value in this class interval is 9.5 and the greatest value is 19.5.

You can draw up a cumulative frequency table using the greatest value in each class interval:

Time (seconds)	Cumulative frequency
up to 9.5	14
up to 19.5	25
up to 29.5	34
up to 39.5	41
up to 49.5	47
up to 60.5	50

You plot the points
(9.5, 14)
(19.5, 25)
and so on and join them with a smooth curve.

Here is the cumulative frequency curve:

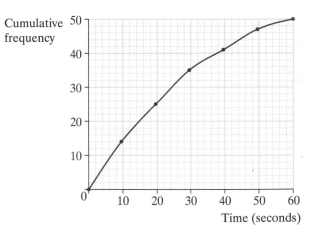

Cumulative frequency diagrams may have different shapes. Here are some examples and how to interpret them:

In this table of data the values increase at a steady rate, giving a straight line in the cumulative frequency diagram:

Class interval	Frequency
0–10	10
11–20	10
21–30	10
31–40	10
41–50	10

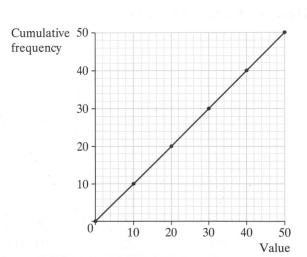

In this table the values are concentrated in the middle. This is shown by the steepness of the curve in the cumulative frequency diagram as it passes through the class interval 21–30:

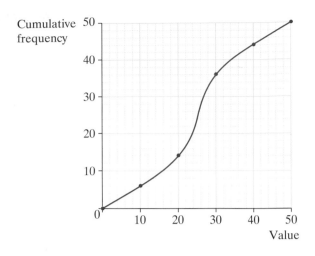

Class interval	Frequency
0–10	6
11–20	8
21–30	22
31–40	8
41–50	6

In this table there are few values in the class interval 21–30. This is shown in the cumulative frequency diagram by a shallow slope:

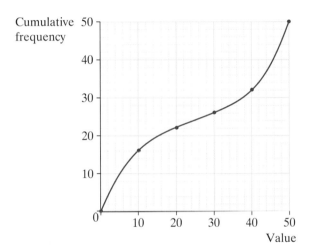

Class interval	Frequency
0–10	16
11–20	6
21–30	4
31–40	6
41–50	16

Exercise 4F

1   The marks in an examination were recorded and are shown in this table. The maximum mark obtainable was 100.

Mark range	Frequency	Mark range	Frequency
1–10	4	51–60	58
11–20	8	61–70	53
21–30	13	71–80	33
31–40	28	81–90	17
41–50	45	91–100	4

(a) Draw a cumulative frequency diagram.

(b) Use your diagram to estimate the percentage of people who scored more than 67 marks.

(c) What mark was exceeded by 60% of people?

2 Jane was performing an experiment to see how good people of various ages were at balancing a small block of wood on their finger. To save time she took measurements from Year 6 pupils at a local primary school and from Year 10 pupils at her own school. The times in seconds are shown in this table:

Primary school

Time (seconds)	Frequency
0 up to but not including 5	8
5 up to but not including 10	7
10 up to but not including 15	9
15 up to but not including 20	6
20 up to but not including 25	3
25 up to but not including 30	1

Secondary school

Time (seconds)	Frequency
0 up to but not including 5	3
5 up to but not including 10	5
10 up to but not including 15	13
15 up to but not including 20	10
20 up to but not including 25	8
25 up to but not including 30	4

(a) Draw cumulative frequency diagrams for each school.

(b) Compare the performances of the two schools.

4.9 Scatter diagrams

Is this hypothesis true:

'People with big feet also have big hands.'

One way to test the hypothesis is to measure the hands and feet of a sample of people. Here is some data from such a sample:

Name	Hand length	Foot length
James	19.0	28.5
Eric	17.7	31.4
Shane	15.1	25.2
Omah	16.8	26.7
Andrew	19.4	30.8
Ian	20.5	30.9
John	16.8	24.7
Ben	17.5	28.8
Atik	18.4	27.9
Hylton	19.9	29.0

You can plot hand length against foot length on a **scatter diagram** like this:

Notice that in general the bigger the foot length the bigger the hand length. This supports the hypothesis that people with big feet also have big hands.

■ **You can use a scatter diagram to show whether two sets of data are related.**

Plot the points
(19.0, 28.5)
(17.7, 31.4)
and so on.

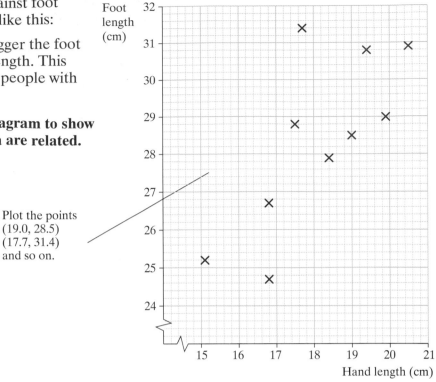

Exercise 4G

1   The demand for water from a pumping station together with the maximum daily temperature are given in this table:

Daily demand (millions of litres)	Maximum daily temperature (°C)
21.5	28
23.6	22
19.9	15
25.9	25
30.0	29
20.7	27
25.6	24
25.7	22
28.6	27

Draw a scatter diagram to illustrate these data. Comment on any association between maximum temperature and demand for water.

2   The table below shows the number of hours of sunshine and the maximum temperature in ten British towns, on one particular day.

Max temp (°C)	13	21	20	19	15	16	12	14	14	17
Number of hours of sunshine	11.7	16.7	15.2	15.4	13.2	11.8	9.8	10.2	12.4	13.7

(a)   Draw a grid like this on cm² graph paper.

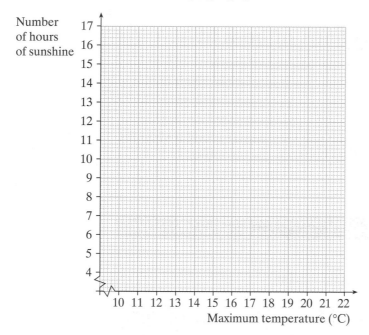

(b)   On the grid, plot the information as a scatter graph.

(c)   What does your diagram tell you about change in the maximum temperature as the number of hours of sunshine increases?   [E]

3   A physicist is experimenting with the resistance in a circuit she is using. She measures and records the resulting current:

Resistance (ohms)	5	10	15	20	25	30	50
Current (amps)	10	3.9	3.2	2.4	1.9	1.7	1.0

(a)   Draw a scatter graph of her results.

(b)   Estimate the current for a resistance of 40 ohms.

(c)   Estimate the resistance for a current of 7.5 amps.

4 This table gives the marks scored by pupils in a French test and in a German test.

French	15	35	34	23	35	27	36	34	23	24	30	40	25	35	20
German	20	37	35	25	33	30	39	36	27	20	33	35	27	32	28

Draw a scatter graph of the marks scored in the French and German tests.

4.10 Correlation and lines of best fit

Scatter diagrams can be used to show whether there is a relationship between two sets of data. Such a relationship is called a **correlation**.

■ **If the points on a scatter graph are very nearly along a straight line there is a high correlation between the variables.**

When the data in one set increases as the data in the other set increases the relationship between them is called **positive correlation**.

When the data in one set increases as the data in the other set decreases the relationship between them is called **negative correlation**.

When there is no linear relationship between the two sets of data there is **no correlation**.

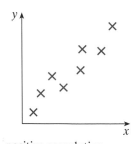

positive correlation

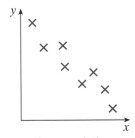

negative correlation

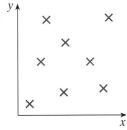

no correlation

Line of best fit

When there is a relationship between two sets of data there is some correlation. You can show this on a scatter diagram by drawing the **line of best fit**. This is a line that shows the general trend of the relationship between the two sets of data. It may or may not pass through any of the data points themselves.

■ **A line that is drawn to pass as close as possible to all the plotted points on a scatter graph is called the line of best fit.**

This scatter diagram (from the data on page 85) shows measurements of hand length plotted against measurements of foot length:

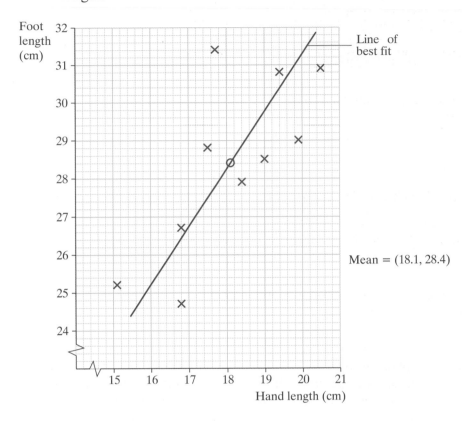

Mean = (18.1, 28.4)

The line of best fit should really be drawn through the point with coordinates (mean of all the hand lengths, mean of all the foot lengths). Using the data on page 85, this point = (18.1, 28.4). The point is circled on the graph. However you would not be expected to do this in your examination. Drawing the line 'by eye' would be suggested.

The line of best fit has been drawn by eye. You should draw the line so that there are about the same number of plotted points on each side of the line. Make the line so that the distances of the plotted points from it on each side balance each other out as near as possible.

The closer the points are to this line of best fit the higher the correlation. The gradient of the line is not important except that a vertical or horizontal line of best fit means that the variables are not connected. Here are some typical examples and their interpretation:

high or strong positive correlation

low positive correlation – not a strong relationship

no correlation – no linear relationship

high or strong negative correlation

low negative correlation – not a strong relationship

Exercise 4H

1   The table shows the marks of eight students in two papers in their mathematics exam.

Paper 1 mark	Paper 2 mark
65	46
83	75
35	40
21	21
56	42
25	20
67	60
75	63

Plot the marks on a scatter diagram. Draw on the line of best fit.
Nabi was ill for Paper 2 but scored 48 for Paper 1. Use the line of best fit to estimate his mark for Paper 2.
Ben was ill for Paper 1 but scored 89 for Paper 2.
Estimate his mark for Paper 1.
Do you think your answer is reasonable?

2   The table shows the volume of ice cream sales and the daily maximum temperature of a shop.

Sales of ice cream (litres)	Temperature °C
110	3
190	5
220	8
230	13
235	10
261	15
285	20
304	25

Which is the independent variable?
Draw a scatter diagram and put on the line of best fit.
Hence, estimate the weekly sales when the average temperature is 17°C.

3 **(a)** The table shows the areas and populations of the Midwestern States of the USA.

State	Area in km^2	Population
Arkansas	137 754	2 399 000
Colorado	269 594	3 470 000
Kansas	213 096	2 523 000
Nebraska	200 349	1 606 000
Ohio	115 998	11 016 000
Oklahoma	181 185	3 212 000
Utah	219 867	1 813 000
Wisconsin	171 496	5 007 000
Wyoming	253 324	466 000

Draw a scatter diagram and the line of best fit.
What information does the slope of the line of best fit provide?

(b) The table shows the areas and populations of eleven Southern States in the USA. Draw a scattergram and a line of best fit.

State	Area (sq km)	Population
Alabama	133 915	4 136 000
Florida	151 939	13 488 000
Georgia	152 576	6 751 000
Louisiana	123 677	4 287 000
Mississippi	123 514	2 614 000
Missouri	180 514	5 193 000
North Carolina	136 412	6 843 000
South Carolina	80 582	3 604 000
Tennessee	109 152	5 024 000
Virginia	105 586	6 377 000
West Virginia	62 758	1 812 000

(c) Compare your answers for **(a)** and **(b)**.

4 Investigate the statistical relationship for one of the following:

(a) hand size and head circumference

(b) the number of cars passing down the high street and the number of other vehicles passing down the high street

(c) attendance at football matches and position in the league

(d) trying to draw by hand a line 5 cm long and trying to draw by hand a line 15 cm long.

5 The scatter diagram shows the number of hours of sunshine
   and the maximum temperature in ten British towns on one day.

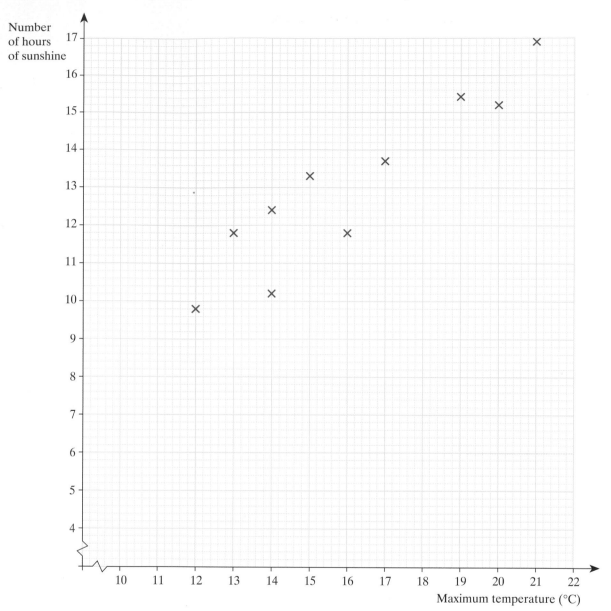

(a) Draw a line of best fit on the scatter diagram.

(b) Use your line of best fit to estimate
   (i)   the number of hours of sunshine when the maximum
         temperature was 18°C
   (ii) the maximum temperature when the number of hours
         of sunshine recorded in the day was 12 hours.   [E]

Summary of key points

1   **Qualitative** data is described using words.

2   **Quantitative** data consists of numbers, and can be discrete or continuous.

3   **Discrete** data can only take particular values. For example, you can buy shoes in these sizes: 6 $6\frac{1}{2}$ 7 $7\frac{1}{2}$ 8 These values are discrete (meaning separate). There are no values in between them. Discrete data has an **exact** value.

4   **Continuous** data can take any value. For example, your foot could be 18 cm long, or 21 cm long, or any value in between such as 19.1573 cm long. Continuous data cannot be measured exactly. The accuracy of a measurement depends on the accuracy of the measuring device.

5   **Class intervals** are groupings of quantitative data.

6   Data that is grouped and continuous can be displayed in a **histogram**:

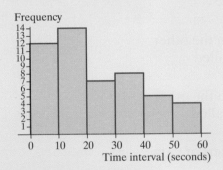

7   When you are writing questions for a questionnaire to collect data:
    ● be clear what you want to find out and what data you need
    ● ask short, simple questions
    ● avoid questions which are too vague, too personal, or which may influence the answer.

8   A **random sample** is one in which each member of the population is equally likely to be selected.

9   A **stratified sample** is one in which the population is divided into groups called strata and each strata is randomly sampled.

10   A **selective sample** is one in which every nth item is chosen.

11 Data that is grouped and discrete or grouped and continuous can be displayed …

… in a frequency polygon:

… in a cumulative frequency diagram:

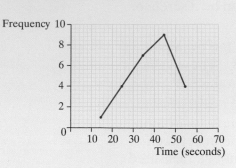

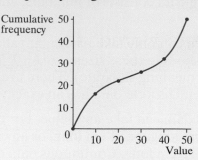

12 **Cumulative frequency** is the running total of the frequency at the end of each class interval.

13 You can use a **scatter diagram** to show whether two sets of data are related.

14 If the points on a scatter diagram are very nearly along a straight line there is a high correlation between the variables.

● When the data in one set increases as the data in the other set increases the relationship between them is called **positive correlation.**

● When the data in one set increases as the data in the other set decreases the relationship between them is called **negative correlation.**

● When there is no relationship between the two sets of data there is no correlation.

15 A line that is drawn to pass as close as possible to all the plotted points on a scatter diagram is called the **line of best fit.**

high or strong positive correlation

low positive correlation — not a strong relationship

no correlation — no linear relationship

high or strong negative correlation

low negative correlation — not a strong relationship

5 Using basic number skills

5.1 Percentage increases

1989 was known as the 'boom year' for house prices in the UK.
Prices went up by an average of 24%.

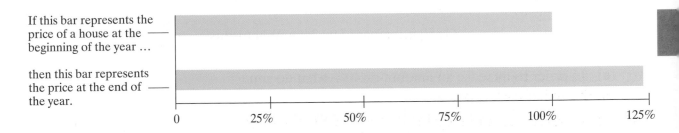

If this bar represents the price of a house at the beginning of the year ...

then this bar represents the price at the end of the year.

As a percentage the new value is:
$$100\% + 24\% = 124\% \text{ of the old value.}$$

Example 1

Jenny earns £26 000 a year. She gets a 12.5% increase.
Calculate her new salary.

If Jenny gets a 12.5% increase her new salary will be $(100 + 12.5)\%$
of her old salary.

112.5% as a decimal is 1.125
Her new salary will be $1.125 \times 26\,000 = £29\,250$

Example 2

Tom gets a 15% rise in pay.

He was earning £9500 a year. What is his new salary?

As a percentage of his old salary his new salary is:
$$(100 + 15)\% = 115\%$$

As a decimal fraction 115% is 1.15
Tom's new salary is $1.15 \times £9500 = £10\,925$

■ **If an amount is increased by $x\%$ the new amount is $(100 + x)\%$
of the original amount.**

Exercise 5A

1   Write the new value as a percentage of the old value if:
    (a) a house increases in value by 12%
    (b) the balance in a bank account increases by 17%
    (c) the weight of a man goes up by 30%
    (d) the cost of a car goes up by 11%
    (e) the cost of a washing machine increases by 8.5%

2   (a) If you want to increase a quantity by 24% what do you multiply by?
    (b) Graham's salary of £1850 is increased by 24%. What is his new salary?

3   (a) In order to increase an amount by 40% what do you multiply by?
    (b) The cost of a theatre ticket is increased by 40% for a special concert. What is the new price if the normal price was £5.40?

4   Gail wants to increase her weight by 4%. What should she multiply her present weight by?

5   (a) Increase £120 by 20%.
    (b) Increase 56 kg by 25%.
    (c) Increase 2.4 m by 16%.
    (d) Increase £1240 by 10.5%.
    (e) Increase 126 cm by 2%.

5.2 Percentage decreases

During 1993 house prices in the UK fell by an average of 15%.

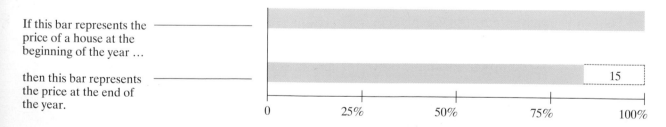

If this bar represents the price of a house at the beginning of the year ...

then this bar represents the price at the end of the year.

15

0   25%   50%   75%   100%

As a percentage the new value is
$$(100 - 15)\% = 85\% \text{ of the old value.}$$

■ **If an amount is decreased by x% the new amount is $(100 - x)$% of the original amount.**

Exercise 5B

1   Write the new value as a percentage of the old value if:
    (a)   a house decreases in value by 16%
    (b)   the balance in a bank account decreases by 11%
    (c)   the weight of a man goes down by 5%
    (d)   the value of a car goes down by 22%

2   **(a)**   If you want to decrease a quantity by 34% what do you multiply by?
    (b)   The value of JJ shares falls by 34%. Before the drop the shares cost £5.25 each. What is the new price?

3   Alan weighs 82 kg before going on a diet. He sets himself a target of losing 5% of his original weight. What is his target weight?

4   **(a)**   Decrease £120 by 20%.
    (b)   Decrease 56 kg by 25%.
    (c)   Decrease 2.4 m by 16%.
    (d)   Decrease £1240 by 10.5%.
    (e)   Decrease 126 cm by 2%.

Exercise 5C   (Mixed questions)

1   Carla invests £100 in a savings account on 1st January 1999. The interest paid is 6% per annum.

    (a)   What do you multiply by to find the amount she will have at the end of the first year?

    (b)   How much money did she have in her account at the end of the first year?

    (c)   Copy and complete this table. Give the amounts to the nearest £.

Date	Amount
1/1/99	£100
1/1/00	
1/1/01	
1/1/02	
1/1/03	

2 A car costing £14000 depreciates (loses its value) by:
   12% in its first year
   18% in its second year
   5% each year thereafter.

   (a) Find the value of the car after 1 year.

   (b) Copy and complete this table giving your
   answers to the nearest £.

New price	£14000
Value after 1 year	
Value after 2 years	
Value after 3 years	
Value after 4 years	
Value after 5 years	

3 The amount in an investment account increases by 10% in the first
   year and a further 8% in the second.

   (a) What should you multiply by to find the value after the first year?

   (b) What should you multiply by to find the value after two
   years? (Be careful: it is not 1.18.)

■ **The percentage increase is** $\dfrac{\textbf{actual increase}}{\textbf{original price}} \times \textbf{100\%}$

5.3 Finding percentage increases

Example 3

In 1994 a box of tissues cost 45p.
In 1996 a similar box cost 60p.
What is the percentage increase?

The actual increase is $60 - 45 = 15$p.

The fractional increase is $\dfrac{\text{actual increase}}{\text{original price}} = \dfrac{15}{45}$

Remember:
To change a decimal to a
percentage multiply by 100.

$\frac{15}{45}$ as a decimal fraction is $0.3\dot{3}$
Percentage increase is $0.3\dot{3} \times 100 = 33\frac{1}{3}\%$.

Exercise 5D

1 In 1999 a box of tissues had increased in price from 45p to 80p. Find:

   (a) the fractional increase from the 1994 cost of 45p

   (b) the fractional increase from the 1996 cost of 60p

   (c) the percentage increase from 1994

   (d) the percentage increase from 1996.

2 Calculate the percentage increase from:

   (a) £24 to £36   (b) 12.5 kg to 20 kg

   (c) 2.45 m to 2.86 m   (d) 50 seconds to 1 min.

3 Before fitting a 'fuel-saver' Jamie's car travelled 36 km on 4.5 litres of petrol. After fitting the fuel-saver it travelled 58.3 km on 5.5 litres. Find:

   (a) the fuel consumption in km/litre before the fuel saver was fitted

   (b) the fuel consumption in km/litre after it was fitted

   (c) the percentage increase in fuel consumption.

4 Del boy Rotter buys an old cottage for £84 000. He spends £10 400 on repairs and renovation, then puts the cottage up for sale for £149 000. Find:

   (a) his actual profit

   (b) his percentage profit to the nearest 1%.

5 A mechanic at a garage draws 1.5 litres of oil from a 24 litre oil drum which was exactly $\frac{1}{4}$ full. Find:

   (a) the amount of oil left in the drum

   (b) the percentage of the original amount that is left.

5.4 Finding percentage decreases

■ **The percentage decrease is** $\dfrac{\textbf{actual decrease}}{\textbf{original price}} \times \textbf{100\%}$

Example 4

Josie invests £800 in a company called Spec-U-Late Ltd. One year later she finds her investment is only worth £680. Find:

(a) her actual loss
(b) her percentage loss

(a) Her actual loss is £800 – £680 = £120.

(b) Her fractional loss is $\frac{120}{800}$
   $\frac{120}{800}$ as a decimal fraction is 0.15
   So her percentage loss is 0.15 × 100 = 15%

Exercise 5E

1 Josie invested £800. One year later this investment was only worth £680. After a further year Josie's investment had fallen to £540. Find:

   (a) her actual loss over the two years

   (b) her percentage loss over the two years

   (c) her percentage loss between the end of the first year and the end of the second year of her investment.

2   Find the percentage decrease to the nearest 1% of:
   (a) a decrease from £48 to £32
   (b) a decrease from 5.2 kg to 3.8 kg
   (c) a decrease from 45 cm to 39.5 cm
   (d) a decrease from 2 min to 110 seconds
   (e) a decrease from 1 metre to 45 cm.

3   This table shows the value of a car after a number of years from new:

New price	£12500
after 1 year	£10000
after 2 years	£8600
after 3 years	£7500
after 4 years	£6600

   (a) Calculate the percentage loss for each year as a percentage of the original price.
   (b) Calculate the percentage loss between the second and third years.

4   Sanjit bought a house in 1984 for £84000. In 1989 he sold it to James for £120000. In 1994 James had to sell the house for £80000. Calculate:
   (a) Sanjit's percentage profit
   (b) James' percentage loss
   (c) the percentage change in the value of the house between 1984 and 1994.

5.5 Mixing percentage increases and decreases

Example 5

In 1987, Mr Peck's house was valued at £60000.
The value of the house increased by 20% during 1988.
During 1989 the value of the house decreased by $12\frac{1}{2}\%$.

Work out the value of the house at the end of 1989.

By the end of 1988, the value of the house, in pounds, was

$$60000 + 20\% \text{ of } 60000$$
or $60000 (1 + 0.2)$
or 60000×1.2
         $= 72000$

By the end of 1989, the value of the house, again in pounds, had fallen by 12.5% of the £72000.

So by the end of 1989, the value of the house was

$$72000 - 12.5\% \text{ of } 72000$$
or $\quad 72000 \times (1 - 0.125)$
or $\quad 72000 \times 0.875$
$$= 63000$$

Exercise 5F

1   The price of a holiday in 2000 was £450.
The same holiday cost an extra £12% in 2001.
In 2002 the same holiday was reduced by 15% of its price in 2001.
Work out the price of this holiday in 2002.

2   Jenny bought a flat in September 1998.
The value of the flat then was £52000.
During the first year, the value of the flat decreased by 8%.
Over the next year the value of the flat increased by 14%.
Work out the value of the flat in September 2000.

3   Erica bought £10000 worth of shares in May 2000.
By May 2001 the value of the shares had increased by 15%.
By May 2002 the value of the shares had decreased by $n\%$ of their value in May 2001.
The value of the shares in May 2002 was again £10000.
Work out the value of n.

5.6 Working out what you save

Example 6

The price of a carpet is reduced by 20% in the sales. It now costs £60. What was the pre-sale price?

The sale price is $(100 - 20)\% = 80\%$ of the pre-sale price
$$= 0.8 \text{ in decimal form}$$

A flow diagram can be used to represent information in the question:

$$\overset{\times\, 0.8}{\text{pre-sale price} \longrightarrow £60}$$

To solve this type of flow diagram problem draw a second flow diagram reversing the direction and the calculation:

$$\overset{\div\, 0.8}{£75 \longleftarrow £60}$$

The pre-sale price of the carpet was £75.

Remember:
In this type of question you are *not* finding 80% of the *given* price, but 80% of the *unknown* price.

Exercise 5G

1. A television is reduced by 10% in a sale. The sale price is £540.
   (a) What percentage of the pre-sale price is the sale price?
   (b) Draw a flow diagram to show this information.
   (c) Find the pre-sale price.

2. The total price of a holiday includes 17.5% VAT. In the brochure the total price is quoted as £320 per person.
   (a) What percentage of the pre VAT price is the total price?
   (b) Draw a flow diagram to show this information.
   (c) Find the price before VAT is added.

3. Due to falling orders, Jonson's Electrical Company decreases its workforce by 8% down to 161 employees. Find:
   (a) the number of employees before the decrease
   (b) the actual decrease in number.

4. After two years the value of a new drilling machine has fallen to £12 000. The depreciation is estimated at 10% per year. Find to the nearest penny:
   (a) the value of the drilling machine after the first year
   (b) the original price.

10% OFF
IF YOU BUY BEFORE
JANUARY 31ST

Sale Price £540

5.7 Compound interest

Tessa invests £250 in a building society account. The account pays an annual interest rate of 6% per annum. She wants to know how much her money will be worth in 6 years' time.

This problem would be easy if the building society paid **simple interest**: 6% of the original amount for each of the 6 years:

6% interest on £250 for 1 year is £15

6% interest on £250 for 6 years is £15 × 6 = £90

Tessa's investment after 6 years would then be worth £250 + £90 = £340.

But building societies usually pay **compound interest**, meaning they pay interest on the interest too. So at the end of the second year Tessa will get 6% interest on both her original investment and on the first year's interest.

This table shows how her investment will grow over 6 years:

Year	Amount at start of year	Interest	Total amount at year end
1	£250	250 × 0.06	£265
2	£265	265 × 0.06	£280.90
3	£280.90	280.90 × 0.06	£297.75
4	£297.75	297.75 × 0.06	£315.62
5	£315.62	315.62 × 0.06	£334.56
6	£334.56	334.56 × 0.06	£354.63

After 6 years' compound interest Tessa's investment will have grown to £354.63 to the nearest penny, compared with £340 if simple interest only were paid.

■ **Compound interest is interest paid on an amount plus the interest already earned.**

Exercise 5H

Using the text above:

1 (a) How much more interest does Tessa receive if compound interest is paid instead of simple interest?

   (b) What percentage of the original investment is the extra interest?

2 Copy the table showing Tessa's compound interest and extend it to find how much the original investment would have grown to after ten years. Use your calculator's memories to help do this.

3 (a) How much more interest does £250 invested at 6% compound interest for 10 years gain than £250 invested at 6% simple interest?

   (b) What percentage of the £250 is the extra interest?

   (c) What annual simple interest rate over the 10 year investment period would give the same amount of interest as the 6% compound interest rate?

4 How many years will it take £250 to double in value when invested:

   (a) at 6% compound interest

   (b) at 6% simple interest?

5 On 1st January Surjit invests £1000 in an investment account that pays an interest rate of 8.2%
Calculate the value at:

(a) the end of the 1st year

(b) the end of the 4th year

(c) the end of the 8th year.

6 Compare the amounts of interest paid on £1000 invested for 8 years at 8.2% per annum in accounts paying:
(a) simple interest   **(b)** compound interest.

7 The Bank of Nirvana pays an annual interest rate such that an investment doubles in value every 2 years.
Claire thinks that the interest rate paid per annum must be 50%. Explain why she must be wrong.

8 (a) Calculate the percentage growth if:

    (i) an interest rate of 50% is paid on £1000 for 2 years

    (ii) an interest rate of 40% is paid on £1000 for 2 years.

  (b) Starting with your answers for parts (i) and (ii) use a trial and improvement method to calculate the annual interest rate paid by the Bank of Nirvana in question **7**. Give your answer to the nearest 0.1%.

9 Colin rents a house at a monthly rent of £320. In the rent agreement it states that the rent will be reviewed annually and will be increased by the annual inflation rate.

Assuming an annual inflation rate of 4% find:

(a) the amount of rent paid in the first year

(b) the rent increase at the end of the first year

(c) the monthly rent paid during the second year

(d) the monthly rent to be paid after the sixth year.

10 Sally takes out a six-year policy with an insurance company. She invests £100 at the start of each year. The company pays a compound interest rate of 8% per annum.

  (a) Copy and complete this table:

Year	Value at start of year	Interest	Total at end of year
1	100	100×0.08	£108
2	108 + 100	208×0.08	£224.64
3	224.64 + 100		
4			
5			
6			

  (b) How much is Sally's policy worth at the end of the tenth year?

11 A car costs £12000 when bought new. It depreciates (loses value) at an annual rate of 10%. Find:
  (a) the value of the car at the end of the first year
  (b) the value at the end of the second year
  (c) the value at the end of the fifth year
  (d) the time it takes for the value of the car to reach £6000.

12 By what factor will an investment increase if paid a compound interest rate of 12.5% per annum over a period of:
  (a) 4 years   (b) 6 years   (c) 10 years.

  Write your answers as decimals or percentages.

13 Alena borrows £4600 pounds to buy a car at an annual interest rate of 16.5%.
  After making all her interest payments Alena works out that she has paid a total of £11500 to the finance company. Using any method you know, calculate how long she took out the loan for.

5.8 Compound measures

Compound measures combine measurements of two or more different types. For example, a speed is a measurement of a *distance* and the *time* taken to travel it. This section shows you how to use a variety of compound measures.

Average speed (rate of travel)

Lisa travelled from Nottingham to London, a distance of 205 km. She started at 07.00 and arrived in London two and a half hours later.

She did not travel at a constant speed throughout the journey, but her average speed can be found from:

■ **Average speed = $\dfrac{\textbf{distance travelled}}{\textbf{time taken}}$**

$$= \frac{205}{2.5} = 82 \text{ km/h}$$

In travel problems times are often given in hours and minutes. To make calculations easier change the minutes into a decimal fraction of an hour.
To do this divide the minutes by 60. For example:

42 minutes = 42 ÷ 60
      = 0.7 hours

Example 7

Jacki takes 36 minutes to walk to school, a distance of 1.8 km. Find her average speed in:

(a) metres per minute (b) km per hour (c) miles per hour.

(a) Distance travelled 1800 m
    Time taken 36 minutes

$$\text{Average speed} = \frac{1800}{36} = 50 \text{ m/min}$$

(b) Distance travelled 1.8 km
Time taken $36 \div 60 = 0.6$ hours

$$\text{Average speed} = \frac{1.8}{0.6} = 3 \text{ km/h}$$

(c) Distance travelled $1.8 \times 0.625 = 1.125$ miles
Time taken $36 \div 60 = 0.6$ hours

$$\text{Average speed} = \frac{1.125}{0.6} = 1.88 \text{ mph}$$

1 km = $\frac{5}{8}$ mile
1 km = 0.625 mile

Exercise 5I

1 *The Flying Scotsman* took 7 hours 15
minutes to travel from Edinburgh to
London, a distance of 632 km.

   (a) Calculate the average speed of the
train in km per hour.

   (b) Given that 1 km is approximately
equal to 0.62 miles, calculate the
average speed in miles per hour.

2 Paul travelled by coach from Derby to
Birmingham, a distance of 45 miles.
The journey took 1 hour 45 minutes.
Calculate the average speed of the coach.

3 Change these minutes into hours, giving your answers correct
to 2 d.p. where appropriate:

   (a) 75 minutes   **(b)** 24 minutes   **(c)** 6 minutes

   (d) 25 minutes   **(e)** 45 minutes   **(f)** 10 minutes.

4 A car travelled 50 km in 2 hours 42 minutes. How long was the
journey in:

   (a) minutes

   (b) hours?

5 Shannon took 1 hour 40 minutes to complete a science
experiment. Jack took 1.6 hours to complete the same
experiment. Work out:

   (a) who took the longest time, and by how much, to complete
the experiment

   (b) the average (mean) time taken by the two students.

To change a decimal fraction
of an hour into minutes
multiply by 60. For example:

0.6 hours = 0.6×60
= 36 minutes

6 Change these hours into hours and minutes:

   (a) 0.4 hours   **(b)** 5.3 hours   **(c)** 3.25 hours

   (d) 6.29 hours   **(e)** 3.9 hours   **(f)** 9.88 hours

7 Ron and Garth set out at 1200 hours for a 25 km walk. They walked for 2 hours 18 minutes, then had a 30 minute rest before completing their walk in a further 2.2 hours. Calculate:

   (a) the time it took to complete the 25 km walk

   (b) the average speed for the journey.

8 A train travelled at 80 km per hour for 2.25 hours, then 60 km per hour for a further 4 hours 36 minutes. Find:

   (a) how far the train travelled altogether

   (b) how long the total journey took

   (c) the average speed for the journey.

Other rates

Example 8

The average fuel consumption of a car is measured in kilometres per litre.

A car uses 17.6 litres (ℓ) of petrol in travelling 220 km. What is its average fuel consumption?

$$\text{Average fuel consumption} = \frac{220}{17.6} = 12.5 \text{ km per litre or km}/\ell$$

Exercise 5 J

1 What is the average fuel consumption of a car which travels 464 km on 32 litres of petrol?

2 Mr Robson sets out with a full tank of petrol on a business trip of 588 km. When he arrives his petrol gauge indicates his tank is $\frac{1}{8}$ full.

   (a) If a full tank holds 48 litres of petrol, calculate:

      (i) how much petrol it took to complete the journey

      (ii) the average fuel consumption.

   (b) If the average fuel consumption remains the same, how much further can he travel before he runs out of petrol?

3 Carol fits a *regular* 4.5V battery into her tape recorder. The battery costs £2.60 and lasts for 320 hours. She replaces it with a *long life* battery which costs £4.68 but lasts for 600 hours.

   (a) Find: (i) the cost per hour for each battery

           (ii) the time per £ for each battery.

   (b) Which type of battery is the best buy and why?

4   Kay got these exchange rates from a newspaper:

£1 was equivalent to 2.25 German marks

£1 was equivalent to 1.58 US dollars

Calculate the exchange rate between the USA and Germany in:

(a) marks per dollar   **(b)** dollars per mark.

5   If this bath is filled from the cold tap only it takes
5 minutes to fill. If filled from the hot tap only it takes
8 minutes to fill.
Calculate:

(a) the flow rate in litres per
minute of each tap

(b) how long it takes to fill the
bath if both taps are
turned fully on.

920 litres

Measuring flow

■   **For liquid flowing through a pipe:**
volume flowing per second = cross-sectional area × flow rate.

Example 9

Roger is watering his vegetable patch with a hosepipe which has a
cross-sectional area of 1.6 cm^2. Water flows out of the pipe at a
speed of 5 cm per second.

Find the amount of water flowing out of the pipe in:

(a) cm^3 per second

(b) litres per hour.

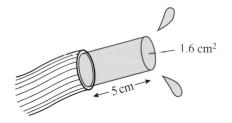
1.6 cm²
5 cm

Answer

(a) In one second a cylinder of water 5 cm long and
1.6 cm^2 in cross-section flows out of the pipe.

Volume flowing per second = cross-sectional area × flow rate
$$= 1.6 \times 5$$
$$= 8 \text{ cm}^3/\text{second}$$

(b) If 8 cm^3 flow per second then

$60 \times 8 \text{ cm}^3$ will flow per minute $= 480 \text{ cm}^3/\text{min}$

and $60 \times 480 \text{ cm}^3$ will flow per hour $= 28\,800 \text{ cm}^3/\text{hour}$

$$= \frac{28\,800}{1000} = 28.8 \text{ litre/hour}$$

$1000 \text{ cm}^3 = 1 \text{ litre}$

Exercise 5K

1   This water butt is leaking at a rate of 0.25 litre/min. How long will it take in hours and minutes for a full butt of water to leak away?

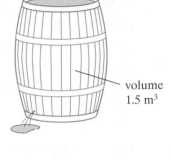

volume
1.5 m³

2   Water flows through a plastic pipe at a rate of 260 cm³ per second. Find:
   (a) the amount of water discharged in:
      (i) cubic metres per second   (ii) litres per minute
   (b) the time it takes to discharge one cubic metre of water
   (c) the time it would take to fill a water tank holding 800 litres of water.

3   A lorry travelling at 40 km per hour uses diesel fuel at an average consumption of 6.25 km per litre.
   (a) What is the consumption:
      (i) in litres per km   (ii) in litres per hour?
   (b) (i) How far will the lorry travel in 15 minutes?
      (ii) How much petrol will it use?

Density

■   **The density of a substance is its mass per unit volume:**

$$\text{density} = \frac{\text{mass}}{\text{volume}}$$

Densities are often given in g/cm³.

Example 10

This block of lead has a volume of 540 cm³ and a mass of 6.156 kg. Find its density.

$$\text{Density} = \frac{\text{mass}}{\text{volume}} = \frac{6156 \text{ g}}{540 \text{ cm}^3} = 11.4 \text{ g/cm}^3$$

Example 11

A cast-iron rod 2 metres long has a cross-sectional area of 14 cm². The mass of the rod is 210 kg. Calculate:
(a) the volume of metal in the rod
(b) the density of cast iron in g/cm³.

Answer

(a) volume = area × length
   $= 14 \text{ cm}^2 \times 200 \text{ cm} = 2800 \text{ cm}^3$

(b) density $= \dfrac{\text{mass}}{\text{volume}} = \dfrac{210\,000}{2800} = 75 \text{ g/cm}^3$

Exercise 5L

1   A cast-iron rod has a mass of 240 kg. Its cross-sectional area is 10 cm^2 and its density is 7.5 g/cm^3. Calculate:

   (a) the volume of the rod

   (b) the length of the rod in metres

2   The density of aluminium is 2590 kg/m^3. Find:

   (a) the mass of a piece of aluminium which has a volume of 2.5 m^3

   (b) the volume of a piece of aluminium whose mass is 1200 kg.

3   This is the cross-section of a prism 12 cm long. It has a mass of 3.5 kg. Calculate:

   (a) its volume

   (b) the density in g/cm^3 of the material from which it is made.

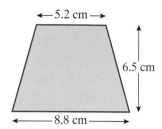

5.9 Ratios

■   **You can use ratios to show how things are divided or shared.**

Example 12

Tony and Kamiljit bought a pack of five blank tapes. They shared the tapes and the cost. Tony took two tapes and Kamiljit took three.

(a) In what ratio did they share the tapes?

(b) What fraction of the tapes did Tony take?

(c) The pack cost £3. What was a fair amount for Tony to pay?

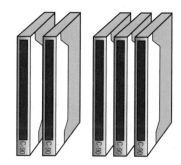

ratio: 2 : 3

fraction: $\frac{2}{5}$ $\frac{3}{5}$

Answer

(a) They shared the tapes in the ratio 2:3

(b) Tony took $\frac{2}{5}$ of the tapes

(c) Tony should pay:

$$\frac{2}{5} \times £3 = £1.20$$

Example 13

Barbara and Peter inherit £4000 in their aunt's Will. The money is to be shared 3:2 in favour of Barbara. How much does each receive?

Sharing in the ratio 3:2 means that the money is divided into 5 parts, with Barbara getting 3 parts and Peter getting 2 parts.

So Barbara gets: $\frac{3}{5} \times £4000 = £2400$

and Peter gets: $\frac{2}{5} \times £4000 = £1600$

(Check this by adding: £2400 + £1600 = £4000)

Exercise 5M

1   Errol, Jane and Karl open a bag of chocolate and find it
    contains 12 bars. They decide to share them in the ratio 1:2:3.
    In turn Errol takes one bar, Jane two and Karl three. They
    keep doing this until they have shared out all the bars.

  (a) How many chocolate bars will:
        (i) Errol have   (ii) Jane have   (iii) Karl have?

  (b) When all the chocolate bars have been shared out, what
        fraction of the total will:
        (i) Errol have   (ii) Jane have   (iii) Karl have?

        Write each fraction in its simplest form.

2   Sally, Brian and Mark want to share £40 in the ratio 2:3:5

  (a) What fraction of the £40 will each person receive?

  (b) How much will each person receive?

3   Donna and Shirley share £270 in the ratio 3:5

  (a) To work out how much each receives, how many parts
        should £270 be divided into?

  (b) What fraction of £270 does Donna receive?

  (c) How much does Donna receive?

  (d) What fraction does Shirley receive?

  (e) How much does Shirley receive?

4   Danny and Melissa work in a bar. Danny works
    five days a week but Melissa only works two
    days a week.

    To be fair they agree to share their tips in the
    ratio of the number of days they work.

  (a) In what ratio do they share their tips?

  (b) What fraction of the tips should
        Danny receive?

  (c) If the tips in one week came to £66.50 how
        much did Melissa receive?

  (d) The following week the tips came to £84. How much did
        Danny receive?

  (e) One week Danny received £25 in tips. How much did
        Melissa receive?

5   In a GCSE maths course 80 of the final marks are for written
    exam papers. The other 20 are from coursework assignments.

  (a) Write the ratio of exam marks to coursework marks in its
        simplest form.

  (b) Daly got a final percentage of 66. The ratio of his exam
        marks to coursework marks was 8:3. How many
        coursework marks did he get?

6   Tom sells three types of soap powder in his shop: New Mold, New Drift and New Purr. In one week he sells 476 boxes of soap powder in the ratio 1:2:4

    (a)   What fraction of the total sales are New Drift?

    (b)   How many boxes of New Mold does he sell?

    (c)   If he makes 8p profit on each box sold, how much profit does he make altogether from the two best-selling powders?

7   Val and Bill run The Grange Residential Home. They held a fund raising morning and raised £360. Val raised £240 from selling raffle tickets. The rest was raised by Bill selling refreshments.

    (a)   How much did Bill raise from the sale of refreshments?

    (b)   What is the ratio of the amount raised from selling raffle tickets to the amount raised from refreshments?

    (c)   The following year a second fund raisingmorning was held and a total of £423 raised. The money raised was in the same ratio as in the previous year.
       How much was raised from selling raffle tickets?

8   Four darts players each threw three rounds of three darts to raise money for charity. The charity received one pound for each point scored. The players scored a total of 1728 between them. Their individual scores were in the ratio 3:4:4:5

    (a)   What score did the best player obtain?

    (b)   What must the best player have scored each round to achieve this total score?

    (c)   What was the lowest total score?

    (d)   What was the difference in score between the highest scorer and the lowest?

9   In Midgrove School the ratio of pupils to teachers is 17.2 to 1.

    (a)   Multiply 17.2 by 5.

    (b)   Rewrite the ratio in the form $m:n$ where m and n are both whole numbers. (Use part **(a)** to help you.)

    (c)   What is the smallest possible number of pupils in the college?

    (d)   If the actual number of pupils and teachers is 1456, how many teachers are there?

Example 14

In a bottle of Pepsea two secret ingredients are added in the ratio 5:8.
If 5.2 mℓ of the first ingredient is added:

(a) how much of the second ingredient is added

(b) how much in total is added?

Answer

(a) Of the total amount added $\frac{5}{13}$ is the first ingredient and $\frac{8}{13}$ the second.

$\frac{5}{13}$ of the total amount added is 5.2 mℓ

Find 1 part from 5 parts:

$\frac{1}{13}$ of the total amount added is $5.2 \div 5 = 1.04$ mℓ

So $\frac{8}{13}$ of the total amount is $8 \times 1.04 = 8.32$ mℓ

8.32 ml of the second ingredient are added.

(b) There are 5.2 mℓ of the first ingredient and 8.32 mℓ of the second so a total of $5.2 + 8.32 = 13.52$ mℓ of the two ingredients are added.

Exercise 5N

1 At Christmas Darren, Nathan and Joe received Christmas cards in the ratio 4:4:5. Joe received 75 cards:

(a) What fraction of the cards did Joe receive?

(b) What fraction did Darren and Nathan receive between them?

(c) How many cards did Darren receive?

(d) How many cards did they receive altogether?

2 A petrol company carried out a fuel consumption test and found that the winter to summer ratio for the same car over the same test track was 3.5:4. The winter fuel consumption rate was 8.2 km per litre. Find the summer consumption rate.

3 When driving in France the speed limit in towns is 60 km/h.
5 miles is approximately equal to 8 km.

(a) Write as a ratio the distance in miles to the distance in kilometres.

(b) Mr Been drives through a French town at 35 mph. Is he breaking the speed limit?

(c) The maximum speed allowed on British roads is 70 mph. What is this speed in km/h?

4 The average mark of three teaching groups at Midgrove School was in the ratio 4:5:7. The average mark of the middle group was 62.5. Find:

(a) the lowest average mark

(b) the difference between the highest and lowest average marks

(c) the ratio of the lowest average mark to the highest average mark expressing your answer in the form $1:n$.

5 Ben and Catherine win the national lottery and have a motoring holiday in Europe. They visit Paris, Rome, Berlin and Oslo, and divide the time spent in each city in the ratio $4:5:7:9$. They spent 12 days in Paris.
Find:

(a) how many days they spent in Oslo

(b) how many days they spent in Berlin

(c) how much longer they spent in Berlin than Rome

(d) how many days they spent away altogether?

6 During one season Blackford Rovers, Muncaster United and Sheffield Thursday scored a total of 270 goals between them. Blackford Rovers scored twice as many as Muncaster United, who scored 3 times as many as Sheffield Thursday.
Find:

(a) the scoring ratio between the three football clubs

(b) how many goals Muncaster United scored

(c) the difference between the number of goals scored by Blackford Rovers and Sheffield Thursday.

7 A gang of four villains robbed a bank but got caught. On passing sentence the judge told the first villain: 'You are the gang leader and will serve the longest sentence.' To the next: 'You were second in charge so you will serve the next longest sentence'. To the fourth robber he said 'You were led astray by the others so I will give you the shortest sentence. I sentence you all to a total of 122 years in prison to be served in the ratio $1:\frac{1}{2}:\frac{1}{3}:\frac{1}{5}$.' Work out how long each person's prison sentence was.

5.10 Standard form

Standard form is a convenient kind of shorthand for writing large and small numbers. To use it you need to know how to write powers of 10 in index form.

For example:

$$10 = 10^1$$
$$100 = 10 \times 10 = 10^2$$
$$1000 = 10 \times 10 \times 10 = 10^3$$

This **power** or **index** means multiply 10 by itself 3 times. It does not mean 3×10

Numbers greater than 1

Here are some numbers greater than 1 written in standard form:

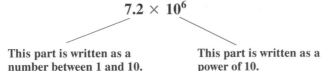

number				in standard form
3000	=	3 × 1000	=	3×10^3
5 000 000	=	5 × 1 000 000	=	5×10^6
7 200 000	=	7.2 × 1 000 000	=	7.2×10^6

■ **A number is in standard form when:**

$$7.2 \times 10^6$$

This part is written as a number between 1 and 10.   **This part is written as a power of 10.**

72×10^5 is *not* in standard form.

This part should be written as a number between 1 and 10.

Astronomers use standard form to record large measurements. The sun's diameter is about 1 392 000 km or 1.392×10^6 km.

Exercise 5O

1 Express these numbers as powers of 10.
(For example $10 \times 10 \times 10 \times 10 = 10^4$.)

  (a) 10 × 10

  (b) 10 × 10 × 10

  (c) 10 × 10 × 10 × 10 × 10 × 10

  (d) 10 × 10 × 10 × 10 × 10 × 10 × 10 × 10 × 10 × 10

You can use the **EXP** key on your calculator to enter numbers in standard form. To enter 6.3×10^7 press:

2 Write each number as a number of millions using decimals where necessary. (For example 8 650 000 is 8.65 million.)

  (a) 6 000 000   **(b)** 3 400 000   **(c)** 7 800 000

  (d) 5 500 000   **(e)** 2 650 000   **(f)** 7 642 000

3 Write these numbers in the form: number $\times 10^6$.
(For example 2 100 000 is 2.1×10^6.)

  (a) 3.1 million   **(b)** 4.3 million   **(c)** 0.5 million

  (d) 2 400 000   **(e)** 7 800 000   **(f)** 8 600 000

  (g) 4 000 000   **(h)** 9 000 000   **(i)** 400 000

4 Write these numbers as powers of 10.
(For example $100 = 10 \times 10 = 10^2$.)

  (a) 1 thousand   **(b)** ten thousand

  (c) ten   **(d)** 1 million

  (e) 1 hundred thousand   **(f)** ten × ten × ten

5 Copy and complete this table. The first one is done for you:

10^6	10^5	10^4	10^3	10^2	10^1	10^0	Standard form
		7	4	0	0	0	7.4×10^4
				2	6	0	
	6	8	0	0	0	0	
					4	5	
9	9	0	0	0	0	0	
					6	2	
						8	

6 Write these numbers in standard form:
(a) 16   (b) 4300   (c) 650000
(d) 87000000   (e) 670   (f) 865
(g) 9870000   (h) 98500   (i) 805000000000

Changing from standard form to a decimal number

Here is how to change 5.2×10^4 from standard form into a decimal number:

$$
\begin{aligned}
5.2 \times 10^4 &= 5.2 \times 10 \times 10 \times 10 \times 10 \\
&= 52 \times 10 \times 10 \times 10 \\
&= 520 \times 10 \times 10 \\
&= 5200 \times 10 \\
&= 52000
\end{aligned}
$$

So $5.2 \times 10^4 = 52000$ as a decimal.

Exercise 5P

1 Change these numbers from standard form to ordinary decimal numbers:
(a) 4.2×10^2 (b) 6.7×10^4 (c) 5.5×10^3
(d) 7.5×10^6 (e) 6.2×10^5 (f) 7.3×10^4
(g) 2.4×10^7 (h) 1.1×10^1 (i) 7.25×10^0

2 Evaluate these expressions giving your answers in standard form:
(a) 25×36 (b) 640×15
(c) 45×900 (d) 25^4

Numbers less than 1

Very small numbers can also be written more conveniently in standard form.

You need to know how to write these as powers of 10:

$$0.1 = \frac{1}{10} = \qquad = 10^{-1}$$

$$0.01 = \frac{1}{100} = \frac{1}{10 \times 10} = 10^{-2}$$

$$0.001 = \frac{1}{1000} = \frac{1}{10 \times 10 \times 10} = 10^{-3}$$

$$0.0001 = \frac{1}{10000} = \frac{1}{10 \times 10 \times 10 \times 10} = 10^{-4}$$

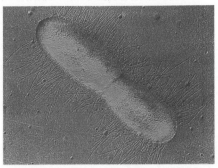

Biologists working with micro-organisms use standard form to record their sizes. This bacterium is 0.000006 m long or 6×10^{-6} m. Magnification: $\times 29400$.

Here is how to write 0.041 in standard form:

First write 0.041 as a fraction with a numerator between 1 and 10.

$$0.041 = \frac{4.1}{100} = 4.1 \times \frac{1}{100} = 4.1 \times \frac{1}{10^2} = 4.1 \times 10^{-2}$$

0.041 in standard form is 4.1×10^{-2}. Here $\times 10^{-2}$ means divide by 10^2 or 100.

Here is how to change a number in standard form back to an ordinary decimal number.

$$2.4 \times 10^{-3} = \frac{2.4}{10^3} = \frac{2.4}{1000} = 0.0024$$

$$10^{-3} = \frac{1}{10^3}$$

Exercise 5Q

1   Copy and complete this table. The first one is done for you:

	10^0	10^{-1}	10^{-2}	10^{-3}	10^{-4}	10^{-5}	**Standard form**
(a)	0	0	0	2	4		2.4×10^{-3}
(b)	0	2					
(c)	0	0	0	0	0	6	
(d)	0	1	5				
(e)	0	0	0	7			
(f)	0	0	0	0	4	5	
(g)	0	0	3	4	6		
(h)	0	0	0	1	2	5	

2 Write these numbers in standard form:

(a) 0.002   (b) 0.15   (c) 0.0004

(d) 0.054   (e) 0.000008   (f) 0.000 000 000 068

(g) 0.346   (h) 0.09   (i) 0.0056

3 Change these numbers back to ordinary decimal number form:

(a) 3.5×10^{-1} (b) 6.0×10^{-2} (c) 7.2×10^{-4}

(d) 2.2×10^{-3} (e) 1.35×10^{-5} (f) 5.33×10^{-6}

(g) 8.8×10^{-10} (h) 4.4×10^{-7} (i) 4.999×10^{-1}

Remember:
The index rule for multiplying powers is:

$$10^m \times 10^n = 10^{m+n}$$

So:

$$10^5 \times 10^3 = 10^8$$

Example 15

Work out the answer to $(2.95 \times 10^5) \times (4.0 \times 10^3)$, giving your answer in standard form.

Rearrange the expression with the powers of 10 on the right:

$$2.95 \times 4.0 \times 10^5 \times 10^3$$

Multiply out the numbers on the left and the powers on the right:

$$11.8 \times 10^8$$

For an answer in standard form this number should be between 1 and 10.

11.8 is 1.18×10 so: $(1.18 \times 10) \times 10^8 = 1.18 \times 10^9$

This is in standard form

Worked examination question

The distance from the Earth to the Moon is 250 000 miles.

(a) Express this number in standard form.

The distance from the Earth to the Sun is 9.3×10^7 miles.

(b) Calculate the value of the expression:

$$\frac{\text{distance from the Earth to the Moon}}{\text{distance from the Earth to the Sun}}$$

giving your answer in standard form.   [E]

(a) $250\,000 = 2.5 \times 10^5$

(b) Substituting the distances in the expression gives: $\dfrac{2.5 \times 10^5}{9.3 \times 10^7}$

Here are two ways of evaluating it:

Method 1	Method 2
$$\frac{2.5 \times 10^5}{9.3 \times 10^7}$$	$$\frac{2.5 \times 10^5}{9.3 \times 10^7}$$
$= \dfrac{250\,000}{93\,000\,000} \;\overset{\div\,10\,000}{\underset{\div\,10\,000}{=}}\; \dfrac{25}{9300}$	$= \dfrac{2.5 \times 1}{9.3 \times 10^2}$
$= 0.002\,69$ (to 3 s.f.)	$= 0.269 \times \dfrac{1}{10^2}$
$= 2.69 \times 10^{-3}$	$= 0.269 \times 10^{-2}$
	$= 2.69 \times 10^{-3}$

Remember:
The index rule for dividing powers is:

$$10^m \div 10^n = 10^{m-n}$$

So:

$$10^5 \div 10^7 = 10^{(5-7)} = 10^{-2}$$

Using Standard Form to make estimates

■ **Standard form can be used to make approximations and estimates.**

Example 16

The distance from the Earth to the Sun is 93 000 000 miles.
The Earth moves around the Sun in an approximately circular orbit.
One complete orbit of the Sun takes 365 days.

Use standard form to help work out an estimate of the average speed of the Earth as it orbits the Sun.

The Earth's orbit of the Sun is a circle of 93 000 000 miles.
This can be approximated as $90\,000\,000 = 9 \times 10^7$ miles.

Assume that the orbit of the Sun is a circle. The distance travelled by the Earth as it makes one orbit of the Sun is equal to the circumference of a circle of radius 9×10^7 miles.
So the total distance travelled by the Earth in 365 days is

$$2 \times \pi \times 9 \times 10^7$$

But π is about 3.1 so $2 \times \pi$ is just over 6.

$$2 \times \pi \times 9 \text{ is about 55.}$$

Hence the distance travelled by the Earth in 365 days is approximately

$$55 \times 10^7 = 5.5 \times 10 \times 10^7$$
$$= 5.5 \times 10^8 \text{ miles.}$$

365 days is 365 × 24 hours, approximately:

$$350 \times 20 = 7000$$
$$= 7 \times 10^3 \text{ hours.}$$

So the Earth travels approximately 5.5×10^8 miles in approximately 7×10^3 hours.

Using the formula

$$\textbf{speed} = \frac{\textbf{distance}}{\textbf{time}}$$

the speed of the Earth as it orbits the Sun is:

$$\text{speed} = \frac{5.5 \times 10^8}{7 \times 10^3}$$

$$= \frac{55 \times 10^7}{7 \times 10^3}$$

$$= \frac{55}{7} \times \frac{10^7}{10^3}$$

$$= \frac{55}{7} \times 10^4$$

This sort of division should be done **mentally**.

$\frac{55}{7}$ is approximately 8

So the speed is approximately 8×10^4 miles per hour.

A good method for approximating $55 \div 7$ mentally is to think of the number closest to 55 in the 7 times table. $7 \times 8 = 56$.

You could use different approximations to get your answer. For example, you could have approximated the distance from the Earth to the Sun as 100 000 000 miles, or 1×10^8 miles. This would give you a different answer.

In your exam you should make sure that you clearly state what approximations you are using.

Exercise 5R

1   Using a calculator evaluate these expressions giving your answers in standard form:

   (a) $(6.4 \times 10^8) \times (1.5 \times 10^5)$ (b) $(8.2 \times 10^7) \times (2.1 \times 10^{-4})$

   (c) $(5.25 \times 10^2) \times (1.6 \times 10^{-5})$ (d) $(4.1 \times 10^{-6}) \times (3.6 \times 10^{-7})$

   (e) $\dfrac{6.4 \times 10^4}{2.2 \times 10^5}$ (f) $\dfrac{5.6 \times 10^{-2}}{3.8 \times 10^7}$ (g) $\dfrac{3.4 \times 10^{24}}{1.2 \times 10^{-6}}$

   (h) $\dfrac{8.8 \times 10^{25}}{8.8 \times 10^{22}}$ (i) $(2.8 \times 10^3)^2$ (j) $(2.2 \times 10^{-2})^3$

2   Evaluate these expressions giving your answers in standard form:

   (a) $500 \times 600 \times 700$

   (b) 0.006×0.004

   (c) $\dfrac{0.08 \times 480}{180}$

   (d) $\dfrac{89\,000 \times 0.0086}{48 \times 0.25}$

   (e) $\dfrac{65 \times 120}{1500}$

   (f) $\dfrac{8.82 \times 5.007}{10\,000}$

   (g) $(12.8)^4$

   (h) $(36.4 \times 24.2)^{-3}$

3   The Earth's diameter is 1.27×10^4 km and the diameter of Mars is 6.79×10^3 km.

   (a) Which planet has the larger diameter?

   (b) What is the difference between their diameters?

   (c) What is the total if the two diameters are added?

   (d) How many times bigger is one planet than the other? (Hint: consider their volumes.)

4   The distance from the Earth to the Sun is approximately $93\,000\,000$ miles.
Light travels at a speed of approximately $300\,000$ kilometres per second.
1 mile is approximately 1.6 kilometres

Use standard form to work out an estimate of the time it takes light to travel from the Sun to the Earth.

Exercise 5S   (Mixed questions)

1   The auction price of a second hand car is 25% less than the retail price of the same car.

   (a) Calculate the auction price of a car which has a retail price of £4200.

   (b) Calculate the retail price of a car which has an auction price of £5400.

2   Asif puts £200 into a building society savings account. He leaves the money in the account for 3 years, without withdrawing any or adding any to his savings.

The building society pays a compound interest rate of 4% per annum.

How much money will Asif have in his account at the end of the three years?

3   On a test drive, a new car travels 307 miles on 8.3 gallons of petrol. Work out the average number of miles the car travels on each gallon of petrol. Give your answer correct to two significant figures.

4   In a competition, the total prize money is £2000.
    This is divided between the winners of the 1st, 2nd and 3rd prizes in the ratio:
    1st : 2nd : 3rd = 5:3:1
    Calculate the value of the 1st prize.

5   Dinosaurs roamed the Earth about 140 million years ago.
    (a) Express 140 million in standard form.
    (b) The average human life span is 72 years. How many average human life spans is 140 million years? Give your answer in standard form, correct to three significant figures.

6   Last year 623 118 students sat GCSE English.
    The entry fee for each GCSE student was £17.50.
    Use standard form to work out an estimate for the total entry fee for all the GCSE English students last year.

7   Approximately $\frac{2}{3}$ of the Earth's surface is covered with water. The Earth can be assumed to be a sphere with a radius of 6350 km.
    Use standard form and the formula

    Surface area of sphere $= 4 \times \pi \times$ radius^2

    to work out an estimate of the area of the surface of the Earth covered by water.

Summary of key points

1   If an amount is increased by $x\%$ the new amount is $(100 + x)\%$ of the original amount.

2   If an amount is decreased by $x\%$ the new amount is $(100 - x)\%$ of the original amount.

3   Compound interest is interest paid on an amount plus the interest already earned.

4   Average speed $= \dfrac{\text{distance travelled}}{\text{time taken}}$ (typical units: km/h)

5   Volume flowing per second = cross-sectional area × flow rate
                                    (typical units: cm^3/s)

6 Density $= \dfrac{\text{mass}}{\text{volume}}$ (typical units: kg/m^3)

7 You can use ratios such as $2:3$ and $5:4:7$ to show how things are divided or shared.

8 Large and small numbers can conveniently be represented in standard form:

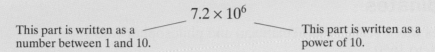

$$7.2 \times 10^6$$

This part is written as a number between 1 and 10.

This part is written as a power of 10.

9 Standard Form can be used to make approximations and estimates.

6 Transformations and loci

6.1 Coordinates

René Descartes was a French mathematician and philosopher who lived from 1596 to 1650.

Descartes developed a method for locating the position of a point by giving its distances from two reference lines called **axes**. These distances are now called the Cartesian coordinates of the point.

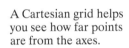

A Cartesian grid helps you see how far points are from the axes.

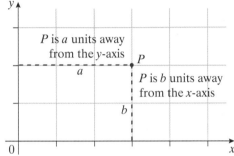

It is said that Descartes found it very difficult to get up in the morning and that he preferred to stay in bed and think. The strain of getting up at 5 o'clock in the morning to teach Queen Christina of Sweden is supposed to have killed him.

The coordinates of point P are (a, b).

The axes can be extended to create four **quadrants**. The quadrant in which a and b are positive numbers is called the first quadrant.

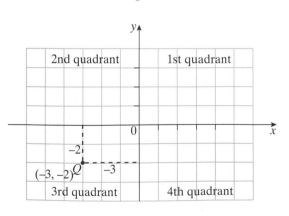

The coordinates of point Q are $(-3, -2)$

■ **Coordinates such as (3, 4) or more generally (x, y) are used to specify location on a Cartesian grid.**

Example 1

Write down the coordinates of the letters shown:

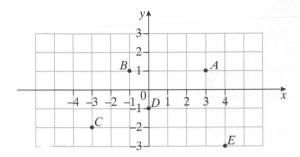

The coordinates are:

$A = (3, 1)$ $B = (-1, 1)$ $C = (-3, -2)$ $D = (0, -1)$ $E = (4, -3)$

Exercise 6A

Use graph paper or squared paper.

1   Draw a set of axes and label each one from –5 to 5. Use the same scale for each axis. Plot these points:
    $A\ (2, -1)$ $B\ (0, -3)$ $C\ (-1, 4)$ $D\ (-4, 1)$ $E\ (-2, -4)$.

2   Draw a set of axes and label each one from –6 to 6. Use the same scale for each axis. Plot these points and join them up in the given order to form a closed shape:
    (a)   (i) $(-4, 5), (-3, 6), (-2, 5), (-3, 2)$
            (ii) $(-6, -4), (-5, -2), (-2, -2), (-3, -4)$
            (iii) $(4, 3), (1, 6), (6, 5)$
            (iv) $(5, -2), (3, -4), (-1, 2)$
    (b)   Draw the diagonals inside the shapes in part **(a)** (i) and (ii). Write the coordinates of the intersection of these diagonals.

3   Give the coordinates of a fourth point D, to complete each rectangle where vertices are at:
    (a) $A\ (3, -2), B\ (7, -2), C\ (7, -4)$
    (b) $A\ (-3, 2), B\ (-5, 2), C\ (-3, -5)$
    (c) $A\ (1, 4), B\ (2, 6), C\ (6, 4)$
    (d) $A\ (-1, 3), B\ (3, 1), C\ (1, -3)$

4   The centre of a square is at $(1, 2)$ and one of the vertices of the square is at $(4, 7)$. Find the coordinates of the other three vertices of the square.

6.2 Bearings

Imagine your boat is in the harbour at point *H*. There is a ship out at sea at point *S*. You can describe the position of the ship by giving its bearing and distance from you:

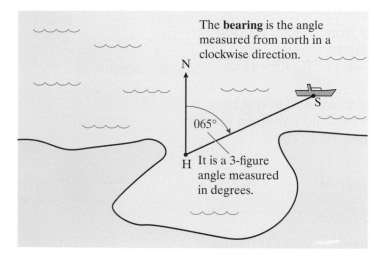

The **bearing** is the angle measured from north in a clockwise direction.

N

065°

S

H It is a 3-figure angle measured in degrees.

The ship's bearing from you is 065°. Giving the direction is not enough. You also need to know that it is 5 km away.

■ **A bearing is an angle expressed in three digits to indicate direction.**

Example 2

Fred lives at *F*. His friend Winston lives 1 km away on a bearing of 210°.

(a) Show this on a diagram.

Fred visits Winston. After a while Fred returns home.

(b) Calculate the bearing Fred takes from Winston's home to arrive back home.

(a)

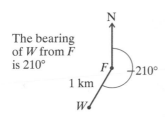

The bearing of W from F is 210°

(b) W is on a bearing of 210° from F. Fred needs to turn through half a turn (180°) to face home again.
The bearing of F from W:

$210° + 180° = 390°$ (This is more than 1 full turn.)
$390° - 360° = 030°$

The bearing of F from W is 030°

Alternative method is to use **alternate angles** (see page 33).

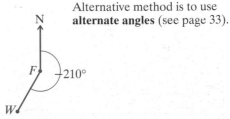

■ **A bearing giving the reverse direction is called a back-bearing.**

Exercise 6B

1 Draw diagrams to show these bearings:
  (a) 016°   **(b)** 200°   **(c)** 325°
  (d) 009°   **(e)** 168°

2 Write the bearings shown in these diagrams for OA, OB, OC.

(a)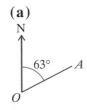

(b)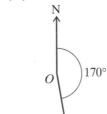

(c)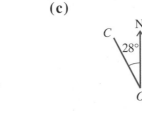

3 Three friends live near each other. Their homes are shown by X, Y and Z.
Y is on a bearing of 055° from X.
Angle $XYZ = 120°$ as shown on the diagram.
Calculate the bearing of:

  (a) X from Y
  (b) Z from Y
  (c) Y from Z.

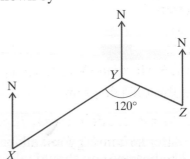

4 This is a plan of the centre of Chichester.

    (a) You are standing at the railway station. Give the bearing and distance to the Market Cross.

    (b) You are standing at the Market Cross. Give the bearing and distance to St. Paul's Church.

    (c) From St. Paul's you want to go swimming. Give the bearing and distance to the swimming pool.

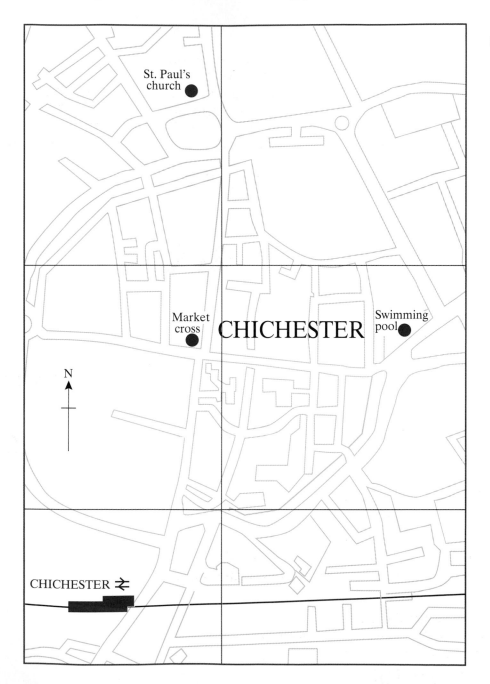

Scale 12.5cm to 1km

    (d) After swimming you walk back to the Market Cross. Write the bearing you should take to return to the railway station.

6.3 Transformations

Changes in an object's position or size are called **transformations**.
There are four kinds of transformations that need to be considered
here.

Translation

The first is translation.

Example 3

Triangle A has been **translated**. Each part of the triangle has moved
an equal distance in the same direction. This transformation is
described by a vector $\begin{pmatrix} x \\ y \end{pmatrix}$. The top number describes the movement
in the x-direction and the lower number describes the movement in
the y-direction. The movement of triangle A to triangle B is given by
$\begin{pmatrix} 2 \\ 1 \end{pmatrix}$. Triangle B is the **image** of triangle A.

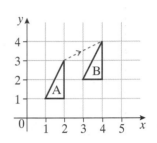

■ **A translation moves every point on a shape the same distance
and direction. A translation is described by a vector $\begin{pmatrix} x \\ y \end{pmatrix}$.**

Exercise 6C

1 Write the vectors describing these translations:
(a) flag A to flag B   (b) flag B to flag A
(c) flag D to flag B   (d) flag C to flag E
(e) flag A to flag E.

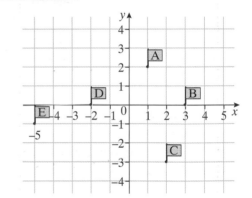

2 Use graph paper or squared paper.
Draw a set of axes and label each one from
–5 to 5. Use the same scale for each axis.
Draw a trapezium A, with vertices at (–5, 3),
(–4, 3), (–3, 2), (–3, 1)
(a) Transform A by the translation $\begin{pmatrix} 7 \\ 1 \end{pmatrix}$
Label the new trapezium B.
(b) Transform B by the translation $\begin{pmatrix} -1 \\ -6 \end{pmatrix}$
Label the new trapezium C.
(c) Write the single translation required to transform A to C.
(d) Write the single translation required to transform C back
to A.

3 Use graph paper or squared paper.
Draw a set of axes and label each one from –5 to 5.
Plot points A (0, 1) and B (3, 2)
T_1 describes a translation of $\begin{pmatrix} 3 \\ 2 \end{pmatrix}$
T_2 describes a translation of $\begin{pmatrix} -2 \\ 1 \end{pmatrix}$

Write down the coordinates of A and B after these translations:

(a) T_1 **(b)** T_1 followed by T_1 again

(c) Write down a single translation for **(b)**

(d) T_1 followed by T_2

(e) T_2 followed by T_1

(f) Write down what you notice about the images in **(d)** and **(e)**.

Reflection

A bounty hunter used reflection to help him locate a World War Two Japanese submarine which had sunk to the bottom of the Pacific Ocean containing £25 million of gold. He used SONAR (**so**und **na**vigation **r**anging) to locate the submarine. Sound pulses sent from a ship on the surface bounce back from the seabed. Changes in the seabed can be detected by the change in time taken for the pulse to return.

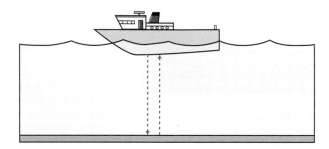

Stand directly in front of a mirror. You see your reflected **image**. The image appears to be the same distance away from the mirror as you are. Move closer to the mirror and the image moves closer. The mirror acts as a line of symmetry. The image of an object can be located by matching corresponding points on the object with points the same distance behind the mirror.

■   **A reflection produces a mirror image in a line of symmetry.**

Example 4

Copy the diagram using a large scale.

(a) Draw the reflection of the shape in the line $y = x$.

(b) Reflect the image back in the line $y = x$.

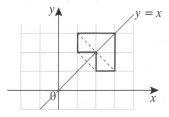

Answer

(a) Take each vertex on the object and locate its image. Imagine a line perpendicular to the line of symmetry. Join the points to produce the image.

(b) The image reflected in the line $y = x$ returns back to the object position.

■ **A repeated reflection in the same line of symmetry returns the image to the original object.**

■ **A reflection of a shape can be described by giving the equation of the line of symmetry.**

Exercise 6D

1 Use graph paper or squared paper.
Draw a set of axes and label each axis from –5 to 5.
Use the same scale for each axis.
Draw a rectangle with vertices (3, 1), (3, 2), (5, 2), (5, 1).
Label the rectangle A.

 (a) (i) Draw the reflection of A in the x-axis.
 Label the image B.

 (ii) Write the coordinates of B.

 (iii) Write how the coordinates of A are changed by the transformation.

 (b) (i) Draw the reflection of A in the y-axis.
 Label the image C.

 (ii) Write the coordinates of C.

 (iii) Write how the coordinates of A are changed by the transformation.

 (c) Draw the reflection of A in the line $y = x$.
 Label the image D.

 (d) Draw the reflection of D in the y-axis.
 Label the image E.

 (e) Write the reflection required to transform E to its image C.

2 (a) Copy this graph.
 Reflect the object in the line $x = 3$.

 (b) Reflect the image in the line $y = 2$.

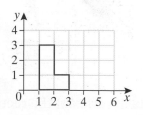

3   Use your results from questions 1 and 2 to complete this table.

Reflection in:	Transforms
x-axis	point (**a**, **b**) to _____
y-axis	point (**a**, **b**) to _____
line $y = x$	point (**a**, **b**) to _____
line $y = -x$	point (**a**, **b**) to _____
line $x = c$	point (**a**, **b**) to _____ **(a < c)**
line $y = d$	point (**a**, **b**) to _____ **(b < d)**

Rotation

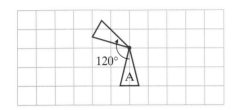

Triangle A is turned 120° clockwise about a **centre of rotation**.

An object can be turned about a point. This point is called the centre of rotation.

Example 5

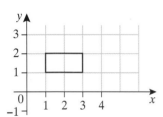

Rotate the rectangle 90° clockwise about the origin.

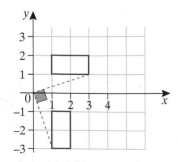

Each vertex of the rectangle has been rotated 90° clockwise.

This transformation could also be described as a rotation of 270° anticlockwise about the origin.

- ■ **A rotation is described by giving:**
  - ● **centre of rotation**
  - ● **amount of turn**
  - ● **direction of turn.**
  (A clockwise direction may be indicated by a negative sign, and an anticlockwise direction by a positive sign.)

- ■ **A rotation is defined by its centre and the angle and direction of turn.**

Exercise 6E

1   Use graph paper or squared paper.
Draw a set of axes and label each axis from –5 to 5.
Use the same scale for each axis.
Draw a triangle with vertices (3, 1), (5, 1), (5, 3).
Label the triangle A.

(a)   Rotate A about the origin through 90° anticlockwise (+90°). Label the image B. Write the coordinates of the vertices of B.

(b)   Rotate B about the origin through 90° anticlockwise (+90°). Label the image C.
Write the coordinates of the vertices of C.

(c)   Write a single rotation to transform A to C.

(d)   Rotate A about the origin through 90° clockwise (–90°). Label the image D. Write the coordinates of the vertices of D.

(e)   Write a single rotation to transform B to D.

(f)   Rotate A about the point (3, 0) through 90° clockwise (–90°). Label the image E.

2   Copy this diagram. Rectangle *ABCD* has been rotated to *A'B'C'D'*. To find the centre of rotation:
Join a pair of corresponding vertices (for example *A A'*).
Construct the perpendicular bisector.
Join another pair of corresponding vertices.
Construct the perpendicular bisector.

The centre of rotation lies on the point of intersection of the bisectors. Check that each vertex of *ABCD* has rotated through the same angle of rotation.

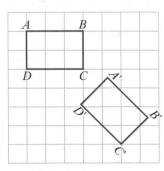

ABCD and *A'B'C'D'* are the **same size** rectangles.

3 The shorthand notation R_0 (90°) means an anticlockwise rotation about the origin through 90°. Applying R_0 (90°) to the point P (2, 1) would result in an image at point Q (−1, 2). Write the coordinates of the images after applying these rotations.

(a) R_0 (180°) to point (3, 4)

(b) R_0 (270°) to point (1, 3)

(c) R_0 (90°) to point (−1, 2)

Enlargement

Different sized photographs can be printed from the same negative. In each photograph the dimensions are in the same proportion as in the original. The images are not distorted. Each length is enlarged or reduced by a **scale factor**.

Example 6

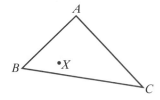

(a) Copy and enlarge the triangle ABC using X as the centre of enlargement, and a scale factor of 2.

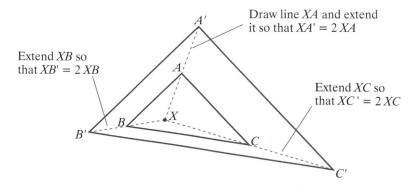

Draw line XA and extend it so that $XA' = 2\,XA$

Extend XB so that $XB' = 2\,XB$

Extend XC so that $XC' = 2\,XC$

The enlarged triangle $A'B'C'$ is similar to triangle ABC. Each side of $A'B'C'$ is twice as long as the corresponding side in ABC. This is because the scale factor of the enlargement is 2. Corresponding angles are identical. Each of the lines in the enlarged shape is parallel to the corresponding original side.

(b) Copy and enlarge the triangle *ABC* using *Y* as the centre of enlargement and a scale factor of 2

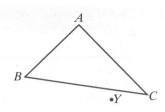

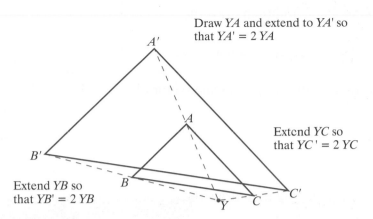

Draw *YA* and extend to *YA'* so that *YA'* = 2 *YA*

Extend *YC* so that *YC'* = 2 *YC*

Extend *YB* so that *YB'* = 2 *YB*

The image triangles in this example are identical in shape and size. The location of the image changes when the position of the centre of the enlargement changes.

Example 7

(a) Enlarge triangle *ABC* with scale factor $\frac{1}{2}$ using *O* as the centre of enlargement.

(b) Enlarge triangle *ABC* with scale factor −1 using *O* as the centre of enlargement.

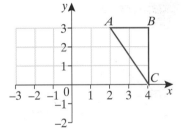

Answer

(a) $OA \times \frac{1}{2} = OA'$ $OB \times \frac{1}{2} = OB'$ $OC \times \frac{1}{2} = OC'$
An enlargement of scale factor $\frac{1}{2}$ has reduced each side of the triangle *ABC*. Triangles *ABC* and *A'B'C'* are similar. The ratio of corresponding sides is $1:\frac{1}{2}$ or 2:1

(b) $OA \times -1 = OA''$ $OB \times -1 = OB''$ $OC \times -1 = OC''$
A negative scale factor indicates measuring from the centre of the enlargement in the opposite direction.

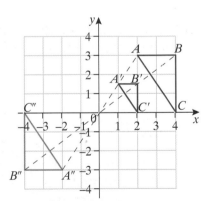

- ■ **An enlargement made with a scale factor smaller than 1 gives a reduced image.**

- ■ **An enlargement made with a negative scale factor indicates that the measuring from the centre of the enlargement must be in the opposite direction.**

- ■ **An enlargement is defined by its centre and scale factor.**

Exercise 6F

1   On squared paper copy this shape and then draw the
    enlargement:

    (a) centre of enlargement at the origin and scale factor of
    enlargement 2

    (b) centre of enlargement at point (2, 1) and scale factor of
    enlargement 2.

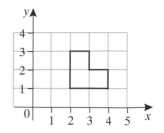

2   On squared paper copy this shape and then draw the
    enlargement:

    (a) centre of enlargement at point (3, 0) and scale factor of
    enlargement 2

    (b) measure the sides of the shape and corresponding sides of
    the enlarged shape. Write the ratio of object side length to
    enlarged side length.

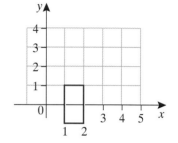

3   On squared paper copy this shape and then draw the
    enlargement:

    centre of enlargement at the origin and scale factor of
    enlargement $\frac{1}{2}$.

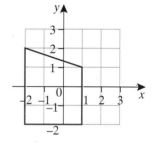

4   On squared paper copy this shape and then draw the
    enlargement:

    centre of enlargement at the origin and scale factor of
    enlargement –3.

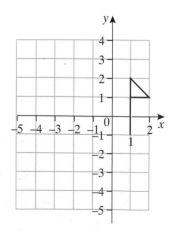

5   State the centre of enlargement and
    scale factor of enlargement for:
    (a) shape A to shape B
    (b) shape A to shape C
    (c) shape C to shape A
    (d) shape A to shape D.

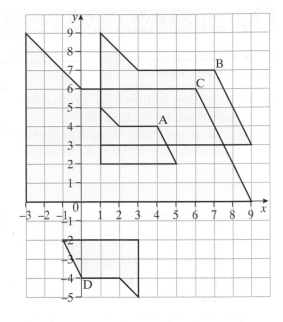

6   These LOGO instructions produce
    a square. Alter these instructions to
    enlarge the square by scale factor 3
    with centre of enlargement at the
    starting position.

```
repeat 4 [ forward 15
           right 90]
```

6.4 Combined transformations

Transformations can be combined by performing one
transformation and then performing another transformation on
the image.

Example 8

(a) Reflect the flag in the y-axis.

(b) Reflect the image in the line $y = x$.

(c) Describe the single transformation to replace (a) and (b).

(d) Reflect the image from (a) in the line $x = 2$.

(e) Describe the single transformation to replace (a) and (d).

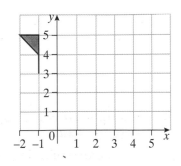

Answer

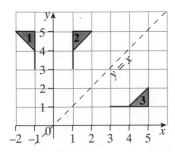

(a) Flag 2 is the image of flag 1.

(b) Flag 3 is the image of flag 2.

(c) Single transformation:
     rotate flag 1 through 90°
     clockwise about the origin.

(d) Flag 2 is the image of flag 1
Flag 4 is the image of flag 2.

(e) Single transformation:
Translate flag 1 by $\binom{4}{0}$.

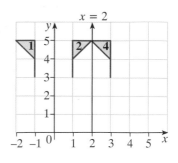

■ **A reflection followed by a reflection can be replaced by the single transformation of:**
- a rotation if the reflection lines are *not* parallel
- a translation if the reflection lines are parallel.

Exercise 6G

1 **(a)** Reflect the triangle in the *x*-axis.

(b) Reflect the image in the *y*-axis.

(c) Describe the single transformation that replaces **(a)** and **(b)**.

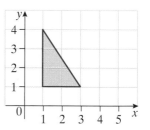

2 **(a)** Reflect the shape in the line $x = 3$.

(b) Reflect the image in the line $x = 6$.

(c) Describe the single transformation that replaces **(a)** and **(b)**.

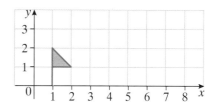

3 **(a)** Reflect the rectangle in the line $y = -x$.

(b) Rotate the image 90° clockwise (−90°) about the origin.

(c) Describe the single transformation that replaces **(a)** and **(b)**.

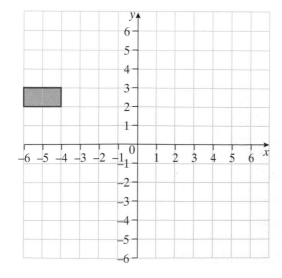

4 (a) Start with the same rectangle as question **3**.
Rotate the rectangle 90° clockwise (–90°) about the origin.

(b) Reflect the image in the line $y = -x$.

(c) Describe the single transformation
that replaces (a) and (b).

5 (a) Enlarge the shape in question 2 with centre of
enlargement at the point (2, 1) and scale factor –1.

(b) Describe this transformation in another way.

6 (a) Rotate the rectangle 90° anticlockwise (+90°)
about the point (1, 0).

(b) Enlarge the image with centre of enlargement
at the point (–1, 3) and scale factor 2.

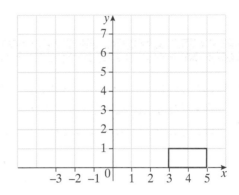

7 (a) Start with the same rectangle as question **6**. Enlarge the
rectangle with centre of enlargement at the point (–1, 3)
and scale factor 2.

(b) Rotate the image 90° anticlockwise (+90°) about the point
(1, 0).

8 Describe the transformations used to produce this pattern.

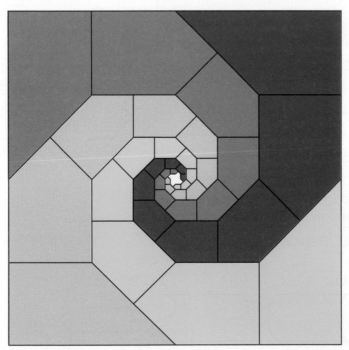

6.5 Tessellations and patterns

■ A single shape or a set of shapes can be used to fill a space. When the shapes fit together without any gaps or overlaps they *tessellate*.

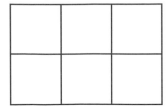

squares

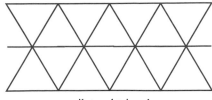

equilateral triangles

4 squares meet at a point of contact.
Interior angle of a square = 90°
$$4 \times 90° = 360°$$

6 equilateral triangles, meet at a point of contact.
Interior angle of an equilateral triangle = 60°
$$6 \times 60° = 360°$$

Regular polygons will tessellate on their own when the sum of the interior angles around the point of contact is 360°.

Exercise 6H

1   Some regular polygons tessellate on their own.
    Complete this table:

Polygon	Number of sides	Interior angle	Tessellates?
Equil. triangle	3	60°	Yes
Square	4	90°	Yes
Pentagon			
Hexagon			
Octagon			
Decagon			

triangle   square

pentagon   hexagon

octagon   decagon

2   More than one shape may be used to tessellate a pattern.
    Create a tessellation using octagons and squares.

3   Use square or square-dotted paper. Repeat these shapes to
    create a pattern which tessellates.

(a)   (b)   (c)   (d)   (e)

4   The designs in question **3** are based on altering a rectangle.
Design a tessellating pattern by altering a square.

5   Patterns which tessellate can be made by transforming a shape.
This tessellation combines a square with its enlargement.

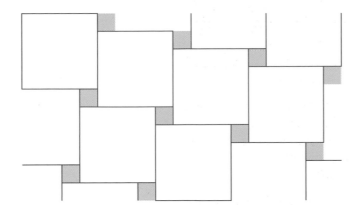

Design a pattern which tessellates by
combining a shape with its enlargement.

6   This pattern was produced by combining an enlargement and
a rotation.

Design a pattern which tessellates by combining an
enlargement and a rotation.

6.6 Scale drawings and scale models

The model of a racing car is a small replica of a real racing car. Each part of the model is a smaller version of the corresponding part on the real car. Each length of 1 cm on the model represents 24 cm on the real car. The model is built to a scale of 1:24. The racing car is an enlargement of the model using a scale factor of 24.

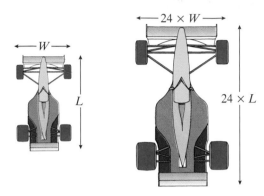

Width of model × 24 = real width

Length of model × 24 = real length

To check the model steering wheel is in the correct scale:

real steering wheel = 28.8 cm

model steering wheel = 1.2 cm

comparing measurements 1.2 : 28.8

1 : 24

$28.8 \div 1.2 = 24$

The model steering wheel is correctly made to a scale 1:24.

■ **Scale drawings or objects are reductions of a real object. The size of reduction is given by the scale factor.**

Example 9

A map has a scale of 1:500 000.

(a) Copy and complete this sentence:
1 cm on the map represents _____

(b) The distance between Arken and Blorcastle is 4 cm on the map. How far is Arken from Blorcastle?

(c) Blorcastle is 12 km from Clarkwood.
How far is this on the map?

Answer

(a) 1:500 000 is shorthand for '1 unit on the map represents 500 000 units'.
1 cm on the map represents 5 km   (500 000 cm = 5 km)

(b) Arken and Blorcastle are 20 km apart $(4 \times 500\,000 \text{ cm} = 20 \text{ km})$

(c) 12 km is represented on the map by 2.4 cm

$$\left(\frac{12}{5} \text{ km} = \frac{1\,200\,000}{500\,000} \text{ cm} \right)$$

Exercise 6I

1   A model boat is made using a scale 1:10.

   (a) The model is 76 cm long.
        How long is the real boat?

   (b) The model is 43 cm wide.
        How wide is the real boat?

2   A model aeroplane is built using a scale 1:50.

   (a) The model is 1.60 m wide.
        How wide is the real aeroplane?

   (b) The real aeroplane is 75 m long.
        How long is the model?

3   This map of the Broads is drawn to a
   scale of 1:600 000.

   (a) Calculate the approximate
        distance in kilometres between
        Acle and Great Yarmouth.

   (b) Calculate the approximate distance
        in kilometres between Norwich and
        Great Yarmouth.

   (c) Lowestoft is 16.25 km from Great
        Yarmouth. How far would this be
        on the map?

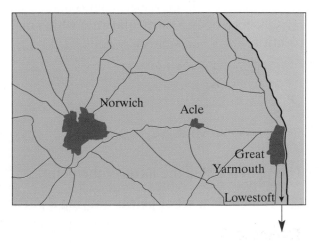

4   This is a scale drawing of Shardia's
   bedroom. She has drawn it so that
   1 cm on the diagram represents 2 m
   in her bedroom.

   (a) Write this as a scale.

   (b) Measure the length and width of
        the bedroom on the plan.

   (c) What is the length and width of the real bedroom?

   (d) How long is the real desk in metres?

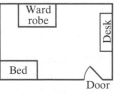

5   This is a sketch of Arfan's bedroom. It is *not* drawn to scale.
   Draw an accurate scale drawing of Arfan's bedroom using a scale
   of 1:50.

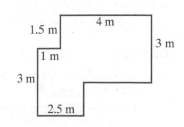

6.7 Locus of a point

Michael threw a cricket ball to his friend Simon.

The ball moved along the path shown in the diagram.

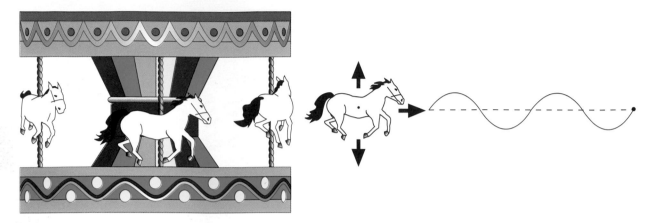

These animals move up and down as the merry-go-round spins. The mechanism on the merry-go-round makes a point on the tip of the horse's nose obey a particular rule which makes the same path every rotation.

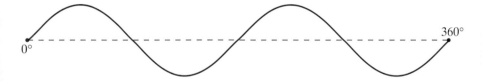

Path of the tip of the horse's nose during one revolution of the merry-go-round.

This path is called the **locus** of the point.

■ **A locus is a set of points obeying a particular rule.**

A locus may be produced by something moving (such as a cricket ball, or a merry-go-round) according to a set of rules, or by a set of points following a mathematical rule.

Using loci to introduce circle theorems

Example 10

A and B are 2 fixed points. Point P can move (vary its position) in the plane but angle $A\hat{P}B = 90°$ always. **Draw the locus of P.**

Step 1 – Mark the two points A and B.

●
A

●
B

Step 2 – Place a set square so that the two edges either side of the 90° pass through A and B. Mark the position of the right-angled corner.

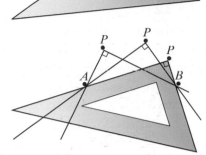

Step 3 – Put the set square in lots of different positions and mark with a dot the right-angled corner, each time.

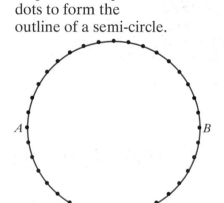

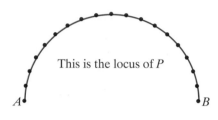

This is the locus of P

Step 4 – Join up all the dots to form the outline of a semi-circle.

If the locus were continued below AB a complete circle would be formed.

Example 11

(a) Draw the locus of a point which moves so that it is always 3 cm from a fixed point.

(b) A garden sprinkler rotates from a fixed position on the lawn. The sprinkler can reach a maximum distance of 4 m as it rotates. Draw a diagram to show the parts of the lawn reached by the sprinkler.

Answer

(a) A is the fixed point. Draw a circle, radius 3 cm centre at A.
The locus of points satisfying the rule is the circumference of the circle.

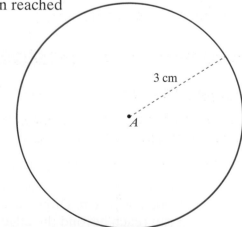

3 cm

(b) *B* is the position of the sprinkler.
The area inside a circle radius 4 m
centre at *B* describes the area of
the lawn reached by the sprinkler.

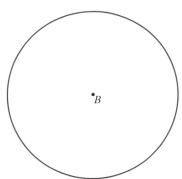

Example 12

(a) Draw the locus of a point which moves so that it
is always 2 cm from a straight line 3 cm long.

(b) A goat is tethered by a 10 m rope. The other end of the rope is
allowed to slide along a horizontal bar 12 m long. Sketch a
diagram to show the parts of the field the goat can graze.

Answer

(a) The locus is two parallel lines (2 cm either side from *AB* and
the same length as *AB*) with a semi-circle radius 2 cm drawn at
either end.

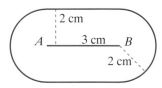

(b) The goat can graze any of
the shaded area.

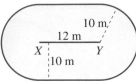

Example 13

Draw the locus of a point which moves so that it is always the same
distance from each of the two fixed points *A* and *B* drawn 5 cm apart.

Draw the perpendicular bisector of the line joining the points *A* and *B*.

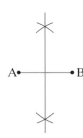

Exercise 6J

1 A running track is designed so that any point on the track is
22.3 m from a fixed line 150 m long.

    (a) Draw the locus of the point.

    (b) Calculate the distance once round the running track.

2 A bicycle has wheels of radius 45 cm. It has a reflector on the
front wheel attached 30 cm from the centre.

    (a) Write the shortest distance the reflector can be from
the road.

    (b) Draw the locus of the reflector as the bicycle
moves forward through one complete wheel turn.

3 Baby Susan is placed inside a rectangular playpen
measuring 1.2 m by 0.8 m. She can reach outside
the playpen by 30 cm. Draw the locus of the points
Susan can reach beyond the edge of the playpen.

4 Daniel lives at the point marked X. There are schools at Ayleton, Bankbury and Corley. Each school allows pupils to attend provided they live in the catchment area. This is a particular distance from the school.

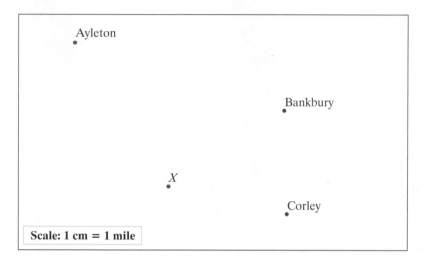

Ayleton accepts pupils who live less than 5 miles away. Bankbury accepts pupils who live less than 4 miles away, and Corley accepts pupils who live less than 3 miles away. Copy the diagram. Draw the catchment area for each school. Write down which school(s) Daniel could attend.

5 Samia enters a rectangular field 40 m by 30 m by a gate in one corner. She crosses the field by taking a path equidistant from the hedges at the side of the gate. Draw the locus of her path.

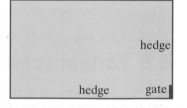

6 Six players stand equally spaced around the circumference of a circle.
Player A passes the ball to player C.
A shorthand rotation for six players passing to the second person along from them could be 6 players step 2.
The locus of the ball is a triangle.
Draw the locus of the ball for these games:

(a) 4 players step 1   **(b)** 8 players step 3

(c) 5 players step 4

7 A ship, S, leaves a harbour, H, and travels 40 km due north to reach a marker buoy, B.
At B the ship turns and travels for a further 30 km on a bearing of 075° from B to reach a lighthouse, L.
At L it turns again and travels back to H in a straight line.

(a) Sketch the locus of the ship's path.

(b) Make an accurate scale drawing of the path taken by the ship.

(c) Measure the bearing of L from H.

(d) Use your scale drawing to find the distance HL.

8 The locus made by a plane cutting a cone horizontally is a circle.
Sketch the locus made by:

(a) **(b)** **(c)**

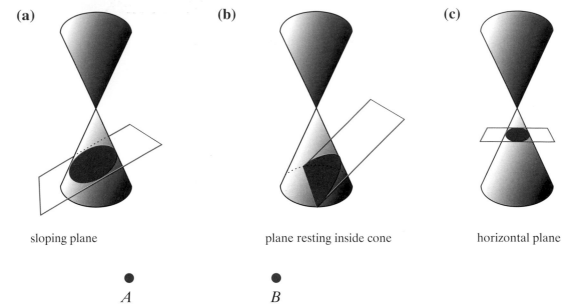

sloping plane   plane resting inside cone   horizontal plane

9

● ●
A *B*

A and *B* are two fixed points.
A point *P* moves in the plane such that $A\hat{P}B = 45°$.
Sketch the complete locus of *P* above and below the line *AB*.

10 Repeat question **9** for $A\hat{P}B = 30°$.

6.8 Formal constructions

■ **A formal construction is an accurate drawing carried out using
a straight edge (ruler), pencil and compasses.**

Constructing a triangle

Example 14

Construct a triangle with sides of length 3 cm, 4 cm and 6 cm.

Draw the longest side using a ruler.

Set the compasses to 4 cm. Draw an arc like this:

Set the compasses to 3 cm. Draw another arc like this:

Join the ends of the line to the point where the arcs cross.

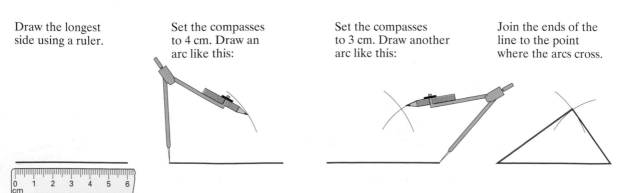

You can measure the angles of your triangle
using a protractor.

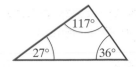

Example 15

Use triangle constructions to draw an angle of 60° without using a
protractor.

An equilateral triangle has angles of 60°.
Construct a triangle keeping your
compasses set to the same length
throughout:

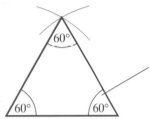

Each angle of the equilateral
triangle is 60°.

Constructing a regular hexagon

Example 16

Construct a regular hexagon.

Keep your compasses set to the same length throughout.

| Draw a circle of any radius: | With the point of the compass on the circumference, draw an arc: | Place your compass point where the first arc crossed the circle. Draw another arc: | Draw four more arcs like this and join them up to make a regular hexagon. |

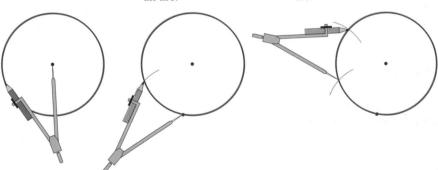

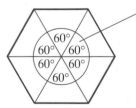

Constructing a regular hexagon is
another way of constructing a
60° angle. Each angle at the
centre of the hexagon is 60°.

Constructing perpendiculars

Perpendicular lines meet at right angles.
The green line is the perpendicular bisector of the black one.

Bisect means 'cut in half'

Example 17

Draw a line segment *AB* of length 8 cm and construct its perpendicular bisector.

You need to set your compasses to more than half the length of the line segment, say 6 cm.

With the compass point at *B*, draw a large arc:	Place the compass point at *A*. Draw this arc:	Join the points where the arcs cross. This is the perpendicular bisector of *AB*.

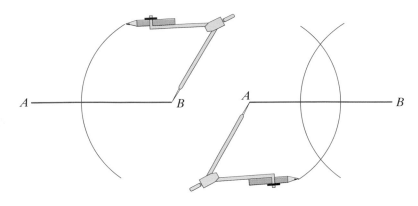
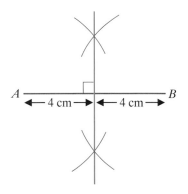

You can construct a rhombus by joining *A* and *B* to the points where the arcs cross. The diagonals of a rhombus always meet at right angles.

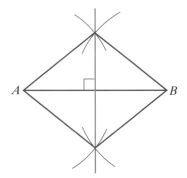

Example 18

Construct a perpendicular to *AB* that passes through the point *P*. This is called constructing the perpendicular **from *P* to *AB***.

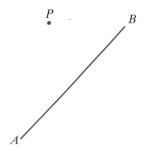

Place your compass point at *P*. Draw two arcs to cut the line *AB*:

With the compass point at the points where the arcs cut the line, draw these arcs:

Join the point where the arcs cross to *P*. This line is the perpendicular from *P* to *AB*:

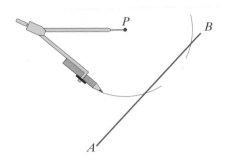 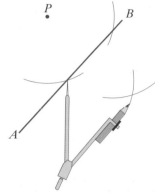 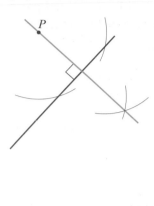

Example 19

Construct a perpendicular to *AB* that passes through the point *P*. This is called constructing the perpendicular **to *AB* at *P***.

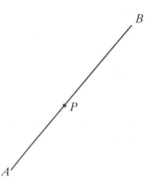

Place your compass point at *P*. Draw two arcs to cut the line *AB*:

With the compass point at the points where the arcs cut the line, draw these arcs:

Join the point where the arcs cross to *P*. This line is the perpendicular to *AB* at *P*:

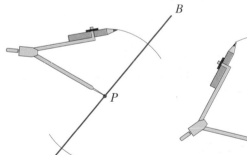 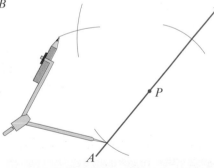 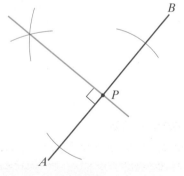

Examples 17, 18 and 19 all showed methods of constructing right-angles. You could also construct a right-angle using your knowledge of Pythagoras' Theorem. (See Chapter 8)

$3^2 + 4^2 = 5^2$, so you can create a right-angle by constructing a triangle with sides of length 3 cm, 4 cm and 5 cm.

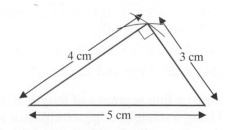

Bisecting an angle

The bisector of an angle is the line that divides
the angle into two equal parts. The green line is
the bisector of this angle.

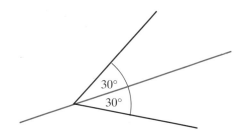

Example 20

Draw a 50° angle and construct its bisector.

Keep your compasses set to the same distance throughout.

With your compass point at
the vertex of the angle,
draw two arcs to cut the
sides of the angle:

Place your compass point
at the points where the arcs
cut the sides of the angle.
Draw these arcs:

Join the point where the
arcs cross to the vertex of
the angle. This line is the
bisector of the angle:

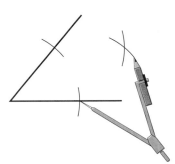

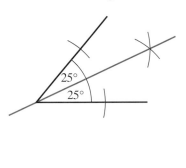

The bisector is a line of symmetry for the angle.
The perpendicular distances from the bisector
to each side of the angle are the same.

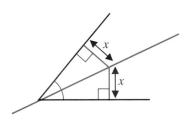

Exercise 6K

1   Using a ruler, compasses and pencil only, construct triangles
    with sides of length:
    (a) 4 cm, 10 cm and 9 cm
    (b) 8 cm, 7 cm and 12 cm.

2   Draw a line segment of length 10 cm.
    Using a straight edge, compasses and pencil only, construct
    the perpendicular bisector of this line segment.

3 Copy this line and construct the perpendicular to *XY* at *P*.

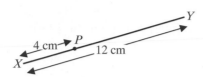

4 Draw a line segment *AB* and a point below it, *Q*.
Construct the perpendicular from *Q* to *AB*.

5 Construct an equilateral triangle with sides of length 8 cm.

6 Draw an angle of any size.
Without using any form of angle measurer, construct the bisector of the angle.

7 The diagram shows a construction of a regular hexagon.

What is the size of:
(a) angle *x*   **(b)** angle *y*?

8 Without using any form of angle measurer, construct an angle of:

(a) 90°   **(b)** 45°   **(c)** 135°

(d) 60°   **(e)** 30°   **(f)** 15°

9 This diagram is a sketch of a triangle *ABC*.

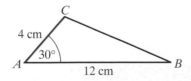

Without using any form of angle measurer,
(a) construct the triangle *ABC*
(b) measure the length of *BC*.

10 Design a method of construction to find the centre of a circle.

11 Without making any measurements at all, use Pythagoras' theorem to construct an angle of 90°

Summary of key points

1   The position of a point P with reference to a fixed point can be identified by:
   - the coordinates of P (x, y) (using a Cartesian grid)

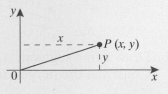

   - the bearing of P (bearings are angles measured clockwise from the north line and expressed in three digits), together with the distance OP.

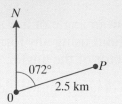

2   A bearing giving the reverse direction is called a back-bearing. It is the outward bearing plus 180°.

3   An object can be transformed by:
   - a translation – this moves every point on a shape the same distance and direction. A translation is given by a vector $\binom{x}{y}$.

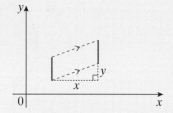

   - a reflection – this produces a mirror image in a line of symmetry. A repeated reflection in the same line of symmetry returns the image to the original object. A reflection can be described by giving the equation of the line of symmetry.

   - a rotation – this is described by giving the centre of rotation, the amount of turn and the direction of turn (– for clockwise, + for anticlockwise). State the centre, angle and direction of rotation.

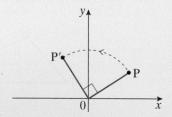

   - an enlargement – this gives a reduced image if made with a scale factor smaller than 1. A negative scale factor indicates that the measuring from the centre of the enlargement must be in the opposite direction. State the centre and the scale factor.

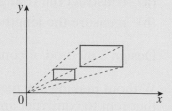

4   A reflection followed by a reflection can be replaced by the single transformation of:
- a rotation if the reflection lines are *not* parallel
- a translation if the reflection lines are parallel.

5   Tessellations are formed by a shape or set of shapes fitting together to fill the whole plane without gaps or overlaps.

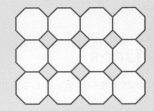

6   Scale drawings and scale models are reductions of a real object. The size of reduction is given by the scale factor, which must be stated, for example 1:50.

7   A locus is a set of points which obey a particular rule.

8   A formal construction is an accurate drawing which is done using a straight edge (ruler), pencil and compasses. Loci can often be drawn using formal constructions.

7 Lines, simultaneous equations and regions

7.1 Lines

The diagram shows the lines with equations $y = 2x + 3$, $y = 2x + 1$ and $y = 2x - 1$.

In all three equations the number in front of x (the **coefficient** of x) is 2 and the lines are **parallel**.

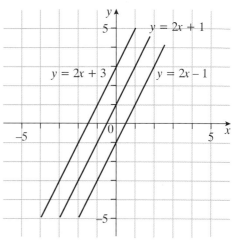

The second number in each equation gives the y-coordinate of the point where the line cuts the y-axis.

For example, the line with equation $y = 2x + 3$ cuts the y-axis at $(0, 3)$.

This point is called its **intercept** on the y-axis. So, the intercept on the y-axis of the line with equation $y = 2x + 1$ is $(0, 1)$ and the intercept on the y-axis of the line with equation $y = 2x - 1$ is $(0, -1)$.

Example 1

(a) Draw the line with equation $y = 3x + 2$.

(b) Write its intercept on the y-axis.

(c) Write the equation of the line parallel to $y = 3x + 2$ whose intercept is $(0, -1)$.

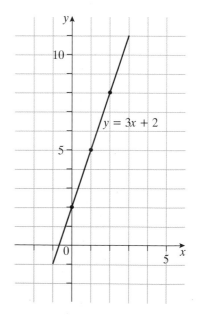

Answer

(a) If you plot two points on the line, you can draw it, but a third point is a useful check.

$$\text{when } x = 0, y = 3 \times 0 + 2 = 0 + 2 = 2$$

$$\text{when } x = 1, y = 3 \times 1 + 2 = 3 + 2 = 5$$

$$\text{when } x = 2, y = 3 \times 2 + 2 = 6 + 2 = 8$$

Table of values

x	0	1	2
y	2	5	8

(b) The intercept on the y-axis is $(0, 2)$.

(c) The line parallel to $y = 3x + 2$ will have the same coefficient of x which is 3.

The equation is $y = 3x - 1$

■ An *intercept* is a point at which a line cuts the y-axis or the x-axis.

Exercise 7A

1 On separate diagrams, draw lines with these equations:
(a) $y = x + 6$ **(b)** $y = 2x + 5$
(c) $y = 3x + 1$ **(d)** $y = 3x - 4$
(e) $y = -2x + 3$ **(f)** $y = \frac{1}{2}x - 2$

2 Write the equations of the two parallel lines in question **1**.

3 **(a)** Without drawing the lines, write the equations of the two parallel lines in:
(i) $y = 4x - 1$ (ii) $y = \frac{1}{4}x + 1$
(iii) $y = x - 4$ (iv) $y = -4x + 1$
(v) $y = \frac{1}{4}x - 1$ (vi) $y = -x + 4$

(b) Explain how you found your answer.

4 Write the intercepts on the y-axis of lines with these equations:
(a) $y = 7x + 5$ **(b)** $y = 5x - 2$
(c) $y = -3x - 5$ **(d)** $y = \frac{1}{4}x + 4$

5 A line is parallel to the line $y = 4x + 5$ and its intercept on the y-axis is $(0, 3)$.
Write the equation of the line.

6 A line is parallel to the line $y = 5x + 7$ and its intercept on the y-axis is $(0, -3)$.
Write the equation of the line.

7 A line is parallel to the line $y = 3x$ and its intercept on the y-axis is $(0, 5)$.
Write the equation of the line.

8 A line is parallel to the line $y = mx$ and its intercept on the y-axis is $(0, c)$.
Write the equation of the line.

9 The point with coordinates $(2, 14)$ lies on the line with equation $y = 5x + c$.
Find the value of c and write the line's intercept on the y-axis.

10 The point with coordinates (8, 1) lies on the line with equation
$y = \frac{1}{2}x + c$.
Find the value of c and write the line's intercept on the y-axis.

11 A line has equation $y = mx + 1$.

(a) Write the line's intercept on the y-axis.

The point with coordinates (3, 16) lies on the line.

(b) Find the value of m.

12 A line is parallel to the line with equation $y = \frac{1}{3}x + 2$ and passes
through the point with coordinates (12, 0).

(a) Find the equation of the line.

(b) Write its intercept on the y-axis.

13 (a) On the same diagram, draw lines with the equations:

   (i) $y = x + 4$ (ii) $y = 2x + 4$
   (iii) $y = 3x + 4$

(b) Write the intercept on the y-axis of each of the lines.

(c) How does the coefficient of x in the equations affect the lines?

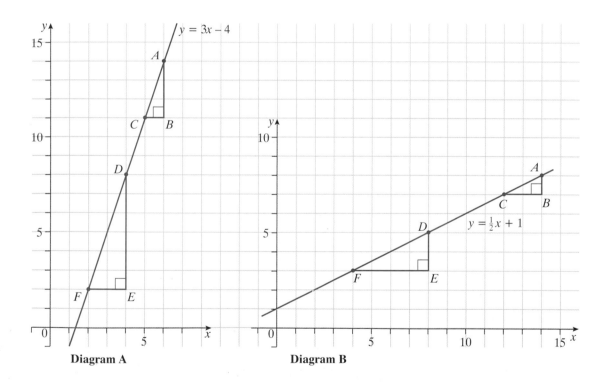

Diagram A   Diagram B

14 Diagram A shows the line $y = 3x - 4$.
Calculate the value of: (a) AB/BC (b) DE/EF
This is the **gradient** of the line.

15 Diagram B shows the line $y = \frac{1}{2}x + 1$.
Calculate the value of: (a) AB/BC (b) DE/EF to find the gradient
of the line.

7.2 $y = mx + c$

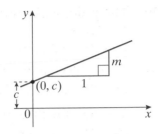

■ The equation of a straight line $y = mx + c$ has a gradient of m and its intercept on the y axis is $(0, c)$.

Example 2

Write down: **(a)** the gradient and **(b)** the intercept on the y-axis of the line with equation $y = 7x - 3$.

(a) gradient $= 7$

(b) intercept on the y-axis is $(0, -3)$.

Example 3

Write down: **(a)** the gradient and **(b)** the intercept on the y-axis of the line with equation $y = -2x + 5$.

(a) gradient $= -2$

(b) intercept on the y-axis is $(0, 5)$.

Diagram C shows the direction of the slope of a line with a **negative** gradient.

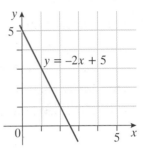

Diagram C

Diagram D shows lines with various gradients.

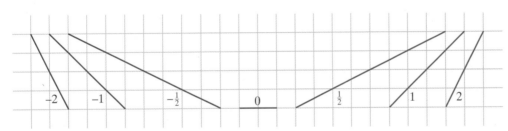

Diagram D

■ Lines with the same gradient (m) are parallel.

Example 4

Find: **(a)** the gradient and **(b)** the intercept on the y-axis of the line with equation $2x - 3y = 18$.

Here the equation must be rearranged in the form $y = mx + c$

$$3y = 2x - 18$$
$$y = \tfrac{2}{3}x - 6$$

We rearrange
$2x - 3y = 18$
to $2x - 18 = 3y$
so $3y = 2x - 18$
then divide by 3.

Answer

(a) gradient $= \tfrac{2}{3}$

(b) intercept on the y-axis is $(0, -6)$.

Example 5

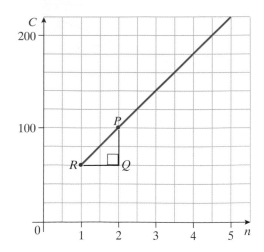

The graph shows the cost, C pounds, of staying in a hotel for n nights.

C is given by the formula $C = an + b$.

Find the values of a and b.

$$\text{Gradient} = PQ/QR = 40/1$$
$$= 40$$
$$\text{so} \quad a = 40.$$

Intercept on C-axis is $(0, 20)$
$$\text{so} \quad b = 20$$

$PQ = £40$

$RQ = 1$ night

Exercise 7B

1   Write: **(a)** the gradient and **(b)** the intercept on the y-axis of lines with the equations:

   (i) $y = 5x + 4$ (ii) $y = 2x - 7$

   (iii) $y = \frac{1}{4}x + 9$ (iv) $y = -4x - 3$

   (v) $y = 8 + 9x$ (vi) $y = 7 - 2x$

2   Find: **(a)** the gradient and **(b)** the intercept on the y-axis of lines with the equations:

   (i) $2x + y = 3$ (ii) $3x - y = 5$

   (iii) $x - 2y = 8$ (iv) $x + 3y = 6$

   (v) $3x - 5y = 10$ (vi) $3x + 4y - 12 = 0$

3   The gradient of a line is 6 and its intercept on the y-axis is $(0, 7)$. Write the equation of the line.

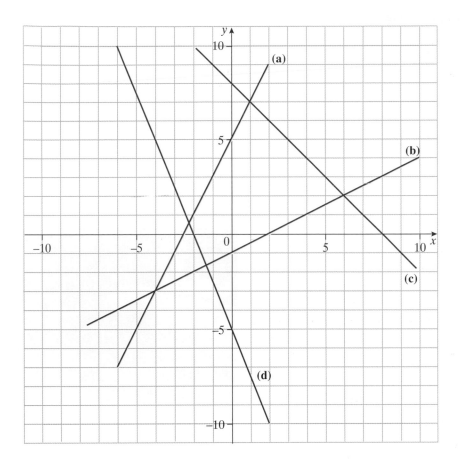

4 Find the equations of the lines shown on the diagram.

5 A line passes through the points with coordinates (0, 2) and
   (2, 8).

   (a) Find the gradient of the line.

   (b) Write the equation of the line.

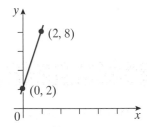

6 The gradient of a line is 4. The point with coordinates (4, 23)
   lies on the line.
   Find the equation of the line.

7 The intercept on the *y*-axis of a line is (0, 11). The point with
   coordinates (–2, 17) lies on the line.
   Find the equation of the line.

8 A line passes through the points with coordinates (2, 11)
   and (5, 23).

   (a) Find the gradient of the line.

   (b) Find the equation of the line.

Draw a diagram.

9

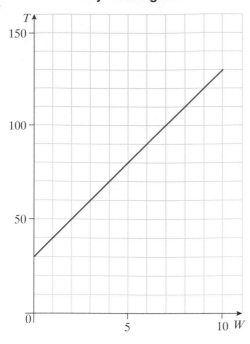

Turkey cooking time

The graph shows the time, T minutes, needed to cook a turkey weighing W lbs.
T is given by the formula $T = mW + c$.
Find the values of m and c.

10

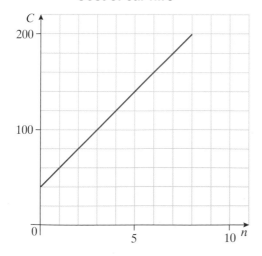

Cost of car hire

The graph shows the cost, C pounds, of hiring a car for n days.
C is given by the formula $C = a + bn$.
Find the values of a and b.

7.3 Perpendicular lines

The diagram shows two pairs of
perpendicular lines.

The gradient of line *a* is 2 and the gradient
of line *b* is $-\frac{1}{2}$.

The gradient of line *c* is $-\frac{2}{3}$ and the gradient
of line *d* is $1\frac{1}{2}$ or $\frac{3}{2}$.

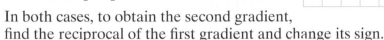

In both cases, to obtain the second gradient,
find the reciprocal of the first gradient and change its sign.

■ **If a line has a gradient of *m*, a line**
perpendicular to it has a gradient of $-\dfrac{1}{m}$.

For example, if a line has a gradient of –3, a line perpendicular to it
has a gradient of $\frac{1}{3}$.

Notice that $2 \times -\frac{1}{2} = -1$, $-\frac{2}{3} \times \frac{3}{2} = -1$ and $-3 \times \frac{1}{3} = -1$

■ **If two lines are perpendicular, the product of their gradients is –1.**

Example 6

A line is perpendicular to the line with equation $y = 4x + 7$ and its
intercept on the *y*-axis is (0, 3). Find the equation of the line.

The line with equation $y = 4x + 7$ has a gradient of 4.

So a line perpendicular to it has a gradient of $-\frac{1}{4}$.

Its intercept on the *y*-axis is (0, 3).

The equation of the line is $y = -\frac{1}{4}x + 3$.

> $y = -\frac{1}{4}x + 3$ can also be
> written as $x + 4y = 3$.

Example 7

A line passes through the point with coordinates (10, –6) and is
perpendicular to the line with equation $5x - 4y = 20$.
Find the equation of the line.

Make *y* the subject of the equation $\quad 5x - 4y = 20$
$$y = \tfrac{5}{4}x - 5$$

The gradient of this line is $\frac{5}{4}$.

The gradient of the perpendicular is $-\frac{4}{5}$.

The equation of the perpendicular line is $y = -\frac{4}{5}x + c$ and it passes
through (10, –6) so $y = -6$ and $x = 10$

$$-6 = -\tfrac{4}{5}(10) + c$$
$$-6 = -8 + c$$
$$c = 2$$

The required equation is $y = -\frac{4}{5} + 2$.

> $y = -\frac{4}{5}x + 2$ can also be
> written as $4x + 5y = 10$.

Exercise 7C

1   Write down the gradient of a line which is perpendicular to lines with each of these gradients.

(a) 5   (b) 2   (c) 1

(d) $\frac{3}{4}$ (e) $-2\frac{1}{2}$ (f) $-\frac{2}{7}$

2   Which of these lines is perpendicular to the line with equation $y = 5x - 3$? Give reasons for your answers.

(a) $y = \frac{1}{5}x + 2$ (b) $y = -\frac{1}{5}x + 2$

(c) $y = -5x - 2$ (d) $y = 5x + 2$

3   A line passes through the origin and is perpendicular to the line with equation $y = 3x - 2$. Find the equation of the line.

4   A line is perpendicular to the line with equation $y = x + 4$ and its intercept on the y-axis is $(0, 6)$. Find the equation of the line.

5   A line passes through the point with coordinates $(4, 1)$ and is perpendicular to the line with equation $y = -\frac{2}{5}x + 3$. Find the equation of the line.

6   Which of these lines is perpendicular to the line with equation $3x + 5y = 15$? Give reasons for your answer.

(a) $3x - 5y = 30$ (b) $5x + 3y = 30$ (c) $5x - 3y = 30$

7   A line passes through the point with coordinates $(6, 1)$ and is perpendicular to the line with equation $2x - 3y = 6$. Find the equation of the line.

8   A line passes through the point with coordinates $(8, 1)$ and is perpendicular to the line with equation $4x + 3y = 12$. Find the equation of the line.

7.4 Simultaneous equations – graphical solutions

An *infinite number* of pairs of values of x and y will make the equation $2x + y = 6$ true.

$x = 0$ and $y = 6$, for example, satisfy this equation.

The coordinates of the points on the line represent these solutions.

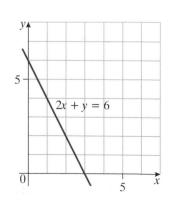

Similarly, an infinite number of pairs of values of x and y satisfy the equation $x + y = 5$.
The coordinates of the points on the line represent these solutions.

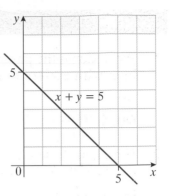

There is only one pair of values which makes both
$2x + y = 6$ and $x + y = 5$ true.

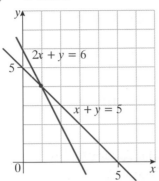

The diagram shows a graphical solution to finding the pair.

The pair of values is found from the x and y coordinates of the point of intersection of the two lines.

$$x = 1$$
$$y = 4$$

Check:

$$2 \times 1 + 4 = 6$$
$$1 + 4 = 5$$
$$2x + y = 6 \quad \text{and} \quad x + y = 5$$

are called **simultaneous equations**
and their solution is $x = 1, y = 4$

Example 8

Solve the simultaneous equations:

$$x + 2y = 10$$
$$x - y = 1$$

Plot three points for each equation and draw straight lines.

The coordinates of the point of intersection of the lines gives the solution.

$$x = 4$$
$$y = 3$$

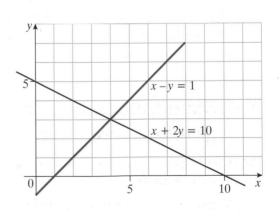

Check:

$$4 + 2 \times 3 = 10$$
$$4 - 3 = 1$$

■ **Simultaneous equations can be solved graphically.**

Exercise 7D

1   Write down four pairs of integer values of x and y which satisfy the equation $x + y = 10$.

2   Which of the following points lies on the line with equation $2x + 3y = 6$:

   (a) $(2, 0)$ **(b)** $(3, 0)$ **(c)** $(6, -2)$ **(d)** $(-3, 4)$

3   Write four pairs of integer values of x and y which satisfy the equation $2x - y = 8$.

4   Which two of the following equations are satisfied by $x = 3$ and $y = 5$:

   (a) $2x + 3y = 20$ **(b)** $3x - y = 4$

   (c) $5x - 2y = 5$ **(d)** $4x + 3y = 25$

5   Use the diagram to solve the simultaneous equations:
$$x + 3y = 6$$
$$x + y = 4$$

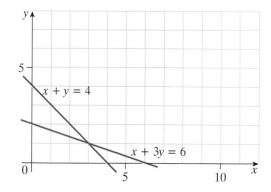

6   On separate diagrams for each part **(a)**, **(b)**, etc., draw appropriate straight lines to solve these simultaneous equations:

   (a) $x + y = 3$ **(b)** $x + y = 6$
              $x - y = 1$ $x + 2y = 8$

   (c) $x + 2y = 6$ **(d)** $2x + 3y = 6$
              $x - y = 3$ $x + 2y = 5$

   (e) $x + y = 2$ **(f)** $3x - y = 9$
              $x - 3y = 6$ $x - 2y = -2$

7.5 Simultaneous equations – algebraic solutions

Example 9

Solve:

$$2x + y = 12$$
$$x - y = 3$$

Here the two equations may be added together in order to eliminate y.

$$2x + y = 12$$
ADD $\quad x - y = 3$

$+ y$ ADD $- y = 0$

$$3x - 0 = 15$$
$$x = 5$$

To find y put $x = 5$ into one of the equations.

$$5 - y = 3$$
$$y = 2$$

The solution is $x = 5, y = 2$.

When the coefficients of x or y in the two equations are equal, you may eliminate one of the unknowns by **subtracting** one equation from the other.

Example 10

Solve:

$$4x + 3y = 5 \qquad (1)$$
$$4x - 5y = 13 \qquad (2)$$

(**Hint:** call the equations (1) and (2).)

In this case adding equations (1) and (2) will not help. Try subtracting:

$$4x + 3y = 5 \qquad (1)$$
SUBTRACT $\quad 4x - 5y = 13 \qquad (2)$

$+ 3y - (-5y)$
$= + 3y + 5y$
$= 8y$

$$8y = -8$$
$$y = -1$$

Substitute $y = -1$ into equation (1)

$$4x - 3 = 5$$
$$4x = 8$$
$$x = 2$$

The solution is $x = 2, y = -1$.

It does not matter which of the two original equations you substitute into (why?) but you should always remember to check that your values fit the other one as well.

■ **Simultaneous equations can be solved algebraically.**

Exercise 7E

Solve these simultaneous equations:

1 $x + 4y = 11$ **2** $3x + 2y = 14$ **3** $5x + 4y = 6$
          $x - y = 1$ $5x - 2y = 18$ $3x - 4y = 10$

4 $3x - 4y = 6$ **5** $4x + 2y = 5$ **6** $2x + 3y = 2$
          $5x - 4y = 2$ $8x - 2y = 1$ $8x + 3y = 17$

7 $x - 6y = 16$ **8** $4x - 2y = 9$ **9** $2x - 3y = 4$
         $2x + 6y = 5$ $4x - 5y = 18$ $2x + 3y = -8$

It is often necessary to multiply one or both of the equations before adding or subtracting.

Example 11

Solve: $6x - 7y = 16$ (1)
                                   $3x - 2y = 5$ (2)

Try multiplying equation (2) by 2. This gives

                          $6x - 4y = 10$ (3)

> This gives the same coefficients of x.

Now subtract (1) from (3)

                        $6x - 4y = 10$ (3)
                        $6x - 7y = 16$ (1)

> $-4y - (-7y) = 3y$

                               $3y = -6$
                                 $y = -2$

Substitute $y = -2$ into (2)

                          $3x + 4 = 5$
                           $3x = 1$
                              $x = \frac{1}{3}$

The solution is $x = \frac{1}{3}, y = -2$.

Example 12

Solve: $2x + 3y = 12$ (1)
                                  $3x - 4y = 1$ (2)

(1) × 4 $8x + 12y = 48$ (3)
(2) × 3 $9x - 12y = 3$ (4)
(3) + (4) $17x = 51$
                                $x = 3$

Substitute in (1) $6 + 3y = 12$
                            $3y = 6$
                              $y = 2$

Solution is $x = 3, y = 2$.

Exercise 7F

Solve these simultaneous equations:

1 $8x + 3y = 35$
$2x - 5y = 3$

2 $7x + 2y = 17$
$5x + 6y = 3$

3 $5x - 4y = 5$
$7x - 12y = 39$

4 $2x + 3y = 12$
$3x - 4y = 1$

5 $6x - 5y = 23$
$4x - 3y = 14$

6 $2x + 7y = 31$
$5x - 3y = 16$

7 $3x - 8y = 11$
$2x - 5y = 6$

8 $9x + 4y = 7$
$8x - 5y = 49$

9 $6x + 7y = 22$
$8x + 9y = 29$

10 $9x + 4y = 8$
$6x + 5y = 3$

7.6 Problems leading to simultaneous equations

Example 13

The total cost of 4 small radiators and 3 large radiators is £159. The total cost of 5 small radiators and 2 large radiators is £134.
Find the cost of one small radiator.

Let the cost of a small radiator be x pounds and the cost of a large radiator be y pounds.

Define any letters you introduce.

The cost, in pounds, of 4 small radiators is $4x$.
The cost, in pounds, of 3 large radiators is $3y$.
The total cost, in pounds, is, therefore, $4x + 3y$.

Express the information in two equations.

So $\qquad\qquad 4x + 3y = 159 \quad (1)$

Similarly $\qquad\quad 5x + 2y = 134 \quad (2)$

$(1) \times 2 \qquad\quad 8x + 6y = 318 \quad (3)$

$(2) \times 3 \qquad 15x + 6y = 402 \quad (4)$

$(4) - (3) \qquad\qquad\quad 7x = 84$

$\qquad\qquad\qquad\qquad x = 12$

Solve the equations.

A small radiator costs £12.

State the answer

Worked examination question

Mrs Rogers bought 3 blouses and 2 scarves. She paid £26. Miss Summers bought 4 blouses and 1 scarf. She paid £28. The cost of a blouse was x pounds. The cost of a scarf was y pounds.

(a) Use the information to write down two equations in x and y.

(b) Solve these equations to find the cost of one blouse.

Answer

(a)
$$3x + 2y = 26 \quad (1)$$
$$4x + y = 28 \quad (2)$$

(b) $(2) \times 2$
$$8x + 2y = 56 \quad (3)$$
$(3) - (1)$
$$5x = 30$$
$$x = 6$$

A blouse costs £6. [E]

Exercise 7G

1 For 6 hours of work at the normal rate of pay and one hour at the overtime rate, Mr Chung is paid £37. For 4 hours at normal rate and 3 hours at the overtime rate, he is paid £41.
Find his normal rate of pay.

2 The cost, C pounds, of staying in a hotel for n nights is given by the formula $C = a + bn$.
It costs £250 for 5 nights and £430 for 9 nights.
Find the values of a and b.

3 The sum of two numbers is 66. Their difference is 8.
Find the numbers.

4 The total cost of a meal and a bottle of wine is £12.50. The meal costs £6.10 more than the bottle of wine.
Find the cost of the meal.
(**Hint:** work in pence.)

5 Heather has 23 coins in her pocket. Some of them are 5p coins and the rest are 10p coins.
The total value of the coins is £2.05.
Find the number of 10p coins.

6 The cost, C pounds, of artificial Christmas trees is given by the formula $C = ah + b$, where h is its height in feet. A 3 foot tree costs £10 and a 6 foot tree costs £25.
Find the values of a and b.

7 Cinema tickets for 2 adults and 3 children cost £12. The cost for 3 adults and 5 children is £19.
Find the cost of an adult's ticket.

8 The total cost of 3 biros and 4 pencils is 98p. The total-cost of 2 biros and 5 pencils is 91p.
Find the cost of a pencil.

9 The total cost of 2 morning papers and 3 evening papers is £1.45. The total cost of 5 morning papers and 4 evening papers is £2.75.
Find the cost, in pence, of a morning paper.

10 The points with coordinates (3, 5) and (5, 12) lie on the line with equation $ax + by = 11$.
   (a) Find the values of a and b.
   (b) Find (i) the gradient of the line and (ii) its intercept on the y-axis.

7.7 Regions

In Unit 2 inequalities were represented on a number line.

■ **Inequalities may also be represented on a graph by a region.**

Example 14

The graph shows the region for $x > 3$. It is the **shaded part** on the right of the line $x = 3$.

Every point on the right of the line $x = 3$ satisfies this inequality.

The region unshaded is the unwanted part for $x > 3$. Because $x = 3$ is also in the unwanted region the line is dotted.

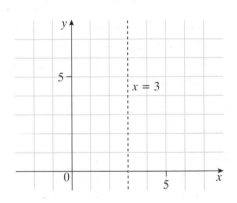

Example 15

$$2x + y \geqslant 6$$

If it is not obvious which side of the line satisfies the inequality, substitute the x and y coordinates of a point, (often the origin), into the inequality.

In this case, letting $x = 0$ and $y = 0$ does not satisfy the inequality and so the origin lies in the unwanted (unshaded) region.

In this case because the inequality is 'greater than **or equal to**', the line is included in the region required. Hence it is drawn as a solid line.

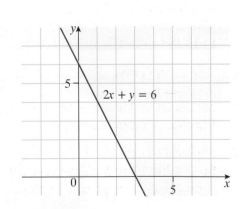

Example 16

$$y < 3x$$

As this line passes through the origin, test with a different point, for example $(1, 0)$. Letting $x = 1$ and $y = 0$ does satisfy the inequality and so $(1, 0)$ lies in the required region.

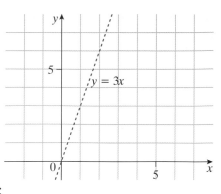

Example 17

Sketch the region which is defined by all of the inequalities:

$$y < x + 1, \quad y < 7 - 2x \quad \text{and } y > 2$$

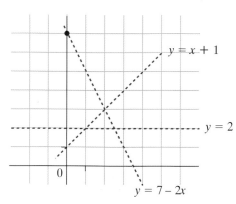

Draw each of the lines:

$$y = x + 1, \quad y = 7 - 2x \quad \text{and } y = 2$$

The region needed is:

      below $y = x + 1$

      below $y = 7 - 2x$

and   above $y = 2$

So it is the shaded region.

Note the region does not include the boundary lines.

(The lines would be included by the use of $y \leqslant x + 1$ etc.)

Exercise 7H

In questions **1–12**, draw diagrams to show the regions which satisfy the inequalities.

1 $x < 4$	**2** $y \geqslant 2$	**3** $x \geqslant 0$
4 $-2 < y \leqslant 3$	**5** $x + y \leqslant 5$	**6** $x + 2y > 6$
7 $3x + 4y < 12$	**8** $2x - 5y < 10$	**9** $y \geqslant 2x$
10 $y \leqslant 3x - 2$	**11** $2x + 3y \geqslant 12$	**12** $3x - 2y > 6$

In questions **13–16**, find inequalities which describe the **unshaded** regions.

13

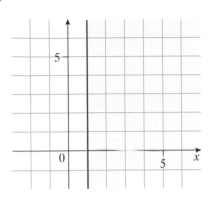

14

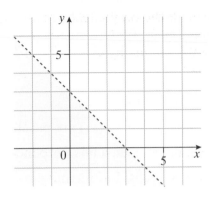

15

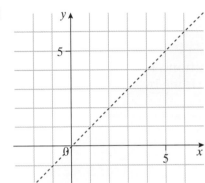

16

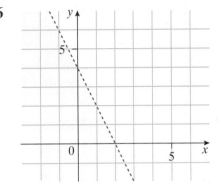

In questions **17–20**, draw diagrams to show the regions which satisfy **all** the inequalities.

17 $x \geqslant 1, y \leqslant 4, y \geqslant x - 1$ **18** $x \geqslant 0, y \geqslant 0, x + y \leqslant 6$

19 $x < 5, y < \frac{1}{2}x, y \geqslant 0$ **20** $x \geqslant 0, y > 2x, 3x + 4y \leqslant 24$

21 Sketch the region defined by all of the inequalities:

$$y > x + 1, \quad y > 7 - 2x, \quad y < 5$$

22 Sketch the region defined by all of the inequalities:

$$y \leqslant 2x + 1, \quad 2x + y \leqslant 9, \quad y \geqslant 3$$

23 Work out the three inequalities which define the shaded region opposite.

Note: the region does not include the boundaries.

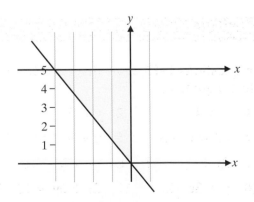

Summary of key points

1   An **intercept** is a point at which a line cuts the y-axis or the x-axis.

2   The general form of the equation of a straight line is
     $y = mx + c$.

     m is its gradient.

     $(0, c)$ is its intercept on the y-axis.

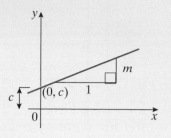

3   Lines with the same gradient (m) are parallel.

4   If a line has a gradient of m, a line perpendicular to it
     has a gradient of $\frac{-1}{m}$.

5   If two lines are perpendicular, the product of their gradients is -1.

6   Simultaneous equations can be solved:

     (a)   graphically, by drawing the straight lines of
             the two equations and finding the coordinates of
             the point of intersection

     (b)   algebraically, by multiplying one or both of the equations, if
             necessary, and then adding or subtracting before dividing by
             the coefficient.

7   Regions on a graph can be used to represent inequalities.

     For example, the shaded region represents $5x + 3y \geqslant 15$.

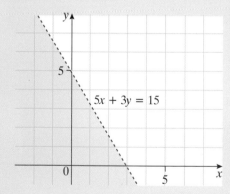

8 Pythagoras' Theorem

8.1 Introducing Pythagoras

Pythagoras was a Greek mathematician and philosopher who lived during the sixth century BC.

The theorem that carries his name, **Pythagoras' Theorem**, is perhaps the best known theorem in the whole of mathematics. It is also extremely useful, very important and a regular feature in GCSE examination papers. It is about finding the lengths of sides in **right-angled** triangles.

Pythagoras' Theorem

The theorem states that:

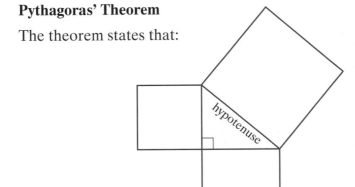

■ **For any right-angled triangle, the area of the square formed on the hypotenuse is equal to the sum of the areas of the squares formed on the other two sides.**

or, in symbols

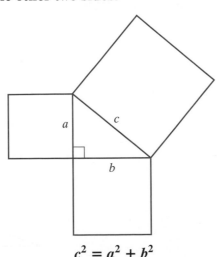

$$c^2 = a^2 + b^2$$

Confirming the theorem

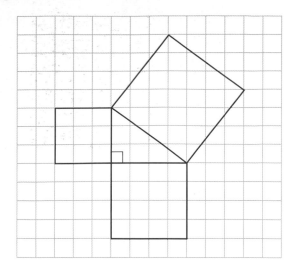

The area of the small square is 9 square units.
The area of the middle square is 16 square units.
The area of the large square is 25 square units.

   Area of small square + area of middle square = area of large square

        9   +   16   =   25

Exercise 8A

1   On squared paper draw a right-angled triangle PQR with the angle at $Q = 90°$ Make $PQ = 4$ units and $QR = 6$ units.

    Draw squares on the three sides of the triangle.

    Confirm, by counting squares, that the area of the square drawn on the hypotenuse PR, that is the largest square, is

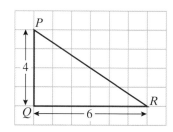

$$4^2 + 6^2 = 16 + 36 = 52 \text{ square units.}$$

2   Repeat question **1** for any right-angled triangle of your own choice.

Finding lengths

Pythagoras' Theorem is often used to find the length of one side of a right-angled triangle, when the other two lengths are known.

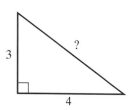

Example 1

Calculate the length of the side marked x.

$$x^2 = 24^2 + 7^2$$
$$x^2 = 576 + 49$$
$$x^2 = 625$$
$$x = \sqrt{625}$$
$$x = 25 \text{ cm}$$

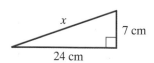

Example 2

A ship leaves a harbour, H, and travels 48 km due north to reach a marker buoy, B. At B the ship turns and travels due west to a lighthouse, L. The straight line distance between H and L is 52 km.

Calculate the distance from B to L.

First draw a diagram.

So
$$HL^2 = HB^2 + BL^2$$
$$52^2 = 48^2 + BL^2$$
$$2704 = 2304 + BL^2$$
$$2704 - 2304 = BL^2$$

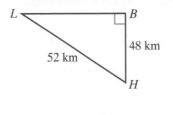

so
$$BL^2 = 400$$
$$BL = \sqrt{400}$$
$$BL = 20 \text{ km}$$

Exercise 8B

1 Calculate the lengths of each of the sides marked with a letter.

(a) 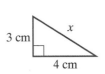   **(b)** 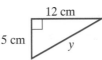   **(c)**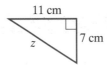

(d)   **(e)**   **(f)**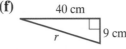

2 A rectangle measures 10 cm by 24 cm. Calculate the length of a diagonal.

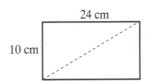

3 The diagonal of a square has a length of 20 cm. Calculate the length of a side of the square.

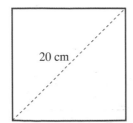

4 A ladder extends to 10 metres in length and is placed 2 metres from a vertical prison wall on horizontal ground. How high up the wall will the ladder reach to the nearest cm?

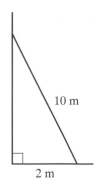

10 m

2 m

5 A yacht leaves a port, P, and travels 43 km due south to a buoy, B. At B the yacht turns due east and travels to a lighthouse, L.

At L, the yacht turns again and travels in a straight line back to P.

The straight line distance from L to P is 54 km.

Calculate the total distance travelled by the yacht.

Sometimes you need to find the length of one side before you can find the length of another side. You may need to apply Pythagoras' Theorem twice.

Example 3 (Pythagoras applied twice)

Calculate the lengths marked x and y.

First
$$x^2 = 3^2 + 4^2$$
$$x^2 = 9 + 16$$
$$x = \sqrt{25}$$
$$x = 5 \text{ cm}$$

Then
$$y^2 + x^2 = 13^2$$
$$y^2 + 5^2 = 13^2$$
$$y^2 + 25 = 169$$
$$y^2 = 169 - 25$$
$$y = \sqrt{144}$$
$$y = 12 \text{ cm}$$

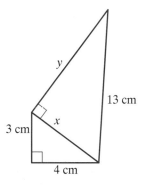

y

13 cm

3 cm

x

4 cm

Exercise 8C

1 In each of these diagrams, all lengths are in centimetres.
Calculate each of the lengths marked with a letter.

(a)

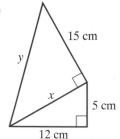

15 cm

y

x

5 cm

12 cm

(b)

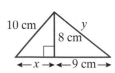

10 cm

y

8 cm

$\leftarrow x \rightarrow \leftarrow$ 9 cm $\rightarrow$

(c)

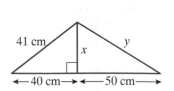

41 cm

x

y

$\leftarrow$ 40 cm $\rightarrow \leftarrow$ 50 cm $\rightarrow$

The distance between two points

Example 4

A is the point with coordinates $(2, 1)$
B is the point with coordinates $(8, 9)$

Find the straight line distance from A to B.

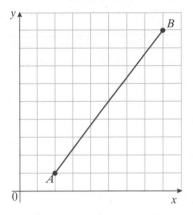

This can be done using Pythagoras.

First we make a triangle ABM where the angle at M is $90°$, M is horizontally across from A and vertically below B.

$AM = 6$ units $\qquad\qquad BM = 8$ units

using Pythagoras
$$AB^2 = AM^2 + BM^2$$
$$AB^2 = 6^2 + 8^2$$
$$AB^2 = 36 + 64$$
$$AB^2 = 100$$
$$AB = 10 \text{ units}$$

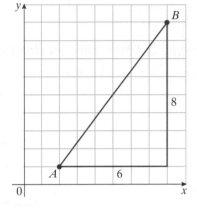

Exercise 8D

1   Calculate the distance between the points in each pair:
    (a) $(1, 2)$ and $(5, 5)$ **(b)** $(3, 5)$ and $(5, 9)$
    (c) $(-1, 4)$ and $(7, 12)$ **(d)** $(-2, 6)$ and $(9, -4)$
    (e) $(8, -3)$ and $(12, -6)$ **(f)** $(-2, -3)$ and $(-7, 8)$

2   Calculate the distance PQ.
    (Each square of the grid is 1 unit by 1 unit.)

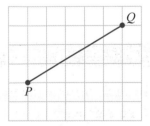

3   Show that the distance between the points with coordinates (x_1, y_1) and (x_2, y_2) is
    $$\sqrt{(x_2 - x_1)^2 + (y_2 - y_1)^2}$$

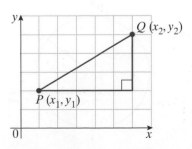

4 A circle of radius 5 units is drawn with its centre at the origin.

Find the coordinates of 8 points which lie on the circle and have integer values for their coordinates.

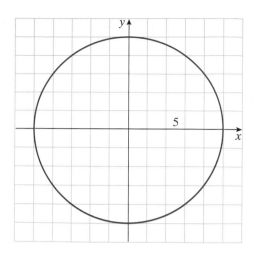

The equation of a circle

Example 5

Find the equation of the circle with centre at the origin (0, 0) and radius 3 units.

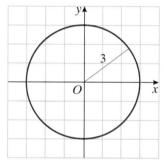

Mark any general point, P, on the circle with coordinates (x, y).

Using Pythagoras' theorem gives

$$x^2 + y^2 = 3^2 \ (= 9)$$

The result will be true for *every* point on the circle.

The *equation* of this circle is

$$x^2 + y^2 = 3^2 \quad \text{or} \quad x^2 + y^2 = 9$$

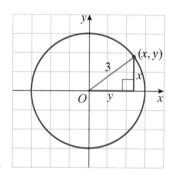

■ **The equation of any circle, centre the origin and radius r units is**
$$x^2 + y^2 = r^2$$

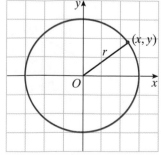

Exercise 8E

1 Write down the equation of each circle.

(a)

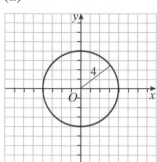

(b)

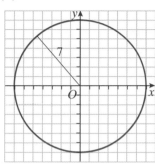

2 Find the equation of the circle
(a) radius 5 units, centre the origin
(b) radius 12 units, centre the origin
(c) radius 15 units, centre the origin.

3 A circle has its centre at the origin.
The circle passes through the point (5, 12).
Find the equation of this circle.

Non right-angled triangles

Example 6

When all angles are acute

The labels L, M and S arc used to indicate
the lengths of the longest side, the middle side
and the shortest side.

The area of the largest square is 9 sq units.
The area of the middle square is 8 sq units.
The area of the smallest square is 5 sq units.

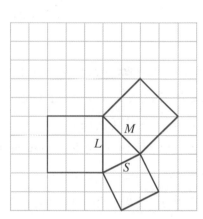

▇ **In the case when all the angles are acute,
then in terms of areas**

> **largest < middle + smallest**

or

$$L^2 < M^2 + S^2$$

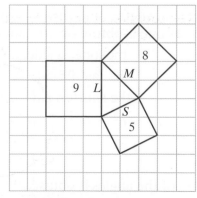

Example 7

When one angle is obtuse

In this diagram one angle is obtuse, and L, the longest side is opposite this angle.

area of largest square = 20 sq units
area of middle square = 9 sq units
area of smallest square = 5 sq units

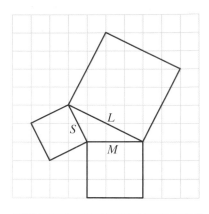

■ **In terms of areas when one angle is obtuse**

largest > middle + smallest

or

$$L^2 > M^2 + S^2$$

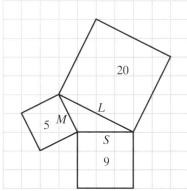

Exercise 8F

1 You will need some squared paper. On it draw:

(a) a triangle where all angles are acute

(b) a triangle where one angle is obtuse.

Using the symbols L, M and S for the longest, middle and shortest sides, convince yourself that in case **(a)** $L^2 < M^2 + S^2$ in case **(b)** $L^2 > M^2 + S^2$

Example 8

The lengths of the sides of a triangle are 3 cm, 5 cm and 6 cm.

Work out whether the angles of this triangle are all acute or whether one is obtuse or whether one is a right angle.

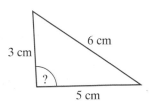

the longest side, $L = 6$ cm
the middle side, $M = 5$ cm
the shortest side, $S = 3$ cm

$$L^2 = 6^2 = 36 \qquad M^2 = 5^2 = 25 \qquad S^2 = 3^2 = 9$$
$$M^2 + S^2 = 25 + 9 = 34$$
$$\text{so } L^2 > M^2 + S^2$$

so one of the angles is obtuse.

Any set of three numbers, usually positive whole numbers (a, b, c), which satisfy (fit into) the relationship

$$c^2 = a^2 + b^2 \text{ (assuming } c \text{ to be the largest)}$$

is known as a **Pythagorean Triple**.

The two most famous Pythagorean Triples

$$(3, 4, 5) \text{ and } (5, 12, 13)$$

are worth remembering.

When:
$L^2 = M^2 + S^2$
the triangle is **right-angled**

Exercise 8G

1   The numbers in the brackets are the lengths in centimetres of the sides of triangles.

In each case, work out whether the angles of the triangle are all acute, whether one angle is obtuse or whether the triangle is right-angled.

   (a) (4, 7, 9)   **(b)** (5, 11, 12)   **(c)** (3, 14, 15)

   (d) (9, 41, 40)   **(e)** (17, 10, 13)   **(f)** (8, 8, 12)

2   Check that (3, 4, 5) and (5, 12, 13) are Pythagorean Triples.

3   Which of the following are Pythagorean Triples?
Show all your working.

   (a) (6, 8, 10)   **(b)** (7, 24, 25)   **(c)** (9, 40, 41)

   (d) (8, 23, 30)   **(e)** (11, 60, 61)   **(f)** (1, 1, 2)

8.2 Pythagoras in three dimensions

It is possible to use Pythagoras' Theorem in 3-D to find the length of the longest diagonal of a cuboid.

Example 8

A cuboid measures 4 cm by 5 cm by 8 cm.
Work out the length of the longest diagonal of this cuboid.

Label the vertices of the cuboid *ABCDEFGH*.

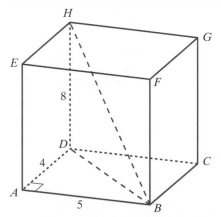

The longest diagonal is HB.

By Pythagoras
$$HB^2 = HD^2 + DB^2$$
$$HD = 8$$

To find the length of DB, look at the base $ABCD$.

Using Pythagoras:
$$AB^2 + AD^2 = DB^2$$
$$5^2 + 4^2 = DB^2$$
$$25 + 16 = DB^2$$
$$DB^2 = 41$$

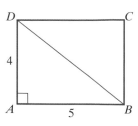

HB can now be found:
$$HB^2 = HD^2 + DB^2$$
$$HB^2 = 8^2 + 41$$
$$HB^2 = 64 + 41$$
$$HB^2 = 105$$
$$HB = \sqrt{105}$$
$$HB = 10.25 \text{ cm (to 2 d.p.)}$$

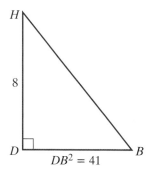

Exercise 8H

1   Calculate the length of the longest diagonal of a cuboid which has sides of length 5 cm, 8 cm and 12 cm.

2   A cuboid measures a cm by b cm by c cm.
Show that the length of its longest diagonal is given by the formula
$$d = \sqrt{(a^2 + b^2 + c^2)}$$

3   $VABC$ is a tetrahedron.

The base ABC is a triangle, right-angled at A.
The vertex V is vertically above A.

$AB = 5$ cm $AC = 12$ cm

and $VA = 17$ cm

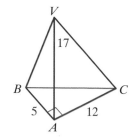

(a) Calculate the lengths of:
(i) BC (ii) VB (iii) VC

(b) Work out whether the angle BVC is acute, obtuse or a right angle.

8.3 Applying Pythagoras

Exercise 8I   (Mixed questions)

1   Calculate the length *PQ*.

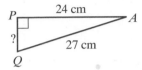

2   The diagram shows a ladder of length 8 metres resting between horizontal ground and a vertical wall.

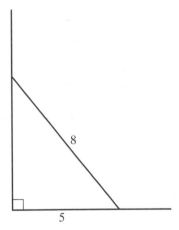

The distance from the base of the wall to the bottom of the ladder is 5 metres. Calculate the height of the top of the ladder above the ground.

3   A ship leaves a harbour, *H*, and travels due north for an unknown distance until it reaches a marker buoy, *B*. At *B* the ship turns due east and travels for 30 km to a lighthouse, *L*. At *L* the ship turns again and travels in a straight line back to *H*. The distance from *L* to *H* is recorded as 43 km.
Calculate the distance from *H* to *B*.

4   Calculate the lengths marked *x* and *y*.

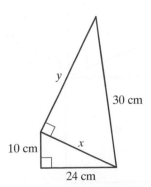

5 Points A and B have coordinates $(-3, 7)$ and $(5, -4)$ respectively.
Calculate the distance AB.

6 ABC is a triangle.

$AB = 7$ cm $AC = 8$ cm   and $BC = 11$ cm

Work out whether the angles of ABC are all acute or whether
one is obtuse or whether the triangle is right-angled.

7 Showing all your working, explain which of these sets are
Pythagorean Triples:

    (a) $(4, 6, 8)$ **(b)** $(10, 24, 26)$ **(c)** $(1, 2, 3)$

8 A cuboid measures 8 cm by 12 cm by 15 cm.

Calculate the length of the longest diagonal of the cuboid.

9 $ABCDEF$ is a prism.

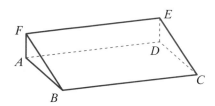

The horizontal base $ABCD$ is a rectangle with $AB = 12$ cm and
$BC = 20$ cm.

The vertical face $ADEF$ is also a rectangle with $AF = 5$ cm.

Calculate the lengths:

    (a) FB **(b)** BD **(c)** FC

10 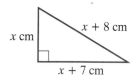   Calculate the value of x.

11 A Special Challenge

This challenge question is outside the
GCSE syllabus for Pythagoras' Theorem.
You can try it if you have completed all
the other work on Pythagoras.

The theorem states that for squares, A, B and C

        Area of C = Area of A + Area of B

but do the shapes drawn on the sides of the
right-angled triangle have to be squares?
Justify your answer.

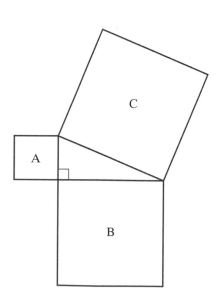

Summary of key points

1   For any right-angled triangle, the area of the square
formed on the hypotenuse is equal to the sum of
the areas of the squares formed on the other two sides.

$$c^2 = a^2 + b^2$$

2   The distance between two points on a coordinate grid is

$$d = \sqrt{(x_2 - x_1)^2 + (y_2 - y_1)^2}$$

3   The equation of any circle, centre the origin and radius
r units is

$$x^2 + y^2 = r^2$$

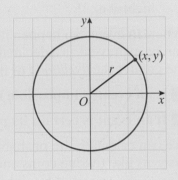

4   In a triangle with largest (L), middle (M), and smallest (S)
lengths of sides:

   if $L^2 < M^2 + S^2$ the angles are acute

   if $L^2 > M^2 + S^2$ one angle is obtuse

   if $L^2 = M^2 + S^2$ one angle is a right-angle

5   Some well-known Pythagorean Triples are:

   (3, 4, 5)   (5, 12, 13)   (6, 8, 10)   (7, 24, 25).

6   The length of the longest diagonal of a cuboid with dimension
a, b, c is:

$$d = \sqrt{a^2 + b^2 + c^2}$$

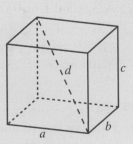

9 Probability

Introduction

Probability is about calculating or estimating what might happen in the future.

■ **The probability of an *event* is expressed as a number from 0 to 1 inclusive.**

When an event can happen its probability is a fraction. That fraction can be expressed as a decimal or a percentage. The larger the fraction the greater the *likelihood* of the event happening.

■ **If an event is impossible its probability is 0.**

■ **If an event is certain its probability is 1.**

We can calculate the probability of events for simple physical systems like tossing coins and rolling dice.

■ **When there are *n* equally likely possibilities then the probability of one of them happening is $\frac{1}{n}$**

Example 1

The numbers 1 to 10 are written on pieces of card which are then placed face downwards on a table. Ian picks up one of the cards at random. Calculate the probability that Ian's card will be:

(a) a multiple of 3　**(b)** an even number　**(c)** a prime number.

Number	1	2	3	4	5	6	7	8	9	10
Multiple of 3	✗	✗	✓	✗	✗	✓	✗	✗	✓	✗
Even	✗	✓	✗	✓	✗	✓	✗	✓	✗	✓
Prime	✗	✓	✓	✗	✓	✗	✓	✗	✗	✗

Answer

(a) Three of the 10 numbers are multiples of 3. The probability of Ian selecting one of these is $\frac{3}{10}$ (0.3 or 30% are equally acceptable).

(b) Five of the 10 numbers are even, so the probability of Ian choosing one of these is $\frac{5}{10}$ ($\frac{1}{2}$ or 0.5 or 50% are equally acceptable).

(c) Four of the 10 numbers are prime, so the probability of Ian obtaining one of these is $\frac{4}{10}$ ($\frac{2}{5}$ or 0.4 or 40% are equally acceptable).

Note: the key word which will indicate that each of the possibilities is equally likely is *random*.

Hint: never hesitate to draw a rough diagram of the possibilities involved in the question. The time that this takes is usually repaid by greater accuracy.

In general terms, the probability of the event that you are considering can be calculated.

■ P(event) = $\dfrac{\text{the number of ways the event can occur}}{\text{the total number of possibilities}}$

> **Hint**: it often helps to understand probabilities if the fractions like $\frac{3}{4}$ are read as 'three out of four possibilities' as well as 'three quarters'.

Example 2

Trevor and Kevin both answer this question:

There are eight snooker balls in a bag. Six of them are red, one is green and the other is yellow. Calculate the probability that a ball taken at random from the bag will be red.

Trevor thinks there are 3 possibilities, red, green and yellow. He then says there are 6 red balls and so his answer is $\frac{6}{3}$ which he cancels down to 2.

Kevin says there are 8 possible balls of which 6 are red. His answer is $\frac{6}{8}$ which he cancels down to $\frac{3}{4}$.

Who is right and who is wrong?

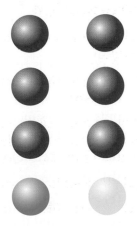

Kevin worked out the correct answer. He realised that each of the red balls is a possible outcome and that altogether there are 8 balls so 8 possible outcomes. He could have left his answer $\frac{6}{8}$ because he was not asked in the question to cancel down his answer. Other correct answers are $\frac{3}{4}$, 0.75 and 75%.

Trevor made the common mistake of just counting the three possible colours. He was correct when he used the fact that there are 6 red balls as *the number of ways the event can occur*. He should have realised with an answer greater than 1 he had made a mistake, because his answer had to be between 0 and 1.

> **Note**: mathematicians use a shorthand way to avoid having to write 'the probability of' all the time. They write P(red) to stand for the probability of selecting a red ball. You should practise reading and writing this way of expressing probabilities in the questions and answers.

Exercise 9A

1 Nine playing cards are numbered 2 to 10. A card is selected from them at random. Calculate the probability that the card will be:

   (a) an odd number

   (b) a multiple of 4.

2 Five strawberry, two orange and three blackcurrant flavoured sweets are placed in a box. One sweet is taken at random from the box. Calculate the probability that the sweet will be:

   (a) blackcurrant flavoured

   (b) not orange flavoured.

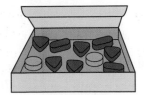

3   A normal die is rolled. Calculate the probability that the number on the uppermost face when it stops rolling will be:

(a) 5   (b) odd   (c) prime   (d) not 6.

A normal die is a cube, is unbiased and has the numbers 1 to 6 on it. The numbers on opposite faces add up to 7.

4   A spinner is made from a regular octagon. It is labelled with three As, two Bs and three Cs. Each of the sides is equally likely to be resting on the table when it stops spinning. Calculate:

(a) P(A resting on the table)

(b) P(B resting on the table)

(c) P(not C resting on the table).

5   A company is trying to prove that chicken flavoured cat food is preferred by most cats. Unfortunately, Alfred the tabby tom cat doesn't know this and selects his food at random. The company puts out 10 saucers of rabbit, 6 saucers of sardine and only 4 saucers of chicken flavoured food. Calculate:

(a) P(Alfred selects sardine flavoured food)

(b) P(Alfred selects rabbit flavoured flood)

(c) P(Alfred selects sardine or rabbit flavoured food)

(d) P(Alfred does not select chicken flavoured food).

The answer to part (c) can be calculated from the answers to parts (a) and (b).

(e) Explain how to answer part (c) from the answers to parts (a) and (b)

(f) Explain how the answer to part (d) can be calculated if you are given that:
P(Alfred selects chicken flavoured food) = p

9.1 Short cuts

In Exercise 9A you might have noticed that it is possible to take short cuts to find some answers. If two events cannot occur at the same time they are **mutually exclusive**.

Events A and B are mutually exclusive when if A happens then B does not and if B happens then A does not.

Example 3

A card is selected from a pack. What is the probability that the chosen card is either a 'king' or a 'ten'?

Choosing a 'king' and choosing a 'ten' are mutually exclusive events.

$$P(K) = \tfrac{4}{52}$$
$$P(10) = \tfrac{4}{52}$$
$$P(K \text{ or } 10) = \tfrac{4}{52} + \tfrac{4}{52} = \tfrac{8}{52}$$

■   **When two events, A and B, are mutually exclusive**
P(A or B) = P(A) + P(B)

Example 4

There are five red, three green and two yellow counters in a bag. A counter is taken at random from the bag. Calculate the probability that the counter will be:

(a) red   **(b)** green   **(c)** yellow   **(d)** red or green

(e) not yellow.

First work out the answers to parts **(a)**, **(b)** and **(c)**:

(a) P(red) = 0.5   **(b)** P(green) = 0.3

(c) P(yellow) = 0.2

(d) A counter cannot be both red and green. So the events 'selecting a red' and 'selecting a green' are **mutually exclusive**.

     P(red or green) = P(red) + P(green) = 0.5 + 0.3 = 0.8

(e) The only possible colours which can occur are red, green or yellow. The sum of their probabilities is 1. You are *certain* to get a red, green or yellow counter.

P(red) + P(green) + P(yellow) = 1

so   P(red) + P(green) = 1 – P(yellow)

so   P(red or green) = 1 – P(yellow)

If the counter selected is red or green then it is not yellow.

so   P(not yellow) = 1 – P(yellow)

                         = 1 – 0.2

                         = 0.8

An obvious case of mutually exclusive events is 'selecting A' and 'not selecting A'. In this situation there is a general statement that is often a useful short cut:

■   **For mutually exclusive events P(not A) = 1 – P(A)**

9.2 Lists and tables

To work out all the possibilities in experiments where two or more objects are used to generate outcomes it is always useful to draw lists or tables.

Example 5

Three coins are tossed. Calculate the probability of getting two heads and one tail.

It is important to consider the three coins separately. Think of them as the first, second and third coins even if they are tossed simultaneously. Draw up a list of possible outcomes:

First coin	H	H	H	H	T	T	T	T
Second coin	H	H	T	T	H	H	T	T
Third coin	H	T*	H*	T	H*	T	H	T

Hint: each of the coins can land in two different ways – head or tail. There are 3 coins. So to check that you have found all the possible outcomes use the fact that there should be 2^3 which is 8. Notice how the 8 different possibilities have been written down systematically. This helps to avoid missing or repeating some of them.

The three starred out of eight possible outcomes are the ones consisting of 2 heads and 1 tail.

$$P(2H \text{ and } 1T) = \tfrac{3}{8}$$

Sometimes the problem involves too many outcomes to list so it is more efficient to draw a table.

Example 6

Two normal dice are rolled and the numbers on the tops are added together. Calculate the probability that the sum will be:

Hint: there are 2 dice and 6 ways each can land. The total number of possibilities is $6^2 = 36$.

(a) 10   (b) a multiple of 5   (c) not 7

Draw a **sample space diagram**. There are 36 different ways that the two dice can land.

Number on the first die

	1	**2**	**3**	**4**	**5**	**6**
1	2	3	4	5	6	7
2	3	4	5	6	7	8
3	4	5	6	7	8	9
4	5	6	7	8	9	10
5	6	7	8	9	10	11
6	7	8	9	10	11	12

Number on the second die

Answer

(a) 10 can occur in 3 ways so $P(10) = \tfrac{3}{36}$

(b) 5 can occur in 4 ways so $P(5) = \tfrac{4}{36}$

$$P(\text{multiple of } 5) = P(10) + P(5) = \tfrac{4}{36} + \tfrac{3}{36} = \tfrac{7}{36}$$

(c) 7 can occur in 6 ways so $P(7) = \tfrac{3}{36}$

$$P(\text{not } 7) = 1 - P(7) = 1 - \tfrac{6}{36} = \tfrac{30}{36}$$

In example 6 it made sense to put the sum of the two scores in the sample space diagram because all the questions concerned sums. This is not always the case.

Example 7

Arif and Moira are working together. Arif has a spinner which has an equal chance of landing on 1, 2, 3, 4 or 5. Moira has a normal die. Arif spins his spinner and Moira rolls her die. Calculate the probability that the two numbers they generate will:

(a) have an odd sum

(b) have a product which is a multiple of three

(c) be consecutive numbers.

First, draw a sample space diagram and:

● put 'o' in every square where the sum of the two scores is an odd number

● put 't' in every square where the *product* of the two scores is a multiple of three

● put 'c' in every square where the two scores are consecutive numbers.

Note: the total number of possibilities is the product of the possibilities on each piece of equipment. So the total number of possibilities is $5 \times 6 = 30$.

Dice score

Spinner score	1	2	3	4	5	6
1		o c	t	o		o t
2	o c		o t c		o	t
3	t	o t c	t	o t c	t	o t
4	o		o t c		o c	t
5		o	t	o c		o t c

Answer

(a) There are 15 squares on the sample space diagram labelled 'o'.

$$P(\text{odd sum}) = \tfrac{15}{30}$$

(b) There are 14 squares labelled 't'.

$$P(\text{product is a multiple of 3}) = \tfrac{14}{30}$$

(c) There are 9 squares labelled 'c'.

$$P(\text{consecutive numbers}) = \tfrac{9}{30}$$

(**Note**: no attempt has been made to cancel down the fractions because the question did not ask for that. It is unwise to cancel down the fractions in probability questions because sometimes you will have to add them to work out other answers. If you have not cancelled down you will retain the common denominators which makes the addition of fractions simpler.)

Exercise 9B

1   A game is played using these two boards. In one 'go' a player spins both arrows. The sum of the numbers where the two arrows point is the score for the 'go'. In the diagram the score is 9.

   (a) Copy and complete the table showing all the possible scores for one 'go'.

   (b) Write the probability that a player will score 11 in one 'go'.   [E]

+	1	3	5	7
2				
4				
6				
8	9			

2 In a game, two normal dice, one red and the other blue, are thrown simultaneously. A score is found by adding together the numbers on the faces which finish uppermost. Four players, Adam, Bethan, Christos and Dean, decide that Adam will win if the red dice shows 6, Bethan will win if the score is 5 or 6, Christos will win if the score is 2 or 3 or 4 and that Dean will win otherwise. Find the probability, expressed as a fraction, that the game will be won by:

(**a**) Adam   (**b**) Bethan   (**c**) Christos (**d**) Dean.   [E]

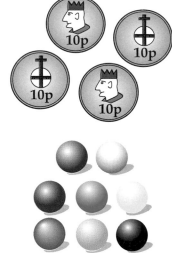

3 Four coins are tossed simultaneously. List all the possible ways in which the coins can land with heads or tails uppermost. Calculate the probability that:

(**a**) four heads will occur

(**b**) an equal number of heads and tails will occur

(**c**) at least one coin will land tails uppermost.

4 A bag contains one red and one white snooker ball. A second bag contains one brown, one green and one yellow ball. A third bag contains one blue, one pink and one black ball. List all the possible sets of three balls which can arise when one ball is selected from each bag at random. Calculate the probability that the three balls selected will include the red and the black ball.

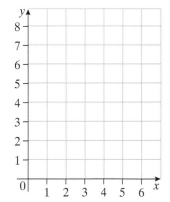

5 Two normal dice are rolled. Calculate the probability that the two numbers on the upper faces will:

(**a**) sum to a prime number   (**b**) have an odd difference

(**c**) have an even product   (**d**) have one which is a factor of the other.

6 A fair octahedral die with the numbers 1 to 8 on its faces will be rolled at the same time as a normal die. The number on the upper face of the normal die will be used as the x-coordinate and the number from the upper face of the octahedral die will be used as the y-coordinate of a point on a grid. Calculate the probability that when the two dice are rolled the point generated will lie on the line:

(**a**) $x = 3$ (**b**) $y = 5$ (**c**) $y = 2x$

(**d**) $x + y = 8$ (**e**) $y = x + 2$ (**f**) $y + 2x = 7$

9.3 Relative frequency

So far we have been able to calculate probabilities because the likelihood of all the possible outcomes was known. Sometimes we can only make estimates of probabilities based on experimental results.

Hannah and Ruth want to estimate the probability of a drawing pin landing point up.

They carry out an experiment which consists of throwing the drawing pin into the air 50 times.

Result	Tally	Frequency
Point up	~~IIII~~ ~~IIII~~ ~~IIII~~ III	18
Point not up	~~IIII~~ ~~IIII~~ ~~IIII~~ ~~IIII~~ ~~IIII~~ ~~IIII~~ II	32

The pin lands point up 18 times out of 50. The **relative frequency** of the pin landing point up $= \frac{18}{50}$.

Hannah thinks that the probability of the pin landing point up is $\frac{1}{3}$. Ruth thinks it is $\frac{2}{5}$.

Both girls make good estimates of the probability from the evidence of the 50 trials. To be sure which is the better estimate they should carry out more trials or combine their results with those of their classmates who have also done the experiment.

To estimate a probability:

- carry out a large number of trials
- calculate the relative frequency

$$\text{relative frequency} = \frac{\textbf{number of times event occurs}}{\textbf{total number of trials}}$$

If a large number of trials is carried out the relative frequency will be approximately equal to the probability.

Example 8

Some plastic trapezia like these are turned into spinners. Ten pairs of pupils each spin their spinner 20 times. They find that the relative frequencies of the 6 cm side stopping on the table are:

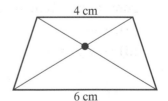

$$\frac{7}{20} \quad \frac{4}{20} \quad \frac{11}{20} \quad \frac{8}{20} \quad \frac{5}{20} \quad \frac{8}{20} \quad \frac{9}{20} \quad \frac{7}{20} \quad \frac{10}{20} \quad \frac{9}{20}$$

Calculate the overall result when these are combined. First add all the numerators to get 78 and then divide by the total number of trials, 200, to obtain 39%. Notice that this is about $\frac{8}{20}$. Two pairs actually got this result but some other pairs got results that were quite different from this. In real experiments, this does happen.

1   Roll an eraser and work out the relative frequency of it finishing in all of its possible positions after 100 trials.

2   Roll an object like a shoe, a video box or a small toy and work out the relative frequencies of it ending up in different positions. Estimate the probability of each of the possibilities and combine your results with classmates' results to refine your estimates.

9.4 Estimating from experience

In some cases even carrying out trials is impossible and so probabilities have to be estimated and justified using a reasonable argument.

Example 9

Hamwick rugby team have won each of their last six home games by at least 20 points. St Debes have lost their last four away games quite heavily. If St Debes play at Hamwick it is much more likely that Hamwick will win. An estimate of Hamwick winning of 80% or more would be reasonable.

Think of some more examples like this one.

9.5 Independent events

James tosses a coin and Sarah rolls a die. James knows that the probability of his coin landing 'head' is $\frac{1}{2}$.

The probability of Sarah's die showing a six is $\frac{1}{6}$.

What is the probability of getting a head and a six?

If we draw the sample space diagram

die

	1	2	3	4	5	6
coin H						✗
T						

We see that there are 12 possible outcomes, and

$$P(\text{H and } 6) = \frac{1}{12}$$
$$P(\text{H}) = \frac{1}{2}, P(6) = \frac{1}{6}$$
$$P(\text{H and } 6) = \frac{1}{2} \times \frac{1}{6} = \frac{1}{12}$$

Whether or not the coin lands 'head' has no effect on whether the die shows a 'six' or any other score.

$$P(H \text{ and } 6) = P(H) \times P(6)$$

■ **If two events, A and B, are independent:**
$$\textbf{P(A and B)} = \textbf{P(A)} \times \textbf{P(B)}$$

The two events, getting a 'head' and scoring a 'six', are **independent**.

Note: P(A and B and C) = P(A) × P(B) × P(C) when the events A, B and C are independent of each other.

Worked examination question

Some college students have to take 3 examination papers, one in English, one in maths and one in science. P(passing English) = 0.7, P(passing maths) = 0.5 and P(passing science) = 0.8. Assume these events are independent. Calculate the probability that a student will:

(a) pass all 3 papers   (b) fail all 3 papers   [E]

(a) P(passing all 3) = 0.7 × 0.5 × 0.8 = 0.28

(b) P(failing all 3) = 0.3 × 0.5 × 0.2 = 0.03

Exercise 9D

1 There are three tame mice, Roger, Susan and Timmy. They can choose to eat at five troughs (A, B, C, D and E) independently and at random.

(a) What is the probability that Roger will eat from trough A?

(b) Find the probability that Susan and Timmy will both eat at
(i) trough A
(ii) at the same trough.   [E]

2 When Nina and Zoe go to the shop the probability of Nina choosing a chocolate bar is $\frac{1}{3}$ and of choosing a toffee bar is $\frac{1}{5}$. The probability of Zoe choosing a chocolate bar is $\frac{1}{4}$ and of choosing a toffee bar is $\frac{1}{2}$. The girls choose independently of each other. Calculate the probability of:

(a) both choosing chocolate bars

(b) both choosing toffee bars

(c) one choosing a chocolate and the other choosing a toffee bar.

9.6 Probability trees

Tree diagrams can be used to illustrate possible outcomes when solving problems involving independent events.

Note: each of the three pupils has 2 choices; to wear or not to wear a pullover. These numbers can be used to work out how many possible paths there are. There are 2 × 2 × 2 or 8 paths.

Example 10

Newton High School pupils can choose to wear or not wear a pullover as part of the uniform. The probability that Ali will choose to wear a pullover is $\frac{1}{3}$. The probability that Bethan will wear one is $\frac{2}{5}$ and the probability that Chris will wear one is $\frac{1}{2}$. Calculate the probability that:

(a) all 3 wear a pullover

(b) exactly 2 of them wear a pullover

(c) at least 2 of them wear a pullover.

(a) The path to A includes the probabilities of each pupil wearing a pullover.
$$P(\text{all wear pullovers}) = \tfrac{1}{3} \times \tfrac{2}{5} \times \tfrac{1}{2}$$
$$= \tfrac{2}{30}$$

(b) Paths B, C or E involve exactly 2 wearing pullovers.

$P(\text{exactly 2 wear pullover})$
$$= (\tfrac{1}{3} \times \tfrac{2}{5} \times \tfrac{1}{2}) + (\tfrac{1}{3} \times \tfrac{3}{5} \times \tfrac{1}{2}) + (\tfrac{2}{3} \times \tfrac{2}{5} \times \tfrac{1}{2})$$
$$= \tfrac{2}{30} + \tfrac{3}{30} + \tfrac{4}{30}$$
$$= \tfrac{9}{30}$$

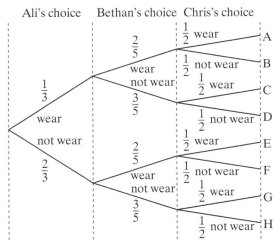

(c) $P(\text{at least 2}) = P(\text{all 3 or exactly 2})$
$$= \tfrac{2}{30} + \tfrac{9}{30}$$
$$= \tfrac{11}{30}$$

(**Note**: it does not matter which choice is put first in the tree because the probabilities are multiplied by each other and multiplication is a commutative operation. Also if you find the products of the probabilities along all the paths their sum must be 1.)

Exercise 9E

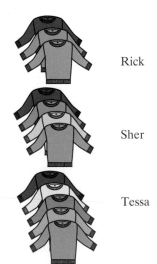

Rick

Sher

Tessa

1 Rick, Sher and Tessa always wear a coloured sweater when they go to the Youth Club. Amongst other colours they each have one, and only one, red sweater. They each pick a sweater at random from the pile in their drawer. Including the red one there are 3 sweaters in Rick's drawer, 4 in Sher's and 5 in Tessa's. Calculate the probability that:

(a) they will all go to the Youth Club in a red sweater

(b) none of them will go in a red sweater

(c) exactly two of them will go in a red sweater

(d) at least 2 of them will go in a red sweater.

2 A game at a school fair involves choosing a plastic cup from each of 3 boxes. The plastic cups each cover a coloured cube which the player cannot see. The first box contains 5 cups covering 3 red and 2 blue cubes. The second box contains 4 cups covering 3 red and 1 blue cube. The third box contains 3 cups covering 2 red and 1 blue cube.

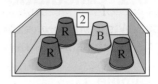
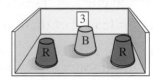

Vicky pays 20p for a go. If she selects 3 reds she loses her money. If she gets exactly 1 blue she gets 10p back. If she uncovers exactly 2 blue cubes she gets 30p back and if all three of her choices cover blue cubes she receives 50p back. Calculate the probability that Vicky will:

(a) receive 50p   (b) receive 30p

(c) have less money at the end of the game than she had before paying her 20p.

Summary of key points

1 Probabilities are measured on a scale from 0 to 1.

2 If an event is impossible its probability is 0.

3 If an event is certain its probability is 1.

4 If there are n equally likely outcomes the probability of one of them occurring is $\frac{1}{n}$.

5 $\text{P(event)} = \dfrac{\text{the number of ways the event can occur}}{\text{the total number of possibilities}}$

6 For mutually exclusive events: P(A or B) = P(A) + P(B).

7 Event A and event 'not A' are mutually exclusive and cover all possibilities and so P(not A) = 1 − P(A).

8 $\text{Relative frequency} = \dfrac{\text{number of times event occurs}}{\text{total number of trials}}$

9 For independent events: P(A and B) = P(A) × P(B)

10 Use lists, sample spaces or trees to aid calculation.

10 Brackets in algebra

10.1 Calculations with brackets

■ **When you are evaluating an expression, calculations inside brackets are always worked out first.**

Example 1

$$8 + (5 \times 4) = 8 + 20$$
$$= 28$$

Even without the brackets $8 + 5 \times 4$ is still 28 as we always do multiplication and division before addition and subtraction. Your calculator should do this automatically.

Example 2

$$10 - (7 - 1) = 10 - 6$$
$$= 4$$

Without brackets $10 - 7 - 1 = 2$ which is incorrect.

Exercise 10A

Evaluate:

1	$9 + (8 \times 2)$	**2**	$12 - (9 - 2)$
3	$5 \times (7 - 3)$	**4**	$(7 - 1) \times 3$
5	$(9 + 7) \div 8$	**6**	$(8 \div 2) + 1$
7	$(3 + 2) \times (9 - 4)$	**8**	$(5 \times 4) - (7 \times 3)$
9	$(6 + 3) \div (1 + 4)$	**10**	$(3 \times 3) \div (5 \times 2)$
11	$(9 - 2) \times (4 - 7)$	**12**	$(4 \times 4) - (8 \times 2)$

Copy the following calculations and insert brackets so that they are correct:

13 $4 \times 6 - 1 = 20$

14 $7 \times 3 - 2 = 19$

15 $9 - 5 - 4 = 8$

16 $7 - 2 \times 6 + 3 = 45$

17 $6 \times 8 - 5 \times 7 = 13$

18 $2 \times 8 + 2 \times 3 = 60$

19 Evaluate and compare:

  (a) $(4 + 5) + 2$ with $4 + (5 + 2)$

  (b) $(9 - 5) - 3$ with $9 - (5 - 3)$

  (c) $(5 \times 2) \times 3$ with $5 \times (2 \times 3)$

  (d) $(12 \div 6) \div 2$ with $12 \div (6 \div 2)$

20 Evaluate and compare:

  (a) $5 \times (6 + 2)$ with $(5 \times 6) + (5 \times 2)$

  (b) $7 \times (9 - 5)$ with $(7 \times 9) - (7 \times 5)$

10.2 Expanding brackets

You need to be able to **expand** algebraic expressions. Here expand means multiplying terms to remove brackets from expressions like $2(3r + 2b)$.

If r represents a red square and b represents a blue square, then $3r + 2b$ represents 3 red squares plus 2 blue squares.

So, an expression like $2(3r + 2b)$ represents 6 red squares plus 4 blue squares:

$$2 \times (3r + 2b) = 6r + 4b$$

$$2 \times (\;\boxed{r \mid r \mid r}\; + \;\boxed{b \mid b}\;) = \boxed{\begin{matrix} r & r & r \\ r & r & r \end{matrix}} + \boxed{\begin{matrix} b & b \\ b & b \end{matrix}}$$

■ **To expand an expression multiply each term inside the brackets by the term outside.**

$$\overset{\times}{\overgroup{2(3r + 2b)}} = 6r + 4b$$

Multiplication can also be seen as repeated addition:

$$3 \times 4 = 4 + 4 + 4 = 12$$
$$3(x + 5) = x + 5 + x + 5 + x + 5 = 3x + 15$$

This diagram helps show that $3(x + 5) = 3x + 15$:

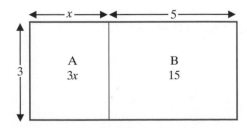

Area of whole rectangle $= 3(x + 5)$

Area of rectangle A $= 3x$

Area of rectangle B $= 3 \times 5 = 15$

Example 3

Expand these algebraic expressions:

(a) $5(x + 2) = 5x + 10$ (b) $4(x - 3) = 4x - 12$

(c) $2(3x + 1) = 6x + 2$ (d) $x(x^2 + y) = x^3 + xy$

(e) $x(5x + 2) = 5x^2 + 2x$ (f) $-2(3x - 4) = -6x + 8$

(g) $3x(4x - 7) = 12x^2 - 21x$ (h) $(6 - x)x^2 = 6x^2 - x^3$

Exercise 10B

Draw diagrams to help show:

1 $5(x + 2) = 5x + 10$ **2** $8(x + 3) = 8x + 24$

Expand these:

3 $7(x + 3)$ **4** $8(x - 2)$ **5** $5(x - 5)$

6 $6(x + 4)$ **7** $6(2x - 5)$ **8** $9(3x + 2)$

9 $5(7x - 1)$ **10** $2(9x + 7)$ **11** $x(x + 4)$

12 $x(x - 8)$ **13** $x(2x + 9)$ **14** $x(5x - 7)$

15 $-5(x + 2)$ **16** $-8(x - 3)$ **17** $-7(3x - 2)$

18 $-6(x + 7)$ **19** $9x(5x - 4)$ **20** $4x(7x + 8)$

21 $-6x(2x - 3)$ **22** $a(x - 8)$ **23** $x(xy - y^2)$

24 $3a(5x - 2)$ **25** $8a(3x + 2a)$ **26** $4a(5ax - 1)$

27 $(3 + x)x$ **28** $(2x - 9)x$ **29** $x^2(x + 7)$

30 $(4 - 3x)x^2$

10.3 Factorizing

Factorizing is the opposite of expanding. You write algebraic expressions in a shorter form using brackets. It also helps you solve some algebra problems.

Example 4

Here are some algebraic expressions before and after factorizing:

	before	after
(a)	$6x + 18$	$6(x + 3)$
(b)	$21x - 28$	$7(3x - 4)$
(c)	$x^2 - 9x$	$x(x - 9)$
(d)	$ax + 4a$	$a(x + 4)$
(e)	$x^2y - xy^2$	$xy(x - y)$
(f)	$3x^2 + 12x$	$3x(x + 4)$

■ **To factorize an expression completely the Highest Common Factors (HCF) must appear outside the brackets.**

In $3x^2 + 12x$ the HCF of $3x^2$ and $12x$ is $3x$. The HCF is the largest term which is a factor of $3x^2$ and $12x$.

$3(x^2 + 4x)$ and $x(3x + 12)$ are both ways of factorizing $3x^2 + 12x$ but they are *not factorized completely* because the HCF $3x$ does not appear outside the brackets.

HCF of $3x^2 + 12x$

$3x^2 = $ | 3 | $\times$ | x | $\times$ | x
$12x = $ | 3 | $\times$ | 4 | $\times$ | x

HCF is $3x$

Example 5

Show that the sum of three consecutive integers is always a multiple of 3.

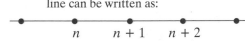

Three consecutive (one after the other) integers on a number line can be written as:

n $n + 1$ $n + 2$

Factorizing can help solve this problem. Let n be the smallest of the three integers. The other integers are $n + 1$ and $n + 2$. The sum of three consecutive integers can be written:

$$\text{Sum} = n + n + 1 + n + 2$$
$$= 3n + 3$$
$$= 3(n + 1)$$

3 is a factor of the sum so it is *always* a multiple of 3.

Exercise 10C

In questions **1–18** factorize the expressions completely:

1	$5x - 20$	**2**	$8x + 24$
3	$12x + 18$	**4**	$20x - 25$
5	$8x^2 - 24$	**6**	$9x^2 + 36$
7	$10x^2 - 15$	**8**	$x^2 + 6x$
9	$x^2 - x$	**10**	$px - 3p$
11	$qx + q^2$	**12**	$5x^2 - 15x$
13	$6x^2 + 9x$	**14**	$15x^2 - 35x$
15	$8x - 12x^2$	**16**	$ax^2 + ax$
17	$4ax - 6a$	**18**	$6ax^2 + 15ax$

19 Show that the sum of five consecutive integers is always a multiple of 5. (**Hint**: Example 5 helps.)

20 Show that the sum of four consecutive integers is always even.

10.4 Simplifying expressions involving brackets

Sometimes you can simplify an algebraic expression by expanding it and collecting like terms together. Here are some examples:

$$7(x + 3) + (x + 4) = 7x + 21 + x + 4 = 8x + 25$$

$$3(x + 2) + 4 (x - 3) = 3x + 6 + 4x - 12 = 7x - 6$$

Collecting like terms
Numbers and terms involving x, x^2 and so on can be collected together like this:

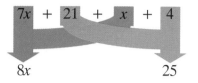

A minus sign in front of the brackets multiplies every term inside by -1. Change the sign of every term inside the brackets then remove them.

$$4(x - 1) - (2x - 3) = 4x - 4 - 2x + 3$$
$$= 2x - 1$$

$- (2x - 3)$ **means**
$-1 \times (2x - 3)$
$= -2x + 3$

$$5(x - 1) - 3(x - 2) = 5x - 5 - 3x + 6$$
$$= 2x + 1$$

$-3(x - 2)$ **means**
$-3 \times (x - 2)$
$= -3x + 6$

$$8 + 3(2x - 3) = 8 + 6x - 9$$
$$= 6x - 1$$

Exercise 10D

Expand and simplify these expressions:

1 $5(x + 3) + 2(x - 7)$ 2 $3(x - 5) + 4(x + 6)$

3 $3(2x - 1) + 2 (3x + 4)$ 4 $4(5x + 2) + 8(2x - 1)$

5 $7(x + 1) - 5(x + 3)$ 6 $3(x + 5) - 2(x - 7)$

7 $4(3x - 2) - 5(2x - 1)$ 8 $7(2x + 1) - 2(5x + 4)$

9 $3(5x - 7) - 7(2x - 3)$ 10 $6(3x + 2) - 9(2x - 1)$

11 $5(x + 4) + (x - 8)$ 12 $8(x - 1) + (3x - 4)$

13 $3(2x - 5) + (4x + 9)$ 14 $2(x - 5) - (x + 3)$

15 $8(x - 1) - (x - 8)$ 16 $5(3x + 4) - (9x - 4)$

17 $7(2x - 1) - (3x - 7)$ 18 $4(2x - 3) - (8x + 5)$

19 $7 + 4(x - 1)$ 20 $5x - 2(3x - 1)$

21 $8(x + 2) + 7(x - 3) - 5(x - 4)$

22 $9 + 6(x - 5) + (3x - 8)$

23 $2(x + 3) - 7(x - 1) - (4x + 15)$

24 $7x - (3x + 2) + (2x - 7) + 10$

25 $6(2x + 5) - 4(3x - 4) - (2x + 7) + 2x$

10.5 Solving equations involving brackets

Sometimes you will need to solve equations involving brackets like

$$3(x + 2) = 10$$

Solve means find the value of x that makes the equation true.

Here are three ways of doing this.

Method 1: Start by expanding the brackets

$$3(x + 2) = 10$$

expanding: $\qquad 3x + 6 = 10$

subtract 6
from each side: $\qquad 3x = 4$

divide both
sides by 3: $\qquad x = \frac{4}{3} = 1\frac{1}{3}$

Method 2: Start by dividing both sides by 3

$$3(x + 2) = 10$$

$\div 3 \qquad\qquad\qquad\qquad \div 3$

$(x + 2) = \frac{10}{3} = 3\frac{1}{3}$

brackets off: $\qquad x + 2 = 3\frac{1}{3}$

subtract 2 from
each side: $\qquad x = 1\frac{1}{3}$

Method 3: Use a flow diagram to build the expression $3(x + 2)$

$$x \quad \xrightarrow{\ +\ 2\ } \quad x + 2 \quad \xrightarrow{\ \times\ 3\ } \quad 3(x + 2)$$

As this is equal to 10 you can use an inverse flow diagram to find the value of x. Start at the right hand side and reverse the steps in the arrows:

$$1\frac{1}{3} \quad \xleftarrow{\ -\ 2\ } \quad 3\frac{1}{3} \quad \xleftarrow{\ \div\ 3\ } \quad 10 \quad \leftarrow \boxed{\text{Start here with the } \textit{value} \text{ of } 3(x + 2)}$$

the inverse   the inverse
of + 2   of × 3

Example 6

Solve $3(x + 2) + 4(x - 3) = 5x - 11$

Expand brackets: $\qquad 3x + 6 + 4x - 12 = 5x - 11$

Collect like terms: $\qquad\qquad 7x - 6 = 5x - 11$

Subtract $5x$ $\qquad 7x - 6 - 5x = -11$

Simplify $\qquad\qquad 2x - 6 = -11$

Add 6 $\qquad\qquad 2x = -11 + 6$

$$2x = -5$$

$$x = -2\frac{1}{2}$$

Exercise 10E

Solve these equations:

1 $5(x + 2) = 30$ **2** $2(x + 7) = 8$

3 $6(x - 2) = 1$ **4** $3(x - 1) = 7$

5 $4(x + 1) = 9x$ **6** $5(x - 4) = 3x - 7$

7 $3(4x - 5) - 5(2x - 3) = 4(x - 1)$

8 $6 + (5x - 7) - 2(3x + 4) = 7(x + 1)$

9 $4(3x - 2) - (7x + 8) = 2(x - 5)$

10 $6(x + 5) - 5(3x - 4) = 11 + 3(2x - 7)$

11 $2(7x - 4) + (5 - 9x) = 1 - 3x$

12 $3(5x + 2) - (9x - 8) = 7(x - 4) + (x + 12)$

13 $8x - 5(2x + 3) - (1 - 6x) = 8 + (4 + 6x)$

14 $5(3x - 7) + 2(9 - 2x) = 8x - (2x + 1) - 16$

15 $3(5x - 4) - 7(2x - 3) = 3(x + 1) + 5$

10.6 Using equations to help solve problems

Example 7

I add 3 to a mystery number and multiply the result by 5. My answer is 55. Find the mystery number.

You can represent this problem as an equation using x as the mystery number. Then:

$$5(x + 3) = 55$$
$$5x + 15 = 55$$
$$5x = 40$$
$$x = 8$$

The mystery number is 8.

What number am I thinking of?

Exercise 10F

Use equations to help you solve these problems:

1 I subtract 9 from a number and multiply the result by 4. My answer is 36. Find the number.

2 I add 2 to a number and multiply the result by 5. The answer is the same as when I add 8 to the number and multiply the result by 3. Find the number.

3 I subtract 7 from a number and multiply the result by 6. The answer is the same as when I multiply the number by 10 and subtract 62 from the result. Find the number.

Steps to solve these problems

1 Use a letter to represent the mystery (unknown) number.
2 Write the problem as an equation.
3 Solve the equation.
4 Write the answer.

4 Mr Benny is now 3 times as old as his son. In 13 years' time he will be twice as old as his son. Find Mr Benny's present age.

5 The sum of the present ages of Mrs Peel and her daughter Emma is 50 years. In 5 years' time Mrs Peel will be 3 times as old as Emma. How old is Emma now?

6 Andrew has 20 coins in his pocket. Some of them are 10p coins and the rest are 5p coins. Their total value is £1.65. How many 5p coins are in his pocket? (Hint: work in pence.)

7 Karen drives 210 miles in 4 hours. Her average speed is 60 mph for part of the time and 40 mph for the rest. For how long is her average speed 60 mph?

8 In a cricket league teams get 5 points for a win, 3 points for a draw and 0 points for a defeat. After 20 matches Mathstown have 80 points. They have had only two defeats. How many of their matches have been draws?

9 One day David works for 10 hours. For part of this time he is paid at the normal rate of £5 per hour. For the rest he is paid at the overtime rate of £7 per hour. He is paid £56 altogether. How many hours overtime has he worked?

10.7 Solving inequalities involving brackets

■ **Inequalities can be solved in the same way as equations except that you cannot multiply or divide both sides by a *negative* number.**

Example 8

Solve $3(x - 2) \leqslant 20$

expand the brackets:	$3x - 6 \leqslant 20$
add 6 to both sides:	$3x \leqslant 26$
divide both sides by 3:	$x \leqslant 8\frac{2}{3}$

> Another way of solving $3(x - 2) \leqslant 20$ is to divide by 3 first.
>
> You can do this because 3 is positive.

Example 9

Solve $2(x + 2) + 3(2x - 3) > 2$

expand the brackets:	$2x + 4 + 6x - 9 > 2$
simplify:	$8x - 5 > 2$
add 5 to both sides:	$8x > 7$
divide both sides by 8:	$x > \frac{7}{8}$

Example 10

Solve $2(3x - 2) + 3(4 - 3x) > 12 - (5x + 7)$

expand the brackets: $\qquad 6x - 4 + 12 - 9x > 12 - 5x - 7$

simplify: $\qquad\qquad\qquad 8 - 3x > 5 - 5x$

add $5x$ to both sides: $\qquad\quad 8 + 2x > 5$

subtract 8 from both sides: $\qquad 2x > -3$

divide both sides by 2: $\qquad\quad x > -1\tfrac{1}{2}$

Exercise 10G

Solve these inequalities:

1. $5(x + 4) > 40$
2. $3(x - 7) > 14$
3. $4(x + 1) \leqslant 2x - 3$
4. $2(3x - 5) \geqslant 8 - 3x$
5. $3(x + 2) - 5(x - 4) < 27$
6. $5 - (3x + 4) \geqslant 4$
7. $7(x - 2) - (5x - 9) > 0$
8. $3x + 4(2x + 3) < 1$
9. $4(2x - 1) \leqslant 3(x + 4)$
10. $3 - (2x - 3) > 6$
11. $6(3x + 1) - 2(5x - 4) < 5(x + 2)$
12. $3(4x + 5) + (8 - 10x) \leqslant 4(x + 6)$
13. $7(3x - 4) - (12x - 5) > 5(2x - 3)$
14. $8x + 7(2 - x) < 9 - 3(x + 1)$
15. $6(x + 4) - 7 - (3x + 10) \geqslant 2(4x - 5)$

10.8 Rearranging formulae

In Unit 2 you looked at rearranging a formula so that a letter such as x is on one side of the equals sign with the rest of the formula on the other side.

This is called **changing the subject** of the formula.

Expanding brackets and factorization can sometimes help you change the subject of a formula.

Example 11

Make x the subject of $a(x + b) = c$.

expand the brackets: $\qquad\quad ax + ab = c$

subtract ab from both sides: $\qquad ax = c - ab$

divide both sides by a : $\qquad\quad x = \dfrac{c - ab}{a}$

Another way of doing this is to divide both sides by a

$$x + b = \frac{c}{a}$$

$$x = \frac{c}{a} - b$$

You could also use a flow diagram as on page 205.

Example 12

Make x the subject of $\dfrac{x}{a} + b = c$.

subtract b from both sides: $\qquad \dfrac{x}{a} = c - b$

multiply both sides by a : $\qquad x = a\,(c - b)$

$$x = ac - ab$$

Example 13

Make x the subject of $a + bx = c - dx$.

add dx to both sides: $\qquad a + bx + dx = c$

subtract a from both sides: $\qquad bx + dx = c - a$

factorize the left side: $\qquad x\,(b + d) = c - a$

divide both sides by $b + d$: $\qquad x = \dfrac{c - a}{b + d}$

Exercise 10H

Make x the subject of each of these:

1   $a\,(x - b) = c$ 2 $ax = bx + c$

3   $a\,(x + b) = cx$ 4 $a + bx = c + dx$

5   $a\,(x + b) = c - dx$ 6 $a - bx = c - dx$

7   $a\,(x + b) = c\,(x + d)$ 8 $\dfrac{x}{a} - b = c$

9   $ax + b\,(x + d) = e$ 10 $ax + \dfrac{x}{b} = c$

11   $\dfrac{x}{a} = bx + c$ 12 $\dfrac{x}{a} = b\,(x + c)$

13   $a\,(x + b) + c\,(x + d) = e$ 14 $a - b\,(x - c) = dx + e$

15   $ax + b\,(x + c) = d\,(x + e)$

10.9 Substituting algebraic expressions

You should know how to substitute numbers in place of terms in an equation. (There is more about this in Unit 14.)

Sometimes you will need to substitute algebraic expressions for terms in an equation. Here's how this can help you solve simultaneous equations:

Example 14

Solve: $3x - 4y = 15$ (equation 1)
$\quad\quad y = 2x - 5$ (equation 2)

Substitute $2x - 5$ for y in equation 1:
$$3x - 4(2x - 5) = 15$$

expand brackets: $\quad\quad 3x - 8x + 20 = 15$

simplify: $\quad\quad\quad\quad\quad -5x + 20 = 15$

add $5x$ to both sides: $\quad\quad\quad 20 = 5x + 15$

subtract 15 from both sides: $\quad 5x = 5$

divide both sides by 5: $\quad\quad\quad x = 1$

substitute x in equation 2: $\quad\quad y = (2 \times 1) - 5$
$$y = -3$$

The solution is $x = 1, y = -3$

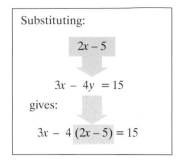

Substituting:

$2x - 5$

$3x - 4y = 15$

gives:

$3x - 4(2x - 5) = 15$

Example 15

Here is another application of substitution:
$$z = 2x + 3y, x = t + 5 \text{ and } y = 5t - 4.$$

Express z in terms of t. $\quad\quad z = 2(t + 5) + 3(5t - 4)$
$$z = 2t + 10 + 15t - 12$$
$$z = 17t - 2$$

Substituting:

$t + 5$ $\quad$ $5t - 4$

$z = 2x \quad + \quad 3$

gives:

$z = 2(t + 5) + 3(5t - 4)$

■ **Simultaneous equations can be solved by substitution.**

Exercise 10 I

In questions **1–6**, solve the simultaneous equations using the method of substitution.

1 $x + 2y = 16$
$\quad y = 3x + 1$

2 $2x - 5y = 7$
$\quad y = 4x - 5$

3 $3x + 4y = 6$
$\quad y = 2x + 7$

4 $4x - 5y = 18$
$\quad y = 5x + 9$

5 $4x + 3y = 5$
$\quad y = -4x + 3$

6 $3x - 2y = 10$
$\quad y = -6x + 5$

In questions **7–10**, express z in terms of t.

7 $z = 4x + 5y, x = 2t - 1 \text{ and } y = 3t - 4$

8 $z = 3x - 2y, x = 5t + 4 \text{ and } y = 7t - 6$

9 $z = 6x - y, x = 2t - 3 \text{ and } y = 5t - 4$

10 $z = 8x + 3y, x = 3t + 1 \text{ and } y = 2 - 7t$

10.10 Multiplying bracketed expressions

Sometimes you will need to multiply bracketed expressions by each other.
For example: $(a + b)(c + d)$. This is called **expanding** the expression.

To help you see what happens when you expand an expression like $(a + b)(c + d)$ look at this rectangle with sides $a + b$ and $c + d$:

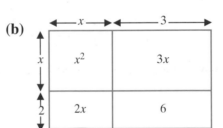

The area of the rectangle is:

$$(a + b)(c + d)$$

The area can also be expressed as the combined area of four smaller rectangles:

$$ac + ad + bc + bd$$

This expression is true for all values of a, b, c and d. It is called an **identity**.

■ **In general:** $(a + b)(c + d) = ac + ad + bc + bd$

Example 16

(a) Expand and simplify $(x + 3)(x + 2)$.

(b) Draw a diagram to illustrate your answer.

Answer

(a) $(x + 3)(x + 2) = x(x + 2) + 3(x + 2)$
$$= x^2 + 2x + 3x + 6$$
$$= x^2 + 5x + 6$$

(b)

Notice that:
$(x + 5)(x + 3)$
$= x^2 + (5 + 3)x + 5 \times 3$

Example 17

(a) $(x + 5)(x + 3) = x(x + 3) + 5(x + 3)$
$$= x^2 + 3x + 5x + 15$$
$$= x^2 + 8x + 15$$

(b) $(x + 4)(x - 2) = x(x - 2) + 4(x - 2)$
$$= x^2 - 2x + 4x - 8$$
$$= x^2 + 2x - 8$$

(c) $(x + 3)(2 - x) = x(2 - x) + 3(2 - x)$
$$= 2x - x^2 + 6 - 3x$$
$$= 6 - x - x^2$$

Exercise 10J

1 **(a)** Expand and simplify these products, drawing diagrams to illustrate your answers.

(i) $(x + 2)(x + 1)$ (ii) $(x + 8)(x + 2)$
(iii) $(x + 3)^2$ (iv) $(x + a)(x + b)$

Expand and simplify these products:

2 $(x + 4)(x + 1)$ **3** $(x + 4)(x - 5)$

4 $(x - 9)(x + 2)$ **5** $(x - 7)(x - 5)$

6 $(x + 8)(x - 3)$ **7** $(x - 1)(x - 6)$

8 $(x - 2)(x + 8)$ **9** $(x + 2)(5 - x)$

10 $(8 - x)(1 - x)$

Multiplying an expression by itself

Example 18

Expand and simplify these:

(a) $(x + 6)^2 = (x + 6)(x + 6)$
$$= x(x + 6) + 6(x + 6)$$
$$= x^2 + 6x + 6x + 36$$
$$= x^2 + 12x + 36$$

(b) $(x - 8)^2 = (x - 8)(x - 8)$
$$= x(x - 8) - 8(x - 8)$$
$$= x^2 - 8x - 8x + 64$$
$$= x^2 - 16x + 64$$

■ **In general:** $(x + a)^2 = x^2 + 2ax + a^2$
$$(x - b)^2 = x^2 - 2bx + b^2$$

Exercise 10K

Expand and simplify these:

1 $(x + 5)^2$ **2** $(x - 3)^2$ **3** $(x + 9)^2$

4 $(x - 2)^2$ **5** $(x + a)^2$ **6** $(x - a)^2$

7 $(3 + x)^2$ **8** $(7 - x)^2$ **9** $(a - x)^2$

10 Copy and complete these by writing appropriate numbers in the boxes:

(a) $(x + \square)^2 = x^2 + \square x + 16$

(b) $(x - \square)^2 = x^2 - 20x + \square$

(c) $(x + \square)^2 = x^2 + 2x + \square$

(d) $(x - \square)^2 = x^2 - \square x + 144$

Multiplying expressions when one sign is different

Example 19

Expand and simplify $(x + 7)(x - 7)$.

$$(x + 7)(x - 7) = x(x - 7) + 7(x - 7)$$
$$= x^2 - 7x + 7x - 49$$
$$= x^2 - 49$$

■ **In general:**
$(x + a)(x - a) = x^2 - a^2$
This is called the difference of two squares.

Exercise 10L

Expand and simplify:

1 $(x + 5)(x - 5)$ **2** $(x - 1)(x + 1)$

3 $(x + 10)(x - 10)$ **4** $(x - 9)(x + 9)$

5 $(x - 7)(x + 7)$ **6** $(x + 3)(x - 3)$

7 $(x + a)(x - a)$ **8** $(x + 6)(6 - x)$

9 $(3 + x)(3 - x)$ **10** $(7 - x)(x + 7)$

Multiplying more complicated expressions

Example 20

Expand and simplify:

(a)
$$(2x + 3)(5x - 4) = 2x(5x - 4) + 3(5x - 4)$$
$$= 10x^2 - 8x + 15x - 12$$
$$= 10x^2 + 7x - 12$$

(b)
$$3(2x - 1)(4x - 3) = 3[2x(4x - 3) - (4x - 3)]$$
$$= 3(8x^2 - 6x - 4x + 3)$$
$$= 3(8x^2 - 10x + 3)$$
$$= 24x^2 - 30x + 9$$

(c)
$$(3x - 5)^2 = (3x - 5)(3x - 5)$$
$$= 3x(3x - 5) - 5(3x - 5)$$
$$= 9x^2 - 15x - 15x + 25$$
$$= 9x^2 - 30x + 25$$

(d)
$$(2x + 7)(2x - 7) = 2x(2x - 7) + 7(2x - 7)$$
$$= 4x^2 - 14x + 14x - 49$$
$$= 4x^2 - 49$$

Exercise 10M

Expand and simplify:

1 $(5x + 2)(3x + 1)$ 2 $(7x - 3)(2x + 5)$

3 $(2x + 3)(8x - 3)$ 4 $(4x - 1)(3x - 5)$

5 $(5x + 1)(3x + 4)$ 6 $(6x + 5)(2x - 1)$

7 $(3x - 7)(2x - 3)$ 8 $(9x - 1)(4x + 5)$

9 $5(x + 2)(x - 6)$ 10 $4(x + 1)(x - 1)$

11 $2(3x + 1)(5x - 1)$ 12 $3(4x - 3)(3x + 2)$

13 $(2x + 7)^2$ 14 $(4x - 3)^2$

15 $(5x - 1)^2$ 16 $(3x + 2)^2$

17 $(3x + 1)(3x - 1)$ 18 $(1 - 4x)(1 + 4x)$

19 $(2x - 5)(2x + 5)$ 20 $(7x + 2)(7x - 2)$

Summary of key points

1 When you are evaluating an expression, calculations inside brackets are always worked out first.
For example: $8 + (5 \times 4) = 8 + 20 = 28$

2 To expand an algebraic expression multiply each term inside the brackets by the term outside.
For example: $2(3r + 2b) = 6r + 4b$

3 Factorizing is the opposite of expanding. To factorize an expression completely, its Highest Common Factor (HCF) must appear outside the brackets.
For example: $3x^2 + 12x = 3x(x + 4)$

4 Inequalities can be solved in the same way as equations except that you cannot multiply or divide both sides by a *negative* number.

5 Simultaneous equations can be solved by substitution.

6 The product of two brackets can be found like this:
$$(x + 2)(x + 4)$$
$$x(x + 4) + 2(x + 4) = x^2 + 4x + 2x + 8$$
$$= x^2 + 6x + 8$$

7 Some useful expansions are
$$(x + a)^2 = x^2 + 2ax + a^2$$
$$(x - b)^2 = x^2 - 2bx + b^2$$

8 The difference of two squares is $(x + a)(x - a) = x^2 - a^2$

11A Using and applying mathematics

Introduction

In this unit you are going to look at an area of mathematics called 'Using and Applying Mathematics'. In your examinations this is most commonly tested through coursework.

A specific investigative task will be used as an aid to showing you the general principles involved in this type of work.

11.1 The specific task

Diagonals

This diagram shows a regular hexagon. A diagonal is a line which joins any two of the vertices (corners) of the shape but which is not an edge of the shape.

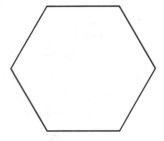

1   Show that the regular hexagon has a total of nine diagonals.
2   Work out the number of diagonals for a regular heptagon, a seven-sided shape.
3   Draw some regular shapes of your own.
    For each shape work out the total number of diagonals.
    Investigate regular shapes to establish the relationship between the number of diagonals and the number of vertices of any shape.
4   Extend your investigation to consider shapes other than regular ones in two dimensions.

The technique for doing investigations is governed by a general process. For assessment purposes this process is sub-divided into three sections:

> A coursework task might start at **3**, or say: 'Investigate …'

Deciding and doing   Showing   Explaining

You should try to get the right balance across these sections.

The process itself can be written as a list of things you should try to do during the investigation and things you should provide as evidence in your final written report.

You should:

● make sure that you understand the problem
● check to see if you have worked on a similar problem and, if you have, try to make use of this experience

- try some simple and special cases
- work in an ordered, strategic way
- record your strategies
- record your observations
- use appropriate diagrams and forms of communication
- record and tabulate any findings and results
- make and test any conjectures
- try to make use of any counter examples
- generalize, especially in symbols if you can
- comment on any generalizations
- explain or justify your generalizations
- try to prove any generalizations.

It is also vitally important during your investigation and in the written report, that you use levels of mathematics appropriate to the tier of entry of your examination. It is also important that you seek advice from your teacher whenever you feel that this is necessary.

We will illustrate this investigative process using the specific task named *Diagonals* to make important points.

11.2 Understanding the problem

You need to understand the main features of the problem. You are being asked to consider regular shapes in two dimensions and examine the number of diagonals of these regular shapes. A diagonal is a line from one corner to the other that is not a side, or edge, of the shape.

An answer to question **1** on page 215, about the regular hexagon, will show that you understand the basic idea of the problem.

The nine diagonals have been drawn on this shape:

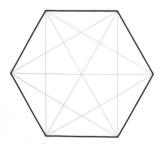

It helps to illustrate your future strategy if you label the six vertices of the regular hexagon as

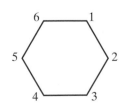

Then you can record the nine diagonals as:

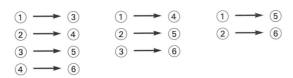

Recording the results for the regular hexagon in this way may help at a later stage.

11.3 Have you worked on a similar problem before?

Only you can answer this but it is quite likely that you have done other work of a similar nature.

Certainly, if instead of writing the diagonals for the regular hexagon as:

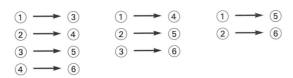

you put an array of dots as:

then there is a chance that you will have seen similar dot patterns when dealing with number patterns such as the triangular numbers or the square numbers. This may be of help later.

Question **2** on page 215, about the regular heptagon, now helps to cement your understanding of the problem.

The diagonals, the recording of them and the dot pattern are shown below.

Diagram

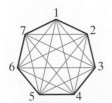

Recording

$\text{①} \longrightarrow \text{③}$ $\text{①} \longrightarrow \text{④}$ $\text{①} \longrightarrow \text{⑤}$ $\text{①} \longrightarrow \text{⑥}$

$\text{②} \longrightarrow \text{④}$ $\text{②} \longrightarrow \text{⑤}$ $\text{②} \longrightarrow \text{⑥}$ $\text{②} \longrightarrow \text{⑦}$

$\text{③} \longrightarrow \text{⑤}$ $\text{③} \longrightarrow \text{⑥}$ $\text{③} \longrightarrow \text{⑦}$

$\text{④} \longrightarrow \text{⑥}$ $\text{④} \longrightarrow \text{⑦}$

$\text{⑤} \longrightarrow \text{⑦}$

Dot pattern

11.4 Try some simple and special cases

It is sensible to be ordered in your way of working at this stage. You should start with a triangle, a square and a pentagon.

These are the results, in terms of diagrams, recording and dot pattern.

	Diagram	Recording	Dot pattern
Triangle		None	
Square		$\text{①} \longrightarrow \text{③}$ $\text{②} \longrightarrow \text{④}$	
Pentagon		$\text{①} \longrightarrow \text{③}$ $\text{①} \longrightarrow \text{③}$ $\text{②} \longrightarrow \text{④}$ $\text{②} \longrightarrow \text{④}$ $\text{③} \longrightarrow \text{⑤}$	

11.5 Strategies

So far, we have worked in an ordered, strategic way. This has been shown in our recording, which suggests that we always start by drawing all of the diagonals that start at vertex 1, then draw all of those that start at vertex 2, then those that start at vertex 3 and so on. Once we have drawn ① – ③ we do not need to draw ③ – ① because this is ① – ③ drawn in the opposite direction. This is an important part of our strategy.

We have also recorded the strategies used. These are shown both in the recording of ① – ③ and so on and in the use we have made of the dot patterns.

It is always worth writing a sentence or two to explain what you have done and how you have done it.

11.6 Recording and tabulating observations and results

We can now record, in a table, the results of our investigation so far.

Shape	Number of vertices	Number of diagonals
Triangle	3	0
Square	4	2
Pentagon	5	5
Hexagon	6	9
Heptagon	7	14

11.7 Forms of communication

In establishing the above table of results we have used diagrams, a labelling system, dot patterns and numbers; so we have used several forms of communication.

We can, and should, be starting to see if we can make sense of what is happening with the sequence for the number of diagonals of these regular shapes.

11.8 Observations and conjectures

The observations we have made can be put in a table:

Vertices	3	4	5	6	7
Diagonals	0	2	5	9	14

The conjectures we can make about the observations are many.

At a very trivial level you might notice that the larger the number of vertices the larger the number of diagonals.

At a slightly more sophisticated level you might notice that the number of diagonals follows a pattern of:

even, even, odd, odd, even

which suggests that for eight vertices there might be an *even* number of diagonals.

One of the best ways of making observations and helping with further work is to use the method of **differences**.

Differences

Vertices	Diagonals	1st difference	2nd difference
3	0		
4	2	2	
5	5	3	1
6	9	4	1
7	14	5	1

(with workings shown: $2-0$, $5-2$, $9-5$, $14-9$ in the 1st difference column and $3-2$, $4-3$, $5-4$ in the 2nd difference column)

Notice that the numbers under the 1st difference column follow a very simple pattern while those under the second difference column are all the same and equal to 1. This useful information **suggests** that the pattern of results can be continued as

Vertices	Diagonals	1st difference	2nd difference
3	0		
4	2	2	
5	5	3	1
6	9	4	1
7	14	5	1
(8)	(20)	(6)	(1)
(9)	(27)	(7)	(1)

(with workings shown: $14+6$, $20+7$ in the Diagonals column and $5+1$, $6+1$ in the 2nd difference column)

This technique allows us to make **conjectures** that the numbers of diagonals for a regular shape with eight vertices is 20 and for a regular shape with nine sides is 27.

It has also told us that, should we be correct, we have established a general pattern.

11.9 Making and testing conjectures

Our **conjectures** at the moment can be put as:

Vertices	Diagonals
8	20
9	27

These can be tested. Our test will be made using the diagrams, recording and dot pattern method we have used previously.

Eight Vertices: Octagon

Diagram   **Recording**   **Dot pattern**

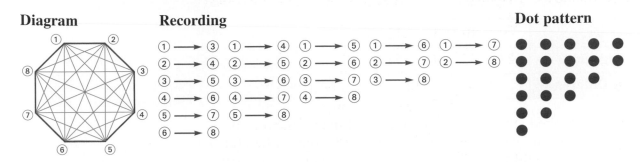

Counting the diagonals confirms the result of 20.

Nine Vertices: Nonagon

Diagram

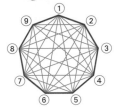

Recording   **Dot pattern**

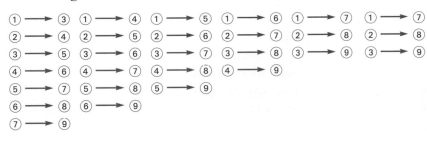
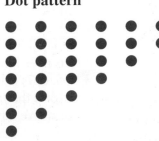

Counting the diagonals confirms the result of 27.

11.10 The use of counter examples

The last two results have helped to confirm the pattern of results for
the number of vertices. The full pattern is:

Vertices	3	4	5	6	7	8	9
Diagonals	0	2	5	9	14	20	27
1st Difference		2	3	4	5	6	7

This is strong evidence in support of the conjecture that the pattern will continue as:

Vertices	9	10	11	12
Diagonals	27	35	44	54
1st Difference		8	9	10

When we looked at the eight- and nine-sided shapes the results confirmed the anticipated pattern. There may be times when further results fail to confirm the anticipated pattern. When this happens these further results are known as **counter examples** and it is necessary to include them with the other results and try to establish another conjecture.

If you work accurately in a systematic way your conjecture should be correct, your testing will confirm it and there will be no need for counter examples.

11.11 Generalizations

Let us look again at our set of results and 1st differences. These are

Vertices	3	4	5	6	7	8	9
Diagonals	0	2	5	9	14	20	27
1st Difference		2	3	4	5	6	7

That the 1st differences follow the sequence of natural numbers (apart from 1, as 2, 3, 4, 5, 6, 7) is itself a generalization, and one which we can make use of to extend the table of results. This process of extending is called extrapolation. We have already seen that it gives the results:

Vertices	10	11
Diagonals	35	44

The next stage in the development of this investigation is to try to put the generalization in symbolic (or algebraic) form. The correct use of symbols is usually an important step on the road to achieving higher grades (C to A*).

For our results, the 1st and 2nd differences are:

Vertices	3	4	5	6	7	8	9
Diagonals	0	2	5	9	14	20	27
1st Difference		2	3	4	5	6	7
2nd Difference			1	1	1	1	1

The 2nd difference being constant means that the general expression is quadratic

Sometimes it is quite easy to spot the rule, but this one is a bit harder. One of the clues here is to see that we have:

Vertices	Diagonals
3	$0 = 3 \times 0$
5	$5 = 5 \times 1$
7	$14 = 7 \times 2$
9	$27 = 9 \times 3$

which suggests that

for 11 vertices the number of diagonals $= 11 \times 4 = 44$

a result we have already conjectured.

Now we look again at the full set of results:

Vertices	Diagonals	Vertices	Diagonals
3	$0 = 3 \times 0 = \dfrac{3 \times 0}{2}$	8	$20 = 8 \times 2\frac{1}{2} = \dfrac{8 \times 5}{2}$
4	$2 = 4 \times \frac{1}{2} = \dfrac{4 \times 1}{2}$	9	$27 = 9 \times 3 = \dfrac{9 \times 6}{2}$
5	$5 = 5 \times 1 = \dfrac{5 \times 2}{2}$	10	$35 = 10 \times 3\frac{1}{2} = \dfrac{10 \times 7}{2}$
6	$9 = 6 \times 1\frac{1}{2} = \dfrac{6 \times 3}{2}$	11	$44 = 11 \times 4 = \dfrac{11 \times 8}{2}$
7	$14 = 7 \times 2 = \dfrac{7 \times 4}{2}$		

This suggests that the **general rule** is:

diagonals = vertices $\times$ ((vertices $- 3$) $\div 2$)

Writing

D for the number of diagonals

and *V* for the number of vertices

the **symbolic form** of the general rule is

$$D = V \times \frac{(V-3)}{2}$$

or

$$D = \tfrac{1}{2}V(V-3)$$

$\frac{1}{2}V(V-3)$ is a quadratic. It is also $\frac{1}{2}V^2 - \frac{3}{2}V$

11.12 Testing the generalization

We can test the generalization in one of two ways.

We can use values of V and D for which we know the result, choosing
$V = 6, D = 9$. This is put in the rule to give:

$$D = \tfrac{1}{2}V(V-3)$$
$$D = \tfrac{1}{2}6(6-3)$$
$$= 3(3) = 9 \quad \text{which is now}$$

confirmed.

Or we can test for some new value of V and check the result geometrically, or otherwise. Choose $V = 12$, for instance

$$D = \tfrac{1}{2}(12)(12-3)$$
$$= 6 \times 9$$
$$= 54$$

If you check either by extending the table of results or by drawing, recording and dot patterns you will find that when $V = 12$, D is equal to 54.

So we have now tested our generalization in two ways.

11.13 Justifying the generalization

Now that the generalization has been expressed in symbols, the next stage is to justify it. We can do this in at least two ways.

Justification using differences

In Unit 14 you will do some work on the use of differences. We shall use this now.

For our set of results, 1st and 2nd differences are:

Vertices (V)	3	4	5	6	7	8	9
Diagonals (D)	0	2	5	9	14	20	27
1st Difference		2	3	4	5	6	7
2nd Difference			1	1	1	1	1

Because the 2nd differences are constant, theory tells us that the symbolic form is quadratic. So it must be of the type:

$$D = AV^2 + BV + C$$

and

$$D = \tfrac{1}{2}V(V-3)$$
$$= \tfrac{1}{2}V^2 - 1\tfrac{1}{2}V$$

is of this type with $A = \tfrac{1}{2}$ and $B = 1\tfrac{1}{2}$

All we need to do with a quadratic is check that it works in three cases. We will try $V = 3, 4$ and 5

V	D
3	$\frac{1}{2}(3)(3-3) \quad = 1\frac{1}{2} \times 0 = 0$ checked
4	$\frac{1}{2}(4)(4-3) \quad = 2 \times 1 = 2$ checked
5	$\frac{1}{2}(5)(5-3) \quad = 2\frac{1}{2} \times 2 = 5$ checked

These 3 successful checks mean that we have justified the general rule as a quadratic and that it is:

$$D = \tfrac{1}{2}V(V-3)$$

The generalization can be justified in a more sophisticated way, leading to a better GCSE grade, by referring back to the geometry.

Justification using geometry

To do this we will look at the dot patterns for six and seven vertices. These are:

6 vertices   7 vertices

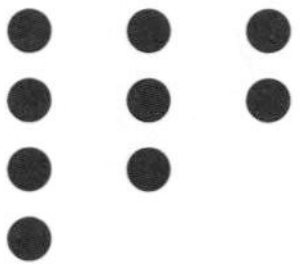

Putting one more dot on the top row of each pattern would give the triangular numbers pattern, so for seven vertices:

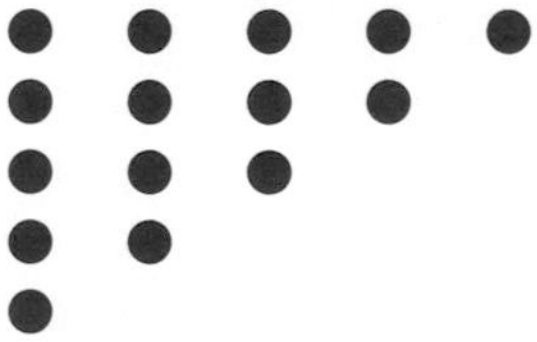

So for seven vertices the number of diagonals is one less than the 5th triangular number.

The expression for the nth triangular number is $\tfrac{1}{2}n(n+1)$.

The dot pattern approach tells us that the rule is:

Note
4th triangular number is:
$1 + 2 + 3 + 4 = 10$
$10 = \tfrac{1}{2}(4 \times 5)$

Vertices	Diagonals
V	$(V-2)$th triangular no. -1

Since the nth triangular number is $\frac{1}{2}n(n+1)$ the $(V-2)$th triangular number is:

$$\frac{1}{2}(V-2)((V-2)+1)$$
$$=\frac{1}{2}(V-2)(V-1)$$
$$=\frac{1}{2}(V^2-V-2V+2)$$
$$=\frac{1}{2}(V^2-3V+2)$$

Replace n by $v-2$

so for V vertices, the number of diagonals D, is:

$$D=\frac{1}{2}(V^2-3V+2)-1$$
$$=\frac{1}{2}V^2-1\tfrac{1}{2}V+1-1$$
$$=\frac{1}{2}V^2-1\tfrac{1}{2}V$$
$$D=\frac{1}{2}V(V-3)$$

There is more about multiplying brackets in Unit 10

This has justified the general rule in a very strong way because it has been related back to the geometry of the physical structure.

11.14 Proving the result

The more or less final stage in the process is to prove the general result. Proofs are never easy.

To do this we will relate back to a picture of a general regular polygon with V vertices, labelled from ① to ⓥ.

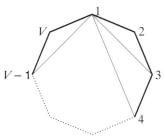

Starting at any vertex, ① for instance, the number of diagonals will be $(V-3)$, because the lines drawn will go to all but ① itself and the two vertices either side of 1, that is ② and ⓥ.

The same will be true for every other vertex. So for each vertex there are $(V-3)$ diagonals and there are V vertices.
So the total number of diagonals is:

$$V(V-3)$$

But this gives the diagonals going in both directions. So it gives double the number of actual diagonals.

Hence the actual number of diagonals is:

$$\frac{1}{2}V(V-3)$$

This **proves** the result.

11.15 Extending the investigation

The process outlined in this chapter is sufficient to guarantee the top grade at GCSE. However, unlike other textbook exercises and examination questions, a mathematical investigation is never finished. One of the things you can always do is extend the investigation.

There are two extensions of *Diagonals* which you might like to see as a future challenge.

Extension 1

What happens in the case when a shape is not regular, looking in particular at concave shapes such as:

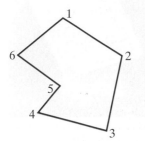

or special cases such as:

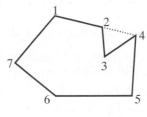

Extension 2

Try to obtain a general rule for shapes in three dimensions, such as:

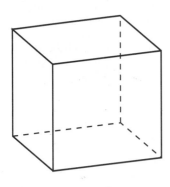

a **cube**

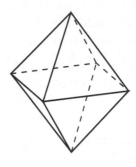

an **octahedron**

Neither of these extensions is easy. Their solutions require mathematical skills beyond the GCSE syllabus – but there is nothing to stop you trying!

11B Handling data

In 2001 a national census was used to collect data about everyone in the country.

You can use mathematics to analyse information or data.

For your GCSE you will need to complete a Handling Data coursework project. This chapter will work through a sample project, showing you how to approach your Handling Data project. Some of the techniques used in this chapter are covered in chapters 15 and 27.

Secondary data suitable for use in the Handling Data project is available to download from Heinemann's website at www.heinemann.co.uk/ mathshandling

Jordan Hill County High School

Jordan Hill County High School is a school for students aged 11 to 16. It is a growing school so the number of students in each year varies. Although fictional, the data is based on a real school.

Data has been provided about all the students at Jordan Hill County High School. Data that has been gathered for you is called **secondary data**. Data you gather yourself is called **primary data**.

The data

This table shows how many boys and girls there are in each year group at Jordan Hill:

Year	Boys	Girls	Total
7	150	150	300
8	145	125	270
9	120	140	260
10	100	100	200
11	84	86	170

Data is provided for each pupil in a number of these categories including:

name, age, year group, IQ, height, hair colour, eye colour, number of brothers/sisters, distance travelled between home and school, and gender (boy or girl).

Lines of enquiry

With so much information it is important to know exactly what you want to find out about Jordan Hill. You will have to choose a **line of enquiry** to investigate. Here are some possibilities:

the hair colour of students

the range of distances travelled to school

the relationship between shoe size and weight or height

the relationship between IQ and Key Stage 3 English results

the height to weight ratio for boys and girls.

You need to choose a line of enquiry that will give you plenty to say without being too difficult. It should also make sense. There wouldn't be any value in investigating the relationship between hair colour and weight, for example. You should ask your teacher whether your line of enquiry is suitable before beginning your project.

The project in this chapter will investigate the relationship between shoe size and height for the students at Jordan Hill.

Collecting data

We will begin by taking a **random sample** of 30 boys and 30 girls from the school register, and recording their shoe sizes and heights.

There are different ways to take a random sample of 30 boys. You could write the names of each of the 599 boys on pieces of paper, put them in a hat, and choose 30 without looking. This would take a very long time. One alternative would be to give each boy a number from 1 to 599 and use the **random number button** on your calculator to choose your sample:

Press **SHIFT** **RAN#** . Your calculator will display a random number between 0 and 1.

You need a number between 1 and 599 so multiply the number by 599. Round your answer to the nearest whole number.

Repeat this process 30 times to select a random sample of 30 boys.

If the calculator selects a number you have already chosen, ignore it and try again.

How big should my sample be?

The bigger a sample the more useful the data will be. If you select a lot of people, your results will be closer to the actual results for the whole school. However, if you choose too many people the data becomes difficult to analyse. A sample of size 25 is an adequate minimum. 30 is a sensible size for a sample because it is bigger than 25 and because it divides 360 exactly. This makes it easier to draw pie charts from your data.

Generate a random number between 0 and 1.

$$0.252$$

Multiply by 599:

$$150.948$$

Round to the nearest whole number:

$$150.948 \approx 151$$

You have selected the 151st boy on the school register.

Here is our sample of 30 boys and 30 girls.

Boys		Girls	
Shoe size	Height (m)	Shoe size	Height (m)
6	1.63	11	1.90
8	1.73	5	1.63
7	1.71	7	1.68
2	1.32	8	1.77
5	1.65	1	1.28
13	1.98	2	1.31
9	1.83	4	1.60
7	1.70	4	1.62
8	1.72	6	1.66
3	1.40	5	1.67
7	1.73	2	1.29
10	1.89	3	1.38
7	1.74	7	1.68
11	1.79	8	1.91
6	1.66	5	1.65
7	1.72	5	1.66
4	1.51	5	1.64
6	1.68	4	1.61
12	1.81	9	1.72
7	1.70	6	1.67
7	1.73	4	1.59
6	1.69	3	1.48
8	1.77	4	1.55
8	1.76	5	1.68
7	1.74	5	1.69
7	1.80	7	1.71
5	1.58	6	1.65
5	1.65	5	1.66
9	1.83	6	1.72
10	1.84	5	1.80

You need a more useful representation of this data. Here are frequency tables for shoe size and height separated into boys and girls:

Boys

Shoe size	Tally	Frequency
1		0
2	|	1
3	|	1
4	|	1
5	|||	3
6	||||	4
7	𝍸 ||||	9
8	||||	4
9	||	2
10	||	2
11	|	1
12	|	1
13	|	1

Height, h (cm)	Tally	Frequency
$120 \leqslant h < 130$		0
$130 \leqslant h < 140$	|	1
$140 \leqslant h < 150$	|	1
$150 \leqslant h < 160$	||	2
$160 \leqslant h < 170$	𝍸 |	6
$170 \leqslant h < 180$	𝍸 𝍸 |||	13
$180 \leqslant h < 190$	𝍸 |	6
$190 \leqslant h < 200$	|	1

In the height column, $130 \leqslant h < 140$ means '130 up to but not including 140'. Any value greater than or equal to 130 but less than 140 would go in this class interval.

Girls

Shoe size	Tally	Frequency
1	|	1
2	||	2
3	||	2
4	𝍸	5
5	𝍸 ||||	9
6	||||	4
7	|||	3
8	||	2
9	|	1
10		0
11	|	1
12		0
13		0

Height, h (cm)	Tally	Frequency
$120 \leqslant h < 130$	||	2
$130 \leqslant h < 140$	||	2
$140 \leqslant h < 150$	|	1
$150 \leqslant h < 160$	||	2
$160 \leqslant h < 170$	𝍸 𝍸 𝍸 |	16
$170 \leqslant h < 180$	||||	4
$180 \leqslant h < 190$	|	1
$190 \leqslant h < 200$	||	2

Shoe sizes

You are now ready to record your results in a diagram. We will begin by analysing the data about shoe sizes, using bar charts to compare the results for boys and girls.

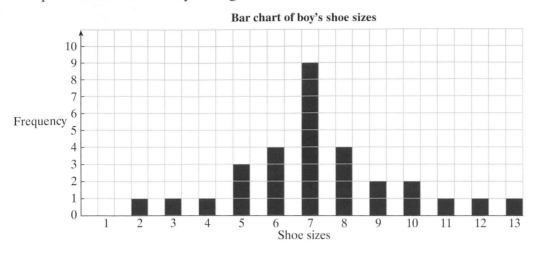

If you want to compare two sets of discrete data you can record them on a **dual bar chart**.

You can use a dual bar chart because there are the same number of boys and girls in your sample. If there were different numbers you could use pie charts to compare the data. In a pie chart you can see the **proportion of 360°** for each shoe size, so you are always comparing like with like.

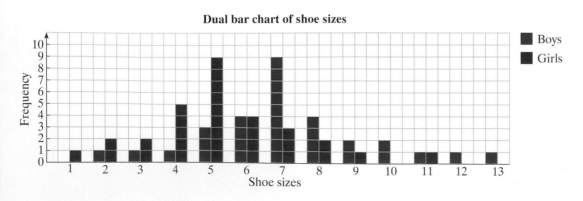

You can now make some simple statements to compare the shoe sizes of boys and girls:

The mode shoe size for the boys in my sample was higher than the mode shoe size for girls

The evidence from the sample suggests that there will be fewer boys with shoe sizes 1 to 6 than girls

You need to gather more information to support your statements. Comparing the **mean, median, mode** and **range** of shoe sizes for boys and girls will give you more evidence.

Mean shoe sizes

You can calculate the mean easily from the frequency tables.

Mean shoe size for boys = 7.23

Mean shoe size for girls = 5.23

Modal shoe sizes

You can read the modes of the shoe sizes for boys and girls straight off your bar charts or frequency tables:

Modal shoe size for boys = 7

Modal shoe size for girls = 5

Median shoe sizes

There are 30 people in each sample, so the median will be half way between the 15th and 16th values:

Median shoe size for boys = 7

Median shoe size for girls = 5

Range of shoe sizes

The range of shoe sizes will show you how spread out your data is:

Range of shoe sizes for boys = 11

Range of shoe sizes for girls = 10

You can summarise these results in a table:

Shoe sizes	Mean	Mode	Median	Range
Boys	7.23	7	7	11
Girls	5.23	5	5	10

You now have more evidence to describe the difference in shoe size between boys and girls:

All three measures of average (mean, median and mode) are greater for boys than for girls. The range of shoe sizes for boys is slightly greater for boys than for girls. In conclusion, although there are a small number of boys with small shoe sizes and girls with large shoe sizes, the evidence suggests that, in general, the shoe sizes for the boys are greater than the shoe sizes for the girls.

You can also use your data to make specific comments:

Evidence from the sample suggests that 17 out of 30, or 57% of boys have a shoe size of either 6, 7 or 8, and that 18 out of 30, or 60% of girls have a shoe size of either 4, 5 or 6.

Height

You can analyse the data about height in exactly the same way. Because height is continuous you need to record it on a histogram.

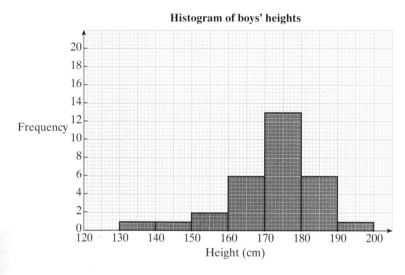

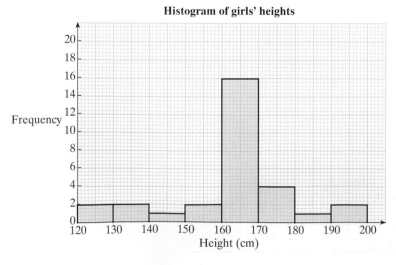

You can compare continuous data by drawing the frequency polygons on the same graph.

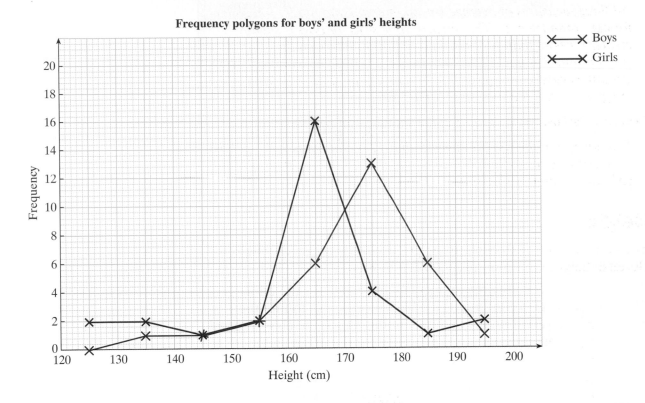

Since the data is grouped into class intervals, it also makes sense to record it in a stem and leaf diagram. This will make it easier to read off the median values.

Boys

Stem	Leaf	Frequency
120		0
130	2	1
140	0	1
150	1, 8	2
160	3, 5, 5, 6, 8, 9	6
170	0, 0, 1, 2, 2, 3, 3, 3, 4, 4, 6, 7, 9	13
180	0, 1, 3, 3, 4, 9	6
190	8	1

Girls

Stem	Leaf	Frequency
120	8, 9	2
130	1, 8	2
140	8	1
150	5, 9	2
160	0, 1, 2, 3, 4, 5, 5, 6, 6, 6, 7, 7, 8, 8, 8, 9	16
170	1, 2, 2, 7	4
180	0	1
190	0, 1	2

Averages

You can also record the mean, median and range for the data. Because the data is continuous it makes more sense to find the **modal class interval** rather than the mode. This is the class interval that contains the most values. The values for the mean and median have been rounded to two decimal places.

Heights (cm)	Mean	Modal class interval	Median	Range
Boys	171	170–180	173	66
Girls	163	160–170	166	63

You can now make comments about the heights of students at Jordan Hill:

All three measures of average in the sample were higher for boys than for girls, though the sample for boys was more spread out, with a range of 0.66 m compared to 0.63 m for the girls. The evidence from the sample suggests that 14 out of 30, or 47% of the boys have a height between 170 and 180 cm, whilst 16 out of 30, or 53% of the girls have a height between 160 and 170 cm. The frequency polygons show that there are fewer boys with heights below 140 cm than girls.

Always support your statements with evidence from your analysis. You should give actual numbers wherever possible.

You need to remember that all your comments are based on a *sample* of the whole school:

These conclusions are based on a sample of only 30 girls and 30 boys. I could extend the sample or repeat the whole exercise to confirm my results.

You shouldn't actually repeat your test, or use a larger sample, but it is very important that you show that you know how to find more evidence to support your statements.

Extending the investigation

You can now extend your line of enquiry and give yourself a hypothesis to test. A hypothesis is a statement that could be true or false. You test a hypothesis by looking at data. We will test the following hypothesis:

In general the larger a person's shoe size, the taller that person is likely to be

To test this hypothesis we need a new random sample of 30 students of any gender.

Shoe size	5	4	2	6	11	3	8	5	2	4	12	4	7	5	4
Height (cm)	160	158	137	163	188	157	177	158	135	158	197	156	172	161	163

Shoe size	7	7	2	5	9	6	1	6	6	3	4	2	8	3	7
Height (cm)	168	171	141	159	180	158	132	166	169	155	160	140	178	152	175

The most sensible way to compare this data is to draw a scatter diagram:

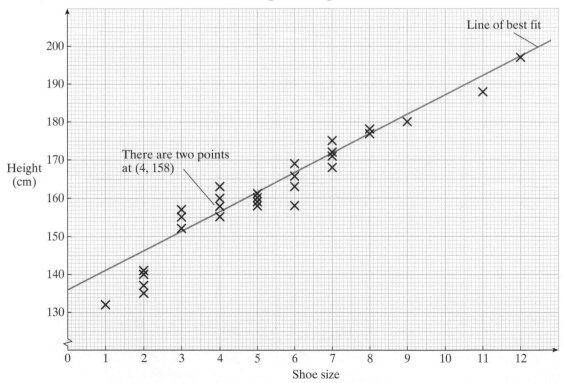

Scatter diagram of height and shoe sizes

You can make simple comments based on your scatter diagram:

There is a positive correlation between shoe size and height. This suggests that the larger a person's shoe size, the taller they will be.

You can use a **line of best fit** to make predictions from your results:

The line of best fit suggests that somebody with shoe size 10 will be 188 cm tall.

Further investigation

In the early part of the investigation you found evidence to suggest that height and shoe size are both affected by gender. A natural next step is to extend our line of enquiry to investigate how the correlation between height and shoe size is affected by gender. We will test the hypothesis:

There will be a better correlation between shoe size and height if we consider boys and girls separately.

We already have random samples of 30 boys and 30 girls that we can use to test this hypothesis. We will plot separate scatter diagrams for the boys and the girls, and one for the whole sample.

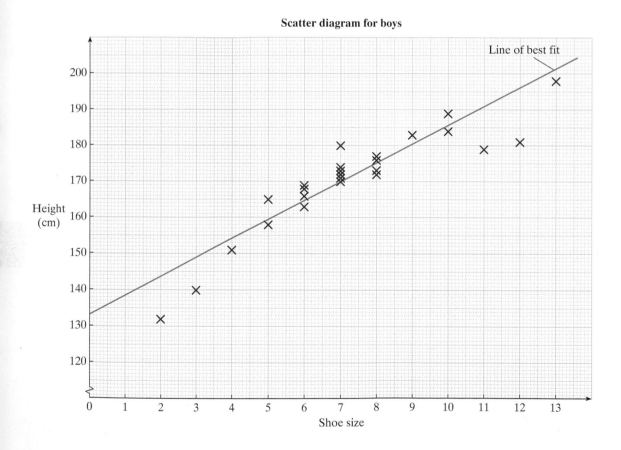

Scatter diagram for boys

Remember the scatter diagram for boys and girls together uses the *same* data as the separate diagrams. In statistics you must always compare **like with like**.

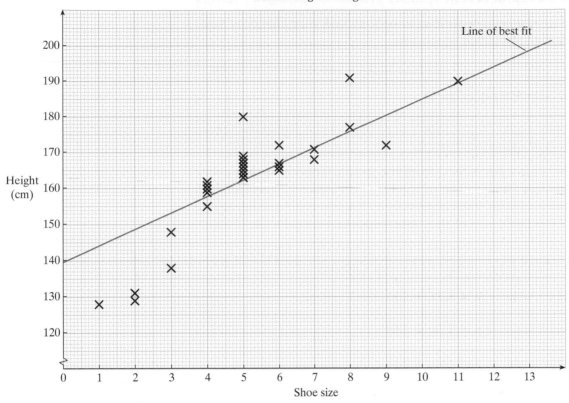

Scatter diagram for girls

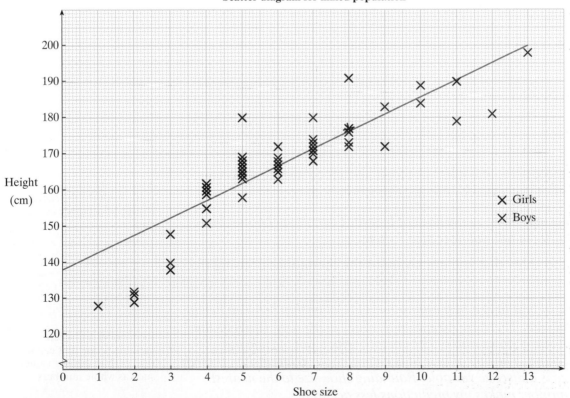

Scatter diagram for mixed population

The evidence supports our hypothesis:

There is a stronger correlation between shoe size and height if boys and girls are considered separately.

The line of best fit on each scatter diagram can be used to make predictions:

The lines of best fit on my diagrams predict that a girl who was 168 cm tall would have a shoe size of 6, wheras a boy of the same height would wear size 7 shoes.

These values have been read from the graph. You know that every straight line has an equation of the form $y = mx + c$. You can find the equations of your lines of best fit by finding their gradients and looking at the point where they intercept the vertical axis.

If y represents height in cm, and x represents shoe size, the equations of the lines of best fit for our data set are:

Boys only: $\qquad y = 5.3x + 133$

Girls only: $\qquad y = 4.5x + 140$

Combined sample: $\quad y = 4.8x + 138$

These equations can be used to make predictions of shoe size when you know height, or height when you know shoe size. For example to predict the shoe size of a boy who is 170 cm tall:

$$y = 5.3x + 133$$

so $\quad x = \dfrac{y - 133}{5.3}$

If $\quad y = 170$ then

$$x = \dfrac{170 - 133}{5.3} = 6.98$$

All the shoe sizes in the data provided are whole numbers, so you should round this value to the nearest whole number:

Using the equations of my lines of best fit, I can predict that a boy who is 170 cm tall will have size 7 shoes.

You need to show that you are aware of the limitations of using the line of best fit.

The line of best fit is a best estimation of relationship between height and shoe size. There are exceptional values in my data (such as the girl with size 8 shoes who is 191 cm tall) which fall outside the general trend. The line of best fit is a continuous relationship, though shoe size is a discrete variable. Rounding shoe size to the nearest whole number makes my predictions less accurate.

The graph with equation $y = mx + c$.

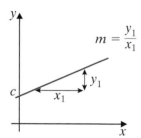

$$m = \frac{y_1}{x_1}$$

There is more about this on page 159.

You should always give your results to the same degree of accuracy as the data.

Remember a **discrete** variable is one that can only take certain distinct values, such as whole numbers.

Cumulative frequency graphs

Cumulative frequency can be a very powerful tool when comparing different data sets. This table shows the cumulative frequency for shoe sizes for boys, girls, and for the mixed sample:

Shoe size (up to and including)	Cumulative frequency		
	Boys	Girls	Mixed
1	0	1	1
2	1	3	4
3	2	5	7
4	3	10	13
5	6	19	25
6	10	23	33
7	19	26	45
8	23	28	51
9	25	29	54
10	27	29	56
11	28	30	58
12	29	30	59
13	30	30	60

If you were to increase the sample size to, for example, 50 boys and 50 girls you could increase the accuracy of your cumulative frequency curve and further justify any inferences made.

The best way of representing this information on a diagram is to draw cumulative frequency curves. If the curves are drawn on the same axis it is easier to compare the results.

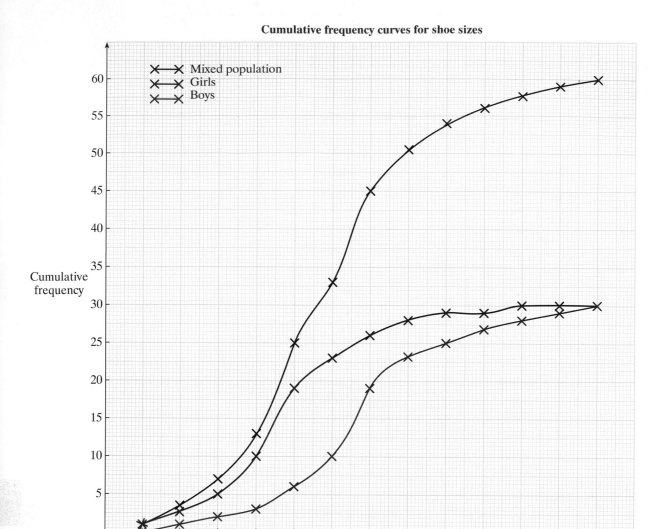

The curves clearly show the trend towards larger shoe sizes amongst boys and girls. However the curve is a continuous measure of cumulative frequency, and shoe sizes are a discrete variable. Cumulative frequency curves would be more appropriate when analysing heights.

Heights (cm)	Cumulative frequency		
	Boys	**Girls**	**Mixed**
<130	0	2	2
<140	1	4	5
<150	2	5	7
<160	4	7	11
<170	10	23	33
<180	24	27	51
<190	29	28	57
<200	30	30	60

Cumulative frequency curves for height

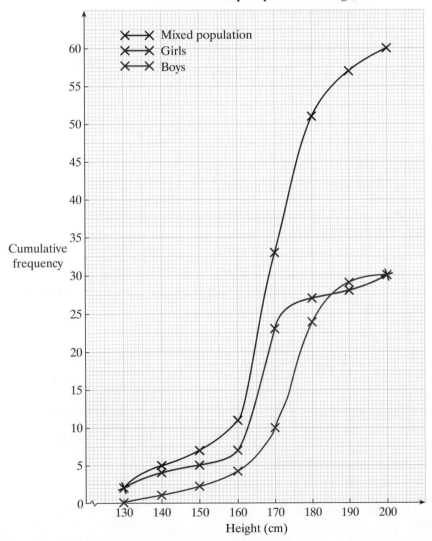

The benefit of drawing cumulative frequency curves for a continuous variable like height is that you can easily read off the median, upper quartile, lower quartile and interquartile range:

Heights (cm)	Median	Lower quartile	Upper quartile	Interquartile range
Mixed	169	163	176	13
Boys	174	167	178	11
Girls	166	161	170	9

You can get a more accurate calculation of the median from a cumulative frequency graph than from a stem and leaf diagram. This is because the cumulative frequency graph is a **continuous approximation** of the distribution of values.

You can also use cumulative frequency curves to predict percentages of students who have a height within a given range. Suppose for example that you wanted to estimate how many boys in the school were between 165 and 180 cm tall. The cumulative frequency curve tells us that 6 boys in the sample had height up to 165 cm, and 24 boys had height up to 180 cm. This means that $24 - 6 = 18$ boys had height between 165 and 180 cm. You can use this figure to estimate that $\frac{18}{30}$ or 60% of boys in the school will be between 165 and 180 cm tall.

You could also say:

If we select a boy at random from the school, our data suggests that the probability of him having a height between 165 and 180 cm is 0.6.

You can also use the information you read from the cumulative frequency curves to draw box-and-whisker diagrams, showing the minimum and maximum values, the median, and the upper and lower quartiles:

Box-and-whisker diagrams for height

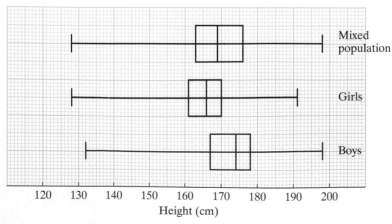

The size of a box-and-whisker diagram depends on the highest and lowest values in the sample. You will need to look back at your stem and leaf diagrams to find these.

These diagrams provide a very clear comparison between the different data sets. For example, you could say:

The box-and-whisker diagrams show that the girls' interquartile range is 2 cm less than the boys'. This suggests that the boys' heights were more spread out than the girls'.

You can use your cumulative frequency graphs to comment on the relationships between the data for boys and the data for girls. The median height for boys is 174 cm. The cumulative frequency curve for girls tells you that 25.5 girls in the sample had height less than 174 cm:

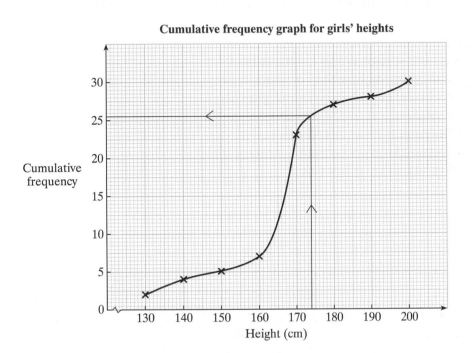

Cumulative frequency graph for girls' heights

So 4.5 out of 30 girls in the sample have a height greater than the median height for boys. You can write this as $\frac{4.5}{30}$ or 15%. You can now make a more sophisticated statement:

Whilst, in general, boys are taller than girls we have evidence to suggest that 15% of the girls have a height greater than the median height of the boys.

Summarizing your results

It is very important to analyse your data, interpret the outcomes and discuss your results and findings. Here is a summary of some of the findings from this investigation – this shows what has been achieved by the investigation, and might produce ideas for further lines of enquiry. You will need to refer back to all of your graphs and comments when you are summarizing your results.

- There is a positive correlation between height and shoe size. In general taller people will have larger shoes than smaller people.
- The points on the scatter diagram for the boys are less dispersed about the line of best fit than those for the girls. This suggests that the correlation is better for boys than for girls and that boys' heights are more predictable.
- The points on the scatter diagrams for boys and girls are less dispersed than the points on the scatter graph for the mixed sample of boys and girls. This suggests that the correlation between shoe size and height is better when boys and girls are considered separately.
- The points on the scatter diagram for the mixed sample of boys and girls are better approximated by a curve than a straight line, suggesting that the overall relationship is not linear.

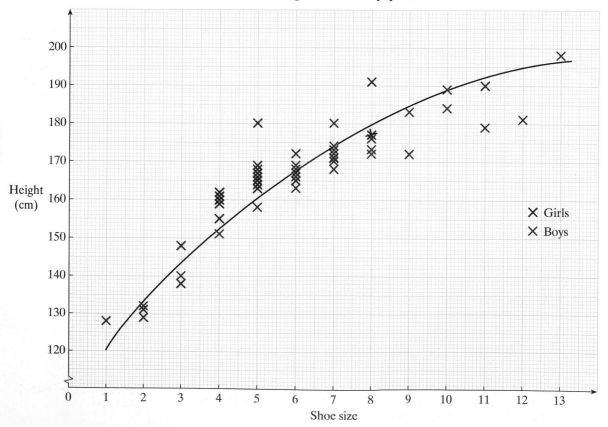

Scatter diagram for mixed population

- The scatter graphs can be used to give reasonable estimates of shoe size and height. This can be done either by reading from the graph or by using equations of lines of best fit.
- Cumulative frequency curves confirm that boys have larger shoe sizes than girls.
- The median height for boys is higher than the median height for girls.
- From the box-and-whisker diagrams we can conclude that, in general, boys are taller than girls, but not exclusively so. The cumulative frequency curves can be used to estimate that 20% of the girls have height greater than 173 cm, the median height for the boys.

You should also make some statements about the limitations of your analysis:

- We could have had greater confidence in our results if we had taken larger samples or given some consideration to the ages of the students in the sample.
- Our predictions are based on general trends observed in the data. In both samples there were exceptional individuals whose results fell outside the general trend.

Considering age

To achieve the best possible marks in your data handling coursework you should address some of the limitations you have identified. Common sense dictates that a person's height and shoe size will be affected by how old they are. We will extend our line of enquiry to consider the relationship between shoe size and height, not only across the gender divide, but also across different age groups.

When you extend your line of enquiry you sometimes call your earlier work a **pre-test**. This is an initial test that gives you enough information to formulate lines of enquiry and hypotheses for your later tests.

When extending your investigation you should always try to refine it and build on the results you have already. Don't start again with a new line of enquiry.

Our analysis so far is a pre-test that allows us to make the following hypothesis

When age is taken into consideration, the correlation between shoe size and height will be better than when age is not considered.

Minimising bias

In our pre-test we sampled 30 students at random from the school. However, the school is growing and there are more students in year 7 than in year 11. This means that your sample is **biased** or unfair. There are likely to be more year 7 students in the sample than year 11 students. To ensure that students from different age groups are equally represented you need to take a **stratified sample**. In a stratified sample you sample values from a particular group in proportion to that group's size within the whole population.

Stratified means layered or grouped.

Here is an example of stratified sampling

There are 100 boys in year 10 and 1200 pupils in the school. So $\frac{1}{12}$ of the school population are year 10 boys. The same proportion of our stratified sample should be year 10 boys. $\frac{1}{12} \times 30 = 2\frac{1}{2}$. There are also 100 girls in year 10 so we should also choose $2\frac{1}{2}$ girls. However it is impossible to sample half a person, so we choose either 2 boys and 3 girls, or 3 boys and two girls (the decision should be made by tossing a coin).

For the students at Jordan Hill County High School a sample of size 30, stratified by year group, should contain:

Year group	Number of boys	Number of girls	Total
7	4	4	8
8	4	3	7
9	3	3	6
10	2	3	5
11	2	2	4

The values have all been rounded to the nearest whole number. In some cases decisions have been made on the toss of a coin.

You can now comment on the fairness of your sample:

By taking a stratified sample I can be as sure as possible that my sample is representative of the whole school. As far as possible, my sample is free from bias caused by gender or age divisions.

Taking a stratified sample using the **RAN#** button on a calculator gives us the following results:

Year	Boys		Girls	
	Shoe size	Height (cm)	Shoe size	Height (cm)
7	4	157	3	152
	3	122	2	130
	5	160	3	146
	4	152	4	155
8	5	159	4	154
	4	151	5	160
	6	177	3	138
	5	174		
9	7	176	5	157
	7	179	4	160
	5	162	6	168
10	8	175	5	165
	7	171	7	168
			5	164
11	10	180	6	166
	7	184	5	163

We can summarise our results for the stratified sample across the whole year group:

	Boys	Girls
Modal shoe size	5 and 7	5
Median shoe size	5	5
Mean shoe size	5.13	4.47
Range of shoe sizes	7	5
Median height (cm)	171	160
Mean height (cm)	165	156
Range of heights (cm)	62	38

Although we can be as certain as possible that these results are representative of the whole school, we don't yet have a big enough sample to make any meaningful statements about the data within each year group. To look in more detail at each year group you can take a **10% sample** for each year group and gender. For example, there are 150 boys in year 7, so you sample 10%, or 15 of them. Here is our sample of year 7 boys:

Shoe size	6	4	5	6	5	2	7	7	5	3	4	5	5	4	6
Height (cm)	165	149	155	159	165	125	162	168	164	154	158	160	163	158	167

Here is a summary of the first analysis of this sample:

Modal shoe size	5
Median shoe size	5
Mean shoe size	4.93
Range of shoe sizes	5
Median height (cm)	159
Mean height (cm)	158
Range of heights (cm)	43

Measures of spread

Earlier in the project we used the range and interquartile range of the data as a measure of spread. Another measure of spread that is useful is the **mean of deviations from the mean**. This is the average distance of data values from the mean. So, for a height of 165 cm, the deviation from the mean is

height − mean height = 165 − 158 = 7 cm.

For a height of 125 cm the deviation from the mean is

$$125 - 158 = -33 \text{ cm.}$$

Before taking the mean you must ignore the minus signs (you are interested in the *magnitude* of the deviation) so this value becomes 33 cm. You can use vertical lines to represent magnitude. For example, $|125 - 158| = 33$.

What happens when you take the mean if you don't ignore the minus sign? Can you prove your answer?

Ignoring minus signs, the mean of the deviations for this sample is:

$$\frac{\Sigma|x - \bar{x}|}{n} = \frac{7 + 9 + 3 + 1 + 7 + 33 + 4 + 10 + 6 + 4 + 0 + 2 + 5 + 0 + 9}{15}$$

$$= \frac{100}{15} = 6.7 \text{ cm}$$

It is important to identify any factors that affect your results. There is one deviation, 33 cm, that is much greater than the others. If we exclude this untypical value, the mean of the deviations is:

$$\frac{67}{14} = 4.8 \text{ cm}$$

We can use this measure of spread to compare our sample of year 7 boys with our stratified sample of all the boys in the school. For all the boys in the stratified sample, the mean of the deviations was 15 cm. The year 7 boy who was 122 cm tall has provided a deviation of 43 cm, which when compared to the rest of the sample is untypical. If we exclude this value the mean deviation becomes 13.5 cm.

We can now comment on the spread of our two samples:

The heights of the boys for the whole school appear to be far more spread out than the heights for boys in year 7 alone. (The mean of the deviations from the mean was 4.8 cm when only year 7 boys were sampled, and 13.5 cm, nearly three times as much, when a stratified sample was taken from the whole school. These values were taken excluding untypical points.)

There is another widely used measure of spread called the **standard deviation** that uses the *square* of the deviations from the mean.

The best way to see the relationship between shoe size and height for our sample of year 7 boys is to draw a scatter diagram as shown at the top of the next page.

The untypical point at (2, 125) causes some concern. We have drawn one line of best fit (in green) that takes this point into account, and one (in red) that excludes it. We have also drawn a curve of best fit that takes all the points into account.

You can analyse lines of best fit to see which is the most suitable by measuring the average distance of the datum points from the line or curve. These distances are called **vertical dispersions**. The vertical dispersions and their mean for the three approximations are shown below the scatter diagram.

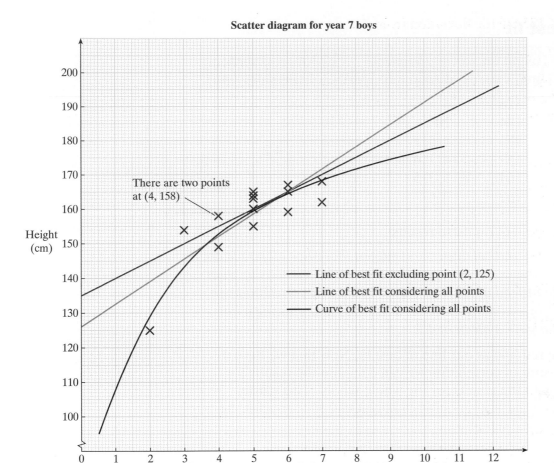

Scatter diagram for year 7 boys

There are two points at (4, 158)

Height (cm)

Shoe size

— Line of best fit excluding point (2, 125)
— Line of best fit considering all points
— Curve of best fit considering all points

Green line of best fit

Shoe size	Height on line (cm)	Vertical dispersions (cm)	Mean of vertical dispersions (cm)
2	139	14	
3	146	8	
4	152	6, 3, 6	
5	159	1, 4, 4, 5, 6	5.3
6	165	0, 2, 6	
7	172	10, 4	

Red line of best fit

Shoe size	Height on line (cm)	Vertical dispersions (cm)	Mean of vertical dispersions (cm)
2	145	20	
3	150	4	
4	155	3, 6, 3	
5	160	0, 5, 3, 4, 5	4.7
6	165	0, 6, 2	
7	170	8, 2	

Curve of best fit

Shoe size	Height on line (cm)	Vertical dispersions (cm)	Mean of vertical dispersions (cm)
2	128	4	
3	144	10	
4	153	4, 5, 5	4.2
5	159	4, 1, 4, 5, 6	
6	164	1, 3, 5	
7	168	0, 6	

You can now comment on these results:

The means of the vertical dispersions suggest that when all points are considered a curve is a better approximation to the relationship than a line (4.2 cm instead of 5.3 cm). However, a strong correlation can also be found by excluding the point at (2, 125) and drawing a line of best fit. This gave a mean of 4.7.

We will use the red line as our line of best fit. Now that we have a numerical measure of correlation we can compare the results for year 7 boys only to the results for the whole school. This scatter graph shows the relationship between height and shoe size for the boys in our stratified sample.

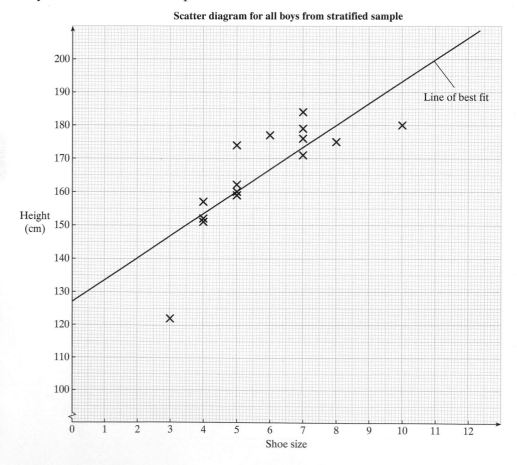

Scatter diagram for all boys from stratified sample

The mean of the vertical dispersions for this graph is 6.7 cm.

We have evidence to suggest that the correlation between shoe size and height may be greater when we restrict ourselves to students from a single year group. The mean of the vertical dispersion for a sample of 15 year 7 boys was 4.7 cm. The mean for a stratified sample of 15 boys from the whole school was 6.7 cm.

Here is a summary of results from a 10% sample taken from each year group:

Year	Gender	Mean height (cm)	Mean of deviations from mean height (cm)	Mean of vertical dispersions on line of best fit (cm)
7	Boys	158	6.7	4.7
	Girls	151	6.8	4.4
8	Boys	162	7.1	5
	Girls	156	6.8	4.7
9	Boys	168	7.2	5.6
	Girls	160	7.1	5.5
10	Boys	171	7.3	5.5
	Girls	163	6.9	5.2
11	Boys	173	7.5	5.2
	Girls	166	6.9	5.1

A final summary

To finish your project your should make some final conclusions. The comments below build on the comments we made in the earlier summary.

- A sample of 30 students stratified over age and gender shows a mean height of 165 cm for boys and 156 cm for girls. However, the range of heights for the boys was considerably greater than that for the girls. This suggests that there will be many boys who are shorter than 156 cm, the girls' mean height.

- A 10% sample of boys in year 7 suggests that this year group of boys has a mean height of 158 cm with a mean deviation about that mean of 6.7 cm. Our evidence suggests that the girls in this year group have a mean height of 151 cm with a mean deviation about this mean of 6.8 cm. Since the deviations are similar we can conclude that the boys in year 7 are in general taller than the girls.

- This conclusion is supported by the evidence gathered from 10% samples in the other years (though the difference in mean height in year 9 is very small). The 10% sample means that the sample sizes for years 9, 10 and 11 were particularly small.

- There is a positive correlation between shoe size and height both across the school as a whole and within each year group. This correlation appears to be stronger when individual year groups and separate genders are considered.

- Over the full range of shoe sizes, there is evidence to suggest that a line of best fit is not the most suitable model to describe the relationship between height and shoe size. Further sampling and analysis could determine whether a curve of best fit would more accurately describe the relationship.

Summary of key points

When completing your Data Handling coursework project you should:

- decide on the line of enquiry for your project

- plan out your line of enquiry, outline its aims and include any hypotheses you will make

- decide on the type of data and how much data you will require

- explain how you intend to collect the data

- record all the data

- use appropriate statistical techniques and diagrams

- explain what you intend to do with any very unusual datum points which seem to be exceptions to the rule

- make appropriate observations and comments on your data

- in cases when you make comparisons remember that a valid comparison usually makes use of a measure of central tendency (or average) such as mean, mode or median and a measure of dispersion (or spread) such as range, interquartile range or percentiles

- provide a summary of your line of enquiry; this should be related to the aims and any hypotheses

- make a final conclusion based on the evidence you have obtained

12 Estimation and approximation

Sometimes you will need to work with approximate numbers, even when a precise value is available. This can help you check calculations mentally.

12.1 Rounding a number

The attendance at a football World Cup final was 112 175, the exact number of tickets sold.

112 175 can be rounded to the nearest ten, hundred, thousand, and so on …

to the nearest ten: to the nearest hundred: to the nearest thousand:
112 180   112 200   112 000

A newspaper would probably report this as:

This gives a good idea of the number of people at the match. The extra 175 in the crowd is very small compared to 112 000.

112,000 supporters attend big match

Notice that 112 175 to the nearest 10 rounds up to 112 180. 175 is halfway between 170 and 180 so we have to choose whether to round up to 180 or down to 170. By convention we round up.

■ **In rounding, halves or 5s are usually rounded upwards.**

Example 1
Round the measurement 655 cm:
(a) to the nearest 10 cm
(b) to the nearest 100 cm.

Answer
(a) 655 cm is exactly halfway between 650 cm and 660 cm. When rounding to the nearest 10 cm, 655 cm rounds upwards to become 660 cm.
(b) 655 cm is between 600 and 700 cm. When rounding to the nearest 100 cm, 655 cm rounds upwards to become 700 cm.

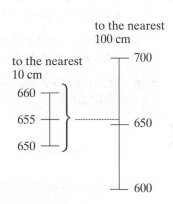

Exercise 12A

1   Round these to the nearest 10:
    (a) 269   **(b)** 129   **(c)** 74   **(d)** 86
    (e) 599   **(f)** 843   **(g)** 7   **(h)** 2

2   Round these to the nearest whole number:
    (a) 2.9   **(b)** 7.4   **(c)** 97.8   **(d)** 0.55
    (e) 0.45   **(f)** 80.4   **(g)** 4.05   **(h)** 5.95

3   Rewrite these with each number rounded to the nearest whole
    number:
    (a) 4.2×6.9 **(b)** 3.9×4.2 **(c)** 9.4×2.9
    (d) 12.4×2.8 **(e)** 5.2×4.8 **(f)** 0.25×5.89

4   Work out the answers to the questions you rewrote in question **3**.

5   49 438 people voted for the New Democrats party in a recent
    election.
    (a) Write this number to:
        (i) the nearest 10   (ii) the nearest 100
        (iii) the nearest 1000
    (b) If you were writing a report on the election which
        'headline' figure would you use and why?

12.2 Rounding to a number of decimal places

Sometimes you will be asked to round (or correct) a decimal
number to a given number of decimal places. For example:

Carl Listie ran 100 m in 10.568 seconds. Round his time to 1
decimal place (1 d.p.).

10.568 has 3 decimal places (digits after the decimal point).

To round to 1 decimal place, count 1 place from the decimal point
and look at the next digit. If it is 5 or more you need to round up.

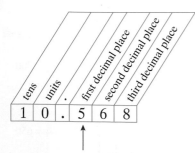

This digit is 6, so you need to round up

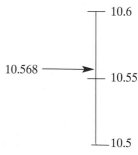

10.568 is closer to 10.6 than to 10.5

10.568 seconds to 1 decimal place (1 d.p.) is 10.6 seconds.

■ **You can round (or correct) numbers to a given number of decimal places (d.p.). The first decimal place is the first number (zero or non-zero) after the decimal point.**

Example 2

Round 46.382 to: **(a)** 1 decimal place **(b)** 2 decimal places.

(a) Rounding to 1 d.p. (1 digit after the decimal point):

46.382
↑
The digit after the first decimal place is 8, so **round up**.

46.382 rounded to 1 d.p. is 46.4

(b) Rounding to 2 d.p.

46.382
↑
The digit after the second decimal place is 2, so **round down**.

Notice that when you round down to 2 d.p. these two decimal places remain the same.
↓↓
46.382 rounded to 2 d.p. is 46.38

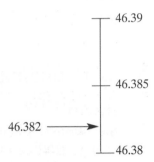

46.382 is closer to 46.38 than to 46.39

Exercise 12B

1 Copy this interval diagram.
   (a) Mark the position of 46.3826 on the diagram.
   (b) Write 46.382 67 rounded to 3 d.p.

2 This table shows the times recorded for the first three athletes in an 800 m race:

Name	Time
R. Grey	1 min 45.4826 s
M. Hobson	1 min 45.4768 s
T. Knight	1 min 45.4817 s

   (a) Write these times correct to 2 d.p.
   (b) Write the names of the athletes in the order in which they finished the race.

3   Write these numbers correct to the approximation given in brackets:

(a)   137.27   (1 d.p.)   (b)   0.67381   (4 d.p.)

(c)   4.999   (1 d.p.)   (d)   8.999   (2 d.p.)

(e)   17.9939   (3 d.p.)   (f)   2.00972   (4 d.p.)

4   Carry out the following calculations and give your answers correct to the number of decimal places asked for:

(a) 24.56×3.87 (correct to 3 d.p.)

(b) 3.764×2.593 (correct to 3 d.p.)

(c) 2.888×3.777 (correct to 2 d.p.)

(d) 13.799×12.752 (correct to 1 d.p.)

12.3 Rounding to a number of significant figures

You will often be asked to round answers to '2 significant figures' (2 s.f.) or '3 significant figures' (3 s.f.). 'Significant' means 'important'.

When you are estimating the number of people at a hockey match you don't need to say that there were exactly 8732 people there. You can give your answer to two significant figures:

There were about 8700 people (to 2 s.f.).

This is called 'correcting to two significant figures'.

■   **You can round (or correct) numbers to a given number of significant figures (s.f.).**
   The first significant figure is the first non-zero digit in the number, counting from the left.

Rounding to a significant figure which is on the right of the decimal point is like the process used in rounding to decimal places. You look at the next digit after the significant figure.

0.0385 correct to 2 s.f. is 0.039.

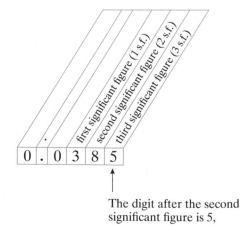

The digit after the second significant figure is 5,

Example 3

Round 642.803:   **(a)** to 1 s.f.   **(b)** to 2 s.f.   **(c)** to 3 s.f.

(d) to 4 s.f.   **(e)** to 5 s.f.

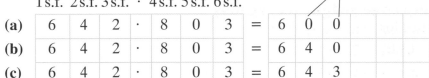

You need the zeros to show the place value of the 6.

	1 s.f.	2 s.f.	3 s.f.	·	4 s.f.	5 s.f.	6 s.f.							
(a)	6	4	2	·	8	0	3	=	6	0	0			(to 1 s.f.)
(b)	6	4	2	·	8	0	3	=	6	4	0			(to 2 s.f.)
(c)	6	4	2	·	8	0	3	=	6	4	3			(to 3 s.f.)
(d)	6	4	2	·	8	0	3	=	6	4	2	·	8	(to 4 s.f.)
(e)	6	4	2	·	8	0	3	=	6	4	2	·	8 0	(to 5 s.f.)

You need the zero to show 5 significant figures.

Exercise 12C

1 Write these numbers correct to the approximation given in brackets.

(a) 3.168   (3 s.f.)   **(b)** 964.8   (3 s.f.)

(c) 15.8   (2 s.f.)   **(d)** 9.9   (1 s.f.)

(e) 55.7639 (5 s.f.)   **(f)** 55.898   (4 s.f.)

2 Using her calculator Samajit works out the answer to 113.12 × 1.7341 as:

$$196.161\,39$$

Write Samajit's answer correct to:

(a) 3 decimal places   **(b)** 3 significant figures

(c) 2 decimal places   **(d)** 2 significant figures

(e) the nearest whole number **(f)** the nearest 10

(g) the nearest 100   **(h)** 7 significant figures.

3 43 672 spectators watched Draycott Town beat Borrowash Rovers in the Cup Final. Write the number of spectators correct to:

(a) 4 s.f.   **(b)** 2 s.f.   **(c)** 1 s.f.

4 Work out the following calculations and give your answers correct to the number of significant figures asked for:

(a) 54 × 36   (3 s.f.)   **(b)** 117 × 38   (3 s.f.)

(c) 148 × 66   (2 s.f.)   **(d)** 235 × 364   (2 s.f.)

(e) 65 × 23   (1 s.f.)   **(f)** 24 × 42 × 59 (1 s.f.)

5 Write these correct to the approximation given:

  (a) 0.0032   (1 s.f.)   **(b)** 0.000876 (2 s.f.)

  (c) 0.00897   (2 s.f.)   **(d)** 0.00091   (1 s.f.)

  (e) 0.0099   (1 s.f.)   **(f)** 0.0004   (1 s.f.)

6 Carry out the following calculations and give your answers correct to the accuracy asked for:

  (a) 0.3×0.023 correct to 1 s.f.

  (b) $3.2 \div 48$ correct to 4 s.f.

  (c) 0.007×0.00041 correct to 2 s.f.

7 1 inch is approximately equal to 2.54 cm. Write:

  (a) 1 cm in inches correct to 2 s.f.

  (b) 1 mm in inches correct to 2 s.f.

  (c) 1 metre in inches correct to 3 s.f.

  (d) 1 km in inches correct to 1 s.f.

12.4 Checking and estimating

Get into the habit of checking the answers to your calculations. Sometimes a wrong answer is obvious: 4.2 metres for the average height of a man in a survey or 0.02 km for the distance travelled by a car on a motorway in a day are clearly incorrect.

Once you notice that an answer is obviously wrong you can repeat the calculation to try and find the correct answer.

Making an estimate

Another way of checking a calculation is to make an estimate. Round all the numbers to one significant figure. Then do the calculation with the rounded numbers.

Example 4

Estimate the answer to:

(a) 5.12×2.79 **(b)** $19.67 \div 2.8$ **(c)** $\dfrac{2.75 \times 8.33}{5.23 + 2.74}$

Answer

(a) Round all numbers to 1 significant figure:
    5.12×2.79 becomes $5 \times 3 = 15$ as an estimate.
    (Actual answer: 14.2848)

(b) Round all numbers to 1 significant figure:
    $20 \div 3 = 6\frac{2}{3} = 7$ (to 1 significant figure).
    (Actual answer: 7.025)

(c) Round all numbers to 1 significant figure:

$$\frac{3 \times 8}{5 + 3} = \frac{24}{8} = 3 \text{ as an estimate.}$$

Actual answer: $22.9075 \div 7.97 = 2.874\ldots$

Worked examination question

$$\frac{10.25 + 29.75}{0.2 \times 45}$$

Show how you would estimate the answer to this expression without using a calculator. Write down your estimate. **[E]**

Round all numbers to 1 s.f.: $\dfrac{10 + 30}{0.2 \times 50} = \dfrac{40}{10} = 4$ as an estimate

■ **To estimate answers round all numbers to 1 significant figure and do the simpler calculation.**

Exercise 12D

1 In each of parts **(a)** to **(f)**:

(i) write down a calculation that could be used to estimate the answer

(ii) work out an estimated answer.

(iii) use a calculator to work out the exact answer.

(a) 8.66×9.56 **(b)** $3.75 \times (2.36 - 0.39)$

(c) 37.12×4.33 **(d)** $(2095 \times 302) + 396$

(e) $\dfrac{187.3 \times 75.4}{47.9}$ **(f)** $\dfrac{0.634 \times 0.0176}{0.0425}$

2 Trevor's calculator is faulty. It doesn't show the decimal point so Trevor has answered the following questions by writing down the digits from the display. By estimating, find the correct answers including the decimal points.

(a) $3.4 \times 2.5 = 85$ **(b)** $5.6 \times 7.23 = 40488$

(c) $2.42 \times 1.93 = 46706$ **(d)** $10.63 \times 12.64 = 1343632$

(e) $3.2 \times 5.1 \times 4.7 = 76704$ **(f)** $2.9 \times 0.32 \times 8.7 = 80736$

3 Estimate the answer to each question giving your answers to the nearest whole number:

(a) $36 \div 5$ **(b)** $47 \div 8$ **(c)** $103 \div 11$

(d) $148 \div 49$ **(e)** $239 \div 31$ **(f)** $197 \div 19$

(g) $281 \div 41$ **(h)** $796 \div 98$ **(i)** $352 \div 58$

(j) $\dfrac{53 \times 3}{49}$ **(k)** $\dfrac{38 \times 5}{19}$ **(l)** $\dfrac{24 \times 11}{59}$

4   Jean is using Pythagoras' Theorem to find the length of the
     missing side of this triangle.
     She uses the calculation:

     $$\sqrt{(3.7)^2 + (9.4)^2}$$

     Estimate, to the nearest whole number, the length of the
     missing side.

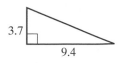

12.5 Describing the accuracy of a measurement

Suppose you use a ruler marked in millimetres to measure a line as
237 mm, correct to the nearest mm.

The true length could be anywhere between 236.5 mm and 237.5
mm.

So the true length could be anywhere in a range of 0.5 mm below
and 0.5 mm above the recorded value:

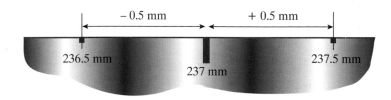

For any measurement you make:

■ **If you make a measurement correct to a given unit the true
value lies in a range that extends half a unit below and half a
unit above the measurement.**

1   Copy and complete this table showing the attendances to the
     nearest hundred at these colleges. The first one is done for you:

School	Attendance to the nearest 100	Least possible attendance	Greatest possible attendance
Southpark	1400	1350	1450
Westown	400		
Eastgrove	800		
Northbury	1200		
Southone	2100		

2 These lengths are all measured to the nearest millimetre. Write down the smallest and the largest possible lengths they could be:

(a) 3.9 cm   (b) 17.620 m   (c) 53.0 cm

(d) 3.010 m   (e) 29 mm

3 Kirsty ran a 1500 metre race. Her time to the nearest 0.01 seconds was 4 minutes 7.56 seconds. Write down the range of times in which her time could lie.

4 The distance between two motorway junctions is given as 17 miles. Write down the range in which the true length could lie.

12.6 Upper and lower bounds

In the measuring example on page 221 the true length could be anywhere between 236.5 mm and 237.5 mm:

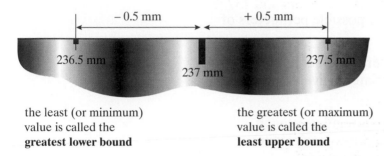

the least (or minimum) value is called the **greatest lower bound**

the greatest (or maximum) value is called the **least upper bound**

Sometimes these are just called the **lower bound** and the **upper bound**.

■ **The greatest lower bound and the least upper bound are the minimum and maximum possible values of a measurement or calculation.**

Example 5

A rectangle has a length of 7.5 cm and width 4.3 cm measured to the nearest 0.1 cm.

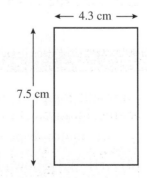

Work out:

(a) the greatest lower bounds of the length and width

(b) the least upper bounds of the length and width

(c) the maximum area (the least upper bound of the area).

Answer

(a) The greatest lower bound of the length is 7.45 cm
The greatest lower bound of the width is 4.25 cm

(b) The least upper bound of the length is 7.55 cm
The least upper bound of the width is 4.35 cm

(c) The maximum area is:

least upper bound of length × least upper bound of width

$$= 7.55 \text{ cm}^2 \times 4.35 \text{ cm}^2$$
$$= 32.8425 \text{ cm}^2$$

Exercise 12F

1 Jafar has a piece of wood that has a length of 30 cm, correct to the nearest centimetre.

(a) Write down the minimum length of the piece of wood.

Fatima has a different piece of wood that has a length of 18.4 cm, correct to the nearest millimetre.

(b) Write down the maximum and minimum lengths between which the length of the piece of wood must lie.

2 The length of each side of a square, correct to 2 significant figures, is 3.7 cm.

(a) Write down the least possible length of each side.

(b) Calculate the greatest and least possible perimeters of this square.

(c) When calculating the perimeter of the square how many significant figures is it appropriate to give in the answer? Explain your answer.

(d) If this question had referred to a regular octagon, instead of a square, would your answer to part (c) have been the same? Explain your answer. [E]

3 Stephanie ran 100 metres.
The distance was correct to the nearest metre.

(a) Write down the shortest distance Stephanie could have run.

Stephanie's time for the run was 14.8 seconds.
Her time was correct to the nearest tenth of a second.

(b) Write down
  (i) her shortest possible time for the run
  (ii) her longest possible time for the run.

(c) Calculate (i) the lower bound (ii) the upper bound for her average speed. Write down all the figures on your calculator display.

(d) (i) Write down her average speed to an appropriate degree of accuracy.
  (ii) Explain how you arrived at your answer. [E]

4

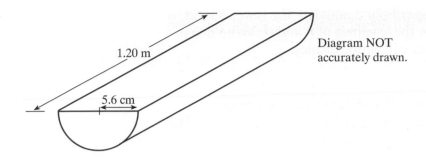

1.20 m

5.6 cm

Diagram NOT
accurately drawn.

The diagram represents a metal girder in the shape of a half
cylinder.

The radius of each semi-circular end is 5.6 centimetres, correct
to the nearest tenth of a centimetre.

(a) Write down the upper and lower bounds of the radius.

The length of the girder is 1.20 metres, correct to the nearest
centimetre.

(b) Write down the upper and lower bounds of the length.

(c) Calculate the upper and lower bounds of the volume of
the girder. Give your answer in cubic centimetres (cm^3).

The girder is melted down and made into a sphere. No metal is
wasted.

(d) Calculate the upper and lower bounds of the radius of the
sphere. Give your answer in centimetres.

5 Brazil has an area of $8\,500\,000$ km^2 correct to the nearest
$100\,000$ km^2.

(a) Write down the limits between which the area of Brazil
must lie.

The population density of a country is the average number of
people per km^2 of the country.

Brazil has a population of 144 million correct to the nearest
million.

(b) Calculate the maximum and minimum values of the
population density of Brazil.   [E]

6 The formula $S = \dfrac{F}{A}$ is used in engineering.

$F = 810$, correct to 2 significant figures.
$A = 2.93$, correct to 3 significant figures.

(a) For the value of F, write down
(i) the upper bound   (ii) the lower bound.

(b) For the value of A, write down
(i) the upper bound   (ii) the lower bound.

(c) Calculate (i) the upper bound and (ii) the lower bound
for the value of S for these values of F and A. Write down
all the figures on your calculator display.
(i) upper bound =
(ii) lower bound =

(d) Write down this value of S correct to an appropriate
number of significant figures. [E]

Summary of key points

1 In rounding, halves or 5s are usually rounded upwards.

2 You can round (or correct) numbers to a given number of
decimal places (d.p.). The first decimal place is the first
number (zero or non-zero) after the decimal point.

3 You can round (or correct) numbers to a given number of
significant figures (s.f.). The first signficant figure is the first
non-zero digit in the number, counting from the left.

4 To estimate answers round all numbers to 1 significant figure
and do the simpler calculation.

5 If you make a measurement correct to a given unit the true
value lies in a range that extends half a unit below and half a
unit above the measurement.

6 The greatest lower bound and the least upper bound are the
minimum and maximum possible values of a measurement or
calculation.

13 Basic trigonometry

The origins of trigonometry

Two of the problems which led to the development of trigonometry were:

creating a device for telling the time by using shadows which vary in length according to the time of day and height of the sun

working out the relationship between the width of a channel in a harbour, the range of a cannon ball and the angle of pivot of the cannon

Today, trigonometry is used by engineers, surveyors, architects and others who need to work out relationships connecting distances and angles. It is also a topic which is always tested in GCSE examinations.

13.1 The mathematical model for trigonometry

Mathematicians took the problems from history and made them into pure mathematical models.

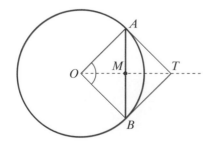

This was their approach.

They drew a circle and investigated the relationships between the length of the chord AB, the distance OM from the centre of the circle to the mid-point of AB and the angle AOB at the centre of the circle. They also examined the length of the tangents AT and BT.

Because of the symmetry of this situation they decided that they only needed to examine the *top half* of the diagram. They made the radius of the circle, *OA*, unit length. For other circles everything would then be in proportion.

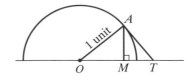

13.2 The three trigonometric functions

Working from the mathematical model, we define three functions, called **trigonometric functions**.

The angle *AOM* is labelled *x* and the radius, *OA*, of the circle is 1 unit in length.

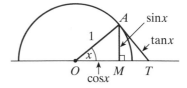

Then the length *MA*, the height of *A* above the horizontal line is called **sine *x*** and often abbreviated to **sin *x***.

The distance *OM* is called **cosine *x*** and often abbreviated to **cos *x***.

The length of the tangent *AT* is called **tangent *x*** and often abbreviated to **tan *x***.

◼ **The three basic trigonometric functions are:**
    sin *x*,   cos *x*   and   tan *x*

Exercise 13A

You need two geostrips of equal length (or similar made from thin card).

The length of each strip is to be regarded as 1.

Keep the black strip *OB* fixed as horizontal. Turn the grey strip *OA* anti-clockwise through an angle (labelled in the diagram as *x*). Vary the angle *x* from 0° to 180° using about 10 different values for *x*.

For each angle *x*, measure and record the distances *MA* and *OM*. (Remember: *OA* = 1 unit.)

$$MA = \sin x \qquad OM = \cos x$$

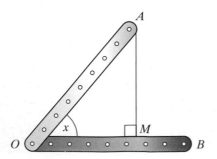

Record your results as:

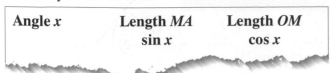

Angle x	Length MA sin x	Length OM cos x

On graph paper plot the graphs of:

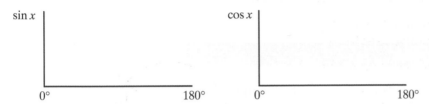

13.3 Using a calculator

We can work out the values of sin x, cos x and tan x for various values of x using a scientific calculator.

Note: different models of calculators work differently. Check with your teacher.

Example 1

Obtain the value of cos 43°.

● Input 43.
● Press the button marked COS.
● Press ▣ .
● The calculator display will be:
  0.731353701

so $\qquad$ cos 43° = 0.731353701

In most cases there is no need to give results to more than 4 decimal places, so we would usually write:

$\qquad$ cos 43° = 0.7314 $\qquad$ having rounded up.

When given the value of any trigonometric function we can also use the calculator to find the value of the angle.

Example 2

Given that tan x = 1.2411, find the value of x

To do this we:

● Press the 2nd (or INV or SHIFT) button
● Press the TAN button.
● Input 1.2411
● Press ▣ .

The calculator display is 51.14032682
So, correct to 2 decimal places (which is usually as accurate as we need), we have:

$\qquad$ the angle whose tangent is 1.2411 is 51.14°

Exercise 13B

Use your calculator to find these:

(a) sin 47°   (b) cos 58°   (c) tan 21°
(d) tan 83°   (e) cos 25°   (f) sin 60°
(g) tan 106°   (h) cos 93°   (i) sin 132°

Answers should be correct to 4 d.p.

Exercise 13C

Use your calculator to find each angle x (correct to 0.1°) when:

(a) $\sin x = 0.3524$ (b) $\cos x = 0.1364$
(c) $\tan x = 1.4142$ (d) $\tan x = 0.4365$
(e) $\cos x = 0.9854$ (f) $\sin x = 0.8856$

and these two, which we shall consider again

(g) $\sin x = -0.7071$ (h) $\cos x = -0.7071$

13.4 Right-angled triangles: the trigonometric ratios

One of the most common uses of trigonometry is in working out lengths and angles in right-angled triangles. This is a topic which is always tested in GCSE examinations and is often related to work on Pythagoras' Theorem, see Unit 8, page 175.

Suppose that we have a right-angled triangle OAM with the angle at O as x and $OA = 1$ unit in length.

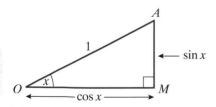

Because $OA = 1$

$\qquad \sin x = MA$ and $\cos x = OM$

and by Pythagoras $(MA^2 + OM^2 = OA^2)$

$$MA^2 + OM^2 = 1$$

and this would make calculations of x, or MA or OM quite easy.

In the general case the right-angled triangle is drawn as:

and we label the

length of BC (the side opposite A) = a

length of AC (the side opposite B) = b

length of AB (the hypotenuse and the side opposite the right angle at C) = c

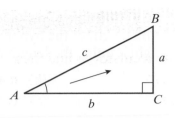

Pythagoras' Theorem gives:

$$a^2 + b^2 = c^2$$

Looking at the simplest case again, where the hypotenuse is 1 unit in length, the vertical height is sin x.

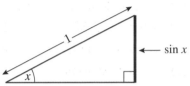

Now, when the length of the hypotenuse is 2 units the vertical height is double the height in the previous diagram, that is, 2 sin x.

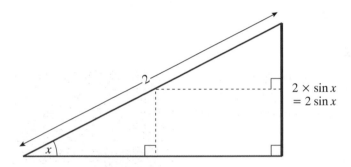

When the length of the hypotenuse is 3 units, then the vertical height is 3 times the original bold height, that is, 3 sin x.

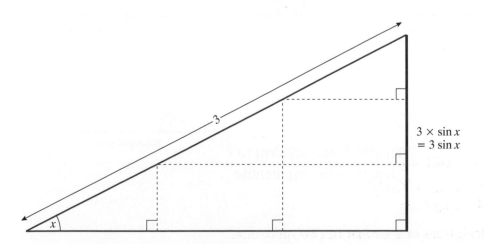

You should now be able to convince yourself that in the general case, with the length of the hypotenuse labelled c units and the vertical height a units, the length of a is given by:

$$a = c \sin x$$

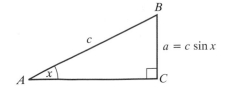

and now making $\sin x$ the subject of this formula

$$\sin x = \frac{a}{c}$$

or

$$\sin x = \frac{\text{length of side opposite } x}{\text{length of the hypotenuse}}$$

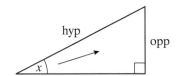

which is often abbreviated to

$$\sin x = \frac{\text{opposite}}{\text{hypotenuse}}$$

or

■ $\sin x = \dfrac{\mathbf{opp}}{\mathbf{hyp}}$

By a similar argument, looking at these triangles:

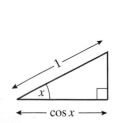
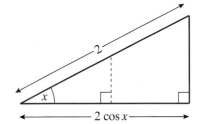
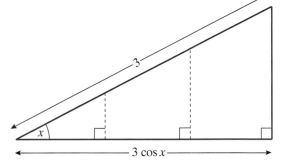

you should also be able to convince yourself that making $\cos x$ the subject of the formula $b = c \cos x$

gives:

$$\cos x = \frac{b}{c}$$

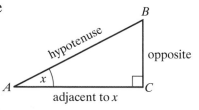

which is often written as:

$$\cos x = \frac{\text{length of side adjacent to } x}{\text{length of the hypotenuse}}$$

or

■ $\cos x = \dfrac{\mathbf{adj}}{\mathbf{hyp}}$

These are the trigonometric ratios for $\sin x$ and $\cos x$.

The tangent ratio is a little different. Look back at the diagram of the original model, on page 268.

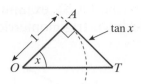

Imagine the triangle OAT is cut out, turned over and re-positioned as:

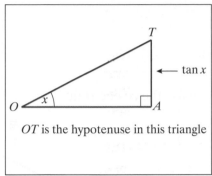

OT is the hypotenuse in this triangle

Note: in this triangle OA is no longer the hypotenuse but is the side adjacent to the angle x.

Now we can look at the sequence of diagrams:

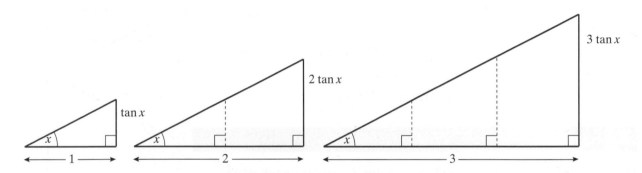

Again, you should be able to convince yourself that in the general case:

$$a = b \tan x$$

and making $\tan x$ the subject of the formula:

$$\tan x = \frac{a}{b}$$

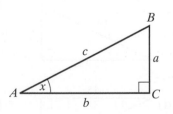

or

$$\tan x = \frac{\text{length of side opposite angle } x}{\text{length of the side adjacent to angle } x}$$

or

 $\tan x = \dfrac{\textbf{opp}}{\textbf{adj}}$

For your GCSE examinations you will be given a formulae sheet which has trigonometrical formulae on it. It will look like this:

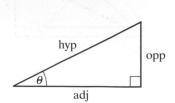

$$\sin \theta = \frac{\text{opp}}{\text{hyp}}$$

$$\cos \theta = \frac{\text{adj}}{\text{hyp}}$$

$$\tan \theta = \frac{\text{opp}}{\text{adj}}$$

You must learn these for your GCSE exam.

Example 3

Calculate the size of the angle at A.

Since 4 is adjacent to A and 5 is the hypotenuse we use the cosine ratio:

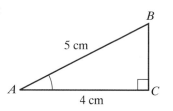

$$\cos A = \frac{\text{adj}}{\text{hyp}} \qquad \text{so} \qquad \cos A = \frac{4}{5}$$

$$\cos A = 0.8$$

so $\qquad A = 36.87°$ (correct to 2 d.p.)

Example 4

Calculate the size of the angle at P

In this triangle 12 is opposite P and 5 is adjacent to P. So we use the tangent ratio:

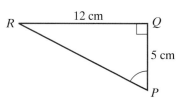

$$\tan P = \frac{\text{opp}}{\text{adj}} \qquad \tan P = \frac{12}{5} \qquad \tan P = 2.4$$

$$P = 67.38° \text{ (correct to 2 d.p.)}$$

Exercise 13D

In this exercise all of the lengths are in cm. Calculate each of the angles marked with a letter, correct to 0.1°.

(a)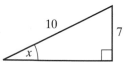

(b)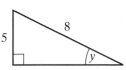

(c)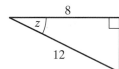

(d)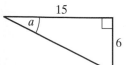

(e)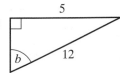

(f)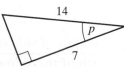

(g)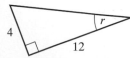

(h)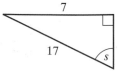

(i)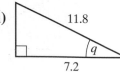

Example 5

Calculate the length of the side marked x.

We are given the angle, 42°.
We are given the *hypotenuse*, 15 cm.
We want to find the side *opposite* the given angle.

So we use the sine ratio:

$$\sin 42° = \frac{\text{opp}}{\text{hyp}}, \quad \text{or} \quad \text{opp} = \text{hyp} \times \sin 42°$$

so
$$x = 15 \times \sin 42°$$
$$x = 15 \times 0.6691$$
$$x = 10.04 \text{ cm (correct to 2 d.p.)}$$

Example 6

Calculate the length of the side marked y.

For the given angle:
y is opposite
12 is adjacent
So we use tangent:

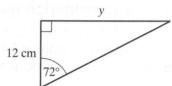

$$\tan 72° = \frac{\text{opp}}{\text{adj}}, \quad \tan 72° = \frac{y}{12}$$

$$y = 12 \times \tan 72°$$
$$y = 12 \times 3.0777$$
$$y = 36.93 \text{ cm (correct to 2 d.p.)}$$

Exercise 13E

In this exercise all lengths are in centimetres. Calculate each length marked with a letter, correct to 2 d.p.

(a)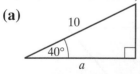

(b)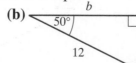

(c)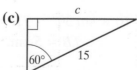

(d)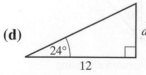

(e)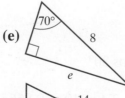

(f)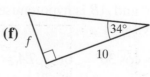

(g)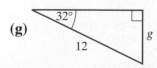

(h)

(i)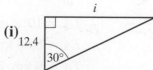

Example 7

A lighthouse, L, is 25 km due north of a port, P. A marker buoy, B, is due east of L and on a bearing of 065° from P. Calculate the distance LB.

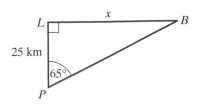

We need a sketch of this situation.

$LB = x$ is opp the 65°
$PL = 25$ is adj the 65°
Use tangent:

$$\tan 65° = \frac{\text{opp}}{\text{adj}} \qquad \tan 65° = \frac{x}{25}$$

so
$$x = 25 \times \tan 65°$$
$$x = 53.61 \text{ km (correct to 2 d.p.)}$$

Worked examination question

The distance from Alton, A, to Burton, B, is 10 km. The bearing of B from A is 030°. A transmitter T, is due north of A and due west of B.

(a) **(i)** Calculate the distance of T from B.
  (ii) Calculate the distance of T from A.
    Give your answer to the nearest kilometre.

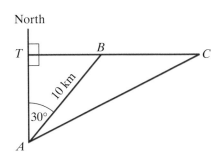

Clowne, C, is 8 km due east of Burton.

(b) Use your answers to **(a)** to calculate angle $T\hat{A}C$. Hence give the 3 figure bearing of C from A, correct to the nearest degree.

(c) A helicopter flies at the same height from A to B to C. If the average speed for the journey is 160 km/h, calculate the time taken. Give your answer correct to the nearest minute. **[E]**

For part **(a)** **(i)**
TB is opposite 30° AB is hypotenuse so we use the sine ratio:

$$\frac{TB}{10} = \sin 30° \qquad \text{or} \qquad TB = 10 \times \sin 30°$$

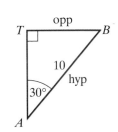

$$TB = 10 \times \sin 30° \qquad TB = 10 \times 0.5$$
so
$$TB = 5 \text{ km.}$$

For part **(a)** **(ii)**
TA is adjacent to 30° and AB is hypotenuse
So we use the cosine ratio:

$$\frac{TA}{10} = \cos 30° \qquad \text{or} \qquad TA = 10 \times \cos 30°$$

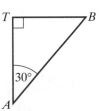

so
$$TA = 10 \times \cos 30° \quad TA = 10 \times 0.866$$
so
$$TA = 8.66 \text{ km}$$
$$TA = 9 \text{ km, correct to the nearest km.}$$

Alternatively, using Pythagoras:
$TA^2 + TB^2 = AB^2$
$TA^2 + 25 = 100$
$TA^2 = 100 - 25 = 75$
$TA^2 = \sqrt{75}$
$TA = 8.66$ km

For part **(b)**

We have the diagram:

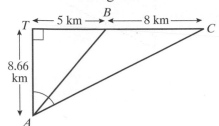 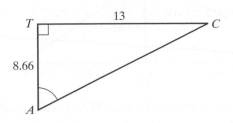

Then: For angle $T\hat{A}C$

$TC = 13 = \text{opp}$
$TA = 8.66 = \text{adj}$

So we use the tangent ratio:

$$\tan TAC = \frac{\text{opp}}{\text{adj}} \qquad \tan TAC = \frac{13}{8.66}$$

$$\tan TAC = 1.50115 \qquad T\hat{A}C = 56.33°$$

So, correct to the nearest degree, the bearing of
C from A is 056°.

For part **(c)**

(*This is not trigonometry but is an example of how trigonometry can be combined with another topic.*)

The total distance from A to B to C is
$$10 + 8 = 18 \text{ km}$$
and $\text{distance} = \text{speed} \times \text{time}$

or $$\text{time} = \frac{\text{distance}}{\text{speed}}$$

so $$\text{time} = \frac{18}{160} = 0.1125 \text{ hours}$$

and 0.1125 hours $= 0.1125 \times 60 = 6.75$ minutes

so time $= 7$ minutes, correct to the nearest min.

Exercise 13F (Mixed questions)

1 ABC is a right-angled triangle.
AB is of length 4 m and BC is of length 13 m.
   (a) Calculate the length of AC.
   (b) Calculate the size of angle $A\hat{B}C$.

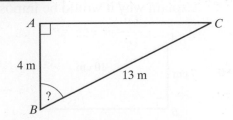

2

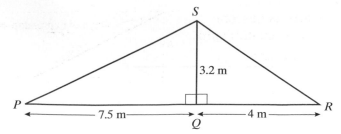

The diagram represents the frame, *PQRS*, of a roof.

$$PQ = 7.5 \text{ m}, \quad QR = 4 \text{ m}, \quad SQ = 3.2 \text{ m}$$

(a) Calculate the length of *PS*.

(b) Calculate the size of the angle $S\hat{R}Q$.

3 The diagram shows a ladder *LD* of length 12 m resting against a vertical wall. The ladder makes an angle of 40° with the horizontal.

Calculate the distance *BD* from the base of the wall to the top of the ladder.

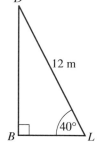

4 A lighthouse, *L*, is 43 km due north of a harbour, *H*. A marker buoy, *B*, is 17 km due east of *L*.

Calculate the bearing of B from H.

5

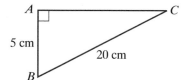

$$AB = 5 \text{ cm}$$
$$BC = 20 \text{ cm}$$
$$\text{angle } BAC = 90°$$

(a) Calculate the length of *AC*.

(b) Calculate the size of the angle *ABC*.

6

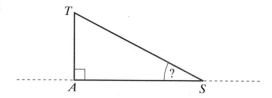

The diagram represents a vertical television aerial *AT*. The aerial casts a shadow *AS* on horizontal ground.

$$AT = 8 \text{ m} \quad AS = 30 \text{ m}$$

Calculate the size of the angle of elevation of the sun ($A\hat{S}T$).

7 Explain why it would be impossible to draw the triangle *ABC*.

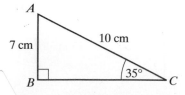

8 Coming in to land, a small aeroplane starts its descent at a vertical height of h metres above the horizontal land.

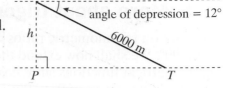

The aeroplane descends along a straight line at a constant angle of depression of 12°.

From starting its descent to touching down, the aeroplane travels through a distance of 6000 m.

(a) Calculate the vertical height, h, at which the aeroplane starts its descent.

At the start of its descent, the aeroplane is vertically above a point P on the ground. It touches down at a point T.

(b) Calculate the distance PT.

9

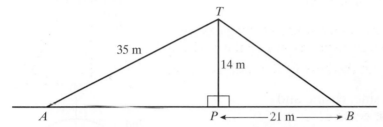

A vertical pole, PT, is held fixed on horizontal ground by two tight ropes AT and BT.

A, P and B are in a straight line.

$$PT = 14 \text{ m}, \quad AT = 35 \text{ m}, \quad PB = 21 \text{ m}.$$

Calculate:

(a) the size of the angle PAT

(b) the length of the rope BT

(c) the size of the angle PBT

(d) the distance from A to B.

10 This is the cross-section, $ABCD$, of a valley

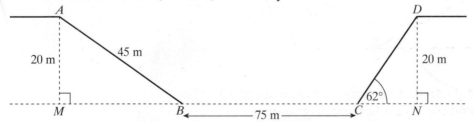

The vertical heights AM and DN above the horizontal base BC of the valley are both equal to 20 metres.

$$AB = 45 \text{ m} \quad \text{angle } NCD = 62° \quad BC = 75 \text{ m}$$

Calculate:

(a) the size of the angle MBA

(b) the distance CN

(c) the distance CD

(d) the distance from A to D across the top of the valley.

13.5 More about trigonometric functions

The work on trigonometric ratios was restricted to angles between 0° and 90°. We shall now extend the original work on the trigonometrical functions $\sin x$, $\cos x$, and $\tan x$ for values of x which are outside the range of 0° to 90°.

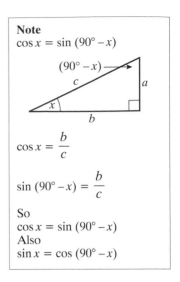

Note

$$\cos x = \sin(90° - x)$$

$$\cos x = \frac{b}{c}$$

$$\sin(90° - x) = \frac{b}{c}$$

So
$$\cos x = \sin(90° - x)$$
Also
$$\sin x = \cos(90° - x)$$

We will go back to the original definitions of the 3 trigonometric functions as:

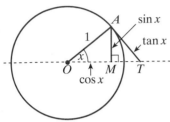

From the diagram we can look at sine, cosine and tangent in all 4 quadrants. That is:

1st quadrant: angles between 0° and 90°

2nd quadrant: angles between 90° and 180°

3rd quadrant: angles between 180° and 270°

4th quadrant: angles between 270° and 360°

We can even consider other angles outside the range 0° to 360°, such as −50° and 410°.

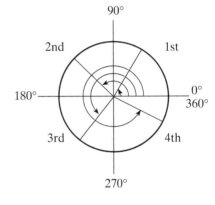

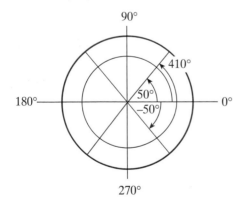

Consider the sine of these angles:

Correct to 4 d.p., your calculator gives:

$$\sin(-50°) = -0.7660$$
$$\sin 130° = 0.7660$$
$$\sin 410° = 0.7660$$

When we compare these with:

$$\sin 50° = 0.7660$$

we can see that:

$$\sin(-50°) = -\sin 50°$$
$$\sin 130° = \sin 50°$$
$$\sin 410° = \sin 50°$$

0.766044443

There are explanations for these results.

Example 8

sin (–50°) and sin 50°

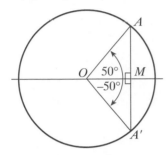

sin 50° is the vertical distance *MA*.

sin (–50°) is the vertical distance *MA'*, which equals *MA* but goes in the opposite direction

■ **sin (– x) = –sin x**

Remember

Angles turning ⤵ are + ve

Angles turning ⤵ are – ve

Example 9

sin 130° and sin 50°

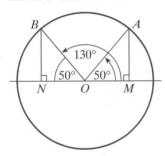

sin 50° is the vertical distance *MA*.

sin 130° is the vertical distance *NB*. By symmetry *NB = MA*, so sin 50° = sin 130°

■ **sin (180° – x) = sin x**

We use the same convention as for coordinate and graph work.

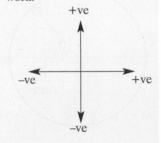

Example 10

sin 410° and sin 50°

In turning through 410° the radius finishes in exactly the same position as in turning through 50°. So sin 50° and sin 410° are both equal to the vertical distance *MA*.

■ **sin (360° + x) = sin x**

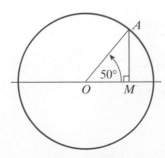

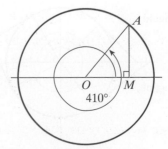

You should be able to convince yourself of the equivalent results for cosine from the following sequence of diagrams.

■ $\cos(-x) = \cos x$

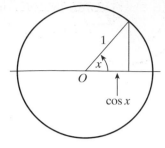 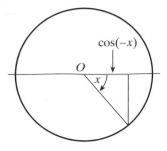

■ $\cos(180° - x) = -\cos x$

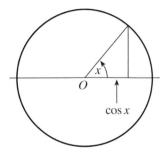 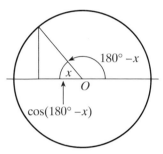

■ $\cos(360° + x) = \cos x$

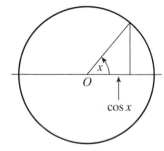 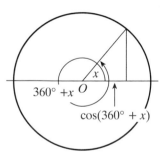

There are similar results for tangent. There is a convention for positive and negative tangents:

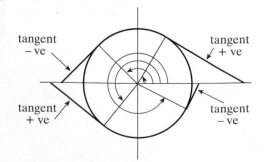

By considering the diagram, you should be able to convince yourself of the results for the tangent function.

■ $\tan(-x) = -\tan x$
$\tan(180° - x) = -\tan x$
$\tan(360° + x) = \tan x$

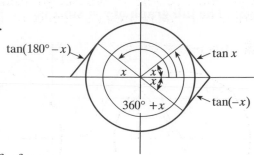

13.6 The graph of sin *x*

We are going to draw the graph of $y = \sin x$ for values of x from 0° to 360°.

Before drawing the graph, it will be useful to look at a few strategically chosen values of x and the resulting values of $\sin x$

$x°$	$\sin x$
0	0
30	0.5
45	0.7071
60	0.8660
90	1
120	0.8660
135	0.7071
150	0.5
180	0
210	−0.5
225	−0.7071
240	−0.8660
270	−1
300	−0.8660
315	−0.7071
330	−0.5
360	0

From this diagram

$\sqrt{2} = \sqrt{1^2 + 1^2}$
(Pythagoras)

$\sin 45° = \dfrac{1}{\sqrt{2}} = 0.707$

From this diagram

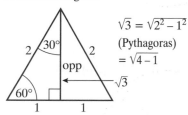

$\sqrt{3} = \sqrt{2^2 - 1^2}$
(Pythagoras)
$= \sqrt{4 - 1}$

$\sin 60° = \dfrac{\sqrt{3}}{2} = 0.8660$

$\sin 30° = \dfrac{1}{2} = 0.5$

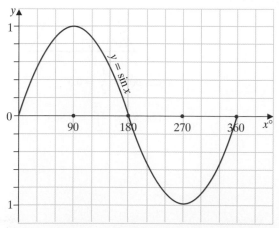

Note:
Graphical calculators are really helpful with this work.

■ **The full graph of $y = \sin x$ is:**

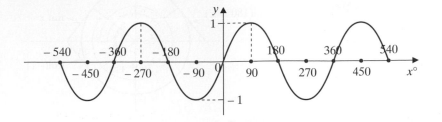

$y = 0$, i.e.
$\sin x = 0$
when
$x = 0°, 180°, -180°$, etc.

The important features of the graph are:

● it repeats itself every 360°, we say that it has a **period** of 360°

● it has maximum values (the largest values of y) of 1 which occur when $x = 90°, 450°, -270°, \ldots$

● it has minimum values (the smallest values of y) of -1 which occur when $x = -90°, 270°, -450°, \ldots$

13.7 The graph of cos x

We will now look at the graph of $y = \cos x$ for values of x from 0° to 360°.

Some important values are:

$x°$	$\cos x$
0	0
30	0.8660
45	0.7071
60	0.5
90	0
120	−0.5
135	−0.7071
150	−0.8660
180	−1
210	−0.8660
225	−0.7071
240	−0.5
270	0
300	0.5
315	0.7071
330	0.8660
360	1

You can confirm this on a graphical calculator.

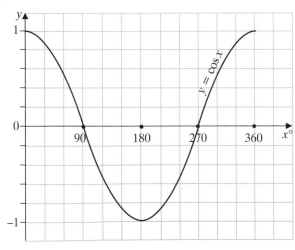

From this diagram

$\cos 45° = \dfrac{1}{\sqrt{2}} = 0.7071$

From this diagram

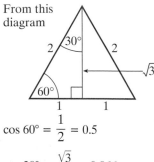

$\cos 60° = \dfrac{1}{2} = 0.5$

$\cos 30° = \dfrac{\sqrt{3}}{2} = 0.866$

■ **The full graph of** $y = \cos x$ **is:**

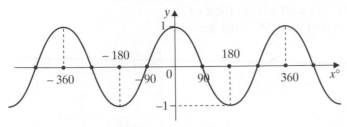

y = 0, i.e.
cos *x* = 0
When
x = 90°, 270°, –90°, etc.

The important features of this graph are:

● it repeats itself every 360°, so it has a period of 360°

● it has a maximum value of 1 which occurs when $x = 0°$, 360°, –360°, …

● it has a minimum value of –1 which occurs when $x = 180°$, –180°, 540°, …

The graph confirms
$\cos (90° - x) = \sin x°$
$\cos x = \sin (90° - x)$

13.8 The graph of tan *x*

We will now look at the graph of $y = \tan x$ for values of *x* from 0° to 360°.

Again we look at some important values:

$x°$	$\tan x$
0	0
30	0.5774
45	1
60	1.7320
90	infinite

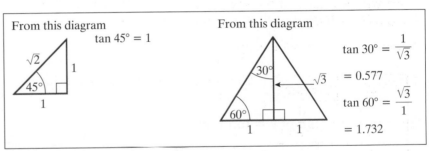

The fact that tan 90° is infinite causes a problem.

In the range of 0° to 90°, the graph of $y = \tan x$ is:

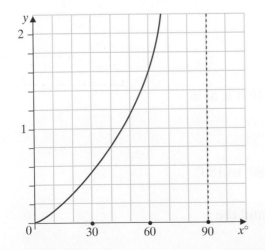

This diagram shows why
tan 90° is infinite

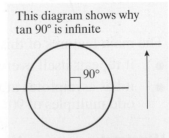

This tangent is
parallel to
the horizontal,
so it will never
meet the *x*-axis.
It is infinitely long.

Angles between 90° and 270°

Taking an angle just greater than 90°, say 95°, we can get an idea of what happens to the tangent graph when x is just over 90° and we look at some important values for x.

In the range 90° to 270° the graph of $y = \tan x$ is:

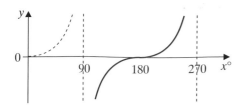

$x°$	$\tan x$
95	–11.43
135	–1
180	0
225	1
270	infinite

Tan 95° is negative and large.

Exercise 13G

1 Show that for the range 270° to 360° the graph of $y = \tan x$ is:

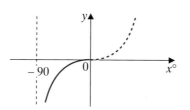

2 Show that the graph of $y = \tan x$ for the range –90° to 0° is:

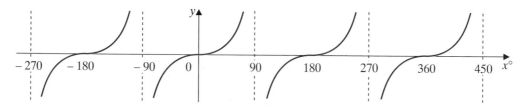

The full graph of $y = \tan x$ is:

The main features of this graph are:

● it repeats itself every 180°, so it has a period of 180°
● it has asymptotes at x values of ±90°, ±270°, ±450°, that is at ± odd multiples of 90°.

Worked examination question

(a) Draw the graph of $y = 3 \sin 2x$ for values of x from –180° to 180°.
(b) State the period of the graph.
(c) State the maximum and minimum values of $3 \sin 2x$ and the values of x at which these occur.

(a) Start with a table of values.

$x°$	$2x$	$3 \sin 2x$
−180	−360	0
−135	−270	3
−90	−180	0
−45	−90	−3
0	0	0
45	90	3
90	180	0
135	270	−3
180	360	0

For simplicity we have chosen intervals of 45°.

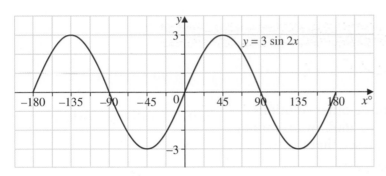

(b) From the drawing we can see that the graph repeats itself every 180°, so the period of the graph is 180°.

(c) The maximum value of $3 \sin 2x$ is 3.
This occurs when $x = -135°$ and $45°$

The minimum value of $3 \sin 2x$ is −3.
This occurs when $x = -45°$ and $135°$

Exercise 13H

1 **(a)** For values of x from −180° to 180°, plot the graph of
$$y = 5 \cos 3x$$
(b) State the period of the graph.
(c) Find the maximum and minimum values of $5 \cos 3x$ and state the values of x at which these occur.

2 **(a)** Plot the graph of $y = 2 \tan x$ for values of x from 0° to 360°.
(b) Find the period of the graph.

3 For values of x from 0° to 360°, plot the graphs of
(a) $y = \sin 4x$ **(b)** $y = \sin \frac{1}{2}x$

Comment on the periods of each of the two graphs.

Example 11 Solving a trigonometric equation

Solve the equation:

$$2 \sin x = 1$$

giving all values of x in the range $-360°$ to $360°$

We have $\qquad\qquad 2 \sin x = 1$

so $\qquad\qquad\qquad \sin x = \frac{1}{2}$

From the calculator:

$$x = 30°$$

This is our first solution and we can use a sketch of the graph of sin $x°$ to find any others.

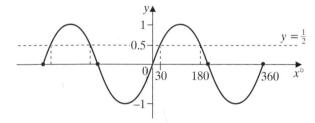

The sketch shows that there are three other solutions to the equation in the range $-360°$ to $360°$. By examining the symmetries of the graph, we can see that the four solutions are:

$$x = 30°, \ 180°-30°, \ -180°-30°, \ -360°+30°$$

So the full set of solutions is:

$$x = 30°, \ 150°, \ -210°, \ -330°$$

In the range $0°$ to $180°$ we have:

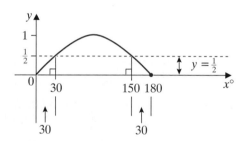

Exercise 13I

1   For values of x in the range $-360°$ to $360°$, solve the equation:

$$2 \cos x = 1$$

2   **(a)** Show that one solution of the equation:

$$5 \sin x = 2$$

is $x = 23.6°$ correct to 1 d.p.

   (b) Hence find all solutions to the equation:

$$5 \sin x = 2$$

which are in the range $0°$ to $720°$, giving your answers correct to 1 d.p.

3   Find all solutions to the equation:

$$\tan x = 2$$

which are in the range $-180°$ to $180°$.

4 **(a)** Sketch the graph of:
$$y = 3 \cos x$$
for values of x from 0° to 360°.

(b) Hence or otherwise obtain all the solutions to the equation:
$$3 \cos x = 1$$
which are between 0° and 360°.

5 **(a)** Show that the equation:
$$5 \sin 3x = 4$$
has a solution:
$$x = 17.71° \text{ correct to 2 decimal places.}$$

(b) Hence or otherwise, obtain all solutions to the equation:
$$5 \sin 3x = 4$$
which are in the range −180° to 180°.

Exercise 13J (Mixed questions)

1 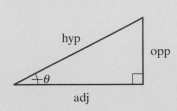 ABC is a triangle.
angle $ABC = 90°$
$AC = 24$ cm
$BC = 13$ cm

(a) Calculate the length of AB.
(b) Calculate the size of the angle ACB.

2 **(a)** Sketch the graph of $y = 2 \cos x°$
for values of x from 0° to 360°.

(b) Hence find all solutions of the equation
$$2 \cos x = -1$$
which are in the range 0° to 360°.

Summary of key points

1 The three basic trigonometric functions are: $\sin x$, $\cos x$, $\tan x$
You should know and be able to use the three trigonometric ratios.

$$\sin \theta = \frac{\text{opp}}{\text{hyp}}$$

$$\cos \theta = \frac{\text{adj}}{\text{hyp}}$$

$$\tan \theta = \frac{\text{opp}}{\text{adj}}$$

and in the forms
$$\text{opp} = \text{hyp} \times \sin \theta \quad \text{adj} = \text{hyp} \times \cos \theta \quad \text{opp} = \text{adj} \times \tan \theta$$

2   For the trigonometric functions you should know why

$$\cos(90° - x) = \sin x \quad \text{and} \quad \sin(90° - x) = \cos x$$
$$\sin(-x) = -\sin x$$
$$\sin(180° - x) = \sin x$$
$$\sin(360° + x) = \sin x$$

$$\cos(-x) = \cos x$$
$$\cos(180° - x) = -\cos x$$
$$\cos(360° + x) = \cos x$$

$$\tan(-x) = -\tan x$$
$$\tan(180° - x) = -\tan x$$
$$\tan(360° + x) = \tan x$$

3   You should be able to draw or sketch the graphs:

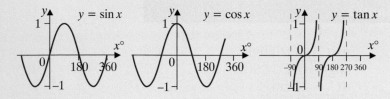

and use them to solve equations of the type

$$a \sin bx = c \quad a \cos bx = c \quad a \tan bx = c$$

14 Sequences and formulae

Unit 1 introduces number patterns such as the triangular numbers:

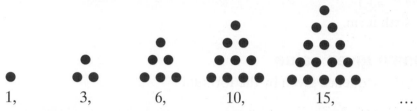

| 1, | 3, | 6, | 10, | 15, | ... |

These numbers are a **sequence** – a succession of numbers formed according to a rule.

Each number in a sequence is called a **term**. 6 is the third term in the sequence of triangular numbers.

14.1 Finding terms of a sequence

Sometimes you will be given an algebraic formula for the nth term of a sequence. You can use this to find any term in the sequence. The nth term is often written as u_n.

■ **When you know a formula for the nth term of a sequence u_n you can calculate any term in the sequence by substituting a value for n in the formula. n must be a positive integer ($n = 1, 2, 3 \ldots$)**

Example 1

The nth term of a sequence is given by: $u_n = 2n - 1$

Find the first two terms of the sequence.

Substituting $n = 1$, the first term $u_1 = (2 \times 1) - 1 = 1$
Substituting $n = 2$, the second term $u_2 = (2 \times 2) - 1 = 3$

The first two terms are 1 and 3.

Example 2

For a given sequence $u_n = 10 - 3n$ Find u_2
Substituting $n = 2$, the second term $u_2 = 10 - (3 \times 2) = 4$

Example 3

For a given sequence $u_n = 2n^2 + 3$ Find the fifth term of the sequence.
Substituting $n = 5$, the fifth term $u_5 = (2 \times 5^2) + 3 = 53$

Exercise 14A

Find the first five terms of each of the following 14 sequences:

1   $u_n = 2n$ 2 $u_n = n + 2$ 3 $u_n = 3n + 7$
4   $u_n = 15 - 2n$ 5 $u_n = n^2 + 5$ 6 $u_n = 3n^2 - 1$
7   $u_n = (n + 1)^2$ 8 $u_n = n^3$ 9 $u_n = 2^n$
10 $u_n = (-2)^n$ 11 $u_n = (\frac{1}{2})^n$ 12 $u_n = \frac{1}{2}n(n + 1)$
13 $u_n = \frac{1}{n}$ 14 $u_n = \frac{n}{n + 1}$

15 $u_n = 4n - 7$.   Find the sixth term.

Finding which term has a given value

The number 105 is a term in the sequence given by the formula
$u_n = 7(n + 4)$

Which term in the sequence has this value?

You can find out by finding the value of n when $u_n = 105$.
Substituting $u_n = 105$ in the formula gives the equation:

$$7(n + 4) = 105$$
$$7n + 28 = 105$$
$$7n = 77$$
$$n = 11$$

So 105 is the 11th term of the sequence.

Example 4

A sequence is given by the formula $u_n = 2n^2 + 5$.
For which value of n does $u_n = 103$?

Substituting $u_n = 103$ in the formula gives the equation:

$$2n^2 + 5 = 103$$
$$2n^2 = 98$$
$$n^2 = 49$$

so $n = 7$

(Notice that $n = -7$ is also a solution of $n^2 = 49$, but for a sequence
n must be a positive integer. There is no -7th term in a sequence.)

Exercise 14B

Find the value of n for which u_n has the given value in these:

1   $u_n = 6n + 7$ $u_n = 55$ 2 $u_n = 7(n + 4)$ $u_n = 112$

3   $u_n = 7 - 3n$ $u_n = -44$ 4 $u_n = \frac{2n + 3}{3n + 1}$ $u_n = \frac{17}{22}$

5   $u_n = n^2 + 5$ $u_n = 149$ 6 $u_n = 3n^2 - 8$ $u_n = 235$
7   $u_n = (n - 8)^2$ $u_n = 169$ 8 $u_n = 2n$ $u_n = 256$
9   $u_n = 3n$ $u_n = 2187$ 10 $u_n = n(n - 1)$ $u_n = 380$

14.2 Finding the formula for a sequence

Arithmetic sequences

The first six terms of a sequence are:

$$9, \quad 14, \quad 19, \quad 24, \quad 29, \quad 34, \quad \ldots$$

By looking for a pattern in the sequence you can find a formula for the **nth term** u_n. This is sometimes called the **general term**.

A good way to start is to find the differences between successive terms:

$$\overset{+5}{9,} \quad \overset{+5}{14,} \quad \overset{+5}{19,} \quad \overset{+5}{24,} \quad \overset{+5}{29,} \quad 34, \quad \ldots$$

The terms go up in fives, so the nth term will include n lots of 5, or $5n$.

For the first term $n = 1$, so $5n = 5$. But the first term is 9, which is 4 more than $5n$. This suggests a formula of the form:

$$u_n = 5n + 4$$

Trying a few values of n gives:

$$u_1 = (5 \times 1) + 4 = 9$$
$$u_2 = (5 \times 2) + 4 = 14$$
$$u_3 = (5 \times 3) + 4 = 19$$
$$u_4 = (5 \times 4) + 4 = 24$$

So the formula $u_n = 5n + 4$ does describe the sequence.

> A sequence in which the differences between successive terms are equal is called an **arithmetic sequence**.

■ **The general term for an arithmetic sequence is of the form $u_n = an + b$. The value of a is the difference between successive terms in the sequence.**

Example 5

The first five terms of a sequence are:

$$5, \quad 3, \quad 1, \quad -1, \quad -3, \quad \ldots$$

Find a formula for the nth number in the sequence u_n.

As each term is 2 less than the previous one, the formula for u_n will include $-2n$.

But $u_n = -2n$ would give the sequence $-2, -4, -6, -8, \ldots$

For the sequence to start at 5, add 7 to each term, giving $5, 3, 1, -1, -3$

The formula that describes the sequence is $u_n = -2n + 7$

Quadratic sequences

The first six terms of a sequence are:

$$2, \quad 6, \quad 12, \quad 20, \quad 30, \quad 42, \quad \ldots$$

$+4 \quad +6 \quad +8 \quad +10 \quad +12 \qquad$ differences

$\quad +2 \quad +2 \quad +2 \quad +2 \qquad$ second differences

The differences between the terms are increasing so you find **second differences**.

The second difference in the sequence above is 2, so the value of a is 1.

■ **A sequence in which the second differences are equal is called a quadratic sequence. The general term of a quadratic sequence is of the form $u_n = an^2 + bn + c$. The value of a is half the second difference.**

To find the value of c you need to work out the **zeroth term** of the sequence.

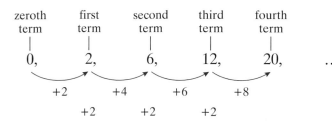

zeroth term	first term	second term	third term	fourth term
0,	2,	6,	12,	20, ...

$+2 \quad +4 \quad +6 \quad +8$

$+2 \quad +2 \quad +2$

Counting back, the zeroth term is 0. So when $n = 0$, $u_n = 0$. You know already that the formula looks like:

$$u_n = n^2 + bn + c$$

When $n = 0$, both n^2 and bn are equal to 0.
So $c = 0$.

You now know that the general term of the sequence is:

$$u_n = n^2 + bn$$

To find the value of b, look at the first term. When $n = 1$, the value of u_n is 2.

But
$$u_n = n^2 + bn$$
$$= 1 + b$$

So $\quad\quad 2 = 1 + b$

So $\quad\quad b = 1$.

The formula for the general term of the sequence is

$$u_n = n^2 + n$$

The quadratic sequence given by $u_n = 3n^2 + 1$ looks like this:

$$4, \quad 13, \quad 28, \quad 49, \quad ...$$

$+9 \quad +15 \quad +21$
$+6 \quad +6$

The second difference is 6 and the value of a is $6 \div 2 = 3$

If $n = 0$ then $bn = 0$ no matter what the value of b is.

Exercise 14C

Find formulae for u_n to describe each of these sequences.
Your answers to Exercise 14A may be helpful.

1 3, 4, 5, 6, 7, ...

2 5, 7, 9, 11, 13, ...

3 7, 11, 15, 19, 23, ...

4 9, 8, 7, 6, 5, ...

5 12, 10, 8, 6, 4, ...

6 7, 4, 1, –2, –5, ...

7 $\frac{1}{2}, \frac{1}{3}, \frac{1}{4}, \ldots$ **8** $2, 1\frac{1}{2}, 1\frac{1}{3}, 1\frac{1}{4}, \ldots$

9 $1, 4, 9, 16, 25, \ldots$ **10** $2, 5, 10, 17, 26, \ldots$

11 $2, 8, 18, 32, 50, \ldots$ **12** $9, 16, 25, 36, \ldots$

13 $4, 1, 0, 1, 4, 9, 16 \ldots$ **14** $5, 25, 125, 625, \ldots$

15 $10, 100, 1000, \ldots$ **16** $-3, 9, -27, 81, \ldots$

17 $0.1, 0.01, 0.001, \ldots$ **18** $1, \frac{1}{3}, \frac{1}{5}, \frac{1}{7}, \frac{1}{9}, \ldots$

19 $1, \frac{2}{3}, \frac{3}{5}, \frac{4}{7}, \frac{5}{9}, \ldots$ **20** $\frac{1}{4}, \frac{1}{9}, \frac{1}{16}, \frac{1}{25}, \ldots$

> **Hint:** What happens to the sequence
> $$u_n = (-1)^n$$

14.3 More about formulae

As well as describing sequences, formulae can also be used to describe other relationships between quantities.

■ **An algebraic formula can be used to describe a relationship between two sets of numbers.**

For example, the formula $C = \pi d$ describes the relationship between the diameter of a circle d and its circumference C.

Notice the difference between a formula and an equation:

$C = \pi \times d$ is a **formula** $C = \pi \times 2.5$ is an **equation**

If you substitute *any* value for d the formula will give you a corresponding value for C. Similarly if you substitute any value for C it will give you a corresponding value for d.

It is only true for a particular value of C ($C = 7.85$ to 2 d.p.)

You need to be able to evaluate formulae by substituting given values into them, including negative numbers and fractions.

Example 9

$$s = pt^2 + q$$

Calculate the value of s when:

(a) $p = 1, q = 2$ and $t = \frac{1}{2}$

(b) $p = 5, q = -1$ and $t = -2$

Answer

(a) $s = (\frac{1}{2})^2 + 2$

      $= \frac{1}{4} + 2$

      $= 2\frac{1}{4}$

(b) $s = 5 \times (-2)^2 - 1$

      $= 5 \times 4 - 1$

      $= 20 - 1$

      $= 19$

Worked examination question

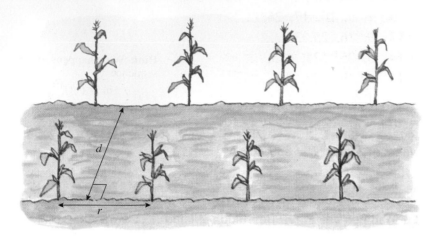

The diagram is taken from a book about growing maize. The distance between the rows of plants is d metres. The spacing between the plants in the rows is r metres.

The number, P, of plants per hectare is given by the formula:

$$P = \frac{10\,000}{dr}$$

$d = 0.8$ and $r = 0.45$
Calculate the value of P.
Give your answer correct to 2 significant figures.

$$P = \frac{10\,000}{0.8 \times 0.45}$$

$$= 27\,777.777$$

$$= 28\,000 \text{ (correct to 2 s.f.)} \qquad [E]$$

Exercise 14D

1   $y = 2x - 5$ Calculate the value of y when $x = -3$

2   $A = LB$ Calculate the value of A when $L = 8\frac{1}{2}$ and $B = 6\frac{1}{4}$

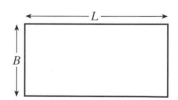

3   $P = 2(L + B)$ Calculate the value of P when $L = 8.3$ and $B = 6.8$

4   $C = \pi d$ Calculate correct to 3 s.f. the value of C when $d = 8.37$

5   $y = 3x^2 + 2$ Calculate the value of y when $x = -1$

6   $A = \pi r^2$ Calculate correct to 3 s.f. the value of A when $r = 3.9$

7   $A = 2\pi r(r + h)$ Calculate correct to 3 s.f. the value of A when $r = 3.7$ and $h = 5.9$

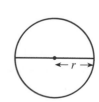

8 $s = \frac{1}{2}gt^2$ — Calculate the value of s when $g = -10$ and $t = 3$

9 $x = L(1 + at)$ Calculate the value of x when $L = 100$, $a = 0.000\ 019$ and $t = 80$

10 $v = u + at$ — Calculate the value of v when $u = 5$, $a = -8$ and $t = 6$

11 $V = \frac{1}{3}\pi r^2 h$ — Calculate correct to 3 s.f. the value of V when $r = 2.6$ and $h = 6.7$

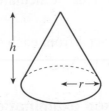

12 $I = \dfrac{PRT}{100}$ — Calculate the value of I when $P = 850$, $R = 5\frac{3}{4}$ and $T = 4$

13 $s = ut + \frac{1}{2}at^2$ Calculate the value of s when $u = -3$, $t = 5$ and $a = -4$

14 $D = v + \dfrac{v^2}{20}$ — Calculate correct to 2 s.f. the value of D when $v = 31.6$

15 $c = \sqrt{a^2 + b^2}$ — Calculate correct to 3 s.f. the value of c when $a = 3.7$ and $b = 8.3$

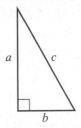

16 $f = \dfrac{uv}{u + v}$ — Calculate the value of f when $u = 3.6$ and $v = -5.4$

17 $r = \sqrt{\dfrac{A}{\pi}}$ — Calculate correct to 3 s.f. the value of r when $A = 37.2$

18 $r = \sqrt{\dfrac{A}{4\pi}}$ — Calculate correct to 3 s.f. the value of r when $A = 21.3$

19 $T = 2\pi\sqrt{\dfrac{L}{g}}$ — Calculate correct to 3 s.f. the value of T when $L = 23.4$ and $g = 9.8$

20 $T = \dfrac{2Mmg}{M + m}$ — Calculate correct to 2 s.f. the value of T when $M = 3.8$, $m = 1.7$ and $g = 9.8$

14.4 Manipulating formulae

Sometimes you will need to rearrange a formula to find a value. Methods of rearranging formulae are introduced in Unit 2 (page 18). The rest of this unit provides you with more practice in manipulating formulae.

Example 10

$C = \pi d$ — Calculate the value of d when $C = 23.2$ giving your answer correct to 3 s.f.

$$23.2 = \pi d$$

Dividing each side by π gives:

$$d = \frac{23.2}{\pi} = 7.38 \text{ (to 3 s.f.)}$$

Example 11

$v = u + at$ Calculate the value of t when $v = 39$, $u = 4$ and $a = 3$

$$39 = 4 + 3t$$

subtract 4 from each side: $35 = 3t$

divide by 3: $t = 11\frac{2}{3}$

Worked examination question

The air temperature $T\,°C$ outside an aircraft flying at a height of h
feet is given by the formula: [E]

$$T = 26 - \frac{h}{500}$$

The air temperature outside an aircraft is $-52\,°C$. Calculate the
height (h) of the aircraft.

$$-52 = 26 - \frac{h}{500}$$

add $\dfrac{h}{500}$ to each side: $\dfrac{h}{500} - 52 = 26$

add 52 to each side: $\dfrac{h}{500} = 78$

multiply by 500: $h = 39\,000$

The height of the aircraft is $39\,000$ feet.

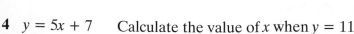

Exercise 14E

1 $C = \pi d$ Calculate correct to 2 s.f. the value of d when
$C = 56.8$

2 $A = LB$ Calculate the value of L when $A = 32.94$
and $B = 6.1$

3 $C = 2\pi rh$ Calculate correct to 2 s.f. the value of h, when
$C = 800$ and $r = 5.9$

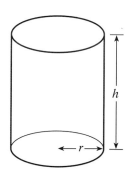

Note: $C = 2\pi rh$
is the formula for the curved
surface area of a cylinder.

4 $y = 5x + 7$ Calculate the value of x when $y = 11$

5 $v = u + at$ Calculate the value of a when $v = 17$, $u = -3$
and $t = 5$

6 $I = \dfrac{PRT}{100}$ Calculate the value of R when $I = 45$, $P = 250$
and $T = 4$

7 $y = 8 + \frac{1}{4}x$ Calculate the value of x when $y = 6$

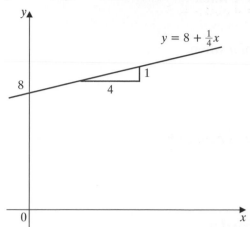

Note: $y = 8 + \frac{1}{4}x$
is the formula – or equation
– of a straight line of
gradient $\frac{1}{4}$ and intercept on
the y-axis of $(0, 8)$.

8 $y = 9 - 4x$ Calculate the value of x when $y = 14$

9 $y = 5 - \frac{1}{2}x$ Calculate the value of x when $y = 2$

10 $P = 2(L + B)$ Calculate the value of B when $P = 42$ and $L = 13\frac{1}{2}$

11 $S = \frac{1}{2}n(a + d)$ Calculate the value of a when $S = 72, d = 3$ and $n = 18$

12 $A = 2\pi r(r + h)$ Calculate correct to 3 s.f. the value of h when $A = 540$ and $r = 4.9$

13 $y = 2x^2$ Calculate the two possible values of x when $y = 98$

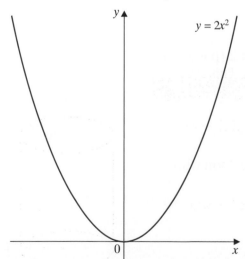

14 $A = \pi r^2$ Calculate correct to 3 s.f. the positive value of r when $A = 50$

15 $s = ut + \frac{1}{2}at^2$ Calculate the value of u when $s = 121, t = 5$ and $a = 8$

16 $c = \sqrt{a^2 + b^2}$ Calculate the positive value of a when $c = 29.9$ and $b = 11.5$

17 $r = \sqrt{\dfrac{A}{\pi}}$ Calculate correct to 3 s.f. the value of A when $r = 8$

18 $V = \frac{1}{3}\pi r^2 h$ Calculate correct to 3 s.f. the value of h when $V = 100$ and $r = 2.5$

19 $y = \dfrac{20}{x + 1}$ Calculate the value of x when $y = 8$

20 $f = \dfrac{uv}{u + v}$ Calculate the value of v when $f = 12$ and $u = -8$.

14.5 Changing the subject of a formula

In the formula

$$C = \pi d$$

C is called the subject of the formula. It appears on its own on one side of the formula. This is a convenient form for the formula when you want to calculate values of C from given values of d.

If instead you want to calculate values of d from given values of C it is more convenient to rearrange the formula so that d is the subject and appears on its own on one side of the formula. You can do this by dividing both sides of the formula by π giving:

$$d = \dfrac{C}{\pi}$$

This is called **changing the subject of the formula**. Unit 2 (page 18) and unit 10 introduce methods of changing the subject. This section provides further practice.

Worked examination question

The formula below gives the Body Mass Index, I, of a person who is h metres tall and weighs W kg.

$$I = \dfrac{W}{h^2}$$

Make h the subject of the formula.

multiply by h^2: $Ih^2 = W$

divide by I: $h^2 = \dfrac{W}{I}$

find square root: $h = \sqrt{\dfrac{W}{I}}$

h is now on its own on one side – it is the subject of the formula.

Worked examination question

The approximate range, R miles, of a radar mounted x feet above sea level is given by the formula:

$$R = \sqrt{2x}$$

Make x the subject of the formula.

square both sides: $\qquad R^2 = 2x$

divide by 2: $\qquad \frac{1}{2}R^2 = x$

x is on its own on one side. The formula can be written as $x = \frac{1}{2}R^2$.

Exercise 14F

In questions **1–30** make the letter in square brackets the subject of the formula:

1 $V = IR$ $\qquad [I]$ $\qquad$ **2** $A = LB$ $\qquad [L]$

3 $C = \pi d$ $\qquad [d]$ $\qquad$ **4** $V = LBH$ $\qquad [B]$

5 $v = u + at$ $\qquad [u]$ $\qquad$ **6** $v = u + at$ $\qquad [t]$

7 $C = 2\pi rh$ $\qquad [h]$ $\qquad$ **8** $I = \dfrac{PRT}{100}$ $\qquad [R]$

9 $P = 2(L + B)$ $\quad [L]$ $\qquad$ **10** $A = 2\pi r(r + h)$ $[h]$

11 $\dfrac{y}{x} = 3$ $\qquad [x]$ $\qquad$ **12** $\dfrac{PV}{T} = k$ $\qquad [T]$

13 $y = 5x^2$ $\qquad [x]$ $\qquad$ **14** $E = kx^2$ $\qquad [x]$

15 $A = \pi r^2$ $\qquad [r]$ $\qquad$ **16** $S = \frac{1}{2}n(a + d)$ $\;[n]$

17 $S = \frac{1}{2}n(a + d)$ $[a]$ $\qquad$ **18** $y = kx^2$ $\qquad [x]$

19 $V = \frac{1}{3}\pi r^2 h$ $\qquad [h]$ $\qquad$ **20** $V = \frac{1}{3}\pi r^2 h$ $\qquad [r]$

21 $c = \sqrt{a^2 + b^2}$ $\;[a]$ $\qquad$ **22** $a = 2\sqrt{b^2 - 2}$ $\;[b]$

23 $t = \sqrt{\dfrac{2s}{g}}$ $\qquad [s]$ $\qquad$ **24** $T = 2\pi\sqrt{\dfrac{L}{g}}$ $\;[L]$

25 $x^2 + y^2 = 1$ $\qquad [y]$ $\qquad$ **26** $v^2 = u^2 + 2as$ $\;[u]$

27 $v = w\sqrt{a^2 - x^2}$ $\;[x]$ $\qquad$ **28** $y = 3\sqrt{1 - x^2}$ $\quad [x]$

29 $y = \dfrac{5 + x}{1 - x}$ $\qquad [x]$ $\qquad$ **30** $f = \dfrac{uv}{u + v}$ $\qquad [u]$

14.6 Substituting in one formula for another

You can sometimes eliminate one of the letters from a formula by substituting an expression from a second formula.

Example 12

Two formulae used when designing electrical circuits are $P = IV$ and $V = IR$.

Find a formula for P in terms of I and R.

In $P = IV$, replace V by IR.

$$P = I \times IR$$
$$P = I^2R$$

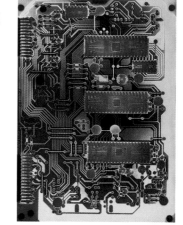

Example 13

The length of a rectangle is l and its area is A.
Find a formula for its perimeter, P, in terms of l and A.

Let b represent the breadth of the rectangle.

$$P = 2l + 2b \qquad (1)$$
$$A = lb \qquad (2)$$

Make b the subject of formula (2)

$$b = \frac{A}{l}$$

In formula (1), replace b by $\dfrac{A}{l}$.

$$P = 2l + \frac{2A}{l}$$

Exercise 14G

1   $P = IV$ and $V = IR$.
    Find a formula for P in terms of V and R.

2   $s = vt$ and $v = \frac{1}{2}gt$.
    Find a formula for s in terms of g and t.

3   $A = \pi r^2$ and $r = \dfrac{d}{2}$.

    Find a formula for A in terms of d.

4   $a = \dfrac{v^2}{r}$ and $v = rw$.

    Find a formula for a in terms of r and w.

5 $y = 4t + 3$ and $x = 2t - 1$.
Find a formula for y in terms of x.

6 The area of a square is A and its perimeter is P.

    (a) Find a formula A in terms of P.

    (b) Find a formula for P in terms of A.

> **Hint:** Let x represent the length of the square's sides and express A and P in terms of x.

7 $x = at^2$ and $y = 2at$.
Find a formula for x in terms of y.

8 $A = 2\pi r^2 + 2\pi rh$ and $V = \pi r^2 h$.
Find a formula for A in terms of r and V.

9

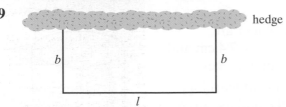

A farmer encloses sheep in a rectangular pen using fencing for three sides and a long hedge for the other side. The length of the pen is l and its breadth is b. The total length of fencing used is F and the area enclosed is A. Find a formula for F in terms of A and l.

10

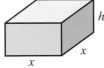

The diagram shows a cuboid with a square base. The length of each side of its base is x and its height is h. The surface area of a cuboid is A and its volume is V. Find a formula for A in terms of V and x.

Exercise 14H   (Mixed questions)

1 A formula used in science is:
$$s = \tfrac{1}{2}gt^2$$
Tracy has to work out the value of s when $g = 9.81$ and $t = 23.67$
Before using her calculator to work out the value of s, Tracy decides to obtain an estimate for s.

    (a)   **(i)** Write down values for g and t which Tracy could use in her calculation for an estimate of s.

        (ii) Calculate the estimate of s that these values would give.

    (b) Use your calculator to work out the actual value of s, giving your answer correct to two decimal places.

2 An estate agent quotes the dimensions of a rectangular floor of a room as 4.3 metres by 3.7 metres, both correct to 10 cm.

    (a) Write down the maximum and minimum values for each of these dimensions.

    (b) Calculate the maximum and minimum values for the area of the floor, giving your answers correct to three significant figures in each case.

3 The formula for the volume V of a cylinder with circular base of radius r and vertical height h is:

There is more about volume formulae on page 391.

$$V = \pi r^2 h$$

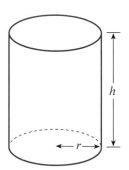

The radius of the base and vertical height of a cylinder are measured to the nearest millimetre and quoted as $r = 7.4$ cm and $h = 10.8$ cm, whilst $\pi = 3.1416$, correct to 4 decimal places.

    (a) Without using a calculator and showing all your working, quote sensible approximations for π, r and h and use these to find an estimate for the volume of the cylinder.

    (b) Use your calculator and the quoted values of π, r and h to calculate the volume of the cylinder, giving your answer in cubic centimetres, correct to three significant figures.

    (c) Keeping π as the quoted figure, find the maximum volume of the cylinder, giving your answer correct to two decimal places.

4 In Florida the speed limit for motor vehicles is 55 miles per hour. The 'on the spot' fine for motorists caught exceeding this limit is 4 dollars for each mile per hour over 55 miles per hour.

    (a) Write down a formula for f, the 'on the spot' fine in dollars, for a motorist caught travelling at v miles per hour. You may assume that $v > 55$.

    (b) Make v the subject of the formula.   [E]

Summary of key points

1 When you know a formula for the nth term of a sequence u_n you can calculate any term in the sequence by substituting a value for n in the formula. n must be a positive integer ($n = 1, 2, 3 \ldots$).

2 The general term for an arithmetic sequence is of the form $u_n = an + b$. The value of a is the difference between successive terms in the sequence.

3 A sequence in which the second differences are equal is called a quadratic sequence. The general term of a quadratic sequence is of the form $u_n = an^2 + bn + c$. The value of a is half the second difference.

4 An algebraic formula can be used to describe a relationship between two sets of numbers.

15 Averages and measures of spread

When you have a lot of data you can find a typical or representative value called an average.

This unit shows you how to calculate averages and measures of the spread of data and draw conclusions from them.

15.1 Averages for discrete data

The mean
The mean for discrete data is defined as:

■ **The mean of a set of data is the sum of the values divided by the number of values.**

$$\text{mean} = \frac{\text{sum of values}}{\text{number of values}}$$

This can also be written using the summation sign Σ:

$$\bar{x} = \frac{\Sigma x}{n}$$

where Σx is the sum of the values and n is the number of values.

The median
■ **The median is the middle value when the data is arranged in order of size.**

If there is an even number of values in the data then the median is the mean of the middle two values.

The mode
■ **The mode of a set of data is the value which occurs most often.**

There can be more than one mode for a set of data.

Example 1

Scott kept a record of the number of points he had scored in his school basketball games.

15   8   7   11   9   5   6   9

Calculate the:

(a) mean　　　　　**(b) median**　　　　**(c) mode**

Answer

(a) Calculate the **mean** by adding all the scores and dividing by the number of games.

$$\text{mean} = \frac{15 + 8 + 7 + 11 + 9 + 5 + 6 + 9}{8} = \frac{70}{8} = 8.75$$

(b) Calculate the **median** by putting the scores in order of increasing size and then selecting either the middle value or the mean of the two middle values.

5   6   7   8   9   9   11   15

$$\text{median} = \frac{8 + 9}{2} = 8.5$$

If in his next game Scott scored 6 points the modes of all the scores would then be 6 and 9.

(c) Calculate the **mode** by selecting the score that appears most often.

mode = 9

15.2 Using appropriate averages

The three different types of average are useful in different situations. The choice of which average you use or quote can be crucial; this is especially true when you do statistical-based coursework.

The **mean** is very useful when you need to quote a 'typical' value, provided that the data is quite closely grouped around the mean.

The mean can, however, be seriously influenced by extreme values. For example, consider the situation below.

> Bob is the Chairman of a small company which employs 10 people, including himself. The salaries of the 10 people are:
>
> | | |
> |---|---|
> | Chairman | £80 000 per year |
> | 7 people earning | £20 000 per year |
> | 2 people earning | £12 000 per year. |

The mean of these salaries is:

$$\frac{80\,000 + 7 \times 20\,000 + 2 \times 12\,000}{10}$$

or

$$\frac{80\,000 + 140\,000 + 24\,000}{10}$$

$$= £24\,400$$

It would be silly to say that a 'typical' salary is £24 400 when in fact nobody at the company earns that amount and only one person earns above, and considerably above, that amount.

In a situation such as this, it would be far more sensible to use the **median** salary as the 'typical' value, i.e. £20 000.

The **mode** is useful in situations such as when you want to know:

Which dress size is most common?
Which brand of dog food is most popular?

The mode is an average which shops use when they calculate the stock they need.

■ The **median and mode** are not influenced by extreme values, but the **mean** is influenced by extremes.

Example 2

There are 10 houses in Streetfield Close.

On Monday, the number of letters delivered to each of the houses is:

$$0, 2, 5, 3, 34, 4, 0, 1, 0, 2$$

Calculate the mean, mode and median of the number of letters.

Comment on your results.

$$\text{mean} = \frac{0 + 2 + 5 + 3 + 34 + 4 + 0 + 1 + 0 + 2}{10}$$

$$= 5.1$$

$$\text{mode} = 0$$

$$\text{median} = 2$$

In this case the mean has been distorted by the large number of letters delivered to one of the houses. It is, therefore, not a good measure of a 'typical' number of letters delivered to any house in the Close.

The mode is also not a good measure of a 'typical' number of letters delivered to a house, since 7 out of the 10 houses do actually receive some letters and the mode is 0.

The median is perhaps the best measure of the 'typical' number of letters delivered to each house, since half of the houses received 2 or more letters and the other half received 2 or fewer letters.

Exercise 15A

1   Ten pupils submitted their English coursework which was marked out of 40.

The marks they obtained were:

$$37, 34, 34, 34, 29, 27, 27, 10, 4, 28$$

(a) For these scores find:
(i) the mean (ii) the mode (iii) the median

(b) Comment on your results.

An external moderator reduced all the marks by 3.

(c) Find the mean, mode and median of the moderated marks.

2 The Matthews family has six members living.
   These people and their ages are:

Grandma Matthews	79
Mr Matthews	43
Mrs Matthews	41
Sarah	16
Peter	12
Lucy	3

   (a) Calculate the mean age of the living members of the
       Matthews family.

   (b) State, with a reason, whether or not the mean age is a
       sensible measure of the age of a 'typical' member of the
       Matthews family.

3 The diagram represents a spinner in the shape of a regular
   hexagon.

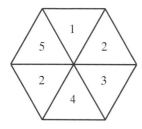

   In an experiment the spinner is spun once and the number it
   lands on is recorded.

   (a) What will be the theoretical mean of the recorded
       numbers?

   (b) Explain your result.

   (c) Explain why the mean number would be a silly measure of
       the 'typical' recorded number.
       The experiment is repeated several times.

   (d) What will be the mode of the recorded numbers?
       Give a reason for your answer.

4 A newspaper headline reads:
       'The average weekly wage in the UK is in excess of £200'
   Explain why this headline could be misleading.

5 Fatima wishes to conduct a survey on the amount of money a
   typical family spends on food each week.
   In her survey she is going to make some quotes about
   'averages'.
   State, with reasons, which 'averages' she should use and how.

Moving averages

This graph shows the variation of job vacancies in a district over
3 years.

■ **A graph showing how a given value changes over time is called
a time series.**

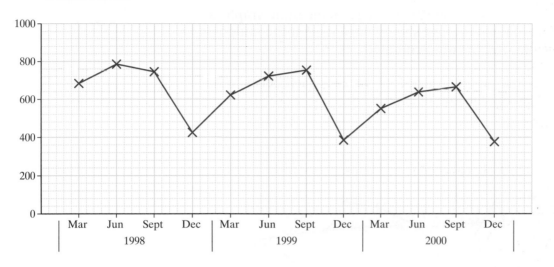

It is difficult from the graph to see whether the number of job
vacancies has risen or fallen over the three years. There is too much
variation within each year.

You can account for this variation by working out a **moving average**.
This takes the average of consecutive pieces of data.

Here is the table of results for the above graph.

Year	March	June	September	December
1998	682	785	742	423
1999	622	722	752	385
2000	550	639	661	376

There are four values for each year, so look at groups of four.

The first moving average, M_1, is the mean of the first four values.

If you consider groups of
four you are finding a **four-
point moving average**.

$$M_1 = \frac{682 + 785 + 742 + 423}{4} = 658$$

The next moving average, M_2, is found by 'moving up one place'. It
is the mean of the values from June 1998 to March 1999:

$$M_2 = \frac{785 + 742 + 423 + 622}{4} = 643$$

The four-point moving average for the whole data set is:

M_1	M_2	M_3	M_4	M_5	M_6	M_7	M_8	M_9
658	643	627	630	620	602	582	559	557

You can plot the four-point moving average on the same graph as your original data. Each average should be plotted at the midpoint of the values from which it was generated.

You can now see that the number of job vacancies fell between the beginning of 1998 and the end of 2000.

A moving average has the effect of 'smoothing out' the data.

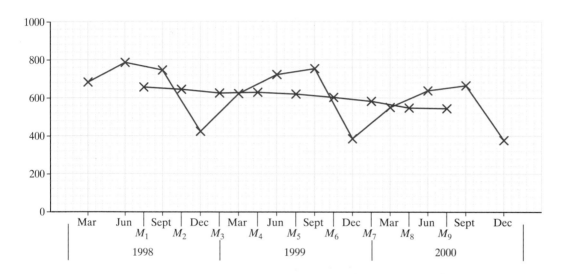

Exercise 15B

1   This table shows the number of students in a school getting at least a grade C in mathematics for the years 1994 to 2001.

The three-point moving average means you need to find the mean for successive groups of three values.

   (a) Represent this data as a time series.

1994	1995	1996	1997	1998	1999	2000	2001
97	118	115	117	121	125	111	125

   (b) Calculate the three-point moving average and plot it on the same graph.

   (c) Are the school's maths results improving?

   (d) Explain why this is not a good way to work out whether the school's results are improving.

2   The takings of a theatre were recorded every quarter for 4 years. The results are shown in this table.

	1997	1998	1999	2000
1st quarter	180 000	200 000	210 000	225 000
2nd quarter	145 000	178 000	195 000	210 000
3rd quarter	135 000	150 000	180 000	198 000
4th quarter	190 000	183 000	192 000	205 000

(a) Work out the four-point moving average for the data.

(b) Plot the original data and the moving average on the same graph.

(c) Comment on how the theatre's takings have changed over the four years.

3   The number of issues of books from a library are given for the years 1991–2001. The numbers are in hundreds of thousands.

1991	1992	1993	1994	1995	1996	1997	1998	1999	2000	2001
125	131	122	118	125	129	132	135	128	123	120

(a) Plot the data and the five-point moving average on the same graph.

(b) Comment on your graph.

15.3 Measures of average for grouped discrete data

The mean

After a science investigation this frequency table was drawn up to show the number of snails per square metre and their frequency.

Number of snails per m^2 (x)	Frequency (f)
0	5
1	6
2	8
3	7
4	0
5	4

This shows that on five occasions zero snails were present, on six occasions one snail was present, and so on.

To calculate the mean you need to add another column to the table showing the frequency $f \times$ number of snails x:

Number of snails (x)	Frequency (f)	$f \times x = fx$
0	5	0 (5 × 0)
1	6	6 (6 × 1)
2	8	16 (8 × 2)
3	7	21 (7 × 3)
4	0	0 (0 × 4)
5	4	20 (4 × 5)
	30 = Σf	63 = Σfx

$$\bar{x} = \frac{\text{sum of } f \times x \text{ column}}{\text{sum of } f \text{ column}} = \frac{\Sigma fx}{\Sigma f} = \frac{63}{30} = 2.1$$

■ **The mean for grouped discrete data is:**

$$\bar{x} = \frac{\Sigma fx}{\Sigma f}$$

The median

In the science investigation the total frequency in the table is $\Sigma f = 30$.

The median value is the middle value.

There are 30 values so the median is the mean of the 15th and 16th values when they are put in order. These occur in the third row of the table above.

So the median is 2.

The mode

The mode is the number of snails with the greatest frequency.

In the table above the greatest frequency is 8 so the mode is 2.

Exercise 15C

1   Find the mean, median and mode of these sets of data:
  (a) 14, 12, 24, 36, 23
  (b) 114, 112, 124, 136, 123
  (c) 2, 3, 4, 5, 30
  (d) $x, 2x, 3x, 4x, 5x$

2   (a) The mean of 5 numbers is 8. Four of the numbers are 7, 9, 11 and 5. Find the fifth number.
  (b) The mean of 4, 8, 9, x and $2x$ is 6. Calculate the value of x.

3 The wage structure of a firm is:

Position	Number of employees	Wage (£)
Manager	1	60 000
Team leader	4	25 000
Operative	25	12 000

(a) Calculate the mean, median and mode of this data.

(b) All the employees in the factory are given a 2.5% pay rise. Calculate the new mean wage.

4 In a class 10 students did a mental arithmetic test and had a mean score of 22.6. The remaining 15 students in the class did the same test later and had a mean of 19.6.

(a) Calculate the mean test score of the whole class.

The median score of the first 10 students was 21 and of the remaining fifteen was 17.

(b) What can you say about the median of the whole class?

5 Five squares have sides of length 1 cm, 2 cm, 3 cm, 4 cm and 5 cm respectively.

(a) Calculate the mean length of a square.
Hence, or otherwise, calculate:

(b) (i) the mean perimeter   (ii) the mean area.

6 Jenna kept a record of the number of goals her hockey team had scored.

Number of goals	Frequency
0	9
1	7
2	8
3	5
4	3
5	1

Calculate the mean, median and mode.

7 The vertical line diagram shows the number of A* to C grades that were obtained in Year 11 of a small school.

(a) Write down the mode.

(b) Calculate the mean and median.

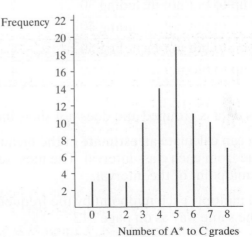

8 The number of goals scored during one weekend in football matches in the league is shown in this table.

Number of goals	Number of games
0	4
1	6
2	11
3	4
4	6
5	3
6	1
7	1

(a) Calculate the mean, median and modal number of goals per game.

(b) Is it true that 2 × team mean = game mean? Justify your answer.

15.4 The mean and the modal class for grouped continuous data

You can also calculate an estimate of the mean for grouped continuous data.

This table gives the times between successive vehicles passing a house:

Class interval (seconds)	Frequency
0 up to but not including 10	12
10 up to but not including 20	14
20 up to but not including 30	7
30 up to but not including 40	8
40 up to but not including 50	5
50 up to but not including 60	4

The class interval with the greatest frequency is called the **modal class**.

This data is grouped and does not show individual data values.

You can calculate an **estimate** of the mean by using a representative value from each class interval. The most sensible value to choose is the midpoint of the interval.

You multiply each midpoint by the frequency for that class interval in the same way as on page 312.

■ The estimate of the mean for grouped continuous data is:

$$\bar{x} = \frac{\Sigma fx}{\Sigma f}$$

where x is the midpoint of the class interval.

Class interval (seconds)	Frequency	Midpoint x	fx
0 up to but not including 10	12	5.0	60
10 up to but not including 20	14	15.0	210
20 up to but not including 30	7	25.0	175
30 up to but not including 40	8	35.0	280
40 up to but not including 50	5	45.0	225
50 up to but not including 60	4	55.0	220
	$\Sigma f = 50$		$\Sigma fx = 1170$

The midpoint of the interval from 0 to 10 is:
$$\frac{0 + 10}{2} = 5$$

Now: $\bar{x} = \dfrac{\Sigma fx}{\Sigma f} = \dfrac{1170}{50} = 23.4$

but this value is an **estimate**.

A slightly different value for the estimate of the mean can be obtained if the data is grouped as 0–9, 10–19, 20–29, and so on. It is now assumed that the 10–19 interval runs from 9.5 to 19.5 with a midpoint of 14.5.

The 0–9 interval really means from 0 up to 9.5 so the midpoint of this interval is:

$$\frac{0 + 9.5}{2} = 4.75$$

Then $\Sigma fx = 1138.25$ so the estimate of the mean is:

$$\frac{1138.25}{50} = 22.8$$

Example 3

A saleswoman kept a record of the mileage she had travelled.
Calculate an estimate for the mean distance travelled from the data
in this table.

Mileage m (miles)	Frequency
$0 \leqslant m < 50$	2
$50 \leqslant m < 100$	8
$100 \leqslant m < 150$	17
$150 \leqslant m < 200$	21
$200 \leqslant m \leqslant 300$	2

First draw up a table of frequencies and midpoints:

Mileage	f	Midpoint x	fx
$0 \leqslant m < 50$	2	25	50
$50 \leqslant m < 100$	8	75	600
$100 \leqslant m < 150$	17	125	2125
$150 \leqslant m < 200$	21	175	3675
$200 \leqslant m \leqslant 300$	2	250	500
	$\Sigma f = 50$		$\Sigma fx = 6950$

The estimate of the mean $\bar{x}$ is:

$$\bar{x} = \frac{\Sigma fx}{\Sigma f}$$

$$= \frac{6950}{50} = 139$$

Exercise 15D

1 Calculate an estimate of the mean for the data in the table below.

Class interval (seconds)	Frequency
0 up to but not including 20	23
20 up to but not including 40	16
40 up to but not including 60	12
60 up to but not including 80	9
80 up to but not including 100	5
100 up to but not including 120	3
120 up to but not including 140	1

2   Most eggs are sold in one of five sizes based on weight, correct to the nearest gram. The table shows the production of eggs on a farm.

Size	Weight (g)	Frequency
1	70 or more	120
2	65 to 69	245
3	60 to 64	318
4	55 to 59	189
5	50 to 54	78

Assuming that the size 1 interval is the same width as the others, calculate an estimate of the mean weight. Suppose that the size 1 interval is twice the width of the other sizes. Calculate an estimate for the mean weight in that case.

3   The lengths of leaves are given in the table below.

Length (cm)	Frequency
0 up to but not including 1	20
1 up to but not including 2	24
2 up to but not including 3	27
3 up to but not including 4	15
4 up to but not including 5	8

Calculate an estimate of the mean length.

4   Ian looked at a passage from a book. He recorded the number of words in each sentence in a frequency table using class intervals of 1–5, 6–10, 11–15, etc.

Class interval	Frequency f	Class interval	Frequency f
1–5	16	26–30	3
6–10	28	31–35	1
11–15	26	36–40	0
16–20	14	41–45	2
21–25	10		

(a)   Write down:
    (i) the modal class interval
    (ii) the class interval in which the median lies.
(b)   Work out an estimate of the mean number of words in a sentence.   [E]

15.5 Using cumulative frequency graphs

A cumulative frequency graph can be used to estimate some useful statistical measures. (There is more about cumulative frequency graphs on page 80.)

Estimating the median

A set of data arranged in class intervals is called a **distribution**.

■ **The median is the middle value of the distribution.**

To estimate the median from a cumulative frequency graph:

Strictly speaking, the middle value of this distribution is the
$$\frac{200 + 1}{2} = 100\tfrac{1}{2}\text{th value.}$$
For an estimate from this graph it is accurate enough to read off the value when the cumulative frequency is 100.

For simplicity the cumulative frequency here to estimate the median is 200. Follow the same procedure for any cumulative frequency.

Estimating the quartiles

■ **The lower quartile is the value one-quarter of the way into the distribution.**

To estimate the lower quartile from a cumulative frequency graph:

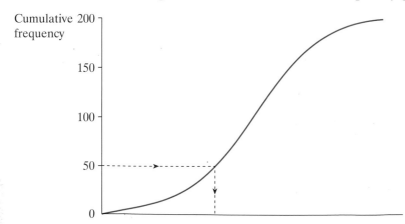

■ **The upper quartile is the value three-quarters of the way into the distribution.**

To estimate the upper quartile use the same method as for the lower quartile but start at the three-quarter way value of the cumulative frequencies. Here it would be the $150\frac{3}{4}$th value so you could use the 151st value for an estimate.

The interquartile range

To estimate the interquartile range (IQR) from a cumulative frequency diagram first estimate the upper and lower quartiles, then use:

■ **interquartile range = upper quartile – lower quartile**

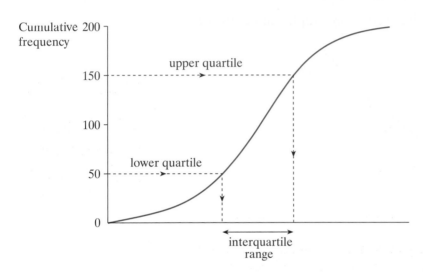

Example 4

This table shows the time in seconds between successive vehicles passing a school. Calculate:

(a) the median

(b)   (i) the upper quartile
     (ii) the lower quartile
     (iii) the interquartile range.

Time in seconds	Frequency
$0 < x \leqslant 10$	12
$10 < x \leqslant 20$	14
$20 < x \leqslant 30$	7
$30 < x \leqslant 40$	8
$40 < x \leqslant 50$	5
$50 < x \leqslant 60$	4

(a) Here is the cumulative frequency table:

Time in seconds	Cumulative frequency
$0 < x \leqslant 10$	12
$0 < x \leqslant 20$	26
$0 < x \leqslant 30$	33
$0 < x \leqslant 40$	41
$0 < x \leqslant 50$	46
$0 < x \leqslant 60$	50

From this table the cumulative frequency graph can be drawn. The median is the middle value. To find it:

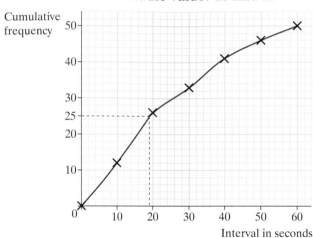

The median is 19 seconds.

(b) **(i)** To find the upper quartile find the value which corresponds to the three-quarter way point on the cumulative frequency axis and draw a horizontal line across to the curve.

From this point draw a line vertically down and read off the upper quartile on the horizontal axis.

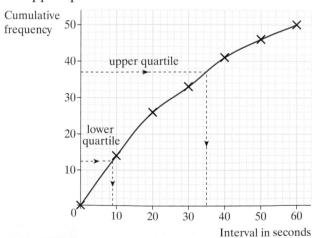

The upper quartile is 35 seconds.

(ii) To find the lower quartile, find the value which corresponds to the quarter way point on the cumulative frequency axis and draw a horizontal line across to the curve.

From this point draw a line vertically down and read off the lower quartile on the horizontal axis.

The lower quartile is 9 seconds.

(iii) To find the interquartile range (IQR), find the difference between the upper and lower quartiles:

$$\text{interquartile range} = \text{upper quartile} - \text{lower quartile}$$
$$= 35 - 9 = 26 \text{ seconds}$$

Worked examination question

This grouped frequency table shows the distribution of the amounts of daily sunshine, in hours in Mathstown in July 1994.

Amounts (s) of daily sunshine in hours	Number of days in this class interval
$0 \leqslant s < 2$	1
$2 \leqslant s < 4$	2
$4 \leqslant s < 6$	3
$6 \leqslant s < 8$	7
$8 \leqslant s < 10$	10
$10 \leqslant s < 12$	6
$12 \leqslant s < 14$	2

(a) Use the data in the table to complete a copy of the cumulative frequency table below.

Amounts (s) of daily sunshine in hours	Cumulative frequency
$0 \leqslant s < 2$	1
$0 \leqslant s < 4$	3
$0 \leqslant s < 6$	
$0 \leqslant s < 8$	
$0 \leqslant s < 10$	
$0 \leqslant s < 12$	
$0 \leqslant s < 14$	

(b) Draw a grid using 1 cm for 1 hour to represent the amounts of daily sunshine (hours) on the horizontal axis and 2 cm for 5 on the vertical cumulative frequency axis. On your grid, construct the cumulative frequency graph for your table.

(c) Use your cumulative frequency graph to find an estimate for the median amount of daily sunshine, in hours, in July 1994.
Make your method clear. [E]

(a) To find the cumulative frequencies add all the previous frequencies.

$1 + 2 = 3$
$1 + 2 + 3 = 6$
$1 + 2 + 3 + 7 = 13$
$1 + 2 + 3 + 7 + 10 = 23$
$1 + 2 + 3 + 7 + 10 + 6 = 29$
$1 + 2 + 3 + 7 + 10 + 6 + 2 = 31$

Amounts (s) of daily sunshine in hours	Cumulative frequency
$0 \leqslant s < 2$	1
$0 \leqslant s < 4$	3
$0 \leqslant s < 6$	6
$0 \leqslant s < 8$	13
$0 \leqslant s < 10$	23
$0 \leqslant s < 12$	29
$0 \leqslant s < 14$	31

(b) The points are plotted with crosses and joined with a smooth curve.

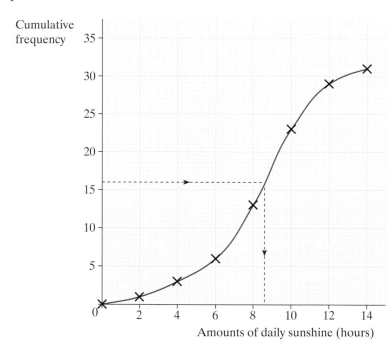

Amounts of daily sunshine (hours)

(c) The median is the $\left(\frac{n+1}{2}\right)$th $= \left(\frac{31+1}{2}\right)$th $= $ 16th value.
Draw a line horizontally where the cumulative frequency is 16. From the point where this line meets the curve draw a line vertically.
The line meets the horizontal axis at 8.6.
So an estimate for the median amount of daily sunshine is 8.6 hours.

If you read off the time when the cumulative frequency is
$$\frac{31}{2} \ (=15\tfrac{1}{2})$$
your answer would be accepted in an exam.

Exercise 15E

1   For the data in this table calculate an estimate of the:
    (a) median   (b) upper quartile (c) interquartile range.

Class interval (seconds)	Frequency
0 up to but not including 10	2
10 up to but not including 20	4
20 up to but not including 30	7
30 up to but not including 40	11
40 up to but not including 50	13
50 up to but not including 60	8
60 up to but not including 70	5

2   The heights of shrubs are given in this table:

Length (cm)	Frequency
0 up to but not including 1	7
1 up to but not including 2	16
2 up to but not including 3	26
3 up to but not including 4	25
4 up to but not including 5	20

Calculate an estimate of the:

(a) median

(b) interquartile range.

3   This table shows the age distribution of workers in a small company:

Age in years	Frequency
22 to 28	18
29 to 35	11
36 to 42	12
43 to 49	15
50 to 56	8
57 to 63	2

(a) Calculate an estimate for the median ages of the workers.

(b) Estimate the percentage of workers who are aged 45 and over.

4 This table shows the distribution of areas of the States of the USA in 1958:

Area in 10 000 km²	Frequency
0 up to but not including 5	10
5 up to but not including 10	4
10 up to but not including 15	14
15 up to but not including 20	8
20 up to but not including 25	5
25 up to but not including 30	5
30 up to but not including 40	2
40 up to but not including 70	2

(a) Calculate estimates for the median area of the States in 1958.

(b) In 1959, Alaska joined the Union. The area of Alaska is 1 531 100 km². Estimate the median area of States after Alaska had joined.

5 The speeds, in miles per hour (mph), of 200 cars travelling on the A320 road were measured.

The results are shown in this table:

Speed (mph)	Cumulative frequency
not exceeding 20	1
not exceeding 25	5
not exceeding 30	14
not exceeding 35	28
not exceeding 40	66
not exceeding 45	113
not exceeding 50	164
not exceeding 55	196
not exceeding 60	200
TOTAL	200

(a) Draw a cumulative frequency graph for these figures.

(b) Use your graph to find an estimate for:
    (i) the median speed (in mph)
    (ii) the interquartile range (in mph)
    (iii) the percentage of cars travelling at less than 48 miles per hour. [E]

6 As part of her project for environmental science, Gina measured the diameter of 64 limpet shells.

  (a) Make a cumulative frequency table.

  (b) Use the table to draw a cumulative frequency diagram.

  (c) Use the cumulative frequency diagram to calculate:
  (i) the median diameter
  (ii) the interquartile range of the diameters.

Diameter (cm)	Frequency
0 up to but not including 0.5	8
0.5 up to but not including 1.0	12
1.0 up to but not including 1.5	15
1.5 up to but not including 2.0	17
2.0 up to but not including 2.5	7
2.5 up to but not including 3.0	5

[E]

7 This cumulative frequency graph gives information on house prices in 1992:

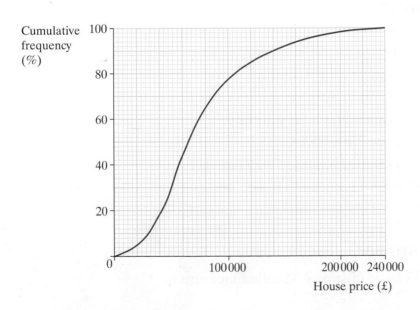

The cumulative frequency is given as a percentage of all houses in England.

  (a) Use the graph to find the percentage of properties valued at less than £60 000.

This grouped frequency table gives the percentage distribution of house prices (p) in England in 1993.

House prices (p) in pounds 1993	Percentage of houses in this class interval
$0 \leqslant p < 40\,000$	26
$40\,000 \leqslant p < 52\,000$	19
$52\,000 \leqslant p < 68\,000$	22
$68\,000 \leqslant p < 88\,000$	15
$88\,000 \leqslant p < 120\,000$	9
$120\,000 \leqslant p < 160\,000$	5
$160\,000 \leqslant p < 220\,000$	4

(b) Use the data above to complete a copy of the cumulative frequency table below.

House prices (p) in pounds 1993	Cumulative frequency (%)
$0 \leqslant p < 40\,000$	
$0 \leqslant p < 52\,000$	
$0 \leqslant p < 68\,000$	
$0 \leqslant p < 88\,000$	
$0 \leqslant p < 120\,000$	
$0 \leqslant p < 160\,000$	
$0 \leqslant p < 220\,000$	

(c) Draw a cumulative frequency graph for your table.

In 1992 the price of a house was £100 000.

(d) Use both cumulative frequency graphs to estimate the price of this house in 1993.

Make your method clear. [E]

15.6 Measures of spread

The interquartile range is a measure of **spread**. Another measure of spread is the **range**.

■ **The range of a set of data is the difference:**

greatest value – least value

The range can be **greatly** affected by extreme values. For example: there are 12 houses in a close. The numbers of letters delivered to the houses one day are, in order:

1 2 2 2 3 3 3 3 5 5 5 30

(One house receives a lot of letters.)

The range of these numbers is 30 – 1 = 29.

The interquartile range is:

upper quartile – lower quartile

= 5 – 2 = 3

If the house that received 30 letters instead received 8 letters then the new range would be 8 – 1 = 7.

This change in the extreme value from 30 to 8 makes no change to the interquartile range.

Because it is not affected by extreme values the interquartile range is a good measure of the spread of the data as a whole, and particularly either side of the median.

A large range tells you very little about how the data is spread.

A large interquartile range tells you that much of the data is widely spread about the median.

A small interquartile range tells you that much of the data is highly concentrated about the median.

Box-and-whisker diagrams

In example 4 you drew a cumulative frequency graph to represent the interval in seconds between successive vehicles passing a school:

Time in seconds	Cumulative frequency
$0 < x \leqslant 10$	12
$0 < x \leqslant 20$	26
$0 < x \leqslant 30$	33
$0 < x \leqslant 40$	41
$0 < x \leqslant 50$	46
$0 < x \leqslant 60$	50

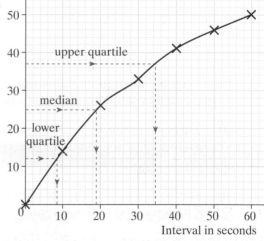

You can represent some of the important features of a cumulative frequency graph in a **box-and-whisker diagram** (or **box-plot**).

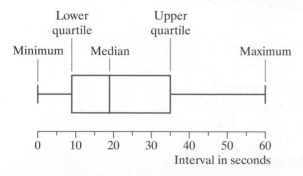

You can draw your box-and-whisker diagrams on graph paper.

■ **Box-and-whisker diagrams are useful for comparing sets of data.**

Example 5

Deenita and Zac are writing party invitations. This table gives information about the time taken in seconds to write an invitation:

	Deenita	**Zac**
Minimum	20	11
Maximum	58	72
Median	35	28
Lower quartile	26	20
Upper quartile	50	55

(a) Draw two box-and-whisker diagrams on the same scale to represent this information.

(b) Compare Deenita's and Zac's results.

(a)

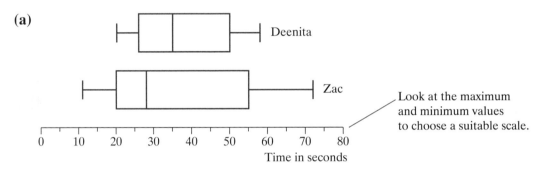

Look at the maximum and minimum values to choose a suitable scale.

(b) Zac had a lower median value but his values were spread over a much wider range. Deenita was much more consistent.

Exercise 15F

1 James collected data on the length of leaves from a bush. The data is summarised in this table.

	Length in centimetres
Minimum	4.2
Maximum	8.7
Median	7.5
Lower quartile	5.6
Upper quartile	8.1

Draw a box-and-whisker diagram to represent this data.

2 Selina collected information about the length of time that people stood in a supermarket queue before being served.
She drew this box-and-whisker diagram.

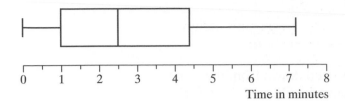

Copy and complete this table.

	time in seconds
Minimum	
Maximum	
Median	
Lower quartile	
Upper quartile	

3 Sibtain asked students at his school to estimate the length of a line he drew. His results are summarised in this table:

Minimum length	2.7 cm
Range	4.8 cm
Median	5.2 cm
Upper quartile	6.5 cm
Interquartile range	2.8 cm

Draw a box-and-whisker diagram to represent Sibtain's results.

4 These box-and-whisker diagrams give information about the results of two different classes from the same test.

Use the diagrams to compare the performance of the two classes.

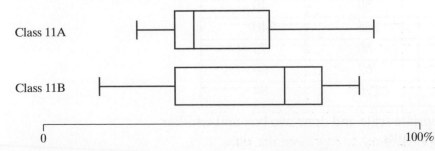

5   Here is a box-and-whisker diagram.

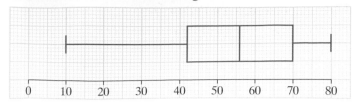

Which of the following histograms has been drawn from the same distribution as the box-and-whisker diagram? Give a reason for your answers.

A
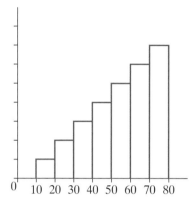

B
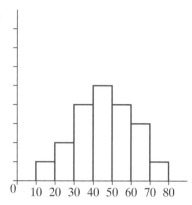

C
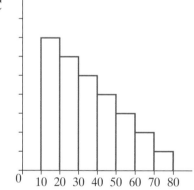

D
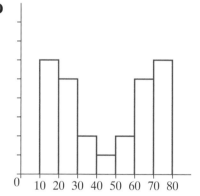

6   A supermarket manager wanted to compare the performance of two checkout assistants. He gathered the following information for the time taken in seconds per customer.

	Assistant A	Assistant B
Minimum	50	40
Maximum	265	740
Median	80	75
Lower quartile	62	60
Upper quartile	170	84

(a) Choose an appropriate scale and draw two box-and-whisker diagrams to these results.

The manager concluded that because Assistant B had a maximum of 740 she was not performing as well as Assistant B.

(b) Explain why the manager's conclusion might have been wrong.

15.7 Drawing conclusions from data

In statistics it is easy to draw incorrect conclusions from data, especially if you do not have sufficient information.

For example: a group of girls and a group of boys took the same test marked out of 20. You are told that:

- **the mean mark for the girls was 10**
- **the mean mark for the boys was 8**

Can you now conclude that the girls have on the whole done better in the test than the boys?

It would be easy to conclude this because the mean mark for the girls is higher than for the boys. But a closer look shows that this need not be true.

Suppose only three girls and three boys took the test and that the marks were:

$$\text{girls:} \quad 9 \quad 10 \quad 11$$
$$\text{boys:} \quad 0 \quad 12 \quad 12$$

The mean for the girls is:
$$\frac{9 + 10 + 11}{3} = 10$$

The mean for the boys is:
$$\frac{0 + 12 + 12}{3} = 8$$

Though the mean for the girls is greater it would not be valid to say that the girls did better than the boys. In fact two-thirds of the boys scored a mark higher than any of the girls.

To say whether the girls did better than the boys further information is needed in addition to the mean mark. You need to know how the marks were distributed or spread out.

Measures of spread

Two measures of spread are used in this book: the range (page 326), the interquartile range (page 319).

■ **To make a judgement when comparing data in situations like the test example above you always need: a measure of average and a measure of spread.**

To help you see why, here is an axis for the scores in the test:

B and G represent the positions of the average marks for Boys and Girls respectively. Here it looks as if the girls have done better than the boys.

But if we add (round) brackets to show the measure of spread of the girls' scores and [square] brackets to show the measure of spread of the boys' scores the diagram now looks like this:

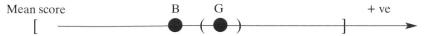

While the average score for the girls is greater than that for the boys, the **measures of spread** mean that many boys may have done better than any of the girls.

If in the test results:

- the mean mark for girls was greater than the mean mark for boys and

- the spread of marks for the girls was about the same as the spread of marks for the boys

you would be justified in concluding that the girls did better in the test than the boys.

Example 6

A group of students took a science examination. The examination consisted of two papers: Paper 1 and Paper 2.

This table gives the median marks and the interquartile ranges of the marks on these papers:

Paper	Median %	Interquartile range %
1	63	12
2	51	38

Comment on these results.

Because the median mark on Paper 1 was greater than the median mark on Paper 2, it looks as if Paper 1 was easier than Paper 2 (or as if more students scored higher marks on Paper 1 than Paper 2).

However, the interquartile range for Paper 2 was much greater than that for Paper 1.

So it could mean that many students obtained a higher mark on Paper 2 than on Paper 1 – and so found Paper 2 to be the easier paper.

Exercise 15G

1 The mean weekly wage at Cromby and Sons is £200.

The mean weekly wage at Drapeway Ltd is £250.

Is it fair to say that the people at Drapeway Ltd are better paid than those at Cromby and Sons? Explain your answer.

2 The Year 11 students at Lucea High School sat an examination in French. The examination consisted of two papers: Paper A and Paper B.

This table shows the median marks and interquartile ranges of the marks on the two papers:

Paper	Median %	Interquartile range
A	72	32
B	60	12

Comment on this data.

3 In the town of Reddshaw there are two golf clubs. This table shows the median ages and interquartile ranges of the ages of the club members:

Club	Median age	Interquartile range of ages
1	34	17
2	42	12

Explain whether or not it would be fair to say that the members of Club 2 are, in general, older than the members of Club 1.

4 The Year 11 students at Russell High School took an examination in mathematics. The examination consisted of three papers: Paper 1, Paper 2 and Paper 3.

This table shows the median marks and interquartile ranges of marks on the three papers:

Paper	Median mark %	Interquartile range%
1	48	32
2	62	17
3	53	21

Make statistical comparisons between the results for:

(a) Paper 1 and Paper 2

(b) Paper 2 and Paper 3

Summary of key points

1 The **mean** of a set of data is the sum of the values divided by the number of values.

$$\text{mean} = \frac{\text{sum of values}}{\text{number of values}}$$

or

$$\bar{x} = \frac{\Sigma x}{n}$$

where Σx is the sum of the values and n is the number of values.

2 The **median** is the middle value when the data is arranged in order of size.

3 The **mode** is the value which occurs most often.

4 The mode and median are *not* affected by extreme values. The mean *is* affected by extreme values.

5 A graph showing how a given value changes over time is called a time series.

6 The mean for grouped discrete data is:

$$\bar{x} = \frac{\Sigma fx}{\Sigma f}$$

7 The estimate of the mean for grouped continuous data is also:

$$\bar{x} = \frac{\Sigma fx}{\Sigma f}$$

but here x is the midpoint of each class interval.

8 The **median** is the middle value of the distribution.

9 The **lower quartile** is the value one-quarter of the way into the distribution.

10 The **upper quartile** is the value three-quarters of the way into the distribution.

11 The **interquartile range** is the difference between the upper and lower quartiles:

   interquartile range = upper quartile − lower quartile

12 The range of a set of data is the difference:

   greatest value − least value

13 Box-and-whisker diagrams are useful for comparing sets of data.

14 To compare two sets of data you should use a measure of average and a measure of spread.

16 Measure and mensuration

Introduction

This unit is about measurement of lengths, areas and volumes – mainly of geometrical shapes.

In the GCSE examination there are a number of formulae that you will need to be able to apply. In some cases these formulae will appear on the examination formulae sheet, in other cases they will not. However, it is best to learn the formulae and understand how they are derived.

16.1 Continuous and discrete data

When someone quotes his or her age as being 15 years, this is really an approximation, where the age is quoted as the number of complete years. It is possible to quote the age more exactly as, say, 15 years, 200 days. But this is still only an approximation, to the nearest day. The age could be quoted even more accurately as, say 15 years, 200 days, 7 hours, 35 minutes and 10 seconds, but this is still only an approximation, now given to the nearest second.

Time passes in a **continuous** way, so if someone quotes his or her age, then they are a little older at the end of the quote than they were at the beginning. No matter how accurate the quote, it is never more than an approximation.

A similar argument is true for a measurement such as length. We might say that the length of a metal rod is 72 cm, but we are dependent on the accuracy of the measuring instrument.

For some data, these arguments do not apply. You might, for instance, be asked to quote the number of GCSE examinations you are taking. This will be a whole number and you can quote it exactly. It is an example of **discrete** data.

The figure does not have to be a whole number. For instance, 10 students could take 76 GCSE examinations between them. Then the mean number of GCSEs taken per person is 76/10 = 7.6, this is an exact number.

16.2 The measurement of continuous data

You will often see examples of continuous data. For instance, a road sign might read:

The distance is quoted to the nearest mile. It could be anywhere between 75.5 and 76.5 miles.

Example 1

The dimensions of a rectangular lawn are given as:

$$\text{width} = 5 \text{ m} \qquad \text{length} = 7 \text{ m}$$

Each measurement is given correct to the nearest metre.

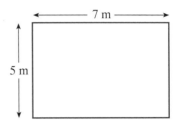

(a) Write the longest and shortest values for the width and the length of the rectangle.

(b) Calculate the largest and smallest values for the area of the lawn.

Answer

(a) For the width, the measurement of 5 m means that it can be anywhere between 4.5 m and 5.5 m.
The shortest width is 4.5 m and the longest width is 5.5 m.

For the length, the measurement of 7 m means that it can be anywhere between 6.5 m and 7.5 m.
The shortest length is 6.5 m and the longest length is 7.5 m.

(b) The area of a rectangle = width × length
To find the smallest area we combine the shortest width with the shortest length.
This gives a smallest area of $4.5 \times 6.5 = 29.25 \text{ m}^2$.
The largest area is found by combining the longest width with the longest length.
This gives a largest area of $5.5 \times 7.5 = 41.25 \text{ m}^2$.

Exercise 16A

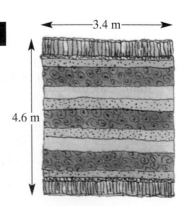

1 A piece of carpet is rectangular. Its dimensions are quoted to the nearest 10 cm as:

$$\text{width} = 3.4 \text{ m} \qquad \text{length} = 4.6 \text{ m}$$

(a) Write the shortest and longest widths of the carpet.

(b) Write the shortest and longest lengths of the carpet.

(c) Calculate the smallest and largest areas of the carpet.

2 A car travels at a constant speed of 70 km per hour for a time of 2 hours and 45 minutes.
Its speed is quoted to the nearest km/h and the time is quoted to the nearest minute.

Calculate:

(a) the maximum possible distance that the car could have travelled

(b) the minimum distance the car could have travelled.

16.3 Lengths, areas and volumes of plane and 3D shapes

You will need to know and apply the formulae for lengths, areas and volumes of many different shapes.

Rectangle

■ **Perimeter = 2 (a + b)**
Area = $a \times b = ab$

Exercise 16B

1

(2x + 3) cm

(x + 1) cm

(a) Find, in its simplest form, an expression in x for the perimeter of this rectangle.

(b) Given that the perimeter is 50 cm, calculate the value of x.

(c) Calculate the area of the rectangle.

2 The area of a square in cm^2 is numerically equal to the perimeter of the square in cm.

(a) Calculate the length of a side of the square.

(b) Calculate the area of the square.

3 A farmer uses exactly 1000 metres of fencing to fence off a square field. Calculate the area of the field.

Triangle

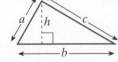

■ **Perimeter = $a + b + c$**
Area = $\frac{1}{2}$ base × height

$= \frac{1}{2}bh$

Exercise 16C

1   Calculate the areas of each of these triangles:

(a)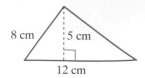

(b)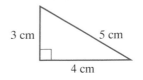

(c)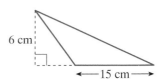

2   **(a)** Write down an expression in x for the perimeter of the triangle.

   (b) If the perimeter of the triangle is 29, calculate the value of x.

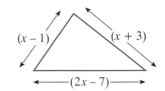

Parallelogram

■ **Perimeter = 2 $(a + b)$**
Area = bh

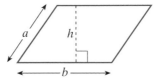

Exercise 16D

1   Calculate the area of each parallelogram:

(a)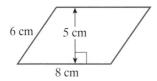

(b)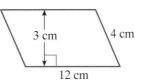

(c)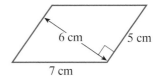

2 $ABCD$ is a parallelogram.

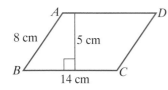

$$BC = AD = 14 \text{ cm} \qquad AB = DC = 8 \text{ cm}$$

The perpendicular distance between AD and BC = 5 cm.

   (a) Calculate the area of the parallelogram.

   (b) Calculate the perpendicular distance between AB and DC.

3

The diagram shows a parallelogram and a square. These two shapes have equal areas.

Calculate the value of x, of the side of the square.

Trapezium

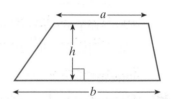

■ **Perimeter = sum of the lengths of the four sides**
Area = $\frac{1}{2}(a + b) \times h$ or $\frac{1}{2}(a + b)h$

Example 2

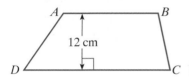

$ABCD$ is a trapezium.
AB is parallel to DC.
The lengths of AB and CD are in the ratio $1:2$.
The perpendicular distance between AB and $CD = 12$ cm.
The area of $ABCD = 72$ cm^2.

Calculate the length of AB.

The statement that the lengths of AB and CD are in the ratio $1:2$ means that $AB = \frac{1}{2}CD$.

We let the length of $AB = x$ cm so then the length of $CD = 2x$ cm.

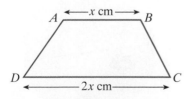

$$\text{Area of } ABCD = \tfrac{1}{2}(AB + CD) \times h = \tfrac{1}{2}(x + 2x) \times 12$$
$$= \tfrac{1}{2}(3x) \times 12 = 18x$$

But the area $= 72$ cm^2

so $\qquad\qquad\qquad\qquad 18x = 72 \quad x = \dfrac{72}{18} = 4 \quad AB = 4$ cm

Exercise 16E

1   *PQRS* is a trapezium.
    The sides *PQ* and *SR* are parallel.
    The perpendicular distance between *PQ* and *SR* = 8 cm.
    PQ = 6.4 cm and *SR* = 8.2 cm.
    Calculate the area of *PQRS*.

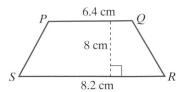

2   *ABCD* is a trapezium with sides *AB* and *DC* parallel.
    The perpendicular distance between *AB* and *DC* = 18 cm.
    The lengths of *AB* and *DC* are in the ratio:
    $$AB : DC = 2 : 3$$
    The area of *ABCD* = 54 cm^2.
    Calculate the length of *AB*.

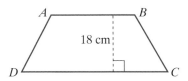

Circle

 Diameter = 2 × radius   or $d = 2r$

 Circumference = $2\pi r = \pi d$

$$\textbf{Area} = \pi r^2 = \frac{\pi d^2}{4}$$

Exercise 16F

1   Calculate the circumference and area of each circle:

    (a)   **(b)**   **(c)**

          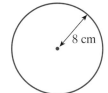

2   Mrs de Silva lays out a circular flower bed of diameter 5 metres.
    (a) Calculate the circumference of the flower bed.
    (b) Calculate the area of the flower bed.

3   The circumference of a circle in centimetres is numerically
    equal to the area of the circle in square centimetres.
    Show that the radius of the circle must be 2 cm.

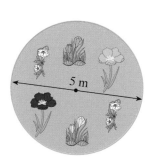

4   The circumference of a circle is 15 cm.
    (a) Calculate the radius of the circle.
    (b) Calculate the area of the circle.

5   The area of a circle is 114 cm^2.
    Calculate the circumference of the circle.

6   Calculate the area and perimeter of the semi-circle
    (a) giving your answer correct to 2 d.p.
    (b) leaving your answer in terms of π.

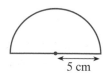

7 The diagram represents a sheet of metal. It consists of a rectangle length 60 cm, width 24 cm, and a semi-circle. Calculate:

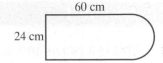

60 cm

24 cm

(a) the perimeter of the sheet of metal

(b) the area of the sheet of metal.

Cuboid

■ **Total length around the edges = 4 ($a + b + c$)**
Surface area = 2 ($ab + ac + bc$)
Volume = abc

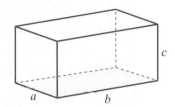

In the special case when the cuboid is a cube then:
$$a = b = c$$

and

total distance around the edges = $12a$

surface area = $6a^2$

volume = a^3

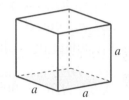

Exercise 16G

1 Calculate the volume of a cube of side length:

(a) 5 cm   (b) 12 cm   (c) 3.8 cm

2 A cube has a volume of 1000 cm^3.
Calculate:

(a) the length of a side of the cube

(b) the surface area of the cube.

3 Calculate the volume of a cuboid with sides of length:

(a) 5 cm, 6 cm and 6 cm   (b) 4.5 cm, 9.2 cm and 11.6 cm.

4 The volume of the cuboid *ABCDEFGH* is 384 cm^3.
The edges *AB*, *BC* and *AF* are in the ratio:
$$AB : BC : AF = 1 : 2 : 3$$

Calculate:

(a) the length of *AB*   (b) the surface area of the cuboid.

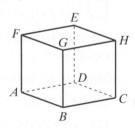

Prisms

■ **A cuboid is a special case of a prism. It is a rectangular prism.**

For a general prism:
Surface area = 2 × area of base + total area of vertical faces

■ **Volume = area of base × vertical height**
= area of base × h

Exercise 16H

1   This is a prism and its base which is a trapezium.

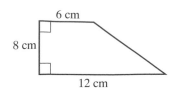
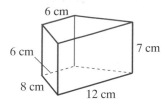

   (a) Calculate the volume of the prism.

   (b) Calculate the surface area of the prism.

2   *ABCDEF* is a triangle-based prism.
   Angle *ABC* = 90°
   AB = 5 cm, *BC* = 8 cm and *CD* = 12 cm
   Calculate the volume of *ABCDEF*.

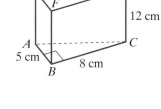

3   *ABCDEF* is a wedge of volume 450 cm³.
   Angle *ABC* = 90°
   AB = 5 cm and *CD* = 15 cm
   Calculate the length of *BC*.

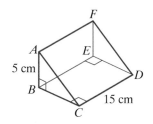

4   This is the base of a prism of vertical height 24 cm.

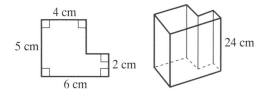

   Calculate the volume of the prism.

16.4 Dimension theory

It can be easy to forget formulae for lengths, areas and volumes or
to confuse them.

For instance, the formulae for the circumference and area of a
circle are:

$$C = 2\pi r \quad A = \pi r^2$$

which are somewhat similar. If you confuse one for the other you
will certainly not be the first person to do so.

Dimension theory is a means of helping you to remember the
formulae and to ensure that you do not confuse them.

Again, the formula for the circumference of a circle is:

$$C = 2\pi r$$

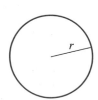

In this formula, r, represents the radius of the circle.
It has units of length.
We say that r has **dimension** $= 1$

In this formula, π is just a number (about equal to 3.14).
It is dimensionless, or it has dimension $= 0$

The 2 in the formula for the circumference is also just a number.
So the 2 is dimensionless, or it has dimension $= 0$

The formula for the area of a circle is:
$$A = \pi r^2$$

Here π is again dimensionless.

The r^2 part is length $\times$ length, or it has dimension of area.
So r^2 has dimension $= 2$

All expressions for length or distance have dimension $= 1$
area have dimension $= 2$
volume have dimension $= 3$
and numbers (like π and 2) are dimensionless.

Example 3

Check the dimensions of $V = abc$ for the volume of a cuboid.
  a, b and c each have dimension $= 1$

So abc has dimension $= 3$, which is correct for volume.

Example 4

Yvonne is using a formula sheet to work out the volume and curved surface area of a cylinder.

Unfortunately a blot on the sheet covers up part of the information.

Her sheet looks like:

$$= \pi r^2 h$$
$$= 2\pi r h$$

Yvonne knows that one of these gives the volume and the other the curved surface area.

How could she tell which is which?

Volume has dimension 3.

In the expression $\pi r^2 h$, π is a number, r is a length, h is a length.
π is dimensionless.
r^2 has dimension 2.
h has dimension 1.
So $\pi r^2 h$ has dimension 3.
So this is the formula for the volume.

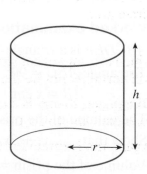

And in $2\pi rh$
2 and π are dimensionless whilst r and h each have dimension 1.
So $2\pi rh$ has dimension 2 and is an expression for an area.

■ **You should be able to use dimension theory to confirm whether formulae are for length, area or volume.**

Exercise 16 I

1 Here are two expressions related to a sphere of radius r:

$$4\pi r^2 \quad \text{and} \quad \frac{4\pi r^3}{3}$$

One of these gives the volume of a sphere and the other the surface area of the sphere.

State, with an explanation, which formula gives the surface area of a sphere.

2 The volume of a solid is given by $V = \lambda a^n b^m$, where λ is dimensionless, a and b have dimensions of length and n and m are non-negative integers.

State the possible values of n and m.

Worked examination question 1

The diagram represents a face of a sheet of metal in the form of a rectangle with a semi-circle cut out from one end.

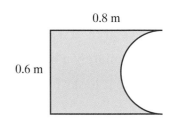

Calculate the area of the face of the sheet of metal.

Radius of semi-circle $= \frac{1}{2} \times 0.6 = 0.3$

Area = area of rectangle – area of semi-circle

$= 0.6 \times 0.8 - \frac{1}{2}\pi(0.3)^2$

$= 0.48 - \frac{1}{2}\pi \times 0.09$

$= 0.48 - 0.141$

$= 0.339 \text{ m}^2$ (correct to 3 d.p.)

Worked examination question 2 (Part question)

$ABCDEF$ is a triangle-based prism.

The angle $A\hat{B}C = 90°$.

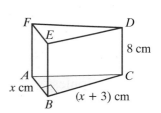

$AB = x$ cm, $BC = (x + 3)$ cm and $CD = 8$ cm.

The volume of the prism $= 40 \text{ cm}^3$.

Show that $x^2 + 3x - 10 = 0$

Volume of the prism = area of base × height

$$\text{Area of base} = \tfrac{1}{2} \times AB \times BC$$
$$= \tfrac{1}{2}x \times (x + 3) = \tfrac{1}{2}x(x + 3)$$

So the volume $= \tfrac{1}{2}x(x + 3) \times 8 = 4x(x + 3)$

But the volume $= 40$, so:

$$4x(x + 3) = 40$$
$$x(x + 3) = 10 \quad \text{(dividing through by 4)}$$
$$x^2 + 3x = 10$$

so, re-arranging, or taking 10 from each side:

$$x^2 + 3x - 10 = 0$$

(Which is what we were asked to show.)

Note: a full question such as this in the examination would almost certainly go on to ask you to solve the equation:

$$x^2 + 3x - 10 = 0$$

which is a quadratic equation and is dealt with in Units 18 and 21.

Worked examination question 3

The diagram shows a solid church door.

The door is made from a rectangle with a semi-circular top.

(a) Work out the area of the cross-section of the door.

The door is made of metal.
The door has a constant thickness of 5 cm.

(b) Work out the volume of the door.

The density of the metal is 7.8 grams per cm^3.

(c) Work out the mass of the door.

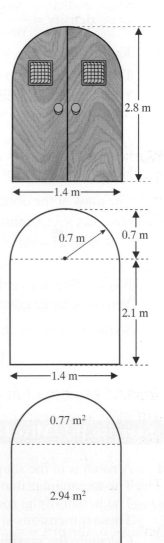

(a) Firstly, you need to split the door into a rectangle and a semi-circle.
The radius of the semi-circle is half of the width of the door, so:

$$\text{radius of semi-circle} = \tfrac{1}{2} \times 1.4 = 0.7 \text{ metres}$$

The rectangle measures 1.4 m by 2.1 m, so:

$$\text{area of rectangular part} = 1.4 \times 2.1 = 2.94 \text{ m}^2$$

The area of the semi-circle is:

$$\frac{\pi r^2}{2} = \frac{\pi \times 0.7^2}{2} = 0.77 \text{ m}^2 \text{ (to 2 decimal places)}$$

So the total area of the cross section of the door is

$$2.94 + 0.77 = 3.7 \text{ m}^2 \text{ (to 2 decimal places)}$$

(b) The door is a prism.
The area of its cross-section is 3.7 m².
The thickness of the door is 5 cm.

The volume of a prism is given by the formula

Volume = Area of Cross-Section × Depth (thickness)

Before applying any formula you need to make sure your units are consistent. It is meaningless to multiply 3.7 m² by 5 cm. You need to convert the cm into m:

5 cm = 0.05 m

Volume of church door = 3.71 × 0.05
= 0.19 m³ (correct to 2 decimal places)

(c) The relationship between mass, volume and density is

$$\text{density} = \frac{\text{mass}}{\text{volume}} \text{ or mass} = \text{density} \times \text{volume}$$

Again you need to be consistent about the units.
The density is given as 7.8 grams per cm³.
The calculated volume is 0.19 m³

1 m = 100 cm
so 1 m³ = 100 × 100 × 100 = 1 000 000 cm³

Hence the volume of the door = 0.19 × 1 000 000 = 190 000 cm³

The mass of the door = density × volume
mass = 7.8 × 190 000
= 1 500 000 grams   (correct to 2
significant figures)

It is sensible to give the most appropriate units.
You can change grams to kilograms by dividing by 1000.

So the mass of the church door is

$$\frac{1\,500\,000}{1000} = 1500 \text{ kg}$$

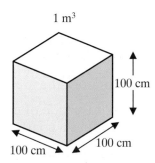

1 m³

100 cm

100 cm

100 cm

Choosing the right degree of accuracy.
The density of the metal was given in the question to 2 significant figures. You should give your answer to the same degree of accuracy.

Exercise 16J (Mixed questions)

1 A room is in the shape of a cuboid.
The rectangular floor has width 4 m and length 5 m.
The height of the room is 3 m.
These dimensions are all quoted correct to the nearest metre.
Calculate the maximum volume of the room.

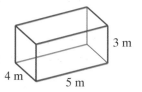

3 m

4 m

5 m

2   The diagram represents the plan of a sports field.
The field is in the form of a rectangle with semi-circular pieces at each end.
Calculate:

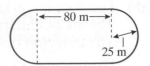

(a)   the perimeter of the field

(b)   the area of the field.

3   Calculate the area of the trapezium *ABCD*.

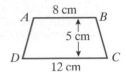

4   The circumference of a circle is 44 cm.
Calculate the area of the circle.

5   Calculate the volume of the wedge *ABCDEF* in which

Angle $A\hat{B}C = 90°$

$AB = 8$ cm, $BC = 12$ cm   and $CD = 20$ cm

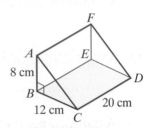

6   The cuboid has a square base of side length x cm.
The height of the cuboid is h cm.
The volume of the cuboid is 50 cm^3.

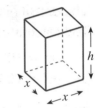

(a)   Show that $x = \sqrt{\dfrac{50}{h}}$

(b)   Calculate the value of x when $h = 4$.

(c)   Calculate the value of h when $x = 5$.

7   In the expression $\mu a^2 h$, μ is dimensionless and a and h have dimensions of length.

Explain whether the expression represents a length, area or volume.

Summary of key points

You should know and be able to apply the following formulae.

1   Perimeter of a rectangle $= 2(a + b)$
    Area of a rectangle $= ab$

2   Perimeter of a triangle $= a + b + c$
    Area of a triangle $= \frac{1}{2}bh$, where b is base and h is height.

3 Perimeter of a parallelogram = $2(a + b)$
Area of a parallelogram = bh, where b is base and h is height.

4 Perimeter of a trapezium = sum of the lengths of the four sides.
Area of a trapezium = $\frac{1}{2}(a + b)h$

5 Diameter of a circle = $d = 2r$, where d is diameter and r is radius.
Circumference of a circle = $2\pi r = \pi d$
Area of a circle = πr^2

6 Surface area of a cuboid = $2(ab + ac + bc)$
Volume of a cuboid = abc

7 Volume of a prism = area of base $\times h$

8 You should be able to use dimension theory
to confirm whether formulae are for length,
area or volume.

17 Proportion

17.1 Direct proportion

Joan volunteers to take part in a charity fund raising walk. She is given 50 pence for each kilometre she walks.

The further she walks the more money she collects. If she doubles the distance she walks then she doubles the money she collects. But if she only walks half the distance then she collects only half the money.

This is an example of direct proportionality. The money collected y is **directly proportional** to the distance walked x.

■ **The symbol $\propto$ means 'is proportional to'.**

Using the symbol:

$$y \propto x$$

■ **$y \propto x$ means y is directly proportional to x.**

A simple rule of thumb method to check whether one quantity is directly proportional to another is to try these two tests:

● If one quantity is zero is the other also zero?
● If one quantity doubles does the other quantity also double?

Exercise 17A

1 Which of the following could be examples of direct proportionality:
  (a) profit made by selling goods
  (b) final examination results and effort put into work
  (c) area of a square and the length of one side
  (d) the area of a rectangle with one side of constant length and the length of the other side
  (e) time and distance travelled at a constant speed
  (f) the height and weight of a student.

2   Given that $w \propto t$ and $w = 8$ when $t = 6$, find:
  (a) t when w doubles in value
  (b) w when t halves in value.

3   Given that $a \propto b$ and $a = 10$ when $b = 8$, find:
  (a) b when a increases to 30
  (b) a when b decreases to 2.

4   Given that $d \propto s$ and $d = 36$ when $s = 16$, find:
  (a) s if d decreases by 9
  (b) d if s increases to 24.

17.2 Graphs that show direct proportion

In a science experiment the time it takes a small pump to fill a water cylinder is recorded. The height of the water in the cylinder is measured at regular time intervals. Here are the results:

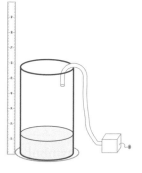

Height of water (cm)	2.1	4.2	6.3	8.4
Time taken (s)	5	10	15	20

This graph shows the result of plotting height against time and joining the points.

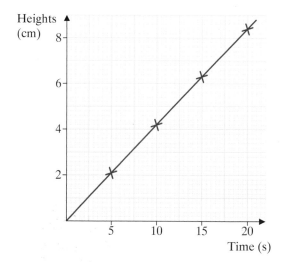

A straight line connects the points. This indicates that the two quantities, height and time taken to reach that height, are in direct proportion.

■ **When a graph of two quantities is a straight line through the origin, one quantity is directly proportional to the other.**

This means that if one quantity changes then the other changes in the same ratio.

Notice that this is consistent with the 'rule of thumb' in the previous section:

● If one quantity is zero the other is also zero.
● If one quantity doubles, the other quantity also doubles.

Example 1

In which of these graphs is *r* directly proportional to *s*?

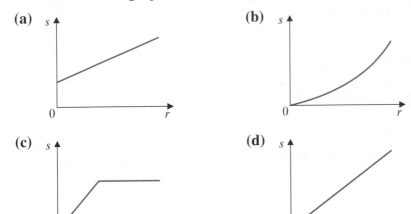

(a)   **(b)**

(c)   **(d)**

Graph **(d)** is a straight line through the origin, showing that *r* is directly proportional to *s*.

17.3 Finding proportionality rules from graphs

You can use a graph to help find the rule connecting quantities that are in direct proportion to one another.

Example 2

The length of a shadow cast by a tree at midday is directly proportional to the height of the tree. At midday a tree 8 metres tall casts a shadow 10 metres long.

(a) Find a relationship connecting the height of the tree and the length of its shadow at midday.

(b) The shadow of another tree at midday is 8 metres long. How tall is the tree?

(c) Calculate the length of the shadow at midday of a tree 10 metres tall.

(a) The relationship between the height and shadow length is directly proportional and can be shown on a straight line graph like this:

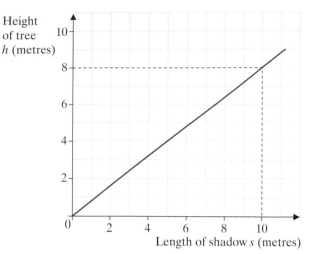

Height of tree h (metres)

Length of shadow s (metres)

The equation of a straight line is:
$$y = mx + c$$

For direction proportion the straight line passes through the origin so:

$c = 0$ and $y = mx$

The gradient m of the line can be used to find the rule connecting h and s. Replacing y with h, and x with s gives:

$$h = ms$$

From the graph, the gradient $m = \frac{8}{10} = 0.8$

So the relationship connecting the height of the tree and the length of the shadow is:

$$h = 0.8s$$

There is more about gradients on page 159.

(b) The relationship $h = 0.8s$ can be used to find the height h of a tree whose shadow is $s = 8$ metres long:

$$h = 0.8 \times 8 = 6.4 \text{ metres}$$

The height of the tree is 6.4 metres when the shadow is 8 metres long.

gradient $m = \frac{8}{10}$

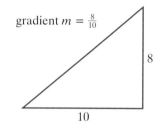

8

10

(c) The relationship can also be used to find the length of shadow for a tree of height $h = 10$ metres:

$$h = 0.8s$$

so: $$10 = 0.8s$$

divide both sides by 0.8 $$\frac{10}{0.8} = s$$

$$12.5 = s$$

The shadow is 12.5 metres long when the height of the tree is 10 metres.

Exercise 17B

1 At 10 am a tree 10 metres high casts a shadow 8 metres long. The length of the shadow (l metres) is directly proportional to the height (h metres) of the tree.

 (a) Sketch a graph to show this information.

 (b) Find a rule connecting h and l.

 (c) Find h when $l = 12.5$.

 (d) Find l when $h = 15$.

2 The area (A cm^2) of a shape is directly proportional to the length (l cm) of one of its sides. When $l = 14$ cm, $A = 42$ cm^3.

    (a) Sketch a graph of this information with area on the vertical axis.

    (b) Work out a rule connecting A and l.

    (c) What is the area of a similar shape with a length of 32.4 cm?

    (d) Find l when $A = 24.6$.

3 In the recipe for Yorkshire puddings, 2 eggs (e) are required to make 24 small Yorkshire puddings (Y).

    (a) Sketch that the graph of $Y = 12e$ for $e = 0$ up to 10.

    (b) Explain why the equation $Y = 12e$ represents the number of eggs to make a certain number of small Yorkshire puddings.

    (c) How many eggs would be needed to make 84 small Yorkshire puddings.

17.4 Finding proportionality rules from ratios

Ratios can also be used to help find the rule connecting quantities that are in direct proportion to one another.

Example 3

The mass of a silver trophy is directly proportional to its height.

A trophy of height 8.2 cm has a mass of 1.148 kg.

(a) Find a rule connecting the mass m and height h of a trophy.

(b) Find the mass of a similar trophy with a height of 15.5 cm. Give your answer correct to 2 d.p.

(a) Instead of drawing a graph you can present this information in a table:

Mass m (kg)	1.148	?
Height h (cm)	8.2	15.5

The mass to height ratio is:

$$1.148 : 8.2 \quad \text{or} \quad \frac{1.148}{8.2} = 0.14$$

The rule connecting the mass m and height h is:

$$m = 0.14h$$

(b) Using the rule, when $h = 15.5$ then $m = 0.14 \times 15.5 = 2.17$

The mass of the second trophy is 2.17 kg.

Exercise 17C

1   These tables each show two variables which are directly proportional to each other.
    For each of these tables find:
    (a)   a rule connecting the two variables
    (b)   the missing values correct to 1 d.p.

w	8.4	?
h	8	12

h	12.0	16.8
s	20.5	?

p	7.8	?
l	6.8	27.6

a	110.4	10.4
p	?	23.4

2   The voltage V across a resistor is directly proportional to the current I flowing through it.
    (a)   Write down the readings of the two meters.
    (b)   Find a rule connecting the voltage and current.
    (c)   Calculate the current when the voltage increases to 11.52 volts.
    (d)   Calculate the voltage when the current is 2.46 amps.

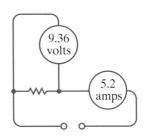

3   Kirstine obtains these readings during an experiment, where V is proportional to I:

Voltage (V)	2.8	4.6	8.4	12.6	22.4	28
Current (I)	1	2	3	4.5	8	10

Unfortunately there is an error in her table.
    (a)   Plot the points and draw a graph of voltage V against current I.
    (b)   Which reading is wrong and what should it be?
    (c)   Find the rule connecting V and I.
    (d)   Find V when $I = 3.5$.
    (e)   Calculate I when $V = 20.5$, giving your answer correct to 3 d.p.

4   Given that $r \propto t$ and $r = 2.4$ when $t = 12.8$, find correct to 2 decimal places:
    (a) r when $t = 8.9$
    (b) t when $r = 3.1$

5   The cost of white correction fluid is directly proportional to the amount of fluid in a bottle. A bottle containing 20 ml of fluid costs 72p. Find:

  (a) a rule connecting the cost and the amount

  (b) the cost of a bottle containing 100 ml of fluid.

6   On a car journey the distance travelled, d, is directly proportional to the time taken, t. After 2 hours the car had travelled 76 miles. Find:

  (a) a rule connecting d and t

  (b) the distance travelled in 5 hours

  (c) the time taken to travel a distance of 418 miles.

17.5 Writing proportionality formulae

■ **When y is directly proportional to x a proportionality statement and a formula connecting y and x can be written:**

$y \propto x$ is the proportionality statement

$y = kx$ is the formula, where k is the constant of proportionality.

Example 4

The extension E of a spring is directly proportional to the force F pulling the spring. The extension is 8 cm when a force of 30 N is pulling it. Calculate the extension when the force is 9 N.

Force is measured in Newtons (N).

As $E \propto F$ the proportionality formula can be written as:

$$E = kF$$

To find the value of the constant of proportionality k, substitute $E = 8$ and $F = 30$ in the formula:

$$8 = k \times 30$$

Divide both sides by 30: $\dfrac{8}{30} = k$

So the proportionality formula is:

$$E = \frac{8}{30} F$$

When the force $F = 9$ N the extension E is:

$$E = \frac{8}{30} \times 9$$

$$= 2.4 \text{ cm}$$

Exercise 17D

1 y is proportional to x so that $y = kx$.
Given that $y = 148$ when $x = 12$, find the value of k.
Calculate the value of y when $x = 7$.

2 p is proportional to q so that $p = kq$.
Given that $p = 34$ when $q = 51$, find the value of k.
Calculate the value of p when $q = 7$.

3 z varies in direct proportion to x.
Write a formula for z in terms of x.
Given that $z = 3\frac{3}{5}$ when $x = 6$, find the value of z when $x = 14$.

4 z varies in direct proportion to w.
Write a formula for z in terms of w.
Given that $z = 315$ when $w = 7$, find the value of z when $w = 105$.

5 y is proportional to x. Given that $y = 12$ when $x = 4$, calculate the value of:

   (a) y when $x = 5$ **(b)** y when $x = 13$ **(c)** x when $y = 10$

6 F varies directly as E. Given that $F = 300$ when $E = 120$, calculate the value of:

   (a) F when $E = 90$

   (b) F when $E = 500$

   (c) E when $F = 180$

7 Given that $y \propto p$, calculate the values missing from this table.

p	1		10
y		4	28

8 The volume v of liquid in a tube is proportional to the height h of the tube. When the height of the liquid in the tube is 7 cm, the volume of liquid is 10 cm^3. Calculate the volume of liquid when the height of liquid is 4 cm.

9 The distance travelled by the tip of a second hand on a clock is proportional to the time elapsed. When the time elapsed is increased by 40%, calculate the increase in the distance travelled by the tip of the second hand, giving your answer as a percentage.

10 The height h of an elephant is directly proportional to the diameter d of its footprint. A baby elephant has a height of 150 cm and the diameter of its footprint is 20 cm.

   (a) Find an equation connecting h and d.

   (b) Calculate the height of an elephant with a footprint of diameter 2.5 cm.

   (c) Calculate the diameter of the footprint of an elephant whose height is 3.2 m.

17.6 Square and cubic proportion

This section introduces proportionality formulae, involving square and cube terms that is $y = kx^2$ and $y = kx^3$.

Sets of Russian dolls are made to fit inside each other. They are exactly the same shape as each other and are mathematically **similar**.

One of the dolls is 8 cm high and has a surface area of 115.2 cm^2.

The surface area S of the dolls is proportional to the square of their height h. This can be written as the proportionality statement:

$$S \propto h^2$$

The proportionality formula can be written as:

$$S = kh^2$$

The constant of proportionality k can be found by substituting information about the 8 cm high doll in the formula:

$$115.2 = k \times 8^2$$
$$64k = 115.2$$
$$k = \frac{115.2}{64}$$
$$k = 1.8$$

so the proportionality formula is:

$$S = 1.8h^2$$

The formula can be used to find the surface of a doll of height 10 cm:

$$S = 1.8 \times 10^2$$
$$= 1800 \text{ cm}^2$$

In this example one quantity (the surface area) is proportional to the **square** of the other quantity (the height).

■ **When y is directly proportional to the square of x:**
$y \propto x^2$ **is the proportionality statement**
$y = kx^2$ **is the proportionality formula, where k is the constant of proportionality.**

Example 5

The volume of a Russian doll is proportional to the cube of its height. The volume of the 8 cm high doll is 64 cm^3. Find the volume of the 10 cm high doll.

The proportionality statement is:

$$V \propto h^3$$

The proportionality formula can be written as:

$$V = kh^3$$

The value of the constant k can be found using the information about the 8 cm high doll:

$$64 = k \times 8^3$$
$$512k = 64$$
$$k = \frac{64}{512} = \tfrac{1}{8}$$
$$k = 0.125$$

So the proportionality formula is:

$$V = 0.125 \times h^3$$

The formula can be used to find the volume of the 10 cm high doll:

$$V = 0.125 \times 10^3$$
$$= 125 \text{ cm}^3$$

In this example one variable is proportional to the **cube** of the other.

■ **When y is directly proportional to the cube of x:**
$y \propto x^3$ is the proportionality statement
$y = kx^3$ is the proportionality formula, where k is the constant
of proportionality.

Exercise 17E

1 y is directly proportional to the square of x so that $y = kx^2$.
   Given that $y = 20$ when $x = 2$, calculate the value of k.
   Calculate the value of y when $x = 4$.

2 z varies in direct proportion to the square of x so that $z = kx^2$.
   Given that $z = 100$ when $x = 4$, calculate the value of:
   (a) the proportionality constant k
   (b) z when $x = 5$
   (c) x when $z = 10$

3 s varies in direct proportion to the cube of t, so that $s = kt^3$.
   Given that $s = 54$ when $t = 3$, calculate the value of:
   (a) the proportionality constant k
   (b) s when $t = 4$
   (c) t when $s = 128$

4 l is directly proportional to the cube of m.
   (a) Write a formula for l in terms of m.
   (b) Given that the value of l when $m = 0.4$ is 3.2:
       (i) calculate the value of l when $m = 0.8$
       (ii) calculate the value of m when $l = 43.2$

5 The resistance R to the motion of a train varies directly as the square of the speed v of the train. The resistance to motion is 100 000 N when the speed is 20 metres per second. Calculate the resistance to motion when the speed of the train is 10 metres per second.

6 A stone is dropped down a well a distance d metres. The value of d is directly proportional to the square of the time of travel t seconds.
When $t = 2$, $d = 20$. Calculate the value of d when:
(a) $t = 3$ **(b)** $t = 4.5$

7 The variables p and q are related so that p is directly proportional to the square of q. Complete this table for values of p and q.

q	0.5	2	
p		12	27

8 z varies in direct proportion to the cube of t. When $t = 5$ the value of z is 0.25. Calculate the value of z when $t = 1$.

9 The length of a pendulum l is directly proportional to the square of the period T of the pendulum. Given that a pendulum which has a period of 3 seconds is 2.25 metres long, calculate the length of a pendulum which has a period of:
(a) 2 seconds   **(b)** 2.5 seconds.

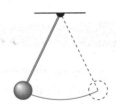

The period of a pendulum is the time taken to complete a full swing and return to its original position.

10 A varies in direct proportion to the square of l. When the value of l is multiplied by 2, by what amount is A multiplied?

11 A set of models of a road bridge is made.
The surface area of the road is directly proportional to the square of the height h of its model.
The height of Model A is 50% greater than the height of Model B. By what percentage is the area of road surface in Model A greater than the area of road surface in Model B?

12 The pressure on a diver under water is directly proportional to the square of her depth below the surface of the water. The diver has reached a depth of 10 metres. How much further must she descend for the pressure to double?

13 The variable z is directly proportional to the cube of the variable w.
Complete the table for these variables.

w	2	1	
z	16		6.75

17.7 Inverse proportion

When one quantity increases while the other decreases proportionally the quantities are in **inverse proportion** to one another. (When one quantity decreases, the other will increase proportionally.)

■ $y \propto \frac{1}{x}$ means *y* is inversely proportional to *x*.

A formula for inverse proportion is $y = k \times \frac{1}{x}$ (or $y = \frac{k}{x}$), where *k* is a constant of inverse proportionality.

Another way of describing such a relationship is to say that *y* is directly proportional to $\frac{1}{x}$.

Example 6

In a physics experiment the pressure *P* and volume *V* of a quantity of gas are measured at constant temperature.

The pressure is inversely proportional to the volume. When the pressure is 2 bar, the volume of the gas is 150 cm³.

Calculate the pressure of the gas when the volume is decreased to 100 cm³.

P is inversely proportional to *V* so:

$$P \propto \frac{1}{V}$$

The formula is:

$$P = \frac{k}{V}$$

The constant *k* is found by substituting the pressure *P* = 2 and the volume *V* = 150 giving the formula:

$$2 = \frac{k}{150}$$

so: $\qquad k = 300$

giving the formula:

$$P = \frac{300}{V}$$

When *V* = 100:

$$P = \frac{300}{100}$$

$$= 3 \text{ bar}$$

Example 7

In Newtonian physics the force of attraction F between two bodies is inversely proportional to the **square** of the distance r between their centres. When the distance between their centres is 1 unit, the force of attraction is 16 units. Calculate the force of attraction when the distance is 16 units. Calculate the force of attraction when the distance between their centres is 2 units.

F is inversely proportional to r^2, so the proportionality statement is:

$$F \propto \frac{1}{r^2}$$

The formula is:

$$F = \frac{k}{r^2}$$

where k is constant for these two bodies only. (The force between the bodies also depends on their masses.)

The constant k is found using the fact that when $r = 1$ unit, the force $F = 16$ units:

$$16 = \frac{k}{1^2}$$

so:

$$16 = k$$

The formula becomes:

$$F = \frac{16}{r^2}$$

When the distance between the centres of the two bodies is 2 units the force is:

$$F = \frac{16}{2^2}$$

$$= 4 \text{ units}$$

> In Newtonian physics the Moon orbits the Earth because there is a force of attraction across space given by the formula:
>
> $$F = \frac{G\, M_E M_M}{r^2}$$
>
> In Einsteinian physics each mass is in free-fall through spacetime which is curved by their presence. This curvature gives rise to the orbit.

Exercise 17F

1   y is inversely proportional to x so that $y = \frac{k}{x}$.
    When $x = 6$ the value of $y = 12$.

   (a) Find the value of k.

   (b) Calculate the value of y when $x = 10$.

2 z is inversely proportional to t so that $z = \frac{k}{t}$.
When $t = 0.3$ the value of $z = 16$.

(a) Find the value of k.

(b) Calculate the value of z when $t = 0.5$.

3 z is inversely proportional to the square of w so that $z = \frac{k}{w^2}$.
When $w = 4$, $z = 32$.

(a) Find the value of k.

(b) Calculate the value of z when $w = 2$.

4 p is inversely proportional to the square of q so that $p = \frac{k}{q^2}$.
When $q = 0.5$, $p = 10$.

(a) Find the value of k.

(b) Calculate the value of p when $q = 0.25$.

5 An essay typed at 50 characters to the line is 372 lines long.
The length of the essay is inversely proportional to the number
of characters to the line. Calculate the number of lines that the
essay will have when it is typed with:

(a) 80 characters to the line

(b) 64 characters to the line.

6 In a mathematics investigation, students draw a series of
rectangles which all have the same area. The length l of each
rectangle is inversely proportional to its width w.

(a) Write the proportionality statement and the formula for
this situation.

(b) What does the constant of proportionality represent?

7 The light intensity I at a distance d from a light source varies
inversely as the square of the distance from the source. At a
distance 1 cm from the light source the light intensity is 64
units. Calculate the light intensity at a distance 4 cm from the
light source.

8 The frequency f of sound varies inversely to the wavelength w.
The frequency of middle C is 256 cycles per second and the
wavelength of this note is 129 cm.

(a) Find the equation connecting frequency f and the
wavelength w.

(b) Calculate the frequency of the note with a wavelength of
86 cm.

(c) Calculate the wavelength of a note whose frequency is
344 cycles per second.

Worked examination question

The distance D moved by an object is proportional to the square of the time t for which it is moving.

(a) Express D in terms of t and a constant of proportionality k.

When $t = 10$ seconds, $D = 500$ metres.

(b) Calculate:

    (i) the value of D when $t = 4$ seconds

    (ii) the value of t when $D = 720$ metres.

Answer

(a) The proportionality statement relating D and t is:

$$D \propto t^2$$

So the expression involving D, t and a constant of proportionality k is the formula:

$$D = kt^2$$

(b) (i) The formula $D = kt^2$ can be used to find the value of k by substituting $t = 10$ seconds and $D = 500$ metres:

$$500 = k \times 10^2$$
$$500 = 100\,k$$

Divide both sides by 100: $5 = k$

So the formula is:

$$D = 5t^2$$

This can be used to find D when $t = 4$ seconds:

$$D = 5 \times 4^2$$
$$= 5 \times 16 = 80 \text{ metres}$$

When $t = 4$ seconds, $D = 80$ metres.

(ii) The same formula can be rearranged to find the value of t when $D = 720$ metres:

$$D = 5t^2$$

divide by 5: $\dfrac{D}{5} = t^2$

take the square root: $\sqrt{\dfrac{D}{5}} = t$

So when $D = 720$:

$$t = \sqrt{\dfrac{720}{5}}$$
$$= \sqrt{144}$$
$$= 12 \text{ seconds}$$

When $D = 720$ metres, $t = 12$ seconds.

Worked examination question

The resistance R of 1 metre of cable of a certain material is inversely proportional to the square of the radius r of the cable. When the radius is 5 mm, the resistance of 1 metre of the cable is 0.06 ohms.

(a) Find a formula connecting R and r.

(b) Calculate the value of R when r is 4 mm.

Answer

(a) The proportionality statement for R and r is:

$$R \propto \frac{1}{r^2}$$

So the formula connecting R and r is:

$$R = \frac{k}{r^2}$$

where k is a constant.

To find R when r is 4, first find the value of k in the formula using the fact that R is 5:

$$R = \frac{k}{r^2}$$
$$0.06 = \frac{k}{5^2}$$

So
$$k = 1.5$$

So the formula is:
$$R = \frac{1.5}{r^2}$$

(b) When $r = 4$:

$$R = \frac{1.5}{4^2}$$
$$R = 0.094 \text{ ohms (correct to}$$
$$\text{2 significant figures)}$$

Exercise 17G (Mixed questions)

1 An electrical heater uses 16 units of electricity in 5 hours. The use of electricity is directly proportional to the time. How much electricity will the heater use in 3 hours?

2 A builder makes concrete paving slabs. All the slabs are the same thickness. The cross-sections of all the slabs are equilateral triangles. The mass of concrete is proportional to the square of the length of one of the sides. A slab with side 60 cm has mass 50 kg. Calculate the mass of a slab with side 20 cm.

3 A spaceship covers a distance of 18 000 kilometres in 3 hours 20 minutes. The distance travelled is directly proportional to the time taken. Calculate the distance that the spaceship will travel in 5 hours 20 minutes.

4 A hang-glider pilot, at a height of h metres above the sea, can see up to a distance of s kilometres. It is known that h is proportional to the square of s.

   (a) Given that $h = 140$ when $s = 16$, find the formula for h in terms of s.

   (b) Calculate the height of the hang-glider when the pilot can just see a lighthouse which is 24 kilometres away.

5 The variables y and x are related by:

$$y \propto \frac{1}{x}$$

When $x = 3, y = 30$. Calculate the value of y when $x = 5$.

6 The height h reached by a ball thrown up into the air varies in direct proportion to the square of the speed v at which the ball is thrown. A ball thrown at a speed of 20 metres per second reaches a height of 20 metres. Calculate the height reached by a ball thrown at a speed of:

   (a) 25 metres per second

   (b) 30 metres per second.

7 The time taken for a journey on a motorway is inversely proportional to the average speed for the journey. The journey takes 1 hour 30 minutes when the average speed is 54 miles per hour. Calculate the time taken, in hours and minutes, for this journey when the average speed is 45 miles per hour.

8 The energy stored in a battery is proportional to the square of the diameter of the battery, for batteries of the same height. One battery has a diameter of 2.5 cm and stores 1.6 units of energy. Another battery of the same height has a diameter of 1.5 cm. Calculate the energy stored in the second battery.

9

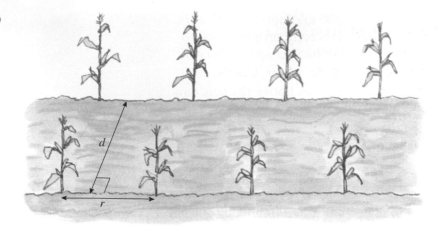

The diagram is taken from a book about growing maize.
The distance between the rows of plants is d metres.
The spacing between the plants in the rows is r metres.
The number, P, of plants per hectare, is given by the formula:

$$P = \frac{10\,000}{dr}$$

$d = 0.9$ and $r = 0.55$.

(a) Calculate the value of p. Give your answer correct to
     2 significant figures.

The value of d is inversely proportional to the value of r and
$d = 0.9$ when $r = 0.4$.

(b)   (i) Calculate the value of r when $d = 1.2$.
      (ii) Calculate the value of r when $r = d$.   [E]

Summary of key points

1 The symbol $\propto$ means 'is proportional to'.

2 $y \propto x$ means y is directly proportional to x.

3 When a graph connecting two quantities is a straight line
   through the origin then one quantity is directly proportional
   to the other.

4 When y is directly proportional to x, a proportionality
   statement and a formula connecting y and x can be written.

   ● $y \propto x$ is the proportionality statement.
   ● $y = kx$ is the proportionality formula, where k is a constant
     of proportionality.

5 When y is directly proportional to the square of x:

   ● $y \propto x^2$ is the proportionality statement
   ● $y = kx^2$ is the proportionality formula, where k is the
     constant of proportionality.

6 When y is directly proportional to the cube of x:

- $y \propto x^3$ is the proportionality statement
- $y = kx^3$ is the proportionality formula, where k is the constant of proportionality.

7 When y is inversely proportional to x:

- $y \propto \frac{1}{x}$ is the proportionality statement

- $y = k \times \frac{1}{x}$ or $y = \dfrac{k}{x}$ are ways of writing the proportionality

formula, where k is the constant of inverse proportionality.

18 Graphs and higher order equations

18.1 Graphs of quadratic functions

■ **A quadratic function is one in which the highest power of x is x^2.**
For example $x^2 - 7$, $2x^2 - 3x + 2$ and $3x^2 + 4x$ are quadratic functions.

The expression $\frac{1}{2}V(V - 3)$ on page 224 is also a quadratic function. There is more about functions in Unit 24.

■ **The graph of a quadratic function is called a *parabola*.**

The simplest parabola has the equation $y = x^2$.

Here is a **table of values** for $y = x^2$:

x	-4	-3	-2	-1	0	1	2	3	4
y	16	9	4	1	0	1	4	9	16

$$(-4)^2 = -4 \times -4$$
$$= 16$$

And here is the parabola; its line of symmetry is the y-axis:

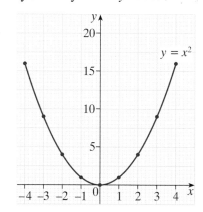

Example 1
Draw the graph of $y = 2x^2 - 4x - 3$, taking values of x from -2 to 4.

Here is a table of values:

x	-2	-1	0	1	2	3	4
y	13	3	-3	-5	-3	3	13

when $x = -2$:
$y = 2 \times (-2)^2 - (4 \times -2) - 3$
$\quad = 2 \times 4 + 8 - 3$
$\quad = 8 + 8 - 3$
$\quad = 13$

And here is the graph:

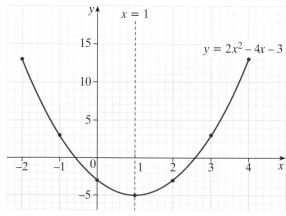

The line of symmetry is $x = 1$

Exercise 18A

(a) Draw graphs with the following equations, taking values of x from -4 to 4.

(b) In each case draw the line of symmetry and write down its equation.

1 $y = x^2 + 5$	**2** $y = x^2 - 10$
3 $y = 3x^2$	**4** $y = \frac{1}{2}x^2$
5 $y = -x^2$	**6** $y = -2x^2$
7 $y = x^2 + 2x$	**8** $y = x^2 + 3x$
9 $y = (x + 1)^2$	**10** $y = (x - 2)^2$
11 $y = x^2 - 4x - 1$	**12** $y = x^2 - 3x - 5$
13 $y = x^2 - x + 4$	**14** $y = 2x^2 + 3x - 5$
15 $y = 3x^2 - 4x + 2$	

18.2 Graphs of cubic functions

■ A *cubic* function is one in which the highest power of x is x^3. For example $x^3 - 3x + 2$, $5x^3$ and $2x^3 + 5x^2 + 4x - 7$ are all cubic functions.

The simplest cubic function has the equation $y = x^3$:

x	-3	-2	-1	0	1	2	3
y	-27	-8	-1	0	1	8	27

when $x = -3$:
$y = (-3)^3$
$\quad = (-3) \times (-3) \times (-3)$
$\quad = -27$

A tracing of this graph will fit on top of itself if you rotate it through $180°$ about the origin $(0, 0)$. It has **rotational symmetry** of order 2 about the origin.

And here is the graph:

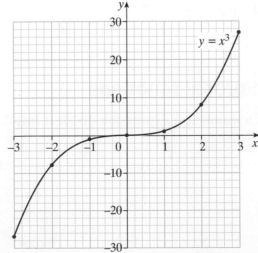

Example 2

Draw the graph of $y = x^3 - 2x^2 - 4x$, taking values of x from -2 to 4.

Here is a table of values for
$y = x^3 - 2x^2 - 4x$:

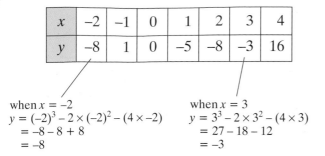

x	-2	-1	0	1	2	3	4
y	-8	1	0	-5	-8	-3	16

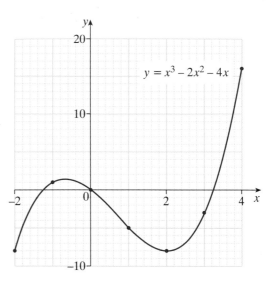

when $x = -2$
$y = (-2)^3 - 2 \times (-2)^2 - (4 \times -2)$
$\quad = -8 - 8 + 8$
$\quad = -8$

when $x = 3$
$y = 3^3 - 2 \times 3^2 - (4 \times 3)$
$\quad = 27 - 18 - 12$
$\quad = -3$

$y = x^3 - 2x^2 - 4x$

Exercise 18B

Draw graphs with the following equations, taking values of x from -3 to 3.

1 $y = x^3 + 5$ 2 $y = x^3 - 10$

3 $y = 2x^3$ 4 $y = \frac{1}{2}x^3$

5 $y = -x^3$ 6 $y = -2x^3$

7 $y = (x - 2)^3$ 8 $y = (x + 1)^3$

9 $y = x^3 - 2x^2$ 10 $y = x^3 + x^2$

11 $y = x^3 + 5x$ 12 $y = x^3 - 5x$

13 $y = x^3 + x^2 - 8x$ 14 $y = x^3 + 3x - 2$

15 $y = x^3 - 2x^2 + 2$

18.3 Graphs of reciprocal functions

■ **To find the reciprocal of a number or expression divide it into 1.**

For example:
the reciprocal of $2 = \frac{1}{2}$
the reciprocal of x is $\frac{1}{x}$.

■ **A line which a graph approaches without touching is called an** *asymptote*.

Example 3

Make a table of values for $y = \frac{1}{x}$ and draw the graph.

Is y undefined for any value of x?

Here is a table of values for $y = \frac{1}{x}$:

x	–3	–2	–1	–0.5	–0.2	0.2	0.5	1	2	3
y	–0.3	–0.5	–1	–2	–5	5	2	1	0.5	0.3

when $x = -3$:
$y = \frac{1}{-3} = -\frac{1}{3}$
$\quad = -0.3$ to 1 d.p.

y is undefined when $x = 0$
as division by 0 is undefined.

When x is very small and +ve
y is $\frac{1}{\text{very small}}$ and +ve
so y is very big and +ve.

There is a
discontinuity (break)
in the graph at $x = 0$

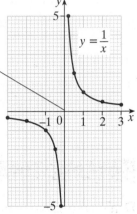

The graph approaches the axes but does not touch them. The axes are asymptotes.

Example 4

Make a table of values $y = 3 - \frac{2}{x}$ and draw the graph.

x	–3	–2	–1	–0.5	–0.2	0.2	0.5	1	2	3
y	3.7	4	5	7	13	–7	–1	1	2	2.3

when $x = -3$
$y = 3 - \frac{2}{-3}$
$\quad = 3 + \frac{2}{3}$
$\quad = 3.7$ to 1d.p.

when $x = -0.5$
$y = 3 - \frac{2}{-0.5}$
$\quad = 3 + 4$
$\quad = 7$

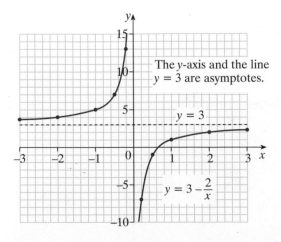

The y-axis and the line $y = 3$ are asymptotes.

Exercise 18C

Draw graphs with the following equations. Use the same x values as in Example 4.

1   $y = \frac{2}{x}$ 2 $y = \frac{-1}{x}$

3   $y = \frac{-3}{x}$ 4 $y = 4 + \frac{1}{x}$

5   $y = 5 - \frac{4}{x}$ 6 $y = \frac{4}{x} - 3$

18.4 Graphs of the form
$$y = ax^3 + bx^2 + cx + d + \frac{e}{x}$$

Sometimes you will need to plot graphs of equations that have a combination of cubic, quadratic, linear and reciprocal terms. For example: $y = ax^3 + bx^2 + cx + d + \frac{e}{x}$

Here a, b, c, d, e are numbers (also called coefficients).

If you have to plot graphs of this form, at least two of the coefficients will be zero.

You could check this using a graphical calculator, computer or graph plotter.

Example 5

Make a table of values for $y = x + \frac{1}{x}$ and draw the graph.

Here is a table of values:

x	0.1	0.2	0.5	1	2	3	4	5
y	10.1	5.2	2.5	2	2.5	3.3	4.3	5.2

when $x = 0.1$
$y = 0.1 + \frac{1}{0.1}$
$\quad = 0.1 + 10$
$\quad = 10.1$

when $x = 0.2$
$y = 0.2 + \frac{1}{0.2}$
$\quad = 0.2 + 5$
$\quad = 5.2$

when $x = 2$
$y = 2 + \frac{1}{2}$
$\quad = 2.5$

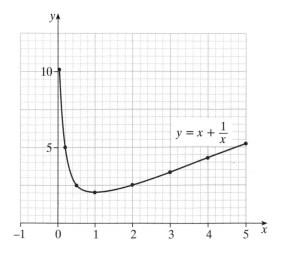

Example 6

Make a table of values for $y = x^2 + 3 + \frac{2}{x}$ and draw the graph.

x	-3	-2	-1	-0.5	-0.2	0.2	0.5	1	2	3
y	11.3	6	2	-0.8	-7.0	13.0	6.3	6	8	12.7

when $x = -3$
$y = (-3)^2 + 3 + \frac{2}{-3}$
  $= 9 + 3 - \frac{2}{3}$
  $= 11\frac{1}{3}$
  $= 11.3$ to 1 d.p.

when $x = 0.2$
$y = 0.2^2 + 3 + \frac{2}{0.2}$
  $= 0.04 + 3 + 10$
  $= 13.0$ to 1 d.p.

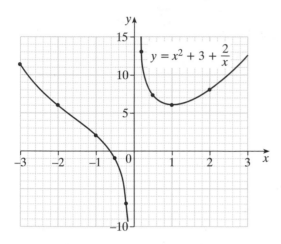

$$y = x^2 + 3 + \frac{2}{x}$$

Exercise 18D

Draw graphs with the following equations. Use the values of x given in brackets.

1. $y = x^2 - 4x + 1$ $(-1$ to $4)$
2. $y = x^3 + 2x^2 - 3$ $(-3$ to $2)$
3. $y = x^3 - x^2 - 2x$ $(-2$ to $3)$
4. $y = x^3 - 5x + 2$ $(-3$ to $3)$
5. $y = 1 + 2x - x^2$ $(-2$ to $4)$
6. $y = 2x + \frac{1}{x}$ (values as in Example 6)
7. $y = x^2 - \frac{1}{x}$ (values as in Example 6)
8. $y = x + 2 + \frac{2}{x}$ (values as in Example 5)
9. $y = x^3 - \frac{1}{x}$ $(0.2, 0.5, 1, 2, 3)$
10. $y = x^2 - x + \frac{2}{x}$ (values as in Example 6)

18.5 Using graphs to solve equations

The solutions to the quadratic equation $x^2 - 4 = 0$ are $x = 2$ and $x = -2$. Notice that these are the values of x where the graph of $y = x^2 - 4$ cuts the x axis.

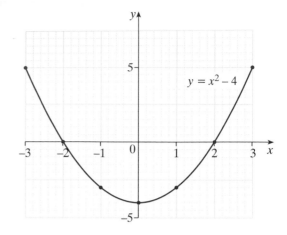

■ **The solutions of a quadratic equation are the values of x where the graph cuts the x axis.**

A similar method can be used to find approximate solutions to higher order equations. There is more about this in Unit 21.

Worked examination question

(a) Make a table of values for $y = x^2 - 2x - 2$.

(b) Plot the points represented by the values in your table on a grid and join them with a smooth curve.

(c) Use your graph to solve the equation $x^2 - 2x - 2 = 0$.
    Give your answers correct to 1d.p.

(a) Here is a table of values:

x	-2	-1	0	1	2	3	4
x^2	4	1	0	1	4	9	16
$-2x$	4	2	0	-2	-4	-6	-8
-2	-2	-2	-2	-2	-2	-2	-2
y	6	1	-2	-3	-2	1	6

(b) Here is the graph:

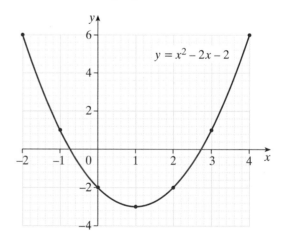

(c) The solutions to the equation $x^2 - 2x - 2 = 0$ are the values of x where the graph cuts the x-axis. These are $x = -0.7$ and $x = 2.7$.

Exercise 18E

Solve the following equations correct to 1 d.p. by drawing appropriate graphs.

Use the values of x given in brackets.

1 $x^2 - 7x + 8 = 0$ (0 to 8)   2 $x^2 - x - 3 = 0$ (−3 to 4)

3 $2x^2 - 3x - 7 = 0$ (−3 to 4)   4 $5x^2 - 8x + 2 = 0$ (−1 to 3)

5 $x^3 - 4x + 1 = 0$ (−3 to 3) [there are 3 solutions]

6 $x^3 + x^2 - 3x = 0$ (−3 to 2)

7 $x^3 - 2x^2 - 1 = 0$ (−2 to 3) [1 solution]
8 $x^3 - 4x^2 + 4x = 0$ (−1 to 4)
9 $x - \frac{5}{x} = 0$ (0.2, 0.5, 1, 2, 3, 4)
10 $x^2 - 3x + \frac{1}{x} = 0$ (0.2, 0.5, 1, 2, 3, 4)

18.6 Trial and improvement methods

You can use trial and improvement methods to solve an equation to any degree of accuracy.

Example 7

Find the positive solution to $x^3 - 3x - 1 = 0$.
Give your answer correct to 1 d.p.

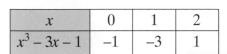

x	0	1	2
$x^3 - 3x - 1$	−1	−3	1

There is a solution to $x^3 - 3x - 1 = 0$ between $x = 1$ and $x = 2$.

It is not obvious whether the solution is nearer to 1 or to 2.

You could try 1.1, 1.2, … until the value of $x^3 - 3x - 1$ changes sign but it often saves time to try the middle value of an interval to establish in which half of the interval the solution lies.

x	1.5
$x^3 - 3x - 1$	−2.125

There is a solution to $x^3 - 3x - 1 = 0$ between $x = 1.5$ and $x = 2$.

x	1.6	1.7	1.8	1.9
$x^3 - 3x - 1$	−1.704	−1.187	−0.568	0.159

… there is a solution to $x^3 - 3x - 1 = 0$ between $x = 1.8$ and $x = 1.9$.

0.159 is nearer than −0.568 to 0. This suggests that the solution is nearer to 1.9 than it is to 1.8, but we must confirm this by trying $x = 1.85$.

x	1.85
$x^3 - 3x - 1$	−0.21838

There is a solution to $x^3 - 3x - 1 = 0$ between $x = 1.85$ and $x = 1.9$.
$$x = 1.9 \quad \text{(to 1 d.p.)}$$

■ **You can use a trial and improvement method to solve an equation by trying a value in the equation and changing it to bring the result closer and closer to the correct figure.**

Worked examination question

Using the method of trial and improvement, or otherwise, find the positive solution of $x^3 + x = 187$.
Give your answer correct to 1 d.p.

x	$x^3 + x$
5	130
6	222
5.5	171.875
5.6	181.216
5.7	190.893
5.65	186.012

$x = 5.7$ (to 1 d.p.)

Exercise 18F

Find the positive solutions to these equations by trial and improvement:

1   $x^3 - x - 1 = 0$ (1 d.p.)

2   $x^3 - x - 7 = 0$ (1 d.p.)

3   $x^3 - 4x - 1 = 0$ (1 d.p.)

4   $x^3 + 2x - 20 = 0$ (2 d.p.)

5   $x^3 + x = 300$ (1 d.p.)

6   $x^3 + x^2 = 700$ (2 d.p.)

7   $x^3 + 5x - 200 = 0$ (2 s.f.)

8   $x^3 - 4x = 300$ (3 s.f.)

9   $x^3 + 2x^2 = 120$ (2 d.p.)

10   $x^2 - \frac{2}{x} = 5$ (2 s.f.)

18.7 Graphs that describe real-life situations

■ **Graphs can be used to describe a wide variety of real-life situations. You may have to interpret or sketch graphs of this type.**

Two important types of graphs are distance-time graphs and speed-time graphs.

Distance-time graphs

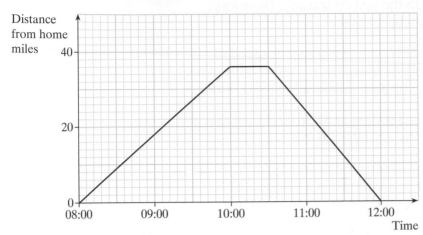

The diagram shows the distance-time graph for a cycle ride.
Between 08:00 and 10:00, the cyclist travels 36 miles. She travels a
constant speed, because this part of the graph is a straight line.

Her constant speed is $\dfrac{36}{2} = 18$ mph, which is the gradient of the line.

Between 10:00 and 10:30 (the horizontal part of the graph), the
cyclist is stationary, having a rest perhaps.

Between 10:30 and 12:00, she travels back home at a constant speed
of $\dfrac{36}{1.5} = 24$ mph.

Speed-time graphs

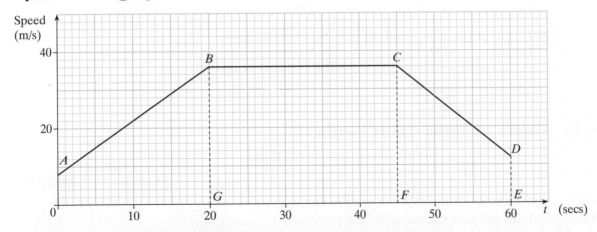

The diagram shows the speed/time graph for one minute of a car's
journey. In the first 20 seconds, its speed increases from 8 m/s to
36 m/s. Because AB is a straight line, the car's speed increases steadily.
The gradient of the line AB gives the rage of increase of speed:

$$\frac{36 - 8}{20} = 1.4 \text{ m/s per second.}$$

Rate of increase of speed is called **acceleration** and the units in this case are m/s per second or m/s^2. In the first 20 seconds, therefore, the car has a constant acceleration of 1.4 m/s^2.

Between $t = 20$ and $t = 45$, the car travels at a constant speed of 36 m/s. Between $t = 45$ and $t = 60$, its speed decreases steadily. The gradient of the line CD gives the constant acceleration as –1.6 m/s^2.

Negative acceleration is called **deceleration** and you could say that the deceleration between $t = 45$ and $t = 60$ is 1.6 m/s^2.

When the car is travelling at a constant speed of 36 m/s for 25 seconds, between $t = 20$ and $t = 45$, it travels a distance of $36 \times 25 = 900$ m. The area of rectangle $BCFG$ represents this distance.

Similarly, the area of trapezium $ABGO$ represents the distance travelled between $t = 0$ and $t = 20$:

$$\tfrac{1}{2}(8 + 36) \times 20 = 440 \text{ m.}$$

The area of trapezium CDEF represents the distance travelled between $t = 45$ and $t = 60$.

$$\tfrac{1}{2}(36 + 12) \times 15 = 360 \text{ m.}$$

So you can work out the total distance travelled by the car in the whole minute:

$$900 + 440 + 360 = 1700 \text{ m}$$

■ **On a distance-time graph, the gradient gives the speed.**

■ **On a speed-time graph, the gradient gives the acceleration.**

■ **The area under a speed-time graph gives the distance travelled.**

Drawing graphs to show water levels

The vases **A**, **B**, **C** and **D** have circular cross-sections and they contain water.

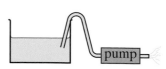

They all start off with the same depth of water. The water is syphoned out of the vases at the same steady rate.

 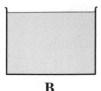

A   **B**   **C**   **D**

The graph shows the relationship between the water level in each vase and the volume of water pumped out of it.

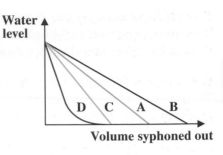

Water level

Volume syphoned out

- Vase **B** has a bigger area of cross-section than **A**. When the same volume has flowed out of each vase the water level in **B** remains higher than in **A**. So the graph for **B** is less steep than the graph for **A**.

- The water level in vase **C** drops more quickly than in **A**, so the graph for **C** is steeper.

- In vase **D** the level drops more quickly and steadily at first, then it gradually drops more slowly. The graph is straight to begin with (steeper than for **C**), but then it is curved to show the changing speed.

Worked examination question

A DJ can control the sound level of the records he plays at a club. The sketch graph is a graph of sound level against time whilst one record was played.

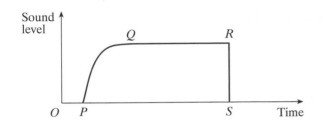

Sound level

(a) Describe how the sound level changed between P and Q on the graph whilst the record was being played.

(b) Give one possible reason for the third part, RS, of the sketch graph.

Answer

(a) The sound level increases quickly at first and then more slowly. It starts to level out.

(b) The sound level suddenly drops to zero. Possible reasons for this are the end of the record or a power cut.

Exercise 18G

1 Sharon and Tracey cycle from Sinton to Coseley and back again. The graph shows their journeys.

(a) Who sets off first?

(b) Describe what happens at A.

(c) Describe what happens at B.

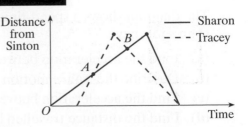

Distance from Sinton

Time

—— Sharon
- - - Tracey

2 Heather walks home from school at a steady speed. Halfway home, she has a rest before continuing her journey at the same speed. Sketch a distance-time graph to show her journey.

3

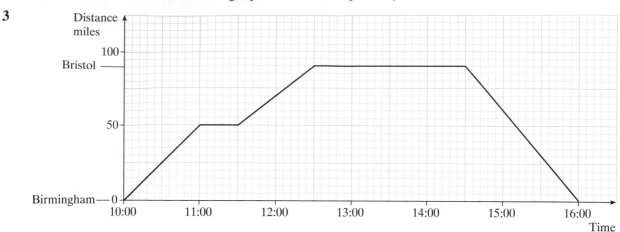

The diagram shows a distance-time graph for a coach trip from Birmingham to Bristol and back again.

(a) At what time did the coach first stop?

(b) At what time did the coach reach Bristol?

(c) For how long did the coach stay in Bristol?

(d) Calculate the speed of the coach on the return journey.

4 A car in traffic travelled for 20 seconds at 15 m/s, stopped for 15 seconds and then travelled for 25 seconds at 10 m/s. Draw a distance-time graph for this 60 second period.

5

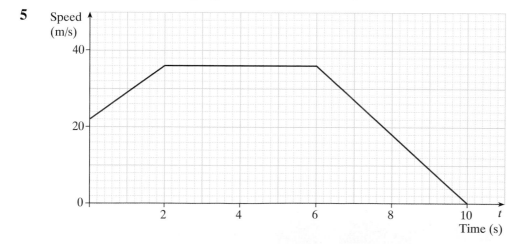

The diagram shows a speed-time graph for 10 seconds of a train's journey

(a) Find the acceleration between $t = 0$ and $t = 2$.

(b) Describe the train's motion between $t = 2$ and $t = 6$.

(c) Find the acceleration between $t = 6$ and $t = 10$.

(d) Find the distance travelled by the train in the 10 seconds.

6 A car accelerates from rest at 5 m/s² for 6 seconds, travels with constant speed for 10 seconds and decelerates at 3 m/s² for 4 seconds.

(**a**) Draw a speed-time graph for this 20 second period.

(**b**) Find the total distance travelled by the car in the 20 seconds.

7 Here are some graphs:

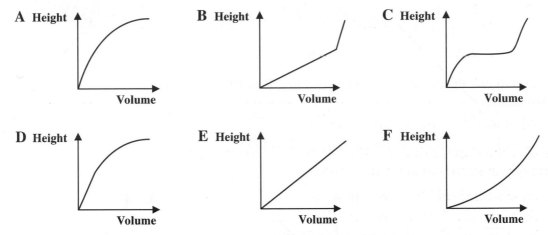

Coloured liquid is poured into these containers at a steady rate. For each one write down the letter of the graph which best illustrates the relationship between the height of the liquid and the volume in the container.

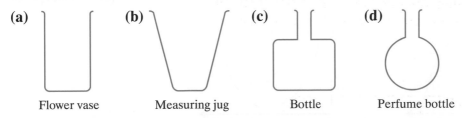

(**a**) Flower vase (**b**) Measuring jug (**c**) Bottle (**d**) Perfume bottle

8 Water is poured into these containers at a steady rate. Sketch a graph to show the relationships between the water level and the volume of water in each container.

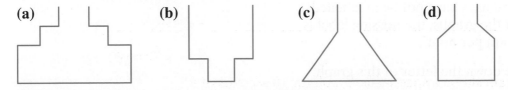

(**a**) (**b**) (**c**) (**d**)

9 Nigel drives 100 miles. Sketch a graph to show the relationship between his average speed and the time he takes.

10 Sketch a graph to show how the value of a new car changes over a number of years.

11 One evening Mr Fish marks a set of maths exam papers. From Graph A describe the three parts of his evening:

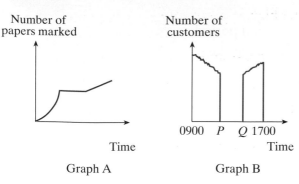

Graph A   Graph B

Graph B shows the number of customers in a Post Office during one day:

(a) What was the busiest part of the day?

(b) What happened in the time between P and Q?

12

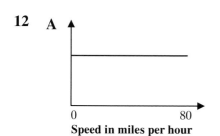

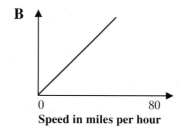

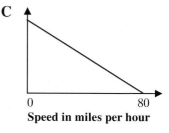

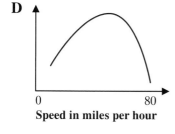

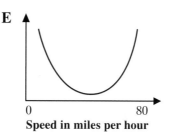

The diagrams show the shapes of five graphs, **A**, **B**, **C**, **D** and **E**.
The vertical axes have not been labelled.
On one of the graphs, the missing label is 'Speed in km per hour'.

(a) Write down the letter of this graph.

On one of the graphs the missing label is 'Petrol consumption in miles per gallon'. It shows that the car travels furthest on 1 gallon of petrol when it is travelling at 56 miles per hour.

(b) Write down the letter of this graph.   [E]

Summary of key points

1   A **quadratic** function is one in which the highest power of x is x^2.

2   The graph of a quadratic function is called a **parabola**.

3   A **cubic** function is one in which the highest power of x is x^3.

4   To find the **reciprocal** of a number or expression divide it into 1. For example, the reciprocal of x is $\frac{1}{x}$.

5   A line which a graph approaches without touching is called an **asymptote**.

6   The solutions of a quadratic equation are the values of x where the graph cuts the x-axis.

7   You can use a trial and improvement method to solve an equation by trying a value in the equation and changing it to bring the result closer and closer to the correct figure.

8   Graphs can be used to describe a wide variety of real-life situations. You may have to interpret, or sketch graphs of this style.

9   On a distance-time graph, the gradient gives the speed.

10   On a speed-time graph, the gradient gives the acceleration.

11   The area under a speed-time graph gives the distance travelled.

19 Advanced mensuration

19.1 Finding arcs and areas of circles

One of the problems that led to the development of trigonometry was finding the relationship between:
the angle a cannon could pivot through,
the distance a cannon ball could travel, and
the width of a passage into a harbour.

There is more about this on page 267.

A natural extension of this work is to look at how to find:

- the length of an arc of a circle for an angle θ:
- the area of a sector:
- the area of a segment:

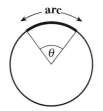

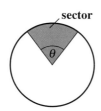

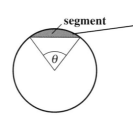

This is called a **minor** segment.

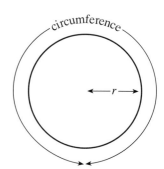

To do this you will need these formulae:

■ **Circumference of a circle = $2\pi r$**

■ **Area of a circle = πr^2**

Finding the length of an arc of a circle

In this diagram the arc length is $\frac{1}{4}$ of the circumference because the angle at the centre is $90°$ which is $\frac{90}{360} = \frac{1}{4}$ of a whole turn.

In this diagram the angle at the centre is $60°$. This is $\frac{60}{360} = \frac{1}{6}$ of a whole turn. So the arc length is $\frac{1}{6}$ of the circumference of the circle.

In this diagram the angle at the centre is $240°$. This is $\frac{240}{360} = \frac{2}{3}$ of a whole turn. So the arc length is now $\frac{2}{3}$ of the circumference of the circle.

■ If the angle at the centre of a circle is θ then:

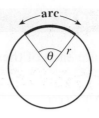

$$\text{arc length} = \frac{\theta}{360} \text{ of the circumference}$$

$$= \frac{\theta}{360} \times 2\pi r$$

$$\text{so: arc length} = \frac{2\pi r\theta}{360} = \frac{\pi r\theta}{180}$$

Example 1

Calculate the length of the arc AB:

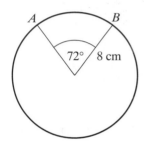

Answer

$$\text{Arc length} = \frac{\pi \times 8 \times 72}{180}$$

$$= 10.05 \text{ cm (correct to 2 d.p.)}$$

Example 2

The length of the arc PQ is 12 cm.
Calculate the angle θ.

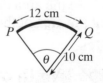

Answer

$$\text{arc length} = \frac{\pi r\theta}{180}$$

so:

$$\frac{\pi r\theta}{180} = 12$$

Rearrange this by multiplying each side by 180 and dividing each side by πr:

$$\theta = \frac{12 \times 180}{\pi r}$$

$$\theta = \frac{12 \times 180}{\pi \times 10}$$

so:

$$\theta = 68.75°$$

Exercise 19A

1 Calculate each of these arc lengths:

(a)

(b)

(c)

(d)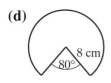

(e)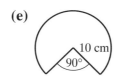

(f)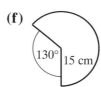

2 Calculate each of the angles marked θ:

(a)

(b)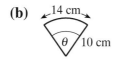

(c)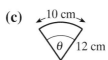

(d)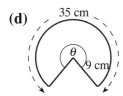

(e)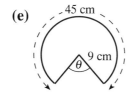

(f)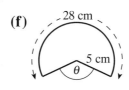

Finding the area of a sector of a circle

In this diagram the sector is $\frac{1}{4}$ of the whole circle, because the angle at the centre is 90°. This is $\frac{1}{4}$ of a whole turn or 360°. So the area of this sector is $\frac{1}{4}$ of the area of the circle.

In this diagram the angle at the centre is 60°. So the sector is $\frac{60}{360}$ or $\frac{1}{6}$ of the area of the whole circle. The area of this sector is $\frac{1}{6}$ of the area of the circle.

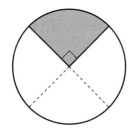

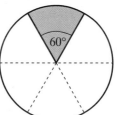

■ **If the angle at the centre of a circle is θ then:**

$$\text{area of a sector} = \frac{\theta}{360} \text{ of area of circle}$$

$$= \frac{\theta \times \pi \times r^2}{360}$$

so: area of a sector $= \dfrac{\pi r^2 \theta}{360}$

Example 3

Calculate the area of this sector:

$$\text{area} = \frac{\pi r^2 \theta}{360}$$

When $r = 8$ then $r^2 = 8 \times 8 = 64$

so:

$$\text{area} = \frac{\pi \times 64 \times 75}{360}$$

$$\text{area} = 41.89 \text{ cm}^2 \text{ (correct to 2 d.p.)}$$

Example 4

AOB is a sector of a circle.
Angle $AOB = 130°$
Area of the sector $AOB = 200 \text{ cm}^2$.

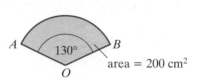

Calculate the radius OA of the circle of which AOB is a sector.

$$\text{area of sector} = \frac{\pi r^2 \theta}{360} = 200$$

Rearranging by multiplying each side by 360 and dividing each side by $\pi\theta$ gives:

$$r^2 = \frac{200 \times 360}{\pi \times \theta} = \frac{200 \times 360}{\pi \times 130}$$

$$= 176.2947$$

so:

$$r = \sqrt{176.2947}$$

$$= 13.28 \text{ cm (correct to 2 d.p.)}$$

Exercise 19B

1 Calculate the area of each of these sectors of circles:

(a)

(b)

(c)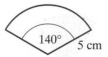

(d)

(e)

(f)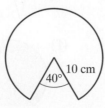

2   *OPQ* is a sector of a circle centre *O* of radius 9 cm.
The area of the sector *OPQ* is 51 cm².
Calculate the size of the angle *POQ*.

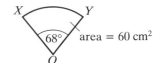
area = 51 cm² 9 cm

3   *OXY* is a sector of a circle centre *O*.
The area of the sector *OXY* is 60 cm².
The angle *XOY* = 68°.
Calculate the length of the radius *OX* of the circle.

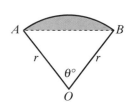
68° area = 60 cm²

Finding the area of a segment of a circle

The area of the sector $OAB = \dfrac{\pi r^2 \theta}{360}$

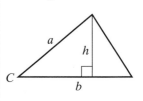

In triangle *OAB*: area of triangle $OAB = \frac{1}{2} r \times r \sin\theta$
$$= \frac{1}{2} r^2 \sin\theta$$

The area of the triangle *OAB* is $\frac{1}{2} r^2 \sin\theta$.

So the area of the shaded segment is the difference between these two areas:

■   **area of segment** $= \dfrac{\boldsymbol{\pi r^2 \theta}}{\mathbf{360}} - \frac{1}{2}\boldsymbol{r}^2 \sin \boldsymbol{\theta}$

> **Note:**
> For any triangle:
> Area $= \frac{1}{2} b \times h$
> $h = a \sin C$
> Area $= \frac{1}{2} b \times a \sin C$
> Area $= \frac{1}{2} ab \sin C$

Example 5

Calculate the area of the shaded segment of the circle:

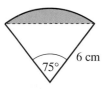
75° 6 cm

area of segment $= \dfrac{\pi r^2 \theta}{360} - \frac{1}{2}r^2 \sin \theta$

$$= \dfrac{\pi \times 6^2 \times 75}{360} - \frac{1}{2} \times 6^2 \times \sin 75°$$

$$= 23.562 - 17.387$$

$$= 6.18 \text{ cm}^2$$

Exercise 19C

1   Calculate the area of each shaded segment:

  (a)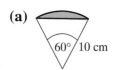
60° 10 cm

  (b)
75° 8 cm

  (c)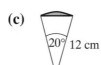
20° 12 cm

  (d)
32° 15 cm

  (e)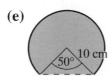
50° 10 cm

  (f)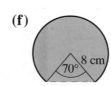
70° 8 cm

Worked examination question

O is the centre of the circle, radius 8 cm.
P and Q are points on the circumference.
Angle $POQ = 108°$.

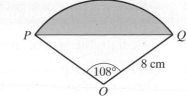

(a) Calculate the area of the sector POQ.

(b) Calculate the difference between the arc length PQ and the length of the chord PQ.

(c) Calculate the area of the shaded segment PQ.

Give your answers correct to three significant figures.

(a) area of sector $= \dfrac{\pi r^2 \theta}{360}$

When $r = 8$ then $r^2 = 64$ and $\theta = 108$

so: area of sector $= \dfrac{\pi \times 64 \times 108}{360} = 60.318\,578\,95$ (by calculator)

$\qquad\qquad\qquad = 60.3$ cm^2 (correct to 3 s.f.)

(b) Length of arc $PQ = \dfrac{\pi r \theta}{180} = \dfrac{\pi \times 8 \times 108}{180} = 15.0796$ cm

To find the length of the chord PQ mark the mid-point of PQ as M:

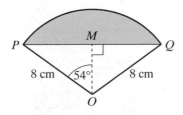

$\dfrac{PM}{8} = \sin 54°$ so $PM = 8 \times \sin 54° = 6.4721$ cm

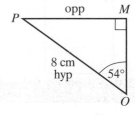

$PQ = 2 \times PM$ so $PQ = 2 \times 6.4721 = 12.9442$ cm

The difference between the arc length PQ and the chord PQ is

$\qquad\qquad 15.0796 - 12.9442 = 2.1354$

$\qquad\qquad\qquad\qquad = 2.14$ cm (correct to 3 s.f.)

(c) The area of the shaded segment PQ $\quad \dfrac{\pi r^2 \theta}{360} - \tfrac{1}{2} r^2 \sin \theta$

$\qquad$ area of shaded segment $=$ result from part (a) $- \tfrac{1}{2} r^2 \sin \theta$

$\qquad\qquad\qquad = 60.3186 - \tfrac{1}{2} \times 64 \times \sin 108°$

$\qquad\qquad\qquad = 60.3186 - 30.4338$

$\qquad\qquad\qquad = 29.8848$

$\qquad\qquad\qquad = 29.9$ cm^2 (correct to 3 s.f.)

1

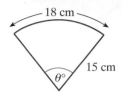

The arc of the circle of radius 15 cm has a length of 18 cm.

Calculate: **(a)** the angle θ **(b)** the area of the sector.

2

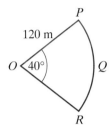

The diagram shows the landing area for a javelin competition. $OPQR$ is a sector of a circle, centre O, of radius 120 metres.

Calculate:

(a) the length of the chord PR **(b)** the length of the arc PQR

(c) the area of the sector $OPQR$ **(d)** the area of the segment PQR.

3

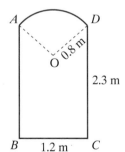

A church door is in the shape of a rectangle $ABCD$ with a sector OAD of a circle.

$DC = AB = 2.3$ m, $BC = AD = 1.2$ m and the radius of the circle is OA where $OA = OD = 0.8$ m.

Calculate:

(a) the perimeter of the door   **(b)** the area of the door.

The door is made of wood which has a uniform thickness of 7 cm.

Calculate:

(c) the volume of the door.

19.2 Finding volumes and surface areas

You need to know how to find the volumes and surface areas of a variety of 3-D objects.

The surface area and volume of a cylinder

This cylinder has a circular base of radius r cm and a height of h cm.

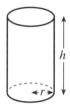

Its surface area is made up of the curved surface area plus the areas of the circular top and base.

The area of the circular base and top are both equal to πr^2. So the combined area of the top and base is $2\pi r^2$.

For the curved surface area, imagine unwrapping the cylinder. This creates a rectangle with length equal to the circumference of the circular base $2\pi r$ and width equal to the height h.

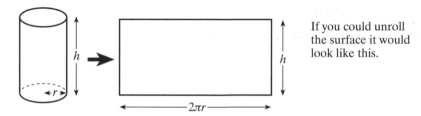

If you could unroll the surface it would look like this.

The curved surface area is $2\pi rh$.

So the total surface area of the cylinder in cm^2 is $2\pi rh + 2\pi r^2$.

■ **For a cylinder of height h with a circular base of radius r
surface area $= 2\pi rh + 2\pi r^2$.**

The volume of a cylinder

■ **The volume of a cylinder is $\pi r^2 h$.**

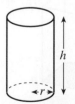

A cylinder is like a prism with a circular base. So:
volume = area of base × height

Example 6

Calculate the surface area and volume of a cylinder with circular base of radius 12 cm and height 30 cm.

$$\text{surface area} = 2\pi rh + 2\pi r^2$$
$$= 2 \times \pi \times 12 \times 30 + 2 \times \pi \times 12^2$$
$$= 2261.947 + 904.77868$$
$$= 3166.725 = 3167 \text{ cm}^2 \text{ (4 s.f.)}$$

$$\text{volume} = \pi r^2 h$$
$$= \pi \times 12^2 \times 30$$
$$= 13572 \text{ cm}^3 \text{ (5 s.f.)}$$

The volume of other prisms

A prism is a 3-D shape which has the same cross-section throughout its height (see pages 43 and 341).

Each of these is also a prism:

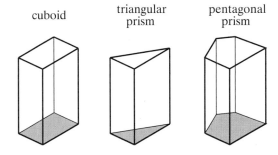

cuboid   triangular prism   pentagonal prism

The volume of any prism is:

■ **volume of a prism
= area of base × vertical height**

or as the base, top and any cross-section are all identical:

■ **volume of a prism
= area of cross-section × vertical height**

Example 7

Here is a pentagonal prism.
The base *ABCDE* is a regular pentagon of side length 8 cm.
The height of the prism is 15 cm.

Calculate the volume of the prism.

First calculate the area of the base *ABCDE*.

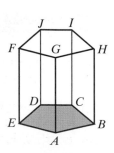

To do this split it into 5 congruent triangles and look at the area of one of these triangles *AOB*.

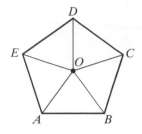

The area of *AOB* is $\frac{1}{2}AB \times p$ where p is the perpendicular distance *OM* from *O* to *AB*.

$$\text{Angle } AOB \text{ is } \frac{360°}{5} = 72°$$

so angle *AOM* is $\frac{1}{2} \times 72° = 36°$

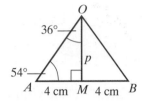

and angle *OAM* is $90° - 36° = 54°$

Using the tangent ratio: $\quad \dfrac{p}{AM} = \tan 54°$ so $p = AM \times \tan 54°$

so:
$$p = 4 \times \tan 54°$$
$$= 5.5055 \text{ cm}$$

so the area of *AOB* is:
$$AM \times p = 4 \times 5.5055 = 22.022 \text{ cm}^2$$

As *ABCDE* is made up of 5 identical triangles:
$$\text{area } ABCDE = 5 \times 22.022$$
$$= 110.11 \text{ cm}^2$$

The volume of the prism is area of base × vertical height

so:
$$\text{volume of prism} = 110.11 \times 15$$
$$= 1651.65 \text{ cm}^3$$
$$= 1652 \text{ cm}^3 \text{ (4 s.f.)}$$

Exercise 19E

1   Find the volume and surface area of a cylinder of height 4 cm and circular base of radius 5 cm (to 3 s.f.).

2   A prism of height 10 cm has the cross-sectional shape of an equilateral triangle of side 6 cm. Find the volume and surface area of the prism (to 3 s.f.).

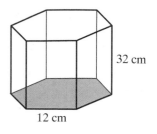

3 A waste bin is the shape of a circular cylinder. Its base has radius 0.5 m and its height is 1.2 m. Find its volume (to 3 s.f.).

4 A hexagonal prism has a vertical height of 32 cm. The base of the prism is a regular hexagon of side length 12 cm.

Calculate the volume of the prism.

5 A prism has a vertical height of 30 cm. The base of the prism is a regular polygon with 10 sides. Each side is of length 6 cm.

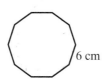

(a) Calculate the volume of the prism.

(b) Calculate the surface area.

The volume of a pyramid

These shapes are all **pyramids**:

tetrahedron

square-based pyramid

hexagonal pyramid

The base of any pyramid is a **polygon**. The other edges are all straight lines which meet at a point, usually called the vertex.

Here is how to find the volume of a pyramid:

The area of a triangle is $\frac{1}{2}$ the area of its smallest surrounding rectangle.

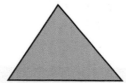
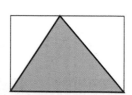

In a similar way the volume of a pyramid is $\frac{1}{3}$ the volume of its smallest surrounding prism.

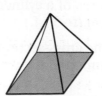
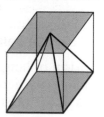

As the volume of a prism is area of base × vertical height:

■ **volume of a pyramid $= \frac{1}{3} \times$ area of base × vertical height**

A **cone** is a pyramid with a circular base. Its smallest surrounding *prism* is a cylinder, so:

volume of a cone
$= \frac{1}{3} \times$ area of base × height
$= \frac{1}{3}\pi r^2 h$

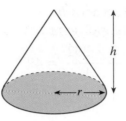
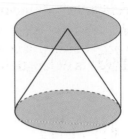

Example 8

A pyramid *VABCD* has a rectangular base *ABCD*.
The vertex *V* is 15 cm vertically above the mid-point of the base. *AB* = 4 cm and *BC* = 9 cm.

Calculate the volume of the pyramid.

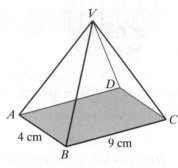

The area of the base is $4 \times 9 = 36$ cm²

so: volume of pyramid $= \frac{1}{3} \times$ area of base × vertical height
$= \frac{1}{3} \times 36 \times 15$
$= 180$ cm³

Example 9

Here is a cone. The circular base has a diameter *AB* of length 10 cm. The **slant height** *AV* is of length 13 cm.

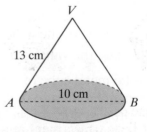

Calculate the volume of the cone.

Before you can work out the volume, you need to work out the vertical height from the mid-point of *AB* to *V*.

Do this by applying Pythagoras' Theorem to the triangular cross-section *ABV*:

$$AV^2 = h^2 + MA^2$$
$$h^2 = AV^2 - MA^2$$
$$h^2 = 13^2 - 5^2$$
$$h^2 = 169 - 25 = 144$$
$$h = 12 \text{ cm}$$

so: volume of cone $= \frac{1}{3}\pi r^2 h$
$= \frac{1}{3} \times \pi \times 5^2 \times 12$
$= \frac{1}{3} \times \pi \times 25 \times 12$
$= 314.2$ cm³ (correct to 1 d.p.)

Example 10

A pyramid has a square base of size x cm and a vertical height 24 cm. The volume of the pyramid is 392 cm³.

Calculate the value of x.

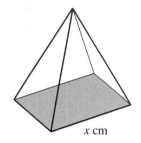

$$\text{volume} = \tfrac{1}{3} \times \text{area of base} \times \text{height}$$
$$= \tfrac{1}{3} \times x^2 \times h$$

so:
$$392 = \tfrac{1}{3} \times x^2 \times 24$$
$$x^2 = \frac{3 \times 392}{24} = 49$$
$$x = 7$$

Exercise 19F

1   $VABCD$ is a square-based pyramid. The vertex V is 20 cm vertically above the mid-point of the horizontal square base $ABCD$, and $AB = 12$ cm.

   Calculate the volume of the pyramid.

2   $VABC$ is a tetrahedron. The vertex V is vertically above the point B. The base, ABC is a triangle right angled at B.
   $AB = 5$ cm, $BC = 7$ cm and $VC = 25$ cm.

   Calculate the volume of the tetrahedron $VABC$.

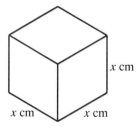

3   A cone has a circular base of radius r cm. The vertical height of the cone is 15 cm. The volume of the cone is 600 cm³.

   Calculate the value of r.

4   A solid metal cube of side x cm is melted down and re-cast to make a cone. During this process none of the metal is lost and all of the metal is used to make the cone. The cone has a circular base of diameter 12 cm and a slant height of length 10 cm.

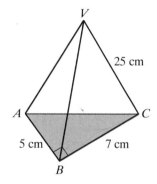

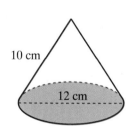

   Calculate:

   (a)   the volume of the cone

   (b)   the value of x.

5   A solid pyramid has a square base of side length 8 cm. The vertical height of the pyramid is 15 cm. The pyramid is made by melting down and re-casting a metal cube of side length y cm. During the melting and re-casting process 10% of the metal is lost but all of the remainder is used to make the cone.

   Calculate the value of y.

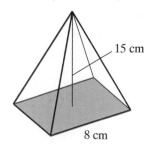

Spheres

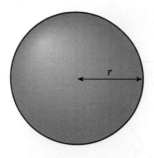

■ **For a sphere of radius r:**

volume of sphere $= \dfrac{4\pi r^3}{3}$

surface area $= 4\pi r^2$

Example 11

Calculate the volume of a sphere of radius 5 cm.

$$\text{volume of sphere} = \frac{4 \times \pi \times 5^3}{3}$$

$$= \frac{4 \times \pi \times 125}{3}$$

$$= 523.6 \text{ cm}^3 \text{ (correct to 1 d.p.)}$$

Example 12

The surface area of a sphere is 2000 cm^2.

Calculate the radius of the sphere.

Using surface area $4\pi r^2$:

$$4\pi r^2 = 2000$$

so: $$r^2 = \frac{2000}{4\pi}$$

$$r^2 = \frac{500}{\pi}$$

$$r^2 = 159.15$$

so: $$r = 12.62 \text{ cm (correct to 2 d.p.)}$$

Exercise 19G

1   Calculate the volume and surface area of a sphere:

   (a) of radius 8 cm

   (b) of diameter 19 cm.

2   A sphere has a volume of 5000 cm^3.
    Calculate the radius of the sphere.

3   A cube of side x cm and a sphere of radius 6 cm have equal volumes.
    Calculate the value of x.

19.3 Areas and volumes of similar shapes

This rectangle measures 1 unit by 2 units: Enlarge it by scale factor 2:

It takes 4 of the smaller rectangles to fill the enlarged one.

   area of enlarged rectangle = 4 × area of first rectangle

Here is a 1 by 2 rectangle: Enlarge it by scale factor 3:

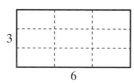

This time it takes 9 (or 3^2) of the smaller rectangle to fill the larger one.

   area of enlarged rectangle = 9 × area of first rectangle

This relationship can be generalized for any enlargement:

■   **When a shape is enlarged by scale factor k the area of the enlarged shape is k^2 × the area of the original shape.**

For example: if the scale factor is 5, the new area will be 5^2 (= 25) times the original area.

A similar result holds for 3-D shapes:

Here is a 1 by 1 by 2 cuboid:

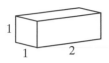

Enlarge it by scale factor 2:

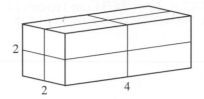

8 (or 2^3) of the smaller cuboids are needed to fill the enlarged shape. So the volume of the enlarged shape is 8 times the volume of the original.

This result generalises to:

■ **When a 3-D shape is enlarged by scale factor k the volume of the enlarged shape is $k^3 \times$ volume of the original shape.**

For example: if the scale factor is 5, the new volume will be 5^3 ($= 125$) times the original volume.

Example 13

The Kola company manufactures two **similar** bottles called *Standard* and *Super*.

The *standard* bottle has volume 1000 cm^3. The *super* bottle has volume 1500 cm^3. The height of the *standard* bottle is 30 cm. Calculate the height of the *super* bottle.

Because the two bottles are similar, *super* is an enlargement of *standard*.

To find the height of *super* you need to calculate the scale factor of the enlargement. Let this scale factor be k.

So: volume of *super* $= k^3 \times$ volume of *standard*

$$k^3 = \frac{\text{volume of } super}{\text{volume of } standard}$$

$$k^3 = \frac{1500}{1000}$$

$$k^3 = 1.5$$

$$k = \sqrt[3]{1.5}$$

$$k = 1.1447$$

Then the height of *super* is scale factor ($k = 1.1447$) × height of *standard*.

The height of *super* is $1.1447 \times 30 = 34.34$ cm (correct to 2 d.p.).

Example 14

The ratio of the radii of two spheres is $1:3$.
Calculate the ratio of:

(a) the surface areas of the spheres

(b) the volumes of the spheres.

In this case the scale factor of the enlargement is 3 because the larger sphere has a radius 3 times that of the smaller sphere. So the surface area of the larger sphere is $3^2 = 9$ times that of the smaller sphere.

So for **(a)** the ratio of the surface areas is $1:9$.

The volume of the larger sphere is $3^3 = 27$ times that of the smaller sphere.

So for **(b)** the ratio of the volumes is $1:27$.

Exercise 19H

1 Washing liquid is sold in two sizes of similar bottles called *Basic* and *Extra*.

The volume of a *Basic* bottle is $\frac{1}{2}$ litre.
The height of a *Basic* bottle is 10 cm.

The volume of an *Extra* bottle is 0.8 litres.

Calculate the height of an *Extra* bottle.

2 A ball is a sphere of radius 12 cm.
The ball is pumped up to increase its radius to 12.5 cm.
Show that the volume of the ball increases by approximately 13%.

3 The heights of two similar cylinders are in the ratio 2:3.
Calculate the ratio of:
  (a) the surface area of the cylinders
  (b) the volumes of the cylinders.

4 A model car is a scale replica of the real car.
The length of the model car is $\frac{1}{50}$ of the length of the real car.

Calculate the fraction $\dfrac{\text{volume of the model car}}{\text{volume of the real car}}$

5 Beans are sold in two similar cylindrical cans called *Standard* and *Large*.
The *Standard* can holds 500 grams of beans.
The *Large* can holds 750 grams of beans.
The height of a *Standard* can is 11 cm.
Calculate:
  (a) the height of a *Large* can
  (b) the ratio of the areas of the circular bases of the cans.

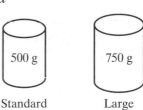

Worked examination question

The diagram represents a box of chocolates.
This is a *standard* size box.

The lid and base of the box is a sector of a circle centre O.
The radius OA of the circle is 18 cm.
The angle $A\hat{O}B = 75°$.
The box is 6 cm deep.

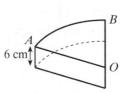

Calculate:

(a) the area of the face of the top of the box

(b) the volume of the box.

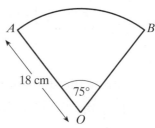

The company manufacturing these boxes decides to manufacture a *large* size box which will be similar in shape to the *standard* size box.

The volume of the *large* size box is to be twice the volume of the *standard* size box.

(c) Calculate the depth of the *large* size box.

Answer

(a)

$$\text{Area of the face of the top} = \frac{\pi \times 18^2 \times 75}{360}$$

$$= \frac{\pi \times 324 \times 75}{360}$$

$$= 212 \text{ cm}^2 \text{ (correct to 3 s.f.)}$$

(b) Volume of box = area of top (or area of base) × height (or depth)

so: volume of the box = 212 × 6

$$= 1272 \text{ cm}^3$$

(c) Let the scale factor for lengths be k. Then we have:

volume of *large* box = k^3 × volume of *standard* box

so:

$$k^3 = \frac{\text{volume of } large \text{ box}}{\text{volume of } standard \text{ box}} = 2 \qquad \text{so:} \quad k^3 = 2$$

$$k = \sqrt[3]{2}$$

$$= 1.2599$$

As the scale factor k is 1.2599:

depth of the *large* box = 1.2599 × depth of *standard* box

$$= 1.2599 \times 6 = 7.56 \text{ cm (correct to 2 d.p.)}$$

19.4 Compound solids

You will need to be able to find the volume of solids made from combining shapes you already know.

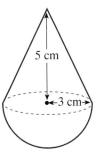

Example 15

This spinning top is made from a hemisphere and a cone. Find its volume. Give your answer to 2 d.p.

Answer

Imagine splitting the shape into two sections.

A sphere of radius 3 cm has volume

Remember: You can leave your answers in terms of π until you have finished your working out.

$$\frac{4\pi(3)^3}{3} = 36\pi \text{ cm}^3$$

The hemisphere is half a sphere so it has volume 18π cm^3.

The volume of the cone is

$$\tfrac{1}{3}\pi(3)^2(6) = 15\pi \text{ cm}^3$$

So the volume of the entire spinning top is

$$18\pi + 18\pi = 36\pi = 113.10 \text{ cm}^3 \text{ (correct to 2 d.p.)}$$

Example 16

This hat is made from a cone with the top chopped off. (This is known as a **truncated cone**.)
Find its volume correct to 2 d.p.

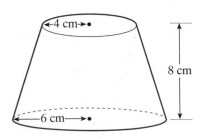

Answer

Imagine the hat was continued up to make a complete cone of vertical height h. If you know the value of h you can find the volume of the hat by subtracting the volume of the smaller cone from the volume of the larger one.

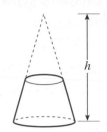

You can find h using similar triangles.

Here is a section from the hat:

Triangle ABE and triangle APD are similar, so

$$\frac{h}{6} = \frac{8}{2}$$

$$h = 24 \text{ cm}$$

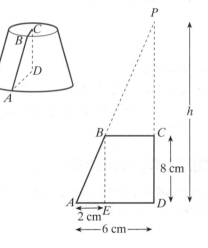

So: volume of larger cone $= \frac{1}{3}\pi(6)^2(h)$

$$= \frac{1}{3}\pi(6)^2(24)$$

$$= 288\pi$$

volume of smaller cone $= \frac{1}{3}\pi(4)^2(h - 8)$

$$= \frac{1}{3}\pi(4)^2(16)$$

$$= \frac{256\pi}{3}$$

So the volume of the hat is

$$288\pi - \frac{256\pi}{3} = \frac{608\pi}{3} = 636.70 \text{ cm}^3 \text{ (2 d.p.)}$$

Exercise 19 I

1   This box is made from a cube of side 8 cm and a pyramid of height 5 cm. Find its volume.

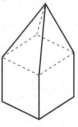

2   This spinning top is made from a cube, a circular prism and a cone. Find its volume, giving your answer to 2 d.p.

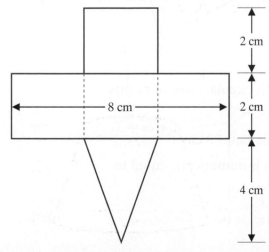

3 Find the volume of this truncated cone.
Give your answer to 2 d.p.

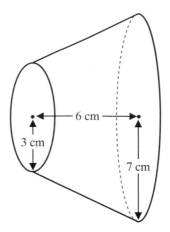

4 The diagram shows a model of
Cleopatra's Needle. It is made
from a truncated pyramid and
a pyramid. Find its volume,
giving your answer to 2 d.p.

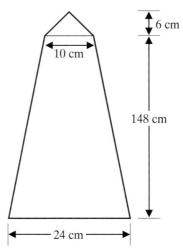

Exercise 19J (Mixed questions)

1 Here is a sector of a circle,
OAPB with centre *O*.

Calculate:

(a) the length of the chord *AB*

(b) the length of the arc *APB*

(c) the area of the sector *OAPB*

(d) the area of the segment *APB*.

2 Calculate the volume of a cylinder with circular base of radius
12 cm and vertical height 20 cm.

3 Calculate the volume of a sphere of diameter 8 cm.

4 The volume of a sphere of radius *r* cm is numerically equal to
the radius of that sphere.

$$\text{Show that } r = \sqrt{\frac{3}{4\pi}}$$

5 *VABCD* is a pyramid with a rectangular base *ABCD*.

$AB = 8$ cm, $BC = 12$ cm,

M is at the centre of *ABCD* and *VM* = 35 cm.

Calculate the volume of *VABCD*.

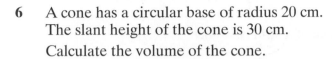

6 A cone has a circular base of radius 20 cm.

The slant height of the cone is 30 cm.

Calculate the volume of the cone.

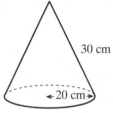

7 Wine is sold in two similar bottles.

The volume of a small bottle of wine is 70 cl.

The volume of a larger bottle of wine is 100 cl.

The height of the larger bottle of wine is 30 cm.

Calculate the height of the smaller bottle of wine.

8 This trophy is made from a truncated cone, a cylinder and a sphere. Find its volume correct to 2 d.p.

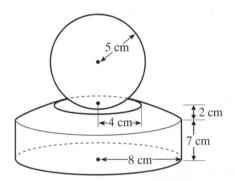

Summary of key points

1 **Circumference** of a circle = $2\pi r$

2 **Area** of a circle = πr^2

3 The formula for the **arc length** of a circle is $\dfrac{\pi r \theta}{180}$

4 The formula for the **area of a sector** of a circle is $\dfrac{\pi r^2 \theta}{360}$

5 The formula for the **area of a segment** is $\dfrac{\pi r^2 \theta}{360} - \dfrac{1}{2} r^2 \sin \theta$

6 For a **cylinder** of height *h* and circular base of radius *r*:

$$\text{surface area} = 2\pi rh + 2\pi r^2$$
$$\text{volume} = \pi r^2 h$$

7 The volume of a **prism** = area of base × vertical height.

8 The volume of a **prism** = area of cross-section × vertical height.

9 The volume of a **pyramid** = $\frac{1}{3}$ × area of base × vertical height.

10 Cylinders and cones are special types of prisms and pyramids, respectively.

11 **Volume of a sphere** = $\dfrac{4\pi r^3}{3}$ **Surface area of a sphere** = $4\pi r^2$

12 When a shape is enlarged by a scale factor k to produce a similar shape:
   - area of enlarged shape = k^2 × area of original shape
   - volume of enlarged shape = k^3 × volume of original shape.

20 Simplifying algebraic expressions

20.1 Terminology you need to know

$x \times x$ is usually written as x^2 and is read as 'x squared'.

$x \times x \times x$ is usually written as x^3 and is read as 'x cubed'.

When x is multiplied by itself any number of times, the word 'power' is used. For example, $x \times x \times x \times x \times x$ is usually written as x^5 and is read as 'x to the power of five' or 'x to the fifth'.

x^1, read as 'x to the power of one', is usually written just as x.

■ **In the expression x^n the number x is called the *base* and the number n is called the *index* or power.**

The plural of index is **indices**.

20.2 Multiplying and dividing expressions involving indices

Multiplying $x^m \times x^n$

Unit 1 shows you how to multiply numbers in index form, for example:

$$4^2 \times 4^3 = 4^5$$

In the same way:

$$x^2 \times x^3 = (x \times x) \times (x \times x \times x)$$
$$= x^5$$

so:
$$x^2 \times x^3 = x^5$$

In general the rule for multiplying algebraic expressions involving powers is:

■ $x^m \times x^n = x^{m+n}$

You may find it easier to remember this rule as 'to multiply numbers which have the same base, add the indices'.

Example 1

Simplify $2x^3 \times 5x^4$

$$2x^3 \times 5x^4 = 2 \times x^3 \times 5 \times x^4$$
$$= 2 \times 5 \times x^3 \times x^4$$
$$= 10 \times x^{3+4}$$
$$= 10x^7$$

A **common error** is to forget that the numbers must have the same base. For example:

$$2^3 \times 3^2 \neq 6^5$$

($\neq$ means 'is **not equal** to').

The rule for multiplying algebraic expressions involving powers can also be used with more complicated expressions like this:

Example 2

Simplify $4a^2b^3 \times 3ab^2 \times 2b$

$$4a^2b^3 \times 3ab^2 \times 2b = 4 \times a^2 \times b^3 \times 3 \times a^1 \times b^2 \times 2 \times b^1$$
$$= 4 \times 3 \times 2 \times a^2 \times a^1 \times b^3 \times b^2 \times b^1$$
$$= 24 \times a^{2+1} \times b^{3+2+1}$$
$$= 24a^3b^6$$

Multiplying out $(x^m)^n$

$(x^2)^3$ means $(x^2) \times (x^2) \times (x^2)$ or $x^2 \times x^2 \times x^2$

so $(x^2)^3 = x^{2+2+2} = x^6 = x^{2 \times 3}$

Similarly:

$(x^3)^4$ means $x^3 \times x^3 \times x^3 \times x^3 = x^{3+3+3+3} = x^{12} = x^{3 \times 4}$

In general the rule for multiplying algebraic expressions involving powers is:

■ $(x^m)^n = x^{m \times n} = x^{mn}$

Example 3

Simplify:

(a) $(x^2)^6$

(b) $(xy^2)^3$

(c) $(2x^3)^5$

(a) $(x^2)^6 = x^{2 \times 6} = x^{12}$

(b) $(xy^2)^3 = (x)^3 \times (y^2)^3 = x^3 \times y^{2 \times 3} = x^3 \times y^6 = x^3y^6$

(c) $(2x^3)^5 = (2)^5 \times (x^3)^5 = 32 \times x^{3 \times 5} = 32 \times x^{15} = 32x^{15}$

Exercise 20A

Simplify:

1 $x^3 \times x^5$ **2** $y^4 \times y^4$ **3** $6x \times x^6$ **4** $a^3 \times 5a$

5 $2a^2 \times 3a \times 4a^3$ **6** $2y^3 \times 4y^2 \times 6$ **7** $2x^2 \times 5x^3y^2$ **8** $3a^2b^3 \times 2a^3 \times 4b^2$

9 $ab \times abc^2$ **10** $2a^2b \times 3b^2c^3 \times 3c^2$ **11** $2x^2 \times 3y^3$ **12** $(x^2)^4$

13 $(3a^2)^2$ **14** $(2x^2)^3$ **15** $2(a^4)^5$ **16** $2(xy^2)^3 \times x$

17 $(3x^4)^2 \times 7$

18 $7(a^2)^3 \times 2a^3b^2 \times (ab)^4$

19 $(x + 2x + 3x)^2$ (Hint: simplify $x + 2x + 3x$ first)

20 $(x \times x + 2x^2)^3$

Dividing $x^m \div x^n$

Unit 1 shows you how to divide numbers in index form, for example:
$$5^6 \div 5^4 = 5^2$$

In the same way:
$$x^6 \div x^4 = \frac{x \times x \times x \times x \times x \times x}{x \times x \times x \times x}$$
$$= x^2$$

In general:

■ $x^m \div x^n = x^{m-n}$

You may find it easier to remember this rule as 'to divide numbers which have the same base, subtract the indices'.

Example 4

Simplify:

(a) $x^6 \div x^2$

(b) $8a^5 \div 4a^3$

(c) $15x^4y^3 \div 3xy^2$

Answer

(a) $x^6 \div x^2 = x^{6-2} = x^4$

(b) $8a^5 \div 4a^3 = \dfrac{8}{4}a^{5-3} = 2a^2$

(c) $15x^4y^3 \div 3xy^2 = 15x^4y^3 \div 3x^1y^2 = \dfrac{15}{3}x^{4-1}y^{3-2} = 5x^3y^1 = 5x^3y$

Exercise 20B

Simplify:

1	$x^7 \div x^4$	**2**	$6a^5 \div 2a^2$	**3**	$18x^4y^3 \div 3x^3$
4	$x^2 \div x^4$	**5**	$a^7 \div a$	**6**	$6a^5 \div 4xa^2$
7	$12x^4 \div 3x^4$	**8**	$x^2 \div x^{-2}$		

20.3 Zero, negative and fractional indices

The zero index

Here is how to find the value of an expression to the power zero.
2^0 can be written as:

$$2^3 \div 2^3 = 2^{3-3} = 2^0$$

but:

$$2^3 \div 2^3 = \frac{2^3}{2^3} = 1$$

so:

$$2^0 = 1$$

Putting $n = m$ in the rule for dividing algebraic expressions involving powers gives:

$$x^m \div x^m = x^{m-m}$$

so:

$$\frac{x^m}{x^m} = x^0$$

$$1 = x^0$$

■ $x^0 = 1$ for all non zero values of x.

Negative indices

You can also find the value of an expression to a negative power.
For example, 6^{-2} can be written as:

$$6^3 \div 6^5 = 6^{3-5} = 6^{-2}$$

so:

$$\frac{6 \times 6 \times 6}{6 \times 6 \times 6 \times 6 \times 6} = 6^{-2}$$

or:

$$\frac{1}{6^2} = 6^{-2}$$

Putting $m = 0$ in the rule for dividing algebraic expressions involving powers gives:

$$x^0 \div x^n = x^{0-n}$$

so:

$$\frac{x^0}{x^n} = x^{-n}$$

which gives:

$$\frac{1}{x^n} = x^{-n}$$

$\boxed{0}$ $\boxed{x^y}$ $\boxed{0}$ $\boxed{=}$

gives 'Error' on most calculators.

What does your calculator display if you input:
(a) a small negative number instead of the first 0
(b) a small positive decimal number instead of the second 0?

Remember:

$$x^0 = 1$$

so $\quad \frac{x^0}{x^n} = \frac{1}{x^n}$

■ $x^{-n} = \dfrac{1}{x^n}$

This result can be written in words as: x^{-n} means the **reciprocal** of x^n.
(Unit 18, page 371, introduces reciprocal functions.)

> The reciprocal of 5 is $\dfrac{1}{5}$.
>
> Multiplying a number by its reciprocal gives 1, for example:
>
> $5 \times \dfrac{1}{5} = 1$ and $x^2 \times \dfrac{1}{x^2} = 1$

Fractional indices

What does a fractional power such as $x^{\frac{1}{3}}$ mean? To see, look at these expressions involving square roots:

$$\sqrt{25} \times \sqrt{25} = 5 \times 5 = 25$$
$$\sqrt{9} \times \sqrt{9} = 3 \times 3 = 9$$

in general: $\quad\quad\quad\quad \sqrt{x} \times \sqrt{x} = x = x^1$

and: $\quad\quad\quad\quad x^{\frac{1}{2}} \times x^{\frac{1}{2}} = x^{\frac{1}{2} + \frac{1}{2}} = x^1$

so: $\quad\quad\quad\quad\quad \sqrt{x} = x^{\frac{1}{2}}$

■ $x^{\frac{1}{2}} = \sqrt{x}$

> **Remember:**
> $\sqrt[3]{}$ means 'cube root'.

Similarly, because:

$$\sqrt[3]{x} \times \sqrt[3]{x} \times \sqrt[3]{x} = x$$

and: $\quad\quad\quad\quad x^{\frac{1}{3}} \times x^{\frac{1}{3}} \times x^{\frac{1}{3}} = x$

you can see that: $\quad\quad\quad \sqrt[3]{x} = x^{\frac{1}{3}}$

■ **In general $x^{\frac{1}{n}} = \sqrt[n]{x}$.**

Example 5

Find the value of $8^{-\frac{1}{3}}$

$$8^{-\frac{1}{3}} = \frac{1}{8^{\frac{1}{3}}} = \frac{1}{\sqrt[3]{8}} = \frac{1}{2}$$

Exercise 20C

Find the value of:

1 $25^{\frac{1}{2}}$ **2** $16^{\frac{1}{4}}$ **3** $64^{\frac{1}{3}}$

4 $49^{-\frac{1}{2}}$ **5** $81^{-\frac{1}{4}}$ **6** $216^{\frac{1}{3}}$

7 $125^{-\frac{1}{3}}$ **8** $100^{-\frac{1}{2}}$ **9** $\left(\frac{1}{4}\right)^{\frac{1}{2}}$

10 $(0.064)^{\frac{1}{3}}$

20.4 Combining indices

You can use the rule for multiplying algebraic expressions involving indices to show that:

$$(\sqrt[n]{x})^m = (x^{\frac{1}{n}})^m = x^{\frac{1}{n} \times m} = x^{\frac{m}{n}}$$

and that:

$$\sqrt[n]{x^m} = (x^m)^{\frac{1}{n}} = x^{m \times \frac{1}{n}} = x^{\frac{m}{n}}$$

$$\frac{1}{n} \times m = \frac{m}{n}$$

■ $x^{\frac{m}{n}} = (\sqrt[n]{x})^m$ **or** $x^{\frac{m}{n}} = \sqrt[n]{x^m}$

Example 6

Find the value of $81^{\frac{3}{4}}$:

Here $m = 3$ and $n = 4$. Substituting in the rule above gives:

$$81^{\frac{3}{4}} = (\sqrt[4]{81})^3 = (3)^3 = 27$$

Exercise 20D

Find the value of:

1 $25^{\frac{3}{2}}$ **2** $16^{\frac{5}{4}}$ **3** $4^{\frac{5}{2}}$

4 $8^{-\frac{2}{3}}$ **5** $(0.125)^{-\frac{5}{3}}$ **6** $125^{\frac{2}{3}}$

7 $(\frac{1}{8})^{\frac{2}{3}}$ **8** $64^{-\frac{5}{6}}$

20.5 Equations involving indices

Example 7

Find the value of k:

(a) $x^k = \dfrac{1}{x^{-2}}$ **(b)** $x^k = \sqrt{x} \div \dfrac{1}{x^3}$

(c) $x^k = \dfrac{x \times \sqrt[3]{x^4}}{(\sqrt[4]{x})^2}$ **(d)** $4^{k+1} = 8$

Answer

(a) $x^k = \dfrac{1}{x^{-2}}$

      $x^k = (x^{-2})^{-1}$

Using the rule for multiplying out $(x^m)^n$ gives:

      $x^k = x^2$

so: $k = 2$ (comparing indices)

(b)
$$x^k = \sqrt{x} \div \frac{1}{x^3}$$
$$x^k = x^{\frac{1}{2}} \div x^{-3}$$
$$x^k = x^{\frac{1}{2} - (-3)}$$
$$x^k = x^{\frac{1}{2} + 3} = x^{3\frac{1}{2}}$$

so: $k = 3\frac{1}{2}$

(c)
$$x^k = \frac{x \times \sqrt[3]{x^4}}{(\sqrt[4]{x})^2}$$
$$x^k = \frac{x^1 \times x^{\frac{4}{3}}}{x^{\frac{2}{4}}}$$
$$x^k = \frac{x^{1 + \frac{4}{3}}}{x^{\frac{1}{2}}}$$
$$x^k = \frac{x^{\frac{7}{3}}}{x^{\frac{1}{2}}}$$
$$x^k = x^{\frac{7}{3} - \frac{1}{2}}$$
$$x^k = x^{\frac{11}{6}}$$

so: $k = \frac{11}{6}$

(d)
$$4^{k + 1} = 8$$
$$4^{k + 1} = 4 \times 2 = 4 \times \sqrt{4}$$
$$4^{k + 1} = 4^1 \times 4^{\frac{1}{2}}$$
$$4^{k + 1} = 4^{1\frac{1}{2}}$$
$$k + 1 = 1\frac{1}{2}$$
$$k = \frac{1}{2}$$

Alternative method
Because 4 and 8 are multiples of 2, you can also answer part **(d)** like this:
$$4^{k + 1} = 8$$
$$(2^2)^{k + 1} = (2)^3$$
$$2^{2k + 2} = 2^3$$
$$2k + 2 = 3$$
$$2k = 1$$
$$k = \frac{1}{2}$$

Example 8

Simplify:

(a) $(25x^2)^{\frac{1}{2}}$ **(b)** $(8x^6)^{-\frac{1}{3}}$

Answer

(a)
$$(25x^2)^{\frac{1}{2}} = (25)^{\frac{1}{2}} (x^2)^{\frac{1}{2}}$$
$$= \sqrt{25}x^{2 \times \frac{1}{2}}$$
$$= 5x^1$$
$$= 5x$$

(b)
$$(8x^6)^{-\frac{1}{3}} = (8)^{-\frac{1}{3}} (x^6)^{-\frac{1}{3}}$$
$$= \frac{1}{\sqrt[3]{8}} (x^{6 \times (-\frac{1}{3})})$$
$$= \frac{1}{2} x^{-2} \quad \text{or} \quad \frac{1}{2x^2}$$

Exercise 20E

1 Find the value of:

(a) $8^{\frac{1}{3}}$ (b) $1000^{-\frac{2}{3}}$ (c) $32^{\frac{2}{5}}$

(d) $(0.25)^{-\frac{5}{2}}$ (e) $27^{\frac{4}{3}}$

2 Find the value of k:

(a) $x^k = \sqrt{x}$ (b) $y^k = 1$ (c) $a^k = 1 \div a^3$

(d) $x^k = \sqrt[3]{x^2}$ (e) $y^k = y^2\sqrt{y^3}$ (f) $a^k = \sqrt{a} \div \dfrac{1}{a^4}$

(g) $x^{k+1} = (x^{-2})^{-3}$ (h) $2^k = 16$ (i) $3^k = 27$

(j) $25^k = 5$ (k) $9^k = 27$ (l) $4^{\frac{k}{2}} = 8$

3 Simplify:

(a) $(3x^0)^2$ (b) $a^{-2} \div \dfrac{1}{a^4}$ (c) $(\sqrt{xy})^3 \times y$

(d) $(9x^2)^{\frac{1}{2}}$ (e) $(8x^{-3})^{\frac{1}{3}}$ (f) $(25a^{-4})^{-\frac{1}{2}}$

(g) $(81x^2)^{\frac{3}{2}}$ (h) $32\sqrt{x} \times (16x^2)^{-\frac{1}{4}}$ (i) $4^{\frac{3}{2}} \times (9y^4)^{\frac{1}{2}}$

20.6 Algebraic fractions

$\dfrac{9x^4}{6x}$ and $\dfrac{3(x+8)}{3x}$ are examples of **algebraic fractions**.

Fractions can be simplified by cancelling *only* if there is a common factor in the numerator and the denominator.

Example 9

Simplify:

(a) $\dfrac{9x^4}{6x}$ (b) $\dfrac{3(x+8)}{3x}$

Answer

(a) $\dfrac{9x^4}{6x} = \dfrac{3 \times \cancel{3} \times \cancel{x} \times x^3}{2 \times \cancel{3} \times \cancel{x}} = \dfrac{3x^3}{2}$

(Check this answer using the rule for dividing indices.)

(b) $\dfrac{3(x+8)}{3x} = \dfrac{\cancel{3} \times (x+8)}{\cancel{3} \times x} = \dfrac{(x+8)}{x}$

You cannot cancel the xs here because x is not a common factor.

Multiplying and dividing algebraic fractions

When you multiply algebraic fractions, you should cancel any factor which is common to both numerator and denominator. When no further cancelling is possible you can then multiply the remaining numerators and denominators.

Example 10

Simplify $\dfrac{x^4}{y} \times \dfrac{xy}{z} \times \dfrac{z}{x^3}$

$$\frac{x^4}{y} \times \frac{xy}{z} \times \frac{z}{x^3} = \frac{x \times \cancel{x^3}}{\cancel{y}} \times \frac{x \times \cancel{y}}{\cancel{z}} \times \frac{\cancel{z}}{\cancel{x^3}}$$

(cancel by x^3, then by y then by z)

$$= \frac{x}{1} \times \frac{x}{1} \times \frac{1}{1} = \frac{x \times x \times 1}{1 \times 1 \times 1} = \frac{x^2}{1} = x^2$$

■ **Dividing by an algebraic fraction is equivalent to multiplying by its reciprocal.**

For example, to divide by $\frac{a}{b}$ you can multiply by $\frac{b}{a}$.

To see why this works, notice that:

$$1 = \frac{a}{b} \times \frac{b}{a}$$

Dividing both sides by $\dfrac{a}{b}$ gives:

$$1 \div \frac{a}{b} = 1 \times \frac{b}{a}$$

The reciprocal.

of $\frac{a}{b}$ is $\frac{b}{a}$.

So dividing 1 by $\frac{a}{b}$ is the same as multiplying 1 by the reciprocal of $\frac{a}{b}$.

Example 11

Simplify $3x \div \frac{a}{b}$

$$3x \div \frac{a}{b} = \frac{3x}{1} \times \frac{b}{a} = \frac{3xb}{a}$$

Example 12

Simplify $\dfrac{x^2(x + 2)}{y^2} \div \dfrac{x(x + 4)}{y^3}$

$$\dfrac{x^2(x + 2)}{y^2} \div \dfrac{x(x + 4)}{y^3} = \dfrac{x^2(x + 2)}{y^2} \times \dfrac{y^3}{x(x + 4)} \qquad \text{(multiply by reciprocal)}$$

$$= \dfrac{\cancel{x} \times x \times (x + 2)}{\cancel{x^2}} \times \dfrac{\cancel{y^2} \times y}{\cancel{x} \times (x + 4)} \qquad \text{(cancel by } x \text{ and then by } y^2\text{)}$$

$$= \dfrac{x(x + 2)}{1} \times \dfrac{y}{(x + 4)} = \dfrac{xy(x + 2)}{(x + 4)} \quad \begin{array}{l}\text{(no further cancelling is}\\\text{possible)}\end{array}$$

Exercise 20F

Simplify:

1 $\dfrac{xy}{4y}$

2 $\dfrac{abc}{acd}$

3 $\dfrac{24x^2}{6x}$

4 $\dfrac{8st}{6sr}$

5 $\dfrac{12a^2}{b^3} \times \dfrac{b^4}{4a}$

6 $\dfrac{4(x - 2)}{2(x - 2)}$

7 $\dfrac{2(x + 2)}{4x}$

8 $\dfrac{9(x + 4)}{3(x + 4)(x - 4)}$

Adding and subtracting algebraic fractions

To add or subtract algebraic fractions you can use the same methods as for ordinary numerical fractions.

Both fractions must have the same denominator. If they do not, find the lowest common denominator (LCD) and change each fraction to an equivalent fraction with this denominator.

Here is a numerical example as a reminder.

To add $\frac{1}{3}$ and $\frac{1}{5}$ find the lowest common denominator of 3 and 5. The LCD is 15.

Change each fraction to an equivalent fraction with this denominator:

$$\overset{\times 5}{\dfrac{1}{3}} = \dfrac{5}{15} \underset{\times 5}{} \qquad \overset{\times 3}{\dfrac{1}{5}} = \dfrac{3}{15} \underset{\times 3}{}$$

So $\frac{1}{3} + \frac{1}{5}$ is equivalent to:

$$\dfrac{5}{15} + \dfrac{3}{15} = \dfrac{8}{15}$$

In general:

$$\blacksquare \quad \frac{1}{n} + \frac{1}{m} = \frac{m+n}{mn}$$

The following examples show you how to add two algebraic fractions, but the same process can be used to add or subtract three or more fractions.

Example 13

Write as a single fraction in its lowest terms:

(a) $\dfrac{2}{3} + \dfrac{1}{4}$ (b) $\dfrac{2}{y} + \dfrac{1}{z}$ (c) $\dfrac{2}{(x+1)} + \dfrac{1}{(x+2)}$

Answers

(a) $\dfrac{2}{3} + \dfrac{1}{4}$ (b) $\dfrac{2}{y} + \dfrac{1}{z}$ (c) $\dfrac{2}{(x+1)} + \dfrac{1}{(x+2)}$

$= \dfrac{2\times4}{3\times4} + \dfrac{1\times3}{4\times3}$ $= \dfrac{2\times z}{y\times z} + \dfrac{1\times y}{z\times y}$ $= \dfrac{2(x+2)}{(x+1)(x+2)} + \dfrac{1(x+1)}{(x+2)(x+1)}$

$= \dfrac{8}{12} + \dfrac{3}{12}$ $= \dfrac{2z}{yz} + \dfrac{y}{zy}$ $= \dfrac{2x+4}{(x+1)(x+2)} + \dfrac{x+1}{(x+2)(x+1)}$

$= \dfrac{8+3}{12}$ $= \dfrac{2z+y}{yz}$ $= \dfrac{2x+4+x+1}{(x+1)(x+2)}$

$= \dfrac{11}{12}$ $= \dfrac{3x+5}{(x+1)(x+2)}$

In the three questions in Example 13 the LCD was found by multiplying each denominator. This will always be the case when the denominators have no common factor.

The next example shows you how to find the LCD when the denominators have a common factor by writing the denominators in terms of their 'prime' factors.

Example 14

Write as a single fraction:

(a) $\dfrac{5}{12} - \dfrac{2}{9}$ (b) $\dfrac{5}{x(x+1)} - \dfrac{2}{x^2}$

In these examples, to find the LCD first split each denominator into its prime factors. Then write the denominator of the first fraction and multiply it by the 'remaining' factors of the other denominators.

Answers

(a) $\dfrac{5}{12} - \dfrac{2}{9}$

$12 = 2 \times 2 \times 3$ and
$9 = 3 \times 3$

the LCD is:

$2 \times 2 \times 3 \times 3 = 36$

$= \dfrac{5 \times 3}{12 \times 3} - \dfrac{2 \times 4}{9 \times 4}$

$= \dfrac{15}{36} - \dfrac{8}{36}$

$= \dfrac{15 - 8}{36}$

$= \dfrac{7}{36}$

(b) $\dfrac{5}{x(x + 1)} - \dfrac{2}{x^2}$

$x(x + 1) = x \times (x + 1)$ and
$x^2 = x \times x$

the LCD is:

$x \times (x + 1) \times x = x^2(x + 1)$

$= \dfrac{5 \times x}{x(x + 1) \times x} - \dfrac{2 \times (x + 1)}{x^2 \times (x + 1)}$

$= \dfrac{5x}{x^2(x + 1)} - \dfrac{2(x + 1)}{x^2(x + 1)}$

$= \dfrac{5x - 2(x + 1)}{x^2(x + 1)}$

$= \dfrac{5x - 2x - 2}{x^2(x + 1)} = \dfrac{3x - 2}{x^2(x + 1)}$

Notice that the fractions in (a) and (b) are the same when $x = 3$.

When $x = 3$, $\dfrac{3x - 2}{x^2(x + 1)} = \dfrac{9 - 2}{3^2(3 + 1)} = \dfrac{7}{9 \times 4} = \dfrac{7}{36}$ which checks.

Checking your solutions

In the GCSE examination you should check your algebra solutions by choosing suitable numbers for the unknown (usually x). When checking work on algebraic fractions do not choose a value of x which makes a denominator zero.

For example, in part (b) of Example 14 you could choose $x = 1$ to check. This makes the question $\frac{5}{2} - 2 = \frac{1}{2}$ which checks with the answer $\dfrac{3 - 2}{1^2(1 + 1)}$.

Exercise 20G

1   Write as a single fraction in its lowest terms:

(a) $\dfrac{x}{5} + \dfrac{2x}{5}$

(b) $\dfrac{1}{x} + \dfrac{1}{y}$

(c) $\dfrac{2}{x} + \dfrac{3}{5x}$

(d) $\dfrac{2x}{3} + \dfrac{x}{4}$

(e) $\dfrac{a - 1}{3} + \dfrac{a + 2}{4}$

(f) $\dfrac{x + 3}{2} - \dfrac{2x - 1}{5}$

(g) $\dfrac{2}{x + 1} - \dfrac{1}{x + 2}$

(h) $\dfrac{1}{x} + \dfrac{1}{2 - x}$

(i) $\dfrac{1}{2} + \dfrac{2}{y - 5}$

2 Find the LCD for the denominators:

(a) 6 and 9   **(b)** x^2 and $x(x-1)$

(c) $4(x-1)(x+1)$ and $6(x+1)$ **(d)** $x^3(1-x)$ and $x^2(1+x)$

3 Write as a single fraction in its lowest terms:

(a) $\dfrac{a}{6} + \dfrac{a}{9}$ **(b)** $\dfrac{2}{x^2} - \dfrac{1}{x(x-1)}$

(c) $\dfrac{x}{4(x-1)(x+1)} - \dfrac{1}{6(x+1)}$ **(d)** $\dfrac{2}{x^3(1-x)} - \dfrac{1}{x^2(1+x)}$

4 Winston was asked to write:

$$\frac{6}{(x-2)} - \frac{4}{(x+2)}$$

as a single fraction in its lowest terms. His solution below shows that he has made two errors. Find where these errors occur and write a correct solution.

$$\frac{6}{(x-2)} - \frac{4}{(x+2)} \qquad\qquad LCD = (x-2)(x+2)$$

$$= \frac{6(x+2)}{(x-2)(x+2)} - \frac{4(x-2)}{(x+2)(x-2)} = \frac{6(x+2)-4(x-2)}{(x-2)(x+2)}$$

$$= \frac{6x+12-4x-8}{(x-2)(x+2)} = \frac{2x+4}{(x-2)(x+2)} = \frac{2\cancel{(x+2)}}{(x-2)\cancel{(x+2)}}$$

$$= \frac{\cancel{2}\,1}{x-\cancel{2}\,1} = \frac{1}{x-1}$$

20.7 Further factorization

Unit 10 (page 200) introduces factorization as the reverse process to expanding brackets. For example, here is how to factorize $2a^2 + 6a$:

$$2a^2 + 6a = \mathbf{2} \times a \times a + \mathbf{2} \times 3 \times a$$

The Highest Common Factor (HCF) of the two terms is $2 \times a$, so $2a$ will go outside the brackets:

$$2a^2 + 6a = 2a(a+3)$$

This section shows you how to factorize more complicated algebraic expressions. In each case the first step is to take out the HCF.

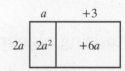

Another way of thinking about this is to see the factors as the lengths of the sides of rectangles. Products such as $2a^2$ are the areas of rectangles.

Factorizing by pairing

An expression such as $ab + cd + bc + ad$ can be factorized by pairing two terms which have a common factor as follows:

$$ab + cd + bc + ad = ab + bc + ad + cd \qquad \text{(put terms in pairs with common factor)}$$
$$= b(a+c) + d(a+c) \qquad \text{(bracket term must be a factor of both pairs)}$$
$$= (a+c)(b+d) \qquad\qquad ((a+c)(b+d) \text{ is the same as } (b+d)(a+c))$$

Example 15

Factorize completely $\quad 5a(x-y) + (y-x)$

$$5a(x-y) + (y-x) = 5a(x-y) - 1(x-y) \qquad (y-x) = -1(x-y)$$
$$= (x-y)(5a-1) \qquad ((x-y) \text{ is a common factor})$$

Example 16

Factorize completely $\quad 2ab - 12cd + 6ac - 4bd$

$$
\begin{aligned}
2ab - 12cd + 6ac - 4bd &= 2[ab - 6cd + 3ac - 2bd] &&(2 \text{ is the HCF})\\
&= 2[ab + 3ac - 2bd - 6cd] &&(\text{group in pairs with common factor})\\
&= 2[a(b+3c) - 2d(b+3c)] &&(\text{factorize in pairs to get } (b+3c) \text{ common})\\
&= 2(b+3c)(a-2d)
\end{aligned}
$$

Before factorizing the second pair, leave a space then write in the bracket term like
$2[a(b+3c) \,?\,?\,(b+3c)]$
That should help you find the correct common factor and correct sign.

Exercise 20H

Factorize completely:

1 $x(p+q) + y(p+q)$
2 $3(x+y) - z(x+y)$
3 $5(x-y) + x(y-x)$
4 $10(a-b) - 2x(b-a)$
5 $xy + 2x + zy + 2z$
6 $ay + 2b + 2ab + y$
7 $x - b - x^2 + bx$
8 $x^2 + yz - xy - zx$
9 $2ab - 10 + 2a - 10b$
10 $3a - 3ab + 6bc - 6c$

Factorizing algebraic quadratic expressions

$$2x^2 - 5x + 4 \qquad 3x^2 - 27 \qquad y^2 + 20y \qquad \text{and} \qquad t^2$$

are all examples of quadratic expressions. These are introduced in Unit 18 (page 368). As a reminder:

■ **An expression of the form $ax^2 + bx + c$, with $a \neq 0$, is called a _quadratic_ in x.**

The expression $3y^2 - 20y$ is a quadratic in y.

■ **In the expression $ax^2 + bx + c$:**
   the coefficient of x^2 is a
   the coefficient of x is b
   the constant term is c.

Example 17

Write the quadratic expression in x in which the coefficient of x^2 is 1, the coefficient of x is –2 and the constant term is –3.

$$1x^2 + (-2)x + (-3) = x^2 - 2x - 3$$

■ **To factorize quadratic expressions of the form $ax^2 + bx$, you take out the HCF.**

For example $3x^2 + 18x = 3x(x + 6)$.

Unit 10 shows you how to expand expressions such as $(x - 3)(2x + 5)$ to get $2x^2 - x - 15$. You can reverse this process to factorize a quadratic.

Steps to factorize quadratic expressions of the form $ax^2 + bx + c$.

Example 18

Factorize $x^2 - 8x + 15$.

Step 1 Check the equation is in the form $ax^2 + bx + c$.

Step 2 Multiply the coefficient of x^2 by the constant term:
$$a \times c = 15$$

Step 3 Find the two factors of this product which add to give the coefficient of x.

Factors are –3 and –5.

Step 4 Rewrite the bx term using these two factors:
$$-3x - 5x$$

Step 5 Proceed as for factorizing by pairing:
$$x^2 - 3x - 5x + 15$$
$$= x(x - 3) - 5(x - 3) = (x - 5)(x - 3)$$

Example 19

Factorize $6x^2 - 45 - 3x$.

Step 1 Take out any HCF and apply the remaining steps to the other factor.

$$= 3[2x^2 - 15 - x]$$

Step 2 Rearrange into the form $ax^2 + bx + c$.

$$= 3[2x^2 - x - 15]$$

Step 3 Multiply the coefficient of x^2 by the **constant term**. Find the two factors of this product which add to give the **coefficient of x**.

The other factor is
$2x^2 - 15 - x$

$2 \times (-15) = -30$
Factors of –30 that add
to –1?
Try:
-30×1 ✗ -15×2 ✗
-10×3 ✗ -6×5 ✔

Step 4 Rewrite the bx term using these two factors. $= 3[2x^2 - 6x + 5x - 15]$

Step 5 Proceed as for factorizing by pairing.
$$= 3[2x(x-3) + 5(x-3)]$$
$$= 3[(x-3)(2x+5)]$$

Example 20

Factorize $14x - 6 - 4x^2$.

Answer

Remember to look for an HCF as a first step to make the rest of the factorization easier.

Step 1 $14x - 6 - 4x^2 = 2[7x - 3 - 2x^2]$ (2 is HCF)

Step 2 $= 2[-2x^2 + 7x - 3]$ (rearrange)

Step 3 $(-2) \times (-3) = (+6)$

You need factors of $+6$ which add to $+7$:
$$(+6) \times (+1) = 6 \text{ ✗}$$
$$(+6) + (+1) = +7 \text{ ✔}$$
$$14x - 6 - 4x^2 = 2[-2x^2 + 7x - 3]$$

Step 4 $= 2[-2x^2 + 6x + 1x - 3]$

Step 5 $= 2[-2x(x-3) + 1(x-3)]$
$$= 2[(x-3)(-2x+1)]$$
or
$$= 2[(x-3)(-1)(2x-1)]$$
$$= -2(x-3)(2x-1)$$

Example 21

Find two numbers whose product P is -60 and whose sum S is -7.

The product is negative, so the two numbers must have different signs.

The sign of the greater number must be negative because their sum is negative.

S is odd so one number must be odd and the other even.

Try: $+1, -60$ ✗ $+3, -20$ ✗

$+4, -15$ ✗ $+5, -12$ ✔

Exercise 20 I

1 Factorize:

(a) $x^2 + 3x$

(b) $8x + 24$

(c) $x^2 + 3x + 8x + 24$

(d) $x^2 + 6x$

(e) $2x + 12$

(f) $x^2 + 6x + 2x + 12$

(g) $x^2 - 5x$

(h) $8x - 40$

(i) $x^2 - 5x + 8x - 40$

(j) $x^2 + 5x + 3x + 15$

(k) $x^2 + 4x - 8x - 32$

(l) $x^2 - 6x - 2x + 12$

(m) $x^2 + 4x - 3x - 12$

(n) $x^2 + 15 + 3x + 5x$

(o) $x^2 + 63 - 7x - 9x$

2   Write the quadratic expression in x in which the coefficient of x^2 is 1, the coefficient of x is -5 and the constant term is -6.

3   In each of the following, find two numbers whose product is P and whose sum is S:

   (a) $P = 6$, $S = -5$ (b) $P = 6$, $S = 7$

   (c) $P = 12$, $S = -8$ (d) $P = 60$, $S = 19$

   (e) $P = -24$, $S = -2$ (f) $P = -24$, $S = 10$

   (g) $P = -48$, $S = -13$ (h) $P = -48$, $S = 8$

4   Factorize:

   (a) $x^2 - 5x + 4$ (b) $x^2 + 7x + 10$ (c) $x^2 - 2x - 15$

   (d) $5x + 6 + x^2$ (e) $10 - 11x + x^2$ (f) $y^2 - y - 12$

5   Factorize completely:

   (a) $2x^2 - 5x + 3$ (b) $6x^2 + 42x + 60$ (c) $2y^2 - y - 10$

   (d) $6a^2 + 10a + 4$ (e) $4x^2 - 5x - 6$ (f) $75 + 35x - 20x^2$

   (g) $5x^2 + 3 - 16x$ (h) $x^4 + 5x^2 + 6$ (Hint: let $x^2 = y$)

   (i) $2x^4 + 14x^2 + 24$

6   Factorize $x^{2n} + 9x^n + 8$ (Hint: let $x^n = y$).

7   Factorize completely $x^{4n + 1} + 6x^{2n + 1} + 8x$.

After further practice, you may be able to factorize some of the quadratics more directly. Diagrams can help, as shown in the next example.

> Question 7 is harder than the others. Only attempt it if you are confident with factorizing expressions.

Example 22

Factorize $2x^2 + 7x - 15$.

You need to find factors of -30 whose sum is $+7$.

These are $(+10$ and $-3)$

$$2x^2 + 7x - 15 = (2x \,\underline{?} \ \ \underline{?}\,)(x \,\underline{?} \ \ \underline{?}\,)$$
$$= (2x \,\underline{?}\, 3)(x \,\underline{?}\, 5)$$
$$= (2x - 3)(x + 5)$$

By the end of your GCSE course you should be able to answer this question more directly. Your solution could be as short as:

$$2x^2 + 7x - 15 = (2x - 3)(x + 5)$$

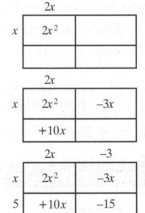

20.8 The difference of two squares

The expression $x^2 - y^2$ is called a difference of two squares. There is an easy way to factorize an expression like this.

Here is a square of side x with a square of side y cut out of the top left corner. The part A which is left has an area $x^2 - y^2$.

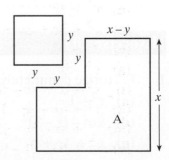

Imagine cutting along the dotted line and moving the shaded piece B around to the bottom end of C.

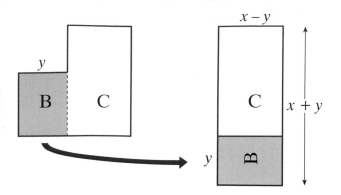

From the diagram you can see that:

$$\text{area A} = \text{area B} + \text{area C}$$
$$= \text{area (B + C)}$$

and:
$$x^2 - y^2 = (x - y)(x + y)$$

So to factorize the difference of two squares you can use:

■ $x^2 - y^2 = (x - y)(x + y)$

Example 23
Factorize completely:
(a) $x^2 - 16$ **(b)** $3x^2 - 27$ **(c)** $18x^2 - 2(x + 5)^2$

Answers
(a) $x^2 - 16$
$= x^2 - 4^2$
$= (x - 4)(x + 4)$

(b) $3x^2 - 27$
$= 3[x^2 - 9]$
$= 3[x^2 - 3^2]$
$= 3(x - 3)(x + 3)$

(c) $18x^2 - 2(x + 5)^2$
$= 2[9x^2 - (x + 5)^2]$
$= 2[(3x)^2 - (x + 5)^2]$
$= 2[3x - (x + 5)][3x + (x + 5)]$
$= 2(2x - 5)(4x + 5)$

Two other results which you should memorize are:

■ $x^2 + 2ax + a^2 = (x + a)^2$

■ $x^2 - 2ax + a^2 = (x - a)^2$

So, for example, putting $a = 6$ in the first you get
$x^2 + 12x + 36 = (x + 6)^2$

Exercise 20J

1 Factorize completely:
(a) $x^2 - 1$ **(b)** $x^2 - 25$ **(c)** $4x^2 - 9$
(d) $9x^2 - (x - 3)^2$ **(e)** $2p^2 - 98$ **(f)** $2x^2 - 8y^2$
(g) $x^2 + 4x + 4$ **(h)** $x^2 - 6x + 9$ **(i)** $2x^2 + 28x + 98$

2   Factorize completely:

(a) $x^2 - 4$ (b) $x^2 - 3x$ (c) $x^2 - 3x - 4$

(d) $4x^2 - 36$ (e) $4x^2 - 70x$ (f) $4x^2 - 70x - 36$

(g) $x^2 + 2x + 1$ (h) $x^2 + 2x$ (i) $2x - 28 + 8x^2$

3   Factorize these expressions in as straightforward a way as you can. Copy (a)–(f) which have been partly done for you and complete the blanks. Check your factors by mentally expanding your answers. (For a reminder on expanding expressions involving brackets see page 201.):

(a) $x^2 - 5x + 4$ Hint: $x^2 - 5x + 4 = (x - 4)(x \underline{\quad})$

(b) $x^2 + 9x - 36$ Hint: $x^2 + 9x - 36 = (x + 12)(x \underline{\quad})$

(c) $x^2 + 4x - 32$ Hint: $x^2 + 4x - 32 = (x + 8)(\underline{\quad})$

(d) $2x^2 - 7x + 5$ Hint: $2x^2 - 7x + 5 = (x - 1)(\underline{\quad})$

(e) $4x^2 - 3x - 7$ Hint: $4x^2 - 3x - 7 = (\underline{\quad} - 7)(\underline{\quad} + 1)$

(f) $5x^2 - 7x - 6$ Hint: $5x^2 - 7x - 6 = (5x \underline{\quad})(\underline{\quad})$

(g) $x^2 - 4x + 3$ (h) $y^2 + 7y + 10$ (i) $x^2 + 2x - 15$

(j) $x^2 - 4x - 21$ (k) $y^2 + 8y + 16$ (l) $4x^2 - 20x + 25$

(m) $5x^2 - 13x + 6$ (n) $4x^2 + x - 5$ (o) $5x^2 + 7x - 6$

20.9 Further fractional algebraic expressions

This section shows you how to combine the skills developed in the previous two sections to simplify expressions which contain more difficult algebraic fractions.

The steps to be taken to do this are:

Step 1 Factorize all expressions completely.

Step 2 Cancel fully.

Step 3 Follow the steps used in **Section 20.6**.

Example 24

Simplify $\dfrac{2x - 4}{x^2 - 5x + 6} \div \dfrac{3x^2 + 9x}{x^2 - 9}$

$$\frac{2x - 4}{x^2 - 5x + 6} \div \frac{3x^2 + 9x}{x^2 - 9} = \frac{2(x - 2)}{(x - 3)(x - 2)} \div \frac{3x(x + 3)}{(x - 3)(x + 3)} \quad \text{(Step 1)}$$

$$= \frac{2}{(x - 3)} \div \frac{3x}{(x - 3)} \quad \text{(Step 2)}$$

$$= \frac{2}{(x - 3)} \times \frac{(x - 3)}{3x} \quad \text{(multiply by reciprocal)}$$

$$= \frac{2}{(x - 3)} \times \frac{(x - 3)}{3x} = \frac{2}{3x} \quad \text{(cancel by } (x - 3))$$

Example 25

Write as a single fraction: $\dfrac{6x - 30}{2x^2 - 50} + \dfrac{21}{x^2 + 3x - 10}$

$$\dfrac{6x - 30}{2x^2 - 50} + \dfrac{21}{x^2 + 3x - 10} = \dfrac{6(x - 5)}{2(x - 5)(x + 5)} + \dfrac{21}{(x + 5)(x - 2)} \quad \text{(factorize fully)}$$

Take out the HCF giving $[2x^2 - 50 = 2(x^2 - 25)]$. Then use the difference of two squares for $x^2 - 25$

$$= \dfrac{3}{(x + 5)} + \dfrac{21}{(x + 5)(x - 2)} \quad \text{(cancel by } 2(x - 5)\text{)}$$

$$= \dfrac{3(x - 2)}{(x + 5)(x - 2)} + \dfrac{21}{(x + 5)(x - 2)} \quad \text{(LCD is } (x + 5)(x - 2)\text{)}$$

$$= \dfrac{3(x - 2) + 21}{(x + 5)(x - 2)}$$

$$= \dfrac{3x + 15}{(x + 5)(x - 2)}$$

$$= \dfrac{3(x + 5)}{(x + 5)(x - 2)} \quad \text{(factorize } 3x + 15\text{)}$$

$$= \dfrac{3}{(x - 2)} \quad \text{(cancel by } (x + 5)\text{)}$$

Exercise 20K

Simplify as fully as possible:

1 $\dfrac{x^2 - 4}{2x - 4}$

2 $\dfrac{x^2 - 5x + 4}{x^2 - 16}$

3 $\dfrac{4x + 8}{4x^2 - 8x} \div \dfrac{x^2 - 4}{x^2 + 2x}$

4 $\dfrac{x^2 + 11x + 18}{x^2 + 3x + 2}$

5 $\dfrac{x^2 + 6x + 8}{x^2 + 4x}$

6 $\dfrac{1}{(x + 1)} - \dfrac{x - 2}{x^2 + 3x + 2}$

7 $\dfrac{x^2 - 5x + 4}{x^2 - 16} + \dfrac{x^2 + 11x + 18}{x^2 + 6x + 8}$

Summary of key points

1 In the expression x^n the number x is called the **base** and the number n is called the **index** or power.

2 $x^m \times x^n = x^{m+n}$

3 $(x^m)^n = x^{mn}$

4 $x^m \div x^n = x^{m-n}$

5 $x^1 = x$

6 $x^0 = 1$ when $x \neq 0$

7 $x^{-n} = \dfrac{1}{x^n}$

8 $x^{\frac{1}{2}} = \sqrt{x}$

9 $x^{\frac{1}{n}} = \sqrt[n]{x}$

10 $x^{\frac{m}{n}} = (\sqrt[n]{x})^m$ or $\sqrt[n]{x^m}$

11 Dividing by an algebraic fraction is equivalent to multiplying by its reciprocal, for example to divide by $\frac{a}{b}$ you multiply by $\frac{b}{a}$.

12 $\dfrac{1}{n} + \dfrac{1}{m} = \dfrac{m+n}{mn}$

13 An expression of the form $ax^2 + bx + c$, with $a \neq 0$, is called a **quadratic** in x.

14 In the expression $ax^2 + bx + c$.
   - the coefficient of x^2 is a
   - the coefficient of x is b
   - the constant term is c.

15 To factorize $ax^2 + bx$, take out the Highest Common Factor (HCF).

16 To factorize $ax^2 + bx + c$, start by looking for two numbers whose product is ac and whose sum is b.

17 $x^2 - y^2 = (x-y)(x+y)$ is called the difference of two squares.

18 $x^2 + 2ax + a^2 = (x+a)^2$

19 $x^2 - 2ax + a^2 = (x-a)^2$

21 Quadratics

Unit 2 (page 18) shows you how to solve equations and inequalities which are **linear**.

Equations of the form $ax^2 + bx + c = 0$, where $a \neq 0$ are called quadratic equations.

Quadratic equations can be used to represent a wide variety of situations such as the diver example shown here.

The solution of $7t^2 - t - 4 = 0$ gives the time in seconds to dive into the pool.

■ **The quadratic equation $ax^2 + bx + c = 0$, with $a \neq 0$ has two solutions (or roots) which may be equal.**

When solving quadratic equations it is helpful to remember that multiplying two numbers to get the answer 0 is only possible if one of the numbers is itself 0. For example:

$$\text{if } 7 \times y = 0 \text{ then } y \text{ must be } 0$$
$$\text{if } x \times 3 = 0 \text{ then } x \text{ must be } 0.$$

■ **In general, if $xy = 0$ then either $x = 0$ or $y = 0$.**

There are three algebraic methods for solving quadratic equations:

by factorizing, by completing the square, and by using a formula.

In each method the first step is to rearrange the quadratic equation into the form $ax^2 + bx + c = 0$.

21.1 Solving quadratic equations by factorizing

Methods of factorizing are introduced on page 202 and page 420. Here they are used to help solve quadratic equations.

Example 1
Solve the equation $x^2 = 8x$.

$$x^2 = 8x$$

rearrange into the form $ax^2 + bx + c = 0$: $\quad x^2 - 8x = 0$

factorize: $\quad\quad\quad\quad\quad\quad\quad\quad\quad x(x - 8) = 0$

either: $\quad\quad\quad\quad\quad\quad\quad\quad\quad\quad\quad\quad x = 0$

or $\quad\quad\quad\quad\quad\quad\quad\quad\quad\quad\quad\quad x - 8 = 0$

so $x = 0$, $x = 8$ are the two solutions or roots of the equation $x^2 = 8x$.

A common error is to divide both sides by x. This loses the $x = 0$ solution. Never divide by any term which can have value 0.

Example 2

Solve the equation $(3x + 2)(2x - 1) = 3$

$$(3x + 2)(2x - 1) = 3$$

expand the brackets: $6x^2 + x - 2 = 3$

rearrange into the form $ax^2 + bx + c = 0$:

$$6x^2 + x - 5 = 0$$

factorize: $(6x - 5)(x + 1) = 0$

either: $6x - 5 = 0$ or $x + 1 = 0$

so: $6x = 5$ or $x = -1$

The two solutions are: $x = \frac{5}{6},$ $x = -1$

Example 3

Solve the equation $y^2 - 5y + 18 = 2 + 3y$.

$$y^2 - 5y + 18 = 2 + 3y$$

rearrange into the form $ay^2 + by + c = 0$:

$$y^2 - 8y + 16 = 0$$

factorize: $(y - 4)(y - 4) = 0$

either: $y - 4 = 0$ or $y - 4 = 0$

so: $y = 4$ or $y = 4$

$y = 4$ is the only solution. In this example the two roots of the quadratic equation are equal. This parabola touches the x-axis ($y = 0$) only once.

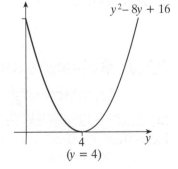

$y^2 - 8y + 16$

4
($y = 4$)

y

Example 4

Solve the equation $2a^2 - 162 = 0$

$$2a^2 - 162 = 0$$

take out the common factor 2: $2(a^2 - 81) = 0$

using the difference of two squares: $2(a - 9)(a + 9) = 0$

for this equation to be 0 either: $a - 9 = 0$ or $a + 9 = 0$

so: $a = 9$ or $a = -9$

Write these roots as $a = \pm 9$. This is read as a equals positive or negative 9.

(As there is only one unknown term (a^2) in the equation, the solution could also be found using a flow diagram method as on page 19.)

Solving equations of the type $y^2 = k$

By taking the square root of both sides you can solve equations of the type $y^2 = k$, but you must remember to write the $\pm$ sign.

■ **If $y^2 = k$ then $y = \pm\sqrt{k}$.**

Example 5

Solve the equation $(3x - 1)^2 = 64$

$$(3x - 1)^2 = 64$$

taking the square root
of both sides: $(3x - 1) = \pm 8$

add 1 to both sides: $3x = 1 \pm 8$

$3x = 1 + 8 \quad \text{or} \quad 3x = 1 - 8$

$3x = 9 \qquad \text{or} \quad 3x = -7$

The solutions are: $x = 3, x = -\frac{7}{3}$

Exercise 21A

1   Solve the equations:
   (a) $(x - 1)(x + 2) = 0$ (b) $x(x - 3) = 0$
   (c) $(x - 4)(2x + 1) = 0$ (d) $2x(2x - 3) = 0$
   (e) $(2x - 1)(3x - 4) = 0$ (f) $2(x - 5)(x - 2) = 0$

2   Solve these quadratic equations in y:
   (a) $y^2 - 3y = 0$ (b) $y^2 - 16 = 0$
   (c) $y^2 - 3y - 4 = 0$ (d) $y^2 + 11y + 28 = 0$
   (e) $2y^2 + 5y - 3 = 0$ (f) $12y^2 - 7y - 12 = 0$
   (g) $4y^2 - 11y + 6 = 0$ (h) $2y^2 - 18y = 0$
   (i) $6y^2 - 11y - 7 = 0$

3   Solve these equations:
   (a) $y^2 = 25$ (b) $3x^2 = x$
   (c) $3y^2 + 5y = 2$ (d) $2x^2 = 8$
   (e) $5x = 2x^2$ (f) $y^2 + 2y = 35$
   (g) $(a - 2)(a + 1) = 10$ (h) $(2x - 1)^2 = 3x^2 - 2$
   (i) $6a^2 = 7 - 11a$ (j) $(3x - 2)^2 = x^2$

4   Explain what is wrong with
   Julie's method to solve the equation
   $6y^2 = 12y$. Write out a correct solution.

$$6y^2 = 12y$$
$$\frac{6y^2}{6y} = \frac{12y}{6y}$$
$$y = 2$$

21.2 Solving quadratic equations by completing the square

The previous section shows how to solve quadratic equations of the type $y^2 = k$. You can solve other quadratic equations by writing them in the form $y^2 = k$. To do this you need to know how to write quadratics in the form $y^2 + r$ where r is a constant. This process is called **completing the square**.

Page 212 shows that:

$$x^2 + 2ax + a^2 = (x + a)^2$$

$x^2 + 2ax + a^2$ is called a **perfect square**.

■ **To make the expression $x^2 + 2ax$ a perfect square you add a^2.**

Here is how to complete the square for the expression $x^2 + 10x$:

$x^2 + 10x$ can be represented by squares and rectangles with sides of length x, 10 and so on:

$$x^2 + 10x = (x + 5)^2 - 25$$

So $x^2 + 10x$ is equivalent to the larger square with a smaller square taken away:

$$x^2 + 10x = (x + 5)^2 - 25$$

In general:

■ **Completing the square:** $x^2 + bx = (x + \frac{b}{2})^2 - (\frac{b}{2})^2$

This formula can be used when the coefficient of x^2 is 1.
The next example shows you what to do if the coefficient is not 1.

Example 6

Write $2x^2 - 12x$ in the form $p(x + q)^2 + r$, where p, q and r are constants to be determined.

Take out the coefficient of x^2: $2x^2 - 12x = 2[x^2 - 6x]$

Complete the square on $x^2 - 6x$ by using $b = -6$ in the completing the square formula:

$$2[x^2 - 6x] = 2[(x - 3)^2 - 9]$$
$$= 2(x - 3)^2 - 18$$

So $p = 2$, $q = -3$ and $r = -18$.

It is possible to write all quadratic expressions in the form $p(x + q)^2 + r$.

Exercise 21B

1   Write the following in the form $(x + q)^2 + r$:

   (a) $x^2 + 4x$ (b) $x^2 - 14x$ (c) $x^2 + 3x$ (d) $x^2 + x$

2   Write the following in the form $p(x + q)^2 + r$:

   (a) $2x^2 + 16x$ (b) $3x^2 - 12x$ (c) $2x^2 + x$ (d) $5x^2 - 15x$

Using the method of completing the square

At first glance it *looks* as though $x^2 + 10x + 18 = 0$ cannot easily be solved by factorizing. (You would need to find two numbers with a product of 18 and a sum of 10.) However, this quadratic can be solved by completing the square. The next example shows you how.

Example 7

Solve the equation $x^2 + 10x + 18 = 0$. Leave your answer in **surd** form.

check that the coefficient of x^2 is 1: $x^2 + 10x + 18 = 0$

subtract 18 to get constant term on RHS: $x^2 + 10x = -18$

complete the square for $x^2 + 10x$: $(x + 5)^2 - 25 = -18$

add 25 to both sides: $(x + 5)^2 = 7$

square root both sides: $x + 5 = \pm\sqrt{7}$

subtract 5 from both sides: $x = -5 \pm\sqrt{7}$

Expressions like $\sqrt{2}$, $\sqrt{3}$ and $\sqrt{7}$ are called **surds**.

The solutions (roots) of $x^2 + 10x + 18 = 0$ are $x = -5 + \sqrt{7}$ and $x = -5 - \sqrt{7}$

Example 8

Solve the equation $2y^2 - 3y - 8 = 0$. Give your answers correct to 2 d.p.

$$2y^2 - 3y - 8 = 0$$

take out the coefficient of y^2: $\quad 2[y^2 - \frac{3}{2}y - 4] = 0$

divide by 2: $\qquad\qquad\qquad\qquad y^2 - \frac{3}{2}y - 4 = 0$

$$y^2 - \frac{3}{2}y = 4$$

complete the square: $\qquad\quad [y - \frac{3}{4}]^2 - (\frac{3}{4})^2 = 4$

$$[y - \frac{3}{4}]^2 = 4 + (\frac{3}{4})^2$$

take the square root of both sides: $\qquad y - \frac{3}{4} = \pm\sqrt{\frac{73}{16}}$

add $\frac{3}{4}$ to both sides: $\qquad\qquad\qquad y = \frac{3}{4} \pm \sqrt{\frac{73}{16}}$

$$= 0.75 \pm 2.136$$

$y = 0.75 + 2.136 \quad$ or $\quad y = 0.75 - 2.136$

so $y = 2.89 \quad$ or $\quad y = -1.39 \qquad$ correct to 2 d.p.

Exercise 21C

1 Solve these equations. Leave your answers in surd form.

   (a) $x^2 + 10x + 3 = 0$ **(b)** $x^2 - 8x = 2$ **(c)** $2x^2 + 18x + 6 = 0$

   (d) $3x^2 - 6x + 1 = 0$ **(e)** $2y^2 - 3y - 4 = 0$ **(f)** $2y^2 = 4y + 7$

2 Solve these equations. Give your answers correct to 2 d.p.

   (a) $x^2 - 4x = 3$ **(b)** $x^2 - 6x + 2 = 0$ **(c)** $2y^2 + 4y = 7$

   (d) $2x^2 - 3x - 3 = 0$ **(e)** $3y^2 - 6y = 1$ **(f)** $4y^2 - y = 8$

21.3 Solving quadratic equations using a formula

The steps shown in the previous section can be applied to the general quadratic equation $ax^2 + bx + c = 0$. This gives a formula which can be used to solve **all quadratic equations**. This section shows a proof of the formula and how to apply it.

Here are two quadratic equations – one with number coefficients, the other a 'general' form of a quadratic with letter coefficients.

The same steps are followed to solve both quadratics. Work your way down to the solutions comparing the effect of the steps at each stage:

$2x^2 - 6x + 1 = 0$ $\qquad\qquad\qquad\qquad\qquad\qquad\qquad ax^2 + bx + c = 0$, where $a \neq 0$

$2[x^2 - 3x + \frac{1}{2}] = 0 \qquad$ take out the coefficient of $x^2 \qquad a\left[x^2 + \frac{b}{a}x + \frac{c}{a}\right] = 0]$

$x^2 - 3x + \frac{1}{2} = 0 \qquad$ divide by the coefficient of $x^2 \qquad x^2 + \frac{b}{a}x + \frac{c}{a} = 0$

$x^2 - 3x = -\frac{1}{2} \qquad$ move non x term to RHS $\qquad x^2 + \frac{b}{a}x = -\frac{c}{a}$

$[x - \frac{3}{2}]^2 - (-\frac{3}{2})^2 = -\frac{1}{2}$ complete the square $\left[x + \frac{b}{2a}\right]^2 - \left(\frac{b}{2a}\right)^2 = -\frac{c}{a}$

$[x - \frac{3}{2}]^2 = (-\frac{3}{2})^2 - \frac{1}{2}$ move non-x terms to RHS $\left[x + \frac{b}{2a}\right]^2 = \left(\frac{b}{2a}\right)^2 - \frac{c}{a}$

$[x - \frac{3}{2}]^2 = \frac{9}{4} - \frac{1}{2}$ simplify RHS to a single fraction $\left[x + \frac{b}{2a}\right]^2 = \frac{b^2}{4a^2} - \frac{c}{a}$

$[x - \frac{3}{2}]^2 = \frac{9-2}{4}$ $\left[x + \frac{b}{2a}\right]^2 = \frac{b^2 - 4ac}{4a^2}$

$[x - \frac{3}{2}]^2 = \frac{7}{4}$ $\left[x + \frac{b}{2a}\right]^2 = \frac{b^2 - 4ac}{4a^2}$

$x - \frac{3}{2} = \pm \sqrt{\frac{7}{4}}$ square root both sides remembering the $\pm$ $x + \frac{b}{2a} = \pm \sqrt{\frac{b^2 - 4ac}{4a^2}}$

$x - \frac{3}{2} = \pm \frac{\sqrt{7}}{\sqrt{4}}$ use $\sqrt{\frac{p}{q}} = \frac{\sqrt{p}}{\sqrt{q}}$ $x + \frac{b}{2a} = \pm \frac{\sqrt{b^2 - 4ac}}{2a}$

$x = \frac{3}{2} \pm \frac{\sqrt{7}}{2}$ get x on its own $x = -\frac{b}{2a} \pm \frac{\sqrt{b^2 - 4ac}}{2a}$

$x = \frac{3 \pm \sqrt{7}}{2}$ write as a single fraction $x = \frac{-b \pm \sqrt{b^2 - 4ac}}{2a}$

■ **The roots of the quadratic equation $ax^2 + bx + c = 0$, where $a \neq 0$, are given by the formula:**

$$x = \frac{-b \pm \sqrt{b^2 - 4ac}}{2a}$$

This formula will be on your examination formulae sheet. You do not have to remember it but you *must* know how to use it.

You can use this formula to prove some interesting results which you may already have noticed as you have worked through this unit:

The roots of $ax^2 + bx + c = 0$ are:

$$\frac{-b + \sqrt{b^2 - 4ac}}{2a} \text{ and } \frac{-b - \sqrt{b^2 - 4ac}}{2a}$$

When you add these roots you get: $\frac{-b}{2a} + \frac{-b}{2a} = \frac{-2b}{2a} = -\frac{b}{a}$

When you multiply the roots you get: $\frac{b^2 - (b^2 - 4ac)}{4a^2} = \frac{4ac}{4a^2} = \frac{c}{a}$

- **The sum of the roots of $ax^2 + bx + c = 0$ is $-\dfrac{b}{a}$**

- **The product of the roots of $ax^2 + bx + c = 0$ is $\dfrac{c}{a}$**

- **If $\sqrt{b^2 - 4ac}$ is an integer then $ax^2 + bx + c$ can be factorized.**

In GCSE examination questions where you are asked to solve equations of the type $ax^2 + bx + c = 0$, $b^2 - 4ac$ will never be negative. If your answer leads to a square root of a negative value, you need to look for an error.

> These two results are useful to check your solutions to quadratic equations.
> In Example 2:
> $$a = 6, b = 1, c = -5$$
> so $-\dfrac{b}{a} = -\dfrac{1}{6}$ and $\dfrac{c}{a} = -\dfrac{5}{6}$
>
> The roots were $\frac{5}{6}$ and -1
> so
> sum of roots $= \frac{5}{6} + -1 = -\frac{1}{6} = -\dfrac{b}{a}$
> and product of roots
> $= \frac{5}{6} \times (-1) = -\frac{5}{6} = \dfrac{c}{a}$

Example 9

Solve the equation $x(x + 3) = 2$. Give your answers correct to 2 d.p.
$$x(x + 3) = 2$$
$$x^2 + 3x = 2$$
$$1x^2 + 3x - 2 = 0$$

Comparing with $ax^2 + bx + c = 0$ gives $a = 1, b = 3, c = -2$.

> A common error is to think that x^2 is $0x^2$ and put $a = 0$. To avoid this error write x^2 as $1x^2$.

Using the formula: $x = \dfrac{-b \pm \sqrt{b^2 - 4ac}}{2a}$

gives the solutions: $x = \dfrac{-3 \pm \sqrt{3^2 - 4(1)(-2)}}{2(1)}$

$$x = \frac{-3 \pm \sqrt{9 + 8}}{2} = \frac{-3 \pm \sqrt{17}}{2}$$

so $\qquad x = \dfrac{-3 + 4.123}{2}$ or $x = \dfrac{-3 - 4.123}{2}$

and $\qquad x = 0.5615\ldots$ or $x = -3.5615\ldots$

The two solutions (roots) are $x = 0.56$ or $x = -3.56$ correct to 2 d.p.

Check the solutions by finding the sum of the roots and the product of the roots and checking against the values found from the expressions on the previous page.

Sum of roots from solutions $= -3.56 + 0.56 = -3$

Sum of roots from expression $= -\dfrac{b}{a} = -\dfrac{3}{1} = -3$. The two results agree.

Product of roots from solutions $= (-3.56) \times (0.56) = -1.9936$

Product of roots from expression: $\dfrac{c}{a} = \dfrac{-2}{1} = -2$. The two results agree.

[The slight discrepancy is due to rounding to 2 d.p.]

Example 10

Solve the equation $2y^2 - y - 4 = 0$. Leave your answers in surd form.

$2y^2 - 1y - 4 = 0$ is a quadratic in y. Comparing with $ay^2 + by + c = 0$ gives $a = 2, b = -1, c = -4$.

Using the formula:

$$y = \frac{-b \pm \sqrt{b^2 - 4ac}}{2a}$$

gives the solutions:

$$y = \frac{-(-1) \pm \sqrt{(-1)^2 - 4(2)(-4)}}{2(2)}$$

$$y = \frac{1 \pm \sqrt{1 + 32}}{4} = \frac{1 \pm \sqrt{33}}{4}$$

In surd form, the roots are $y = \dfrac{1 \pm \sqrt{33}}{4}$

Check: sum of roots $= \frac{1}{4} + \dfrac{(\sqrt{33})}{4} + \frac{1}{4} - \dfrac{(\sqrt{33})}{4} = \frac{1}{2} = -\dfrac{b}{a}$

Product of root $= \dfrac{(1 + \sqrt{33})(1 - \sqrt{33})}{16} = \dfrac{(1 - 33)}{16} = -2 = \dfrac{c}{a}$

On page 432 the quadratic $x^2 + 10x + 18 = 0$ was introduced. We said that at first glance it could not easily be solved by factorizing because two numbers with product 18 and sum 10 would need to be found. In fact two such numbers can be found. The next example shows you how.

Example 11

Find two numbers whose sum is 10 and product is 18.

Half the sum is 5.

Let one number be $5 - x$ then the other must be $5 + x$ so that they add up to 10.

The product of the numbers is 18 so:

$$(5 - x)(5 + x) = 18$$
$$25 - x^2 = 18$$

giving $x^2 = 7$ so $x = \sqrt{7}$ or $x = -\sqrt{7}$

The two numbers are $5 - \sqrt{7}$ and $5 + \sqrt{7}$

Exercise 21D

In this exercise use the formula $x = \dfrac{-b \pm \sqrt{b^2 - 4ac}}{2a}$

1   Write down the values of $b^2 - 4ac$ in these:

(a) $x^2 + 3x + 1 = 0$ (b) $x^2 - 2x - 1 = 0$

(c) $2x^2 + 6x - 1 = 0$ (d) $8x^2 - 9 = 0$

(e) $4 - 3x - 2x^2 = 0$ (f) $2x^2 = 2x + 3$

2 Solve the equations in question 1. Give your answers correct to 2 d.p.

3 Solve the following equations. Give your answers correct to 2 d.p.

(a) $x^2 - 4x + 1 = 0$ (b) $x^2 - 5x + 1 = 0$

(c) $4x^2 + 9x + 1 = 0$ (d) $4x^2 - 2x = 3$

(e) $1 = x^2 - 8x + 2$ (f) $(x + 4)^2 = 2(x + 7)$

4 Find two numbers whose sum is 18 and product 21.

5 Find two numbers whose difference is 6 and product 15.

21.4 Equations involving algebraic fractions

Unit 20 (page 407) shows you how to simplify algebraic fractions. This section shows you how to apply this skill to algebraic equations. For example:

If $\frac{x}{4} = \frac{3}{4}$ then $x = 3$.

■ **In general, if $\frac{p}{r} = \frac{q}{r}$ then $p = q$.**

To solve equations involving algebraic fractions:

Step 1 Write both sides of the equation with the same denominator.

Step 2 Use the numerators to write an equation.

Step 3 Solve the resulting equation.

Example 12

Solve the equation:

$$\frac{2}{y + 1} - \frac{3}{2y + 3} = \frac{1}{2}$$

Give your answer correct to 3 d.p.

$$\frac{2}{y + 1} - \frac{3}{2y + 3} = \frac{1}{2}$$

The LCD of the three denominators is $2(y + 1)(2y + 3)$.

Multiply each term so that this LCD is the denominator for the whole equation:

$$\frac{2 \times 2 (2y + 3)}{2(y + 1)(2y + 3)} - \frac{3 \times 2 (y + 1)}{2(y + 1)(2y + 3)} = \frac{(y + 1)(2y + 3)}{2(y + 1)(2y + 3)}$$

Now the numerators can be simplified, ignoring the denominators because they are the same across the equation.

$$4 (2y + 3) - 6(y + 1) = (y + 1)(2y + 3)$$
$$8y + 12 - 6y - 6 = 2y^2 + 5y + 3$$
$$0 = 2y^2 + 3y - 3$$

Comparing this equation with the general quadratic $0 = ay^2 + by + c$ gives $a = 2, b = 3, c = -3$.

Using the formula:

$$y = \frac{-3 \pm \sqrt{3^2 - 4(2)(-3)}}{2(2)}$$

$$= \frac{-3 \pm \sqrt{33}}{4} = \frac{-3 \pm 5.74456}{4}$$

So $y = 0.686$ or $y = -2.186$.

Check:

sum of roots $= -2.186 + 0.686 = -1.5 = -\frac{3}{2} = -\frac{b}{a}$

product of roots $= (-2.186) \times (0.686) = -1.499596$ which is approximately equal to $\frac{c}{a}$

Exercise 21E

Solve the following equations. Give fractional answers exactly. In all other cases give your answers correct to 2 d.p.

1 $\dfrac{1}{x} + \dfrac{1}{x+1} = \dfrac{7}{12}$ **2** $2x + \dfrac{1}{x} = 5$

3 $\dfrac{5}{2x+1} + \dfrac{6}{x+1} = 3$ **4** $\dfrac{1}{x-1} - \dfrac{1}{x} = 8$

5 $\dfrac{1}{x-3} - \dfrac{3}{x+2} = \dfrac{1}{2}$ **6** $\dfrac{2}{y+1} + \dfrac{3}{2y+3} = 1$

21.5 Problems leading to quadratic equations

The equation for question 1 in Exercise 21E would be the correct one to solve the following problem.

Find consecutive integers whose reciprocals add up to $\frac{7}{12}$.

When answering questions in the GCSE examination, candidates usually find it harder to obtain the quadratic equation than to solve it. You will find these steps useful.

Finding the equation to represent a problem

Step 1 Where relevant, draw a diagram and put all the information on it.

Step 2 Use x to represent the unknown which you have been asked to find.

In the diagonals investigation on page 216 the generalization for the number of diagonals of a regular polygon with n sides is $[\frac{1}{2}n(n-3)]$. Suppose you are told that an n-sided regular polygon has 35 diagonals and are asked to find n. This gives:

$$\frac{1}{2}n(n-3) = 35$$
$$n(n-3) = 70$$
$$n^2 - 3n = 70$$
$$n^2 - 3n - 70 = 0$$

This is a quadratic equation.

Step 3 Use other letters to identify any other relevant unknowns.

Step 4 Look for information given in the question which links these letters to x and write them down.

Step 5 Try simple numbers for the unknowns and see if this helps you to find a method.

Step 6 Make sure that the units on each side of your equation are the same.

When you have solved the quadratic equation you *must* check your two roots to see whether they are meaningful in the context of the problem. Often one of the roots will not be a solution of the problem.

Example 13

In a right-angled triangle, the hypotenuse is 6 cm longer than the shortest side. The third side is 2 cm shorter than the hypotenuse. Find a quadratic equation in the form $ax^2 + bx + c = 0$ which when solved leads to the length of the shortest side.

Let the shortest side be x cm.

Mark the other sides y cm and z cm like this:

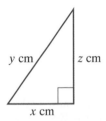

The hypotenuse is 6 cm longer than x so $y = x + 6$

The third side is 2 cm shorter than y so $z = y - 2$ or $z = x + 6 - 2 = x + 4$

The units of length are the same.

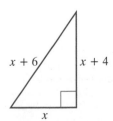

Test: if $x = 1$ then $y = 1 + 6 = 7$ and $z = 1 + 4 = 5$. This confirms the relative size of each side.

Using Pythagoras' Theorem:
$$x^2 + (x + 4)^2 = (x + 6)^2$$
$$x^2 + x^2 + 8x + 16 = x^2 + 12x + 36$$
$$x^2 - 4x - 20 = 0$$

Solving this quadratic gives $x = 6.899$ or -2.899. As length is positive the only answer is 6.9 to 2 s.f.

Example 14

Lisa cycled from Norton to Sufton, a distance of 34 km, and then cycled the same distance back.

Lisa's average speed on the outward journey was x km/h.

On the return journey the wind was behind her so she cycled 2 km/h faster and completed the return journey 16 minutes quicker than on the outward journey.

(a) Write down, in terms of x, the time in hours taken for:

(i) the outward journey

(ii) the return journey.

(b) Show that $x^2 + 2x - 255 = 0$.

(c) Calculate Lisa's average speed on the return journey.

(a) Using:
$$\text{time taken} = \frac{\text{distance}}{\text{average speed}}$$

(i) The time for the outward journey is $\frac{34}{x}$ hours.

(ii) Let Lisa's average speed on the return journey be y km/h.
y is 2 more than x so $y = x + 2$.

The time for the return journey is $\frac{34}{y} = \frac{34}{x+2}$ hours.

(b) The time taken for the outward journey is 16 minutes more than for the return journey.

Time for outward journey – time for return journey = 16 minutes

$$\frac{34}{x} - \frac{34}{x+2} = \frac{4}{15}$$

The units must be the same
16 mins $= \frac{16}{60} = \frac{4}{15}$ hours

$$\frac{34 \times 15(x+2) - 34 \times 15(x)}{15x(x+2)} = \frac{4x(x+2)}{15x(x+2)}$$

$$510x + 1020 - 510x = 4x^2 + 8x$$
$$0 = 4x^2 + 8x - 1020$$
$$0 = 4(x^2 + 2x - 255)$$

So $x^2 + 2x - 255 = 0$

(c) You need to find the value of y, which equals $x + 2$.

From part **(b)** $x^2 + 2x - 255 = 0$

$$(x + 17)(x - 15) = 0$$
$$\text{So } x + 17 = 0 \text{ or } x - 15 = 0$$
$$\text{and } x = -17 \text{ or } x = 15$$

Lisa's speed cannot be negative, so $x = 15$.

Lisa's average speed on the return journey is
$x + 2 = 15 + 2 = 17$ km/h

Exercise 21F

1 Find two consecutive integers whose reciprocals add up to $\frac{9}{20}$.

2 The hypotenuse of a right-angled triangle is $2x$ cm.

The lengths of the other two sides are $(x + 1)$ cm and $(x + 3)$ cm. Calculate the value of x.

3 The sum of the square of a number and five times the number itself is 24. By solving a quadratic equation find the two possible values of the number.

4 The length of a rectangular piece of carpet is 4 m longer than its width. The area of the carpet is 16 m². Calculate the width of the carpet. Give your answer correct to the nearest centimetre.

5 One week a syndicate of x people won £400 on the National Lottery.

(a) Write down, in terms of x, how much each person should receive, assuming that each gets an equal share.

If the prize had been won the previous week when there were $(x + 2)$ people in the syndicate, each person would have received £10 less.

(b) Show that $x^2 + 2x - 80 = 0$.

(c) Calculate the amount of money given to each person in the syndicate.

6 A right-angled triangle is cut from the corner of a rectangular piece of card 15 cm by 8 cm.

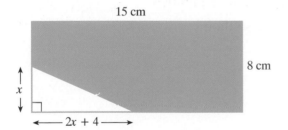

The base of the triangle is $(2x + 4)$ cm and its height is x cm.

The remaining piece of card (green) has an area of 89 cm².
Calculate the value of x, giving your answer correct to 2 s.f.

7 Joe is training for a long-distance cycle race. One day he cycles for x hours in the morning and travels a distance of 84 km.

(a) Write down, in terms of x, Joe's average speed in km/h.

In the afternoon he cycles for 1 hour more to travel the same distance and his average speed is 2 km/h slower than in the morning.

(b) Show that $x^2 + x - 42 = 0$.

(c) Calculate the number of hours Joe cycles in the afternoon.

21.6 Solving linear and quadratic equations simultaneously

In unit 7 you found the coordinates of the point of intersection of two straight lines by solving the equations of the lines simultaneously.

In this section you will find the coordinates of the points of intersection of a straight line and a quadratic curve.

Let us consider the curve $y = x^2$ and three different lines.

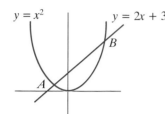 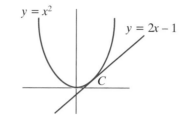 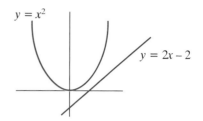

Diagram 1: the line $y = 2x + 3$ cuts the curve at two points, A and B.
Diagram 2: the line $y = 2x - 1$ just touches at the curve at the point C.
Diagram 3: the line $y = 2x - 2$ misses the curve completely.

The following three examples, correspond to the three cases above.

Example 15

Find the coordinates of the points of intersection A and B of the line $y = 2x + 3$ and the curve $y = x^2$.

Answer

The points A and B both lie on the curve and the line so their coordinates must satisfy both.

$$y = x^2 \qquad (1)$$
$$\text{and} \qquad y = 2x + 3 \qquad (2)$$

Eliminate y by substituting (1) into (2) to give

$$x^2 = 2x + 3$$

Rearrange: $\qquad x^2 - 2x - 3 = 0$

Factorize: $\qquad (x + 1)(x - 3) = 0$

So $\qquad\qquad x + 1 = 0 \text{ or } x - 3 = 0$

which gives $\qquad\qquad x = -1 \text{ or } x = 3$

Substitute these values of x in the linear equation (2):

When $x = -1$, $\qquad\qquad y = 2(-1) + 3 = 1$
When $x = 3$, $\qquad\qquad y = 2(3) + 3 = 9$

So the line $y = 2x + 3$ intersects the curve $y = x^2$ at the points $A(-1, 1)$ and $B(3, 9)$.

Check there is a solution:
$a = 1, b = -2$ and $c = -3$,
so
$b^2 - 4ac = 4 + 12 = 16 > 0$

The linear equation is the equation of the straight line.

From the sketch the x-coordinate of A is negative.

Example 16

(a) Solve the simultaneous equations $y = x^2$ and $y = 2x - 1$.

(b) Interpret your solution to part **(a)** geometrically.

Answer

(a)
$$y = x^2 \qquad (1)$$
and
$$y = 2x - 1 \qquad (2)$$

Eliminate y by substituting (1) into (2) to give

$$x^2 = 2x - 1$$

Rearrange: $x^2 - 2x + 1 = 0$

Factorize: $(x - 1)(x - 1) = 0$

So $x - 1 = 0$ repeated

which gives the single value $x = 1$

> Check there is a solution:
> $a = 1, b = -2$ and $c = 1$, so $b^2 - 4ac = 0$.

Substitute this value of x in the linear equation (2):
When $x = 1$, $y = 2(1) - 1 = 1$.

(b) The line $y = 2x - 1$ meets the curve $y = x^2$ in just one point, $(1, 1)$. We say that the line $y = 2x - 1$ is a tangent to the curve $y = x^2$ at the point $(1, 1)$.

Example 17

(a) Show that the x-coordinate of any point of intersection of the line $y = 2x - 2$ and the curve $y = x^2$ would need to satisfy the equation $(x - 1)^2 = -1$.

(b) Interpret the result in part **(a)** geometrically.

Answer

(a) (1) $y = x^2$ and (2) $y = 2x - 2$

Eliminate y by substituting (1) into (2) to give

$$x^2 = 2x - 2$$

Rearrange: $x^2 - 2x + 2 = 0$

Complete the square: $x^2 - 2x + 1 = -2 + 1$

Factorize: $(x - 1)^2 = -1$

> $a = 1, b = -2$ and $c = 2$, so $b^2 - 4ac = 4 - 8 - 4 < 0$.

The x-coordinate would need to satisfy this equation.

(b) For real numbers, $(x - 1)^2$ is always $\geqslant 0$. So there are no real number solutions to the equation and the line $y = 2x - 2$ never intersects the curve $y = x^2$.

> You can use **complex numbers** to solve this type of equation. You can learn about complex numbers at A-level.

Exercise 21G

1 Solve the simultaneous equations:

    (a) $y = x^2$ and $y = 16$ **(b)** $y = x^2$ and $y = 12x - 36$

    (c) $y = x^2$ and $y = 2x + 35$ **(d)** $y = x^2$ and $7x + 18$

    (e) $y = 2x^2$ and $y = x + 3$ **(f)** $y = 3x^2$ and $y = 7x + 6$

    (g) $y = 2 - x^2$ and $y = x - 4$ **(h)** $y = 12x^2 - 5$ and $y = 7 - 7x$

2 Find the coordinates of the points of intersection of these lines and curves:

    (a) $y = 3 - 2x$ and $y = x^2$

    (b) $y = 3x + 10$ and $y = 4x^2$

    (c) $y = x + 5$ and $y = x^2 - 2x - 5$

    (d) $y = 3x + 7$ and $y = 7 - 6x^2$

3 In each case determine the number of points of intersection of the line and the curve and find the coordinates of any points of intersection.

    (a) $y = x^2 + 3$ and $y = 2$ **(b)** $y = x^2$ and $y = 10x - 25$

    (c) $y = 2x^2$ and $y = 5x + 7$ **(d)** $y = x^2 + 3x + 6$ and $y = 1 - x$

    (e) $y = 5x^2$ and $y = 14x + 3$ **(f)** $y = 4x^2$ and $y = 10x + 6$

    (g) $y = 3x - x^2$ and $y = 4 - x$ **(h)** $y = 12x^2 - 5x$ and $y = 6x + 5$

The intersection of a line and a circle

In section 8.1 you were shown that the equation of a circle with centre $(0, 0)$ and radius r is $x^2 + y^2 = r^2$.

Example 18

(a) Show that the x-coordinate of the points of intersection of the line $y = 2x + 2$ and the circle $x^2 + y^2 = 8$ must satisfy the equation $5x^2 + 8x - 4 = 0$.

(b) Hence find the coordinates of the points where the line $y = 2x + 2$ cuts the circle $x^2 + y^2 = 8$.

Answer

(a) The points of intersection both line on the circle and the line so their coordinates must satisfy both.

$$x^2 + y^2 = 8 \qquad (1)$$

and
$$y = 2x + 2 \qquad (2)$$

Eliminate y by substituting (2) into (1) to give

$$x^2 + (2x + 2)^2 = 8$$

Expand the brackets $x^2 + 4x^2 + 8x + 4 = 8$

Rearrange: $5x^2 + 8x - 4 = 0$ as required.

(b) Factorize: $(5x - 2)(x + 2) = 0$

So $5x - 2 = 0$ or $x + 2 = 0$

which gives $x = \frac{2}{5}$ or $x = -2$

Substitute these values of x in the linear equation (2):

When $x = \frac{2}{5}$, $y = 2(\frac{2}{5}) + 2 = \frac{14}{5}$

When $x = -2$, $y = 2(-2) + 2 = -2$

The line $y = 2x + 2$ cuts the circle $x^2 + y^2 = 8$ at the points $(\frac{2}{5}, 2\frac{4}{5})$ and $(-2, -2)$.

Example 19

(a) Solve the simultaneous equations $4y + 3x = 25$ and $x^2 + y^2 = 25$.

(b) Interpret your solution to part **(a)** geometrically.

Answer

(a) (1) $4y + 3x = 25$ and (2) $x^2 + y^2 = 25$

From (1) $\qquad\qquad y = \dfrac{25 - 3x}{4}$ $\qquad$ (3)

Eliminate y by substituting (3) into (2) to give

$$x^2 + \frac{(25 - 3x)^2}{16} = 25$$

Multiply each term by 16: $\quad 16x^2 + (25 - 3x)^2 = 400$

Expand the brackets: $\quad 16x^2 + 625 - 150x + 9x^2 = 400$

Rearrange: $\quad 25x^2 - 150x + 225 = 0$

Factorize: $\quad 25(x^2 - 6x + 9) = \Rightarrow 25(x - 3)^2 = 0$

so $\quad x - 3 = 0$ repeated

which gives $\quad x = 3$

Substitute in (3): $\quad$ when $x = 3$, $y = \dfrac{25 - 9}{4} = 4$

$x = 3, y = 4$ is the only solution of the simultaneous equations.

(b) The line $4y + 3x = 25$ is a tangent to the circle $x^2 + y^2 = 25$ at the point $(3, 4)$.

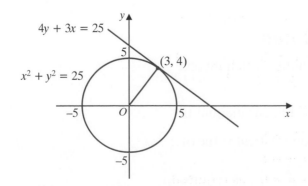

The gradient of the radius from $(0, 0)$ to $(3, 4)$ is $\frac{4}{3}$.

The gradient of the line $4y + 3x = 25$ or $y = -\frac{3}{4}$ $x + \frac{25}{4}$ is $-\frac{3}{4}$.

From section 7.3, since $\frac{4}{3} \times -\frac{3}{4} = -1$ the radius and the line $4y + 3x = 25$ are perpendicular. A radius is always perpendicular to a tangent.

1 In each part solve the simultaneous equations and interpret your solution geometrically.

    (a) $y = 3$ and $x^2 + y^2 = 25$

    (b) $y = x + 5$ and $x^2 + y^2 = 25$

    (c) $x^2 + y^2 = 25$ and $y = 3x + 13$

    (d) $x^2 + y^2 = 25$ and $y + x + 1 = 0$

    (e) $x^2 + y^2 = 50$ and $y = x - 10$

    (f) $x^2 + y^2 = 50$ and $y + 8x = 1$

    (g) $y = 3x - 10$ and $x^2 + y^2 = 10$

    (h) $y + 2x = 5$ and $x^2 + y^2 = 5$

    (i) $x^2 + y^2 = 2$ and $y + 7x + 10 = 0$

    (j) $3y + 4x = 6$ and $x^2 + y^2 = 4$

2 The line $y = x - 3$ intersects the circle $x^2 + y^2 = 25$ at the points A and B.

    (a) Show that the x-coordinates of A and B satisfy the equation $x^2 - 3x - 8 = 0$.

    (b) Hence show that the coordinates of A and B are

$$\left(\frac{3 + \sqrt{41}}{2}, \frac{-3 + \sqrt{41}}{2} \right) \text{ and } \left(\frac{3 - \sqrt{41}}{2}, \frac{-3 - \sqrt{41}}{2} \right)$$

3 **(a)** Show that the line $y = x + 4$ is a tangent to the circle $x^2 + y^2 = 8$.

    (b) (i) Find the equation of the tangent to the circle $x^2 + y^2 = 8$ that is parallel to $y = x + 4$ and write down the coordinates of the point where it touches the circle.

        (ii) Find the distance between these two parallel tangents.

    (c) Give a general result for the distance between two parallel tangents to any circle of the form $x^2 + y^2 = r^2$.

21.7 Solving quadratic inequalities

Unit 2 (page 28) shows you how to solve inequalities which involve linear functions.

$2(x - 2) \leqslant 20$, $2(3x - 5) > 8 - 3x$ are examples of linear inequalities.

The solution of a linear inequality involves just one **critical value** of x.

For example, the solution of $2(x - 2) \leqslant 20$ is $x \leqslant 12$.

We say that $x = 12$ is the critical value. The critical value splits the number line into two regions.

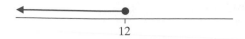

This section shows you how to solve, algebraically, inequalities which involve some types of quadratic functions.

Examples of quadratic inequalities:

$$x^2 - 9 > 0, \quad 2x^2 + 3 \leqslant 53, \quad \frac{(1 - 2x)^2}{4} \geqslant 4$$

The solution of a quadratic inequality involves **two critical values**.

■ **Find the two critical values by replacing the inequality sign by = and solving the quadratic equation.**

These two critical values split the number line into three regions.

■ **The solution of the quadratic inequality will be either the region between the two critical values or the other two regions. If the inequality is not strict, that is either $\leqslant$ or $\geqslant$, the critical values will form part of the solution.**

Example 20

Solve the inequality $x^2 < 9$.

$$x^2 < 9$$

To find the critical values replace $<$ by $=$: $\quad x^2 = 9$

Take the square root of both sides: $\quad\quad x = \pm 3$

The inequality is strict so the critical values -3 and $+3$ are **not** part of the solution.

Pick a value of x between these two critical values and use it to test the inequality $x^2 < 9$.

For values of x between -3 and $+3$, for example $x = 0$, x^2 is < 9 so x lies in the region between -3 and $+3$.

We write the solution of the inequality as $-3 < x < 3$.

Example 21

Solve the inequality $2x^2 + 3 \geqslant 53$

To find the critical values replace $\geqslant$ by $=$:

$$2x^2 + 3 \geqslant 53$$
$$2x^2 + 3 = 53$$
$$2x^2 = 50$$
$$x^2 = 25$$

Take the square root of both sides: $\qquad x = \pm 5$

The inequality is not strict so the critical values -5 and $+5$ **are** part of the solution.

Choosing 0 as a value for x between the critical values you can see that when $x = 0$, $2x^2 + 3 = 3$.

So for $-5 < x < 5$, $2x^2 + 3$ is **not** $\geqslant 53$, therefore x must lie **outside** the region $-5 < x < 5$.
We write the solution of $2x^2 + 3 \geqslant 53$ as $x \leqslant -5$ or $x \geqslant 5$.

Example 22

Solve the inequality $\dfrac{(1 - 2x)^2}{4} \leqslant 9$.

$$\frac{(1 - 2x)^2}{4} \leqslant 9$$

To find the critical values replace $\leqslant$ by $=$: $\qquad \dfrac{(1 - 2x)^2}{4} = 9$

Multiply both sides by 4: $\qquad (1 - 2x)^2 = 36$

Take the square root of both sides $\qquad 1 - 2x = \pm 6$
Subtract 1 from both sides $\qquad -2x = 6 - 1 \;$ or $\; -6 - 1$
So the critical values are $x = -2\frac{1}{2}$ and $x = 3\frac{1}{2}$.

The inequality is not strict so the critical values $-2\frac{1}{2}$ and $3\frac{1}{2}$ **are** part of the solution.

Choosing 0 as a value for x between the critical values you can see that when $x = 0$, $\frac{(1 - 2x)^2}{4} = \frac{1}{4}$ which is $\leqslant 9$ as required.

So x must lie between the critical values.

We write the solution of $\dfrac{(1-2x)^2}{4} \leqslant 9$ as $-2\frac{1}{2} \leqslant x \leqslant 3\frac{1}{2}$.

Exercise 21I

Solve these inequalities:

1 $x^2 \leqslant 25$ **2** $x^2 > 49$ **3** $x^2 < 2.25$

4 $x^2 + 3 \leqslant 19$ **5** $x^2 - 4 \geqslant 5$ **6** $4 + x^2 > 5$

7 $3x^2 \geqslant 27$ **8** $4x^2 < 49$ **9** $3x^2 > 6\frac{3}{4}$

10 $2x^2 - 3 < 47$ **11** $4 + 9x^2 > 5$ **12** $\frac{x^2}{4} \leqslant 16$

13 $1 + \frac{x^2}{6} \geqslant 2\frac{1}{2}$ **14** $10 - x^2 < 1$ **15** $4(13 - x^2) \geqslant 3$

16 $(x + 1)^2 \leqslant 25$ **17** $(2 - x)^2 > 49$ **18** $\frac{(2x + 1)^2}{9} < 16$

21.8 Solving equations graphically

Work on graphical solutions is introduced in Unit 18 (page 368).

Unless graphs have been drawn for you, a lot of valuable time in examinations can be used up drawing them accurately enough to solve equations. Also, graphical solutions are less accurate so you should only use a graphical approach to solve a quadratic equation if asked to do so in a question.

Although most cubic and higher power equations have to be solved graphically at GCSE level rather than algebraically, you should still consider an algebraic approach if x is a common factor. For example:

A sketch of $y = x^3 - 4x^2 + 4x$

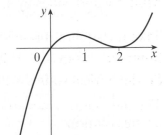

Solve the equation $x^3 - 4x^2 + 4x = 0$

$$x^3 - 4x^2 + 4x = 0$$
$$x(x^2 - 4x + 4) = 0$$
$$x(x - 2)(x - 2) = 0$$

which leads to the solutions $x = 0$ or $x = 2$

Notice that the 'double' solution at $x = 2$ indicates that the graph of $y = x^3 - 4x^2 + 4x$ just touches the x-axis at $x = 2$. The x-axis is a **tangent** to this curve at $x = 2$.

Using a given graph to solve equations

Sometimes you will need to rearrange the equation to be solved to make use of a given graph.

Example 23

The graph of $y = 2 + x - x^2$, for $-3 \leqslant x \leqslant 3$ is drawn below. This shape is a **parabola**.

(a) By drawing lines on the graph, solve these equations giving your answers to (i) correct to 1 d.p.

   (i) $1 + x - x^2 = 0$ (ii) $2 - x - x^2 = 0$

(b) Find the equation of the line you would draw on the graph to solve the equation $2x^2 - x = 6$.

(c) Explain how the graph can be used to show that the equation $x^2 - x + 3 = 0$ has no real solutions.

(d) By drawing a suitable curve on the graph, solve the equation $x^3 - x^2 - 2x + 6 = 0$.

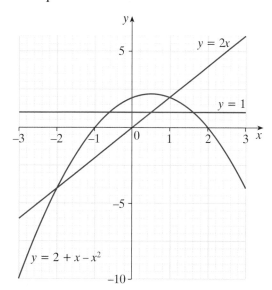

Answer

(a) (i) To solve $1 + x - x^2 = 0$ using the graph:

      add 1 giving: $2 + x - x^2 = 1$

      The graph shows: $2 + x - x^2 = y$

      Comparing gives: $y = 1$

      So draw the horizontal line: $y = 1$

This line meets the parabola at $x = -0.6$ and $x = 1.6$, so the solutions of $1 + x - x^2 = 0$ are $x = -0.6$ and $x = 1.6$ to 1 d.p.

(ii) To solve $2 - x - x^2 = 0$ using the graph:

      add 2x giving: $2 + x - x^2 = 2x$

      The graph shows: $2 + x - x^2 = y$

      Comparing gives: $y = 2x$

      So draw the line: $y = 2x$

This line meets the parabola at $x = -2$ and $x = 1$ so the solutions of $2 - x - x^2 = 0$ are $x = -2$ and $x = 1$.

(b) To solve $2x^2 - x = 6$ using the graph:

rearrange:	$0 = 6 + x - 2x^2$
divide by 2:	$0 = 3 + \frac{1}{2}x - x^2$
add $\frac{1}{2}x$:	$\frac{1}{2}x = 3 + x - x^2$
subtract 1:	$\frac{1}{2}x - 1 = 2 + x - x^2$
The graph shows:	$y = 2 + x - x^2$
Comparing gives:	$y = \frac{1}{2}x - 1$

So you need to draw the line $y = \frac{1}{2}x - 1$ and find where it meets the parabola. The x-coordinates of the points of intersection give the solutions of the equation $2x^2 - x = 6$.

(c) To solve $x^2 - x + 3 = 0$ using the graph:

rearrange:	$0 = -3 + x - x^2$
add 5 gives:	$5 = 2 + x - x^2$
The graph shows:	$y = 2 + x - x^2$
Comparing gives:	$y = 5$

The horizontal line $y = 5$ does not meet the parabola so you can deduce that $x^2 - x + 3 = 0$ has no real solutions.

(d) To solve $x^3 - x^2 - 2x + 6 = 0$ using the graph:

rearrange:	$6 = 2x + x^2 - x^3$
factorize:	$6 = x(2 + x - x^2)$
divide by x:	$\frac{6}{x} = 2 + x - x^2$
The graph shows:	$y = 2 + x - x^2$
Comparing gives:	$y = \frac{6}{x}$

Alternative method:
You could rearrange $x^3 - x^2 - 2x + 6 = 0$ to get $2 + x - x^2 = 3x - x^3 - 4$ Then draw the curve $y = 3x - x^3 - 4$ and find its point of intersection with the parabola. Although valid, we aim to rearrange so that the easiest graph to draw is obtained, preferably straight lines.

Draw the graph of the reciprocal function $y = \frac{6}{x}$. Notice that it only meets the parabola at one point. This means that there is only one real solution of the equation $x^3 - x^2 - 2x + 6 = 0$.

The x-coordinate of this point of intersection gives the solution. So the one real solution of $x^3 - x^2 - 2x + 6 = 0$ is $x = -1.8$ to 1 d.p.

Exercise 21J

1 **(a)** On graph paper draw the graph of $y = x^2 - 4$ for $-3 \leqslant x \leqslant 3$.

(b) On the same axes, draw the graph of $y = 2 + \frac{1}{x}$ for $-3 \leqslant x \leqslant -\frac{1}{2}$ and $\frac{1}{2} \leqslant x \leqslant 3$.

(c) Use your graphs to find approximate solutions for:

(i) $x^2 = 2$ (ii) $2 + \frac{1}{x} = 0.8$ (iii) $2 + \frac{1}{x} = x^2 - 4$

2 Part of the graph of the cubic function
$y = x^3 - 4x$ is shown.

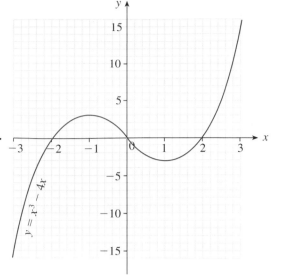

(a) Find the equation of the line which needs
to be drawn on the graph to solve:

(i) $x^3 - 4x = 1$ (ii) $x^3 - 4x + 2 = 0$
(iii) $x^3 - 6x = 1$ (iv) $x^3 - 2x = 4$

(b) Which one of the equations in (a) does not
have three real solutions? Explain your answer.

(c) The graph of $y = \frac{1}{x}$ (not drawn) intersects
the given cubic curve $y = x^3 - 4x$ at points
A and B. Find, in a form which includes
an x^4 term, the equation which has the
x-coordinates of A and B as two of its
solutions.

3 (a) On graph paper, draw the graph of the
parabola

$y = (6-x)(2+x)$ for $-2 \leqslant x \leqslant 6$

A container is in the shape of a cuboid of height x m.
The area of the base of the container is $(6-x)(2+x)$ m^2.
The volume of the container is 24 m^3.

(b) By drawing the graph of an appropriate curve on your axes
in (a) find, correct to the nearest tenth of a metre, the
possible values for the height of the container.

4 (a) Copy and complete the table of values for $y = x^3 - 6x + 1$

x	-3	-2	-1	0	1	2
y	-8					

(b) On graph paper draw and label appropriate axes. Plot the
points represented by the values in your table. Join them
with a smooth curve.

(c) (i) On the same axes, draw the line with equation $y = 2x + 5$.

(ii) Write down the x-coordinates of the two points of
intersection of this line and the curve you have drawn
in (b).

(d) Find the equation, expressed as simply as possible, which
may be solved to give the x-coordinates of each point of
intersection of the curve and the line.

A line L is drawn parallel to the line $y = 2x + 5$ so that L
intersects the curve $y = x^3 - 6x + 1$ at the point on the y-axis.

(e) Write down the equation of line L.

(f) By forming and solving an algebraic equation, find the
x-coordinates of the three points of intersection of L and the
curve $y = x^3 - 6x + 1$.

Summary of key points

1   The quadratic equation $ax^2 + bx + c = 0$, with $a \neq 0$ has two solutions (or roots) which may be equal.

2   If $xy = 0$ then either $x = 0$ or $y = 0$.

3   If $y^2 = k$ then $y = \pm\sqrt{k}$.

4   To make the expression $x^2 + 2ax$ a perfect square you add a^2.

5   Completing the square: $x^2 + bx = (x + \frac{b}{2})^2 - (\frac{b}{2})^2$.

6   The roots of the quadratic equation $ax^2 + bx + c = 0$, where $a \neq 0$
are given by the formula $x = \dfrac{-b \pm \sqrt{b^2 - 4ac}}{2a}$

7   If $\sqrt{b^2 - 4ac}$ is an integer then $ax^2 + bx + c$ can be factorized.

8   The sum of the roots of $ax^2 + bx + c = 0$ is $-\frac{b}{a}$.

9   The product of the roots of $ax^2 + bx + c = 0$ is $\frac{c}{a}$.

10   If $\frac{p}{r} = \frac{q}{r}$ then $p = q$.

11   When applied to problem solving, one of the roots of the quadratic equation is often not a solution to the problem and must be abandoned with an explanation.

12   Solving a linear equation ($y = px + q$) and a quadratic equation ($y = ax^2 + bx + c$) simultaneously:
   (i)   Find y in terms of x from the linear equation (or x in terms of y).
   (ii) Substitute for y (or x) in the quadratic equation.
   (iii) Solve the resulting quadratic equation for x (or y).
   (iv) Substitute the values of x (or y) into the linear equation to find y (or x).
   If the roots of the quadratic equation are equal the line will be a tangent to the curve.

13   Solving quadratic inequalities:
   (i)   Find the two critical values by replacing the inequality sign by $=$ and solving the quadratic equation.
   (ii) Indicate the two critical values on a number line using open circles for strict inequalities, closed circles otherwise.
   (iii) Pick a value between the two critical values and test the given inequality. If satisfied, then the solutions of the inequality lie between the two critical values. If the inequality is not satisfied then the required solutions lie in the other two regions not between the two critical values.

14   Solving equations from a given graph (or one to be drawn):
   (i)   Rearrange the equation to be solved to match the equation of the graph. Try to keep the rearrangement simple.
   (ii) Compare to find y. Note that y may be of form $\frac{k}{x}$ if the equation to be solved has a higher power of x than the given graph function.
   (iii) Draw the line (or curve) from (ii) on the given axes.
   (iv) Read off the values of x at the points of intersection.

22 Advanced trigonometry

This chapter extends the work on Pythagoras' Theorem (Unit 8) and basic trigonometry (Unit 13). In this unit you will learn how to find areas, lengths and angles in triangles which do not contain a right angle. You will also learn to apply Pythagoras' Theorem and trigonometry in three dimensions.

22.1 The area of a triangle

The diagram below shows a general triangle ABC. The lengths of its sides are a, b, and c units.

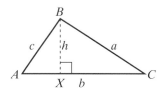
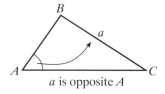

a is opposite A

The perpendicular from B to AC is drawn.
It meets AC at X.
Its length is h.

The area of a triangle is $\frac{1}{2}$ base $\times$ height.
So the area of a triangle ABC is $\frac{1}{2}bh$.

Using the **sine ratio**

$$\sin C = \frac{h}{a} \quad \text{or } h = a \sin C$$

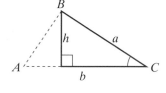

Substituting $h = a \sin C$ into area $ABC = \frac{1}{2}bh$ gives:
$\qquad$ area $ABC = \frac{1}{2}b\,(a \sin C)$, which we can write as:

- **area $ABC = \frac{1}{2}ab \sin C$**

Similarly, it can be shown that:

- **area $ABC = \frac{1}{2}ac \sin B$**

and

- **area $ABC = \frac{1}{2}bc \sin A$**

Note: in all cases, to find the area of a triangle you need the *length of two sides* of the triangle and the *size of the angle between these sides*. Then the area is:

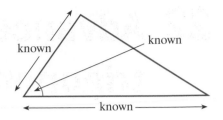

■   **area of triangle = length of first side × length of second side × sine of angle between these sides**

Example 1

Find the area of the triangle *PQR*.

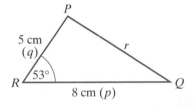

$$\text{Using area } PQR = \tfrac{1}{2}pq \sin R$$
$$\text{area } PQR = \tfrac{1}{2} \times 5 \times 8 \times \sin 53°$$
$$= 20 \times 0.7986$$
$$= 15.97 \text{ cm}^2$$

Example 2

A builder fences off a triangular plot of land *XYZ*.
$$XY = 43 \text{ m}, \quad XZ = 58 \text{ m} \quad \text{and the angle } Y\hat{X}Z = 125°$$

Calculate the area of the plot.

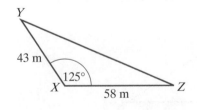
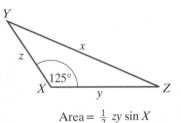

$$\text{Area} = \tfrac{1}{2} zy \sin X$$

$$\text{Using area } XYZ = \tfrac{1}{2}zy \sin X$$
$$\text{area } XYZ = \tfrac{1}{2} \times 43 \times 58 \times \sin 125°$$
$$= \tfrac{1}{2} \times 43 \times 58 \times 0.8192$$
$$= 1021.5 \text{ m}^2$$

Example 3

The area of a triangle *ABC* is 52 cm^2.
All three angles of this triangle are acute.
$$AB = 14 \text{ cm}, \quad AC = 12 \text{ cm}$$

Calculate the angle A.

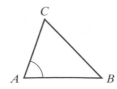

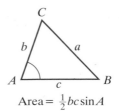

Area $= \frac{1}{2} bc \sin A$

$$\text{Area} = \frac{1}{2} \times 14 \times 12 \times \sin A$$
$$52 = 84 \times \sin A$$

so, dividing both sides by 84:

$$\sin A = \frac{52}{84}$$
$$\sin A = 0.6190$$

so:
$$A = 38.25°$$

Exercise 22A

1   Calculate the area of each of the triangles. (All lengths in cm.)

(a)

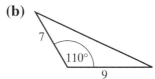

(b)

(c)

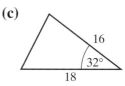

(d)

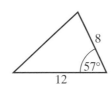

(e)

(f)

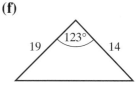

2   A farmer fences off a triangular field PQR.
$PQ = 30$ metres,
$PR = 34$ metres and the angle at P is $75°$.
Calculate the area of the field.

3   Calculate the area of the parallelogram below.

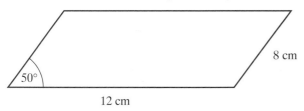

(Hint: the parallelogram can be seen as two triangles.)

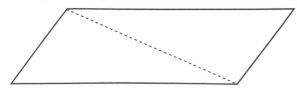

4 A triangle ABC has an area of 60 cm^2.

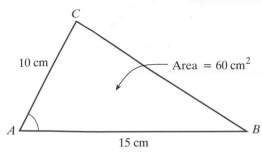

$$AB = 15 \text{ cm}, \quad AC = 10 \text{ cm}.$$

Calculate the size of the acute angle at A.

5 A triangle PQR has an area of 40 cm^2.

$$PQ = 12 \text{ cm}, \quad PR = 8 \text{ cm}$$

The angle at P is obtuse.

Calculate the angle at P.

(Note: you may need to refer to Unit 13, page 281 which shows that $\sin(180° - x) = \sin x$.)

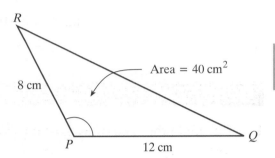

22.2 The sine rule

The diagram shows a general triangle ABC.
The lengths of its sides are a, b and c.
The perpendicular from C to AB is drawn and its length is labelled h. It meets AB at X.

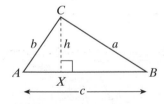

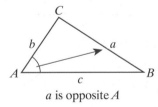

a is opposite A

Now, using the sine ratio in triangle AXC:

$$\sin A = \frac{h}{b} \quad \text{or } h = b \sin A$$

also:

$$\sin B = \frac{h}{a} \quad \text{or } h = a \sin B$$

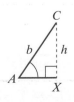

so:

$$a \sin B = b \sin A$$

so: $\qquad \dfrac{a}{\sin A} = \dfrac{b}{\sin B} \quad$ or $\quad \dfrac{\sin A}{a} = \dfrac{\sin B}{b}$

This result is known as the **sine rule**.

Exercise 22B

Draw a triangle ABC.

Draw a perpendicular from B to AC.

Hence show that $\dfrac{a}{\sin A} = \dfrac{c}{\sin C}$

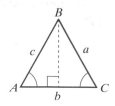

■ **The full version of the sine rule is:**

$$\frac{a}{\sin A} = \frac{b}{\sin B} = \frac{c}{\sin C}$$

or $\dfrac{\sin A}{a} = \dfrac{\sin B}{b} = \dfrac{\sin C}{c}$

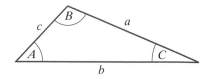

Example 4

Calculate the length of the side marked p.

By the sine rule:

$$\frac{p}{\sin P} = \frac{q}{\sin Q}$$

$$\frac{p}{\sin 37°} = \frac{12}{\sin 58°}$$

$$p = \frac{12 \times \sin 37°}{\sin 58°}$$

$$p = \frac{12 \times 0.6018}{0.8480}$$

$$p = 8.52 \text{ cm (correct to 3 s.f.)}$$

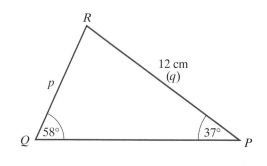

Example 5

The angles in the triangle ABC are all acute.

Calculate the size of the angle at A.

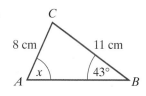

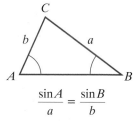

$$\frac{\sin A}{a} = \frac{\sin B}{b}$$

Using the sine rule:

$$\frac{11}{\sin x} = \frac{8}{\sin 43°} \quad \text{or} \quad \frac{\sin x}{11} = \frac{\sin 43°}{8}$$

so:

$$\sin x = \frac{11 \times \sin 43°}{8}$$

$$\sin x = \frac{11 \times 0.6820}{8}$$

$$\sin x = 0.9377$$

$$x = 69.7° \text{ (correct to 3 s.f.)}$$

Example 6

A lighthouse, L, lies 40 km due north of a harbour, H.

A speedboat leaves H and travels on a bearing of 053° from H until it reaches a point P.

The point, P, lies on a bearing of 075° from L.

Calculate the distance travelled by the speedboat.

First we need to make a sketch.

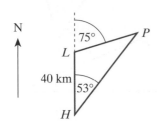 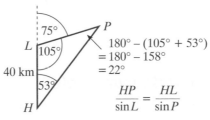

So the angle at P is $180° - (105° + 53°) = 22°$.

Now we can use the sine rule:

$$\frac{HP}{\sin 105°} = \frac{40}{\sin 22°} \quad \text{so} \quad HP = \frac{40 \times \sin 105°}{0.3746}$$

$$HP = \frac{40 \times 0.9659}{0.3746}$$

$$HP = 103.14 \text{ km (to 2 d.p.)}$$

Exercise 22C

1 Calculate the lengths of the sides marked with letters.
   All lengths are in cm.

   (a)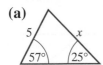

   (b)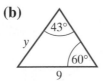

   (c)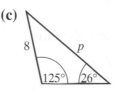

   (d)

   (e)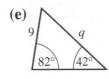

   (f)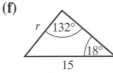

2 Calculate the angles marked with letters.
   All of these angles are acute.

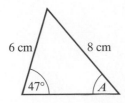 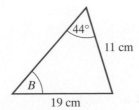 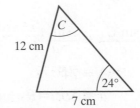

3 The diagram shows the relative positions of a port, *P*, a marker buoy, *B*, and a lighthouse, *L*.

The bearing of *L* from P is 110°.
The distance *PL* is 56 km.
The bearing of *B* from *P* is 147°.
The bearing of *L* from *B* is 030°.

Calculate:

(a) the distance *PB*

(b) the distance *BL*.

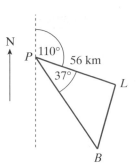

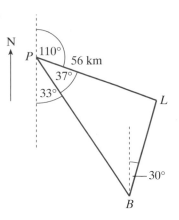

The ambiguous case (for an unknown angle)

In a triangle *ABC*, *AB* = 9 cm, *BC* = 5 cm and the angle at *A* = 24°.

(a) Show that there are two possible values for the angle at *C*.

(b) Use the sine rule to calculate each value.

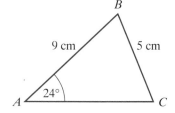

Suppose we draw the base line which is to be *AC* and *AB* = 9 cm, with the angle at *A* as 24°, then we have:

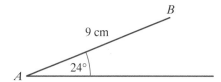

Imagine putting a compass point at *B*, then drawing a circle centre *B* of radius 5 cm. This circle will intersect *AC* at two possible points, marked as *C* and *C'*.

This shows that we have two possible angles at *B*, $A\hat{B}C$ or $A\hat{B}C'$.

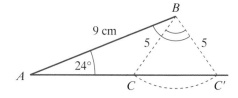

By drawing a general triangle *ABC* as:

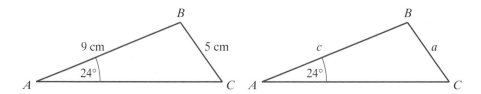

By the sine rule:

$$\frac{\sin C}{c} = \frac{\sin A}{a}$$

$$A = 24°$$

$$a = 5, c = 9$$

We could use the sine rule as: $\dfrac{\sin C}{9} = \dfrac{\sin 24°}{5}$

so:
$$\sin C = \frac{9 \times \sin 24°}{5}$$
$$\sin C = 0.7321$$
$$C = 47.06°$$

$\sin(180° - x) = \sin x$, so we could also have:
$$C = 180° - 47.06°$$
$$C = 132.94°$$

Exercise 22D

In a triangle PQR, $PQ = 10$ cm, $QR = 8$ cm and the angle $QPR = 48°$.

Calculate the two possible values for the angle PQR.

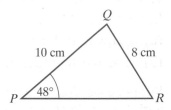

22.3 The cosine rule

ABC is a general triangle with sides a, b and c.
Side a is opposite angle A, and so on.
The perpendicular from B to AC meets AC at P.
This perpendicular has length h units.
We will let $CP = x$ units, so $PA = (b - x)$ units.

In triangle BPA, Pythagoras' Theorem gives:
$$c^2 = h^2 + (b - x)^2$$
$$c^2 = h^2 + b^2 - 2xb + x^2$$
or:
$$c^2 = h^2 + x^2 + b^2 - 2xb \quad \text{①}$$

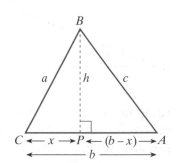

but using Pythagoras' Theorem again in triangle BPC gives:
$$a^2 = h^2 + x^2 \quad \text{②}$$
we have:
$$c^2 = h^2 + x^2 + b^2 - 2xb \quad \text{①}$$
and:
$$a^2 = h^2 + x^2 \quad \text{②}$$

Substitute a^2 for $h^2 + x^2$ in ①

so:
$$c^2 = a^2 + b^2 - 2xb \quad \text{③}$$

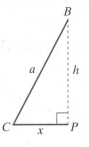

But using cosine ratio on triangle BPC gives:
$$\cos C = \frac{x}{a} \quad \text{or} \quad x = a \cos C \quad \text{④}$$

Substituting $a \cos C$ for x in ③ gives:

■ $\quad c^2 = a^2 + b^2 - 2ab \cos C$

This is the **cosine rule**.

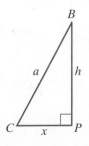

We can rearrange this to give:

$$2ab \cos C = a^2 + b^2 - c^2 \quad \text{or}$$

■ $\cos C = \dfrac{a^2 + b^2 - c^2}{2ab}$

In the GCSE examination the cosine rule will be given on a formula sheet. However, it might help you to remember it as:

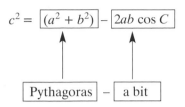

Example 7

Calculate the length of the side marked y.

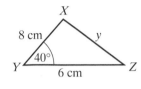
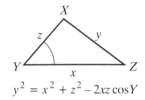

$$y^2 = x^2 + z^2 - 2xz \cos Y$$

Using the cosine rule:

$$y^2 = 6^2 + 8^2 - 2 \times 6 \times 8 \times \cos 40°$$
$$y^2 = 36 + 64 - 96 \times \cos 40°$$
$$y^2 = 100 - 96 \times 0.766$$
$$y^2 = 100 - 73.54 \quad \text{or} \quad y^2 = 26.46$$
$$y = \sqrt{26.46} \quad y = 5.14 \text{ cm (to 3 s.f.)}$$

Example 8

Calculate the angle at A.

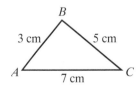
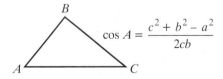

$$\cos A = \dfrac{c^2 + b^2 - a^2}{2cb}$$

Using the cosine rule:

$$\cos A = \dfrac{3^2 + 7^2 - 5^2}{2 \times 3 \times 7}$$

$$\cos A = \dfrac{9 + 49 - 25}{42} = \dfrac{33}{42} = 0.7857$$

so: $\qquad\qquad A = 38.2° \text{ (to 1 d.p.)}$

Exercise 22E

1 Calculate the lengths marked with letters. All lengths are in centimetres.

(a)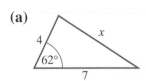

(b)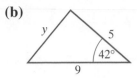

(c)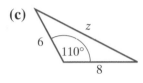

(d)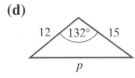

2 Calculate the angles marked with letters. All lengths are in centimetres.

(a)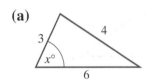

(b)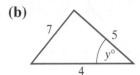

(c)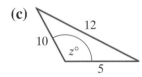

(d)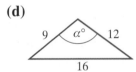

3 A builder ropes off a triangular plot of ground, PQR. The length of $PQ = 42$ m and the length of $PR = 50$ m. The angle $QPR = 72°$. Calculate the length of rope needed by the builder.

4 A ship, S, sets sail from a harbour, H, and travels 32 km due north to a lighthouse L. At L the ship turns onto a bearing of 056° from L and travels for a further 45 km until it reaches a marker buoy, B. At B the ship turns again and travels back in a straight line to H.

Calculate the total distance travelled by the ship.

The following exercises are a mixture of the area of a triangle, sine rule and cosine rule.

Exercise 22F (Mixed questions)

1 Calculate:

  (a) the area of ABC

  (b) the length of BC

  (c) the angle BCA.

2 A yacht, Y, leaves a harbour, H, and travels 45 km due north until it reaches a marker buoy, B. At B the ship turns onto a bearing of 290° from B and travels for a further 56 km until it reaches a lighthouse, L. At L it turns again and travels in a straight line back to H.

  Calculate:

  (a) the total distance travelled by the yacht

  (b) the bearing of L from H

  (c) the shortest distance between Y and B on the yacht's return journey from L to H.

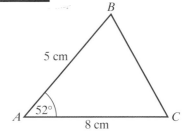

3 A man walks from his home H on a bearing of 060° from H for 3.2 miles until he reaches his friend's house, F. At F he turns onto a bearing of 280° from F and travels a further 4.1 miles to his sister's house, S. At S he turns again and walks in a straight line back home.

  Calculate:

  (a) the distance he walks

  (b) the bearing of H from S

  (c) the shortest distance between the man and F on the man's return journey from S to H.

22.4 Trigonometry and Pythagoras' Theorem in 3 dimensions

Example 9

The diagram shows a tetrahedron $VABC$ with V vertically above A. The base ABC is a triangle right angled at A.

$$AB = 8 \text{ cm}, \quad AC = 6 \text{ cm} \quad \text{and} \quad VA = 15 \text{ cm}$$

Calculate:

(a) the length BC (b) the length VB

(c) the length VC (d) angle $V\hat{B}A$

(e) angle $A\hat{V}C$ (f) angle $B\hat{V}C$

(a) length BC (by Pythagoras' Theorem $BC^2 = AB^2 + AC^2$)

$$BC^2 = 8^2 + 6^2$$
$$BC^2 = 64 + 36$$
$$BC^2 = 100$$
$$BC = \sqrt{100}$$
$$BC = 10 \text{ cm}$$

(b) length VB $(VB^2 = VA^2 + AB^2)$

$$VB^2 = 15^2 + 8^2$$
$$VB^2 = 225 + 64$$
$$VB^2 = 289$$
$$VB = \sqrt{289}$$
$$VB = 17 \text{ cm}$$

(c) length VC $(VC^2 = VA^2 + AC^2)$

$$VC^2 = 15^2 + 6^2$$
$$VC^2 = 225 + 36$$
$$VC^2 = 261$$
$$VC = \sqrt{261}$$
$$VC = 16.16 \text{ cm}$$

(d) angle $V\hat{B}A$ (using the tangent ratio)

$$\tan V\hat{B}A = \frac{15}{8}$$
$$\tan V\hat{B}A = 1.875$$
$$V\hat{B}A = 61.92°$$

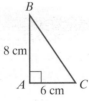

$\tan V\hat{B}A = \frac{\text{opp}}{\text{adj}} = \frac{AV}{AB} = \frac{15}{8}$

(e) angle $A\hat{V}C$ (using the tangent ratio)

$$\tan A\hat{V}C = \frac{6}{15}$$
$$\tan A\hat{V}C = 0.4$$
$$A\hat{V}C = 21.80°$$

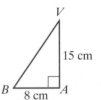

$\tan A\hat{V}C = \frac{\text{opp}}{\text{adj}} = \frac{AC}{AV} = \frac{6}{15}$

(f) angle BVC (using the cosine rule)

$$\cos V = \frac{17^2 + 16.15^2 - 10^2}{2 \times 17 \times 16.15}$$

$$\cos V = \frac{289 + 260.82 - 100}{549.1}$$

$$\cos V = \frac{449.82}{549.1}$$

$$\cos V = 0.8192$$

$$V = 35°$$

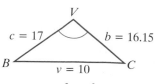

By the Cosine Rule $\cos V = \dfrac{c^2 + b^2 - v^2}{2cb}$

■ **You need to apply Pythagoras' Theorem and trigonometry in 3-D.**

Exercise 22G

1 Here is a wedge $ABCDEF$. The base, $BCDE$, is a rectangle. The back face, $ABEF$, is also a rectangle. The angle between these two rectangles is 90°. M is the mid-point of CD.

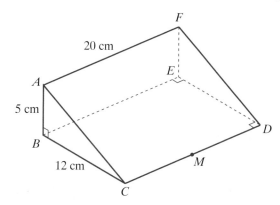

Calculate:

(a) the length AC **(b)** the length FC

(c) the length EC **(d)** angle ACB

(e) angle ECF **(f)** angle EMF.

2 $VABCD$ is a square-based pyramid. The vertex V is vertically above the mid-point, M, of the square base.

$$AB = 10 \text{ cm} \qquad VA = 16 \text{ cm}$$

Calculate:

(a) the length of BD

(b) the length of VM

(c) the angle VAM

(d) the angle AVM.

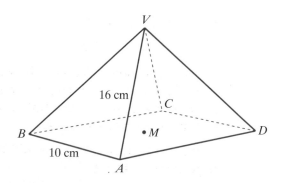

3 *VPQR* is a tetrahedron. The vertex *V* is vertically above *P* with *VP* = 16 cm.

In the triangular base, $Q\hat{P}R = 120°$.

Calculate:

(a) the length *QR*

(b) the length *VQ*

(c) the angle $Q\hat{V}R$.

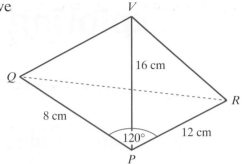

Summary of key points

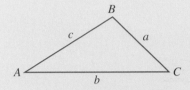

1 For any triangle *ABC*, as above,

the **area of the triangle** is:

- area $= \frac{1}{2}ab \sin C$
- area $ABC = \frac{1}{2}ac \sin B$
- area $ABC = \frac{1}{2}bc \sin A$

2 Area of triangle = length of first side × length of second side × sine of angle between these sides.

3 The **sine rule** is:

$$\frac{a}{\sin A} = \frac{b}{\sin B} = \frac{c}{\sin C} \quad \text{or} \quad \frac{\sin A}{a} = \frac{\sin B}{b} = \frac{\sin C}{c}$$

4 The **cosine rule** gives:

$$c^2 = a^2 + b^2 - 2ab \cos C$$

$c^2 = (a^2 + b^2) - 2ab \cos C$
$c^2 = $ Pythagoras – a bit

and

$$\cos C = \frac{a^2 + b^2 - c^2}{2ab}$$

These formulae appear on the formulae sheet which you will be given in the GCSE examination.

Remember also that

$$\sin x = \sin (180° - x)$$

a fact which is not given on the sheet but which you should be able to remember and might need for certain ambiguous cases related to the sine rule or for the area of a triangle.

5 You need to apply Pythagoras' Theorem and trigonometry in 3-D.

23 Exploring numbers 2

23.1 Terminating and recurring decimals

When a fraction of the type $\frac{a}{b}$ is converted into a decimal, there are two different types of answer. The result will either be a **terminating decimal** such as 0.75 or 0.174 or a **recurring decimal** such as 0.3333… or 0.34343434…

Terminating decimals

When the denominator, b, of the fraction $\frac{a}{b}$ is a factor of 10^n for some integer n (e.g. 10 or 100 or 1000 and so on), the division to evaluate the decimal will always finish at some stage. There will be an exact answer.

■ **A fraction is equivalent to a terminating decimal if the denominator is a factor of 10^n for some integer n.**

Example 1

Change $\frac{1}{4}$ and $\frac{13}{64}$ into decimals. Although 4 is not a factor of 10 it is a factor of 100.

$$\frac{1}{4} = \frac{25}{100} = 0.25$$

And 64 is a factor of 1 000 000.

$$64 \times 15\,625 = 1\,000\,000$$
$$\frac{13}{64} = 0.203125$$

```
        0.203125
64 )13.000000
    12 8
       200
       192
        80
        64
       160
       128
       320
       320
```

We can reverse this process. To change 0.25 to a fraction we multiply 0.25 by 100 and then divide this result by 100.

So:
$$0.25 = \frac{0.25 \times 100}{100} = \frac{25}{100}$$

$$= \frac{1}{4} \quad \text{(cancel by 25)}$$

In general, to convert a decimal which terminates after **2** places into a fraction we multiply the decimal by 10^2 and then divide this result by 10^2.

For a decimal which terminates after **3** decimal places we do the same but multiply and divide by 10^3.

For example:

$$0.125 = \frac{0.125 \times 1000}{1000} = \frac{125}{1000} = \frac{1}{8} \quad \text{(cancel by 125)}$$

Any terminating decimal with n decimal places can be converted into a fraction by multiplying the decimal by 10^n and dividing the result by 10^n. The fraction is then cancelled down.

Recurring decimals

The decimal expansion of $\frac{1}{7}$ is:

$\frac{1}{7} = 0.142857142857\ldots$

This pattern **recurs** every **6** decimal places.

The decimal expansion of $\frac{1}{13}$ is:

$\frac{1}{13} = 0.076923076923076\ldots$

This pattern also **recurs** every 6 decimal places.

But

$\frac{1}{12} = 0.083333\ldots$

Once this pattern starts to recur, it does so every **1** decimal place.

When the denominator is not a factor of any number of the form 10^n, the division to find the equivalent decimal will never finish. However, at some stage, it will start repeating itself to form a pattern. When the denominator is an integer, m, the repetition will start in the first $m - 1$ decimal places.

■ **A recurring decimal occurs when the denominator divides the numerator giving a repeating pattern.**

Example 2

$\frac{1}{3} = 0.33333\ldots$

$\frac{1}{9} = 0.11111\ldots$

$\frac{1}{6} = 0.16666\ldots$

$\frac{1}{11} = 0.090909\ldots$ (a pattern every 2 decimal places)

$\frac{10}{21} = 0.476190476190\ldots$ (a pattern every 6 decimal places)

> 6 is a factor of 7 − 1

> 6 is a factor of 13 − 1

> 6 is not a factor of 12 − 1

Exercise 23A

1   Convert the following fractions into decimals, and indicate which of the decimals are terminating and which are recurring:

$$\frac{3}{4}, \ \frac{3}{5}, \ \frac{3}{7}, \ \frac{3}{8}, \ \frac{3}{10}, \ \frac{3}{11}$$

2   Without converting these fractions into decimals, state, with a reason, which have equivalent decimals that terminate and which recur:

$$\frac{5}{6}, \ \frac{5}{7}, \ \frac{5}{8}, \ \frac{5}{9}, \ \frac{5}{12}, \ \frac{5}{13}, \ \frac{5}{16}$$

3   Write out the recurring decimals which are equivalent to the fractions:

$$\frac{1}{7}, \ \frac{2}{7}, \ \frac{3}{7}, \ \frac{4}{7}, \ \frac{5}{7}, \ \frac{6}{7}$$

Explain the relationship between the pattern of the recurring decimals.

4   How many digits are there in the recurring pattern of decimals of the fraction $\frac{3}{17}$?

5   Find the recurring decimals equivalent to the fractions:

$$\frac{1}{13}, \ \frac{2}{13}, \ \frac{3}{13}, \ \dots \ \frac{10}{13}, \ \frac{11}{13}, \ \frac{12}{13}$$

Separate the recurring decimals into two distinct sets of numbers. Explain the connection between them.

6   By considering the process of long division to find a recurring decimal equivalent to a fraction, $\frac{1}{n}$, such as $\frac{1}{17}$, explain:

(a) why the decimal will recur

(b) why the number of digits in the recurring pattern is fewer than n.

23.2 Finding a fraction equivalent to a recurring decimal

The method is best illustrated with three examples.

Example 3

Find a fraction equivalent to 0.55555…

Let $\qquad\qquad\qquad\qquad x = 0.55555\dots \qquad\qquad\qquad\qquad$ **A**

multiply equation **A** by 10:

$$10x = 5.55555\dots \qquad\qquad\qquad\qquad \textbf{B}$$

subtract equation **A** from equation **B**:

$$10x - x = 5$$

so: $\qquad\qquad\qquad\qquad 9x = 5$

so: $\qquad\qquad\qquad\qquad x = \frac{5}{9} \qquad$ the equivalent fraction

Example 4

Find a fraction equivalent to 0.636363...

Let $$x = 0.636363...$$

multiply by 100:

$$100x = 63.6363...$$

subtract: $$100x - x = 63$$

$$99x = 63$$

$$x = \frac{63}{99} = \frac{7}{11}$$

Example 5

Find a fraction equivalent to 0.103103103...

Let $$x = 0.103103103...$$

multiply by 1000:

$$1000x = 103.103103103...$$

so: $$999x = 103$$

so: $$x = \frac{103}{999}$$

Exercise 23B

1   Find the fractions which are equivalent to the recurring decimals:

   (a) 0.666666...   **(b)** 0.777777...

   (c) 0.34343434...   **(d)** 0.91919191...

   (e) 0.181818...   **(f)** 0.125125125...

   (g) 0.513513513...   **(h)** 0.100110011001...

   (i) 0.127912791279...   **(j)** 0.089108910891...

   (k) 1.4133333...   **(l)** 4.72121212...

2   Find the fraction which is equivalent to the recurring decimal 0.9999999... Explain the significance of your result.

23.3 Surds

The area of this square garden is 20 m^2.
The length of each side is x m.

$$x^2 = 20$$
$$x = \sqrt{20}.$$

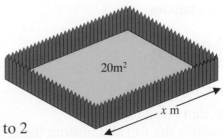

Using a calculator, you could write x as 4.47. This is correct to 2 decimal places but it is not an exact answer. If you wanted to say exactly what the number was you could just write $\sqrt{20}$.

■   **A number written exactly using square roots is called a surd. For example $\sqrt{3}$ and $2 - \sqrt{5}$ are in surd form.**

Unless you are told otherwise in the question, it is perfectly acceptable to leave your answers as surds.

Example 6

Solve the equation $x^2 - 8x + 11 = 0$, leaving your answer in surd form.

Complete the square:

$$(x - 4)^2 - 16 + 11 = 0$$
$$(x - 4)^2 - 5 = 0$$
$$(x - 4)^2 = 5$$
$$x - 4 = \pm\sqrt{5}$$

$x = 4 + \sqrt{5}$ or $x = 4 - \sqrt{5}$

Remember: $\sqrt{5}$ is the positive square root of 5. You have to use a $\pm$ sign.

Manipulating surds

Surds can be added, subtracted, multiplied and divided.

■ $\sqrt{a \times b} = \sqrt{a} \times \sqrt{b}$

■ $\sqrt{\dfrac{a}{b}} = \dfrac{\sqrt{a}}{\sqrt{b}}$

Example 7

Simplify $\sqrt{12}$.

$$\sqrt{12} = \sqrt{4 \times 3}$$
$$= \sqrt{4} \times \sqrt{3}$$
$$= 2\sqrt{3}$$

You can work these rules out using the rules of indices:

$(ab)^n = a^n b^n$

If $n = \frac{1}{2}$: $(ab)^{\frac{1}{2}} = a^{\frac{1}{2}} b^{\frac{1}{2}}$

$\sqrt{ab} = \sqrt{a} \sqrt{b}$

You will sometimes be asked to simplify a fraction with a surd as the denominator.

Example 8

Simplify $\dfrac{1}{\sqrt{2}}$.

Multiply the top and bottom by $\sqrt{2}$:

$$\frac{1}{\sqrt{2}} \times \frac{\sqrt{2}}{\sqrt{2}}$$

$$= \frac{\sqrt{2}}{2}$$

Example 9

Simplify $\dfrac{1}{3\sqrt{5}}$.

Multiply the top and bottom by $\sqrt{5}$.

$$\frac{1}{3\sqrt{5}} = \frac{1}{3\sqrt{5}} \times \frac{\sqrt{5}}{\sqrt{5}}$$

$$= \frac{\sqrt{5}}{3 \times 5} = \frac{\sqrt{5}}{15}$$

■ **Simplified surds should never have a square root in the denominator.**

Exercise 23C

1   Solve the equation $x^2 = 30$, leaving your answer in surd form.

2   The area of a square is $40\,\text{cm}^2$. Find the length of one side of the square. Give your answer in its most simplified form.

3   The lengths of the sides of a rectangle are:

   $3 + \sqrt{5}$ and $3 - \sqrt{5}$ units.

   Work out, in their most simplified forms,
   (a) the perimeter of the rectangle
   (b) the area of the rectangle.

4   Simplify each of the following:

   (a) $\dfrac{1}{\sqrt{7}}$ **(b)** $\dfrac{3}{\sqrt{5}}$ **(c)** $\dfrac{1}{\sqrt{17}}$

   (d) $\dfrac{1}{\sqrt{11}}$ **(e)** $\dfrac{\sqrt{32}}{\sqrt{2}}$

5   Solve these equations, leaving your answer in surd form:
   (a) $x^2 - 6x + 2 = 0$ **(b)** $x^2 + 10x + 14 = 0$

6   Show that

   $$\frac{1}{2\sqrt{17}} = \frac{\sqrt{17}}{34}$$

7   $ABCD$ is a rectangle
   $AB = 2\sqrt{2}$ units $BC = \sqrt{3}$ units
   Work out, in surd form, the length of the diagonal AC.

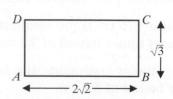

8   The diagram represents a right angled triangle ABC

   $AB = \sqrt{7} + 2$ units $AC = \sqrt{7} - 2$ units

   Work out, leaving any appropriate answers in surd form
   (a)   area of ABC **(b)** the length of BC.

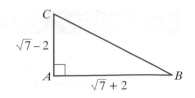

9   Simplify $\dfrac{\sqrt{3}}{\sqrt{2}}$

23.4 Rounding off

Consider the problem of rounding the three measurements 2.14 cm, 2.15 cm and 2.16 cm to 1 decimal place. 2.14 cm **rounds down** to 2.1 cm and 2.16 cm **rounds up** to 2.2 cm. However, although you know there is a convention that 2.15 cm rounds up to 2.2 cm, you know that 2.15 cm is the same distance from both 2.1 cm and 2.2 cm.

What is the largest measurement that rounds down to 2.1 cm?

Try 2.1499999… (recurring).

But what is its value?

Let $x = 2.149999…$

Then using the method of **Section 23.1**:

$$10x = 21.49999…$$

And: $9x = 19.35$

$$x = \frac{19.35}{9}$$

$$= 2.15$$

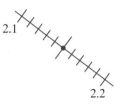

So
2.1499… (recurring) is *the same* as 2.15.

23.5 Upper bounds and lower bounds

These numbers round down to 3.1 cm correct to 1 d.p.:

   3.11 cm, 3.136 cm, 3.1475 cm, 3.14 cm, 3.149 cm

A number is an upper bound of 3.1 cm if it is greater than or equal to all the numbers which round down to 3.1 cm. So upper bounds of 3.1 cm include the measurements 3.2 cm, 3.16 cm, 3.155 cm, 3.152 cm, 3.151 cm and 3.15 cm. They are all greater than the numbers that round down to 3.1 cm.

Because 3.15 cm is the **smallest measurement** in this list, it is called the **least upper bound** of 3.1 cm.

Using similar arguments, the measurement 3.05 cm is the **greatest lower bound** of 3.1 cm.

Exercise 23D

Write down the greatest lower bound and the least upper bound for these measurements, to the given degree of accuracy.

1 To the nearest unit:
4, 14, 104, 10, 100

2 To the nearest 10:
30, 50, 180, 3020, 10, 100

3 To 1 decimal place:
4.7, 2.9, 13.6, 0.3, 157.5, 10.0, 100.0

4 To 2 s.f.:
37, 50, 180, 3.2, 9.5, 9400, 10, 100

5 To the nearest 0.5 unit:
4.5, 7.5, 16.5, 3.0, 15.5, 10, 100

6 To the nearest 0.2 unit:
3.4, 3.2, 4.0, 9.4, 12.2, 24.6, 10.0, 100.0

7 To the nearest quarter unit:
3.75, 3.5, 4.25, 6.0, 15.5, 10.0, 100.0

8 Three children, James, Sunita and Sara, all try to find the size of their classroom.

    (a) James estimates the walls of his classroom to be 7 m by 6 m to the nearest metre. Write the least upper bound and the greatest lower bound for the lengths of the classroom walls.

    (b) Sunita measures the lengths of the classroom walls with her ruler. She writes 730 cm and 590 cm correct to 2 significant figures. Write the least upper bound and the greatest lower bound for the lengths of the classroom walls.

    (c) Sara measures the walls with a tape measure and obtains the lengths 7.32 m and 5.94 m correct to 2 decimal places. Write the least upper bound and the greatest lower bound for the lengths of the classroom walls.

    (d) Explain whether the children's measurements are consistent.

9 The lap record for a racing circuit is 74.36 seconds correct to the nearest $\frac{1}{100}$ of a second.

    (a) Write down the least upper bound and the greatest lower bound for the lap record.

    (b) A racing driver has a recorded lap time of 74.357 seconds. Explain whether this is a new record.

23.6 Calculations involving upper and lower bounds – addition and multiplication

In addition and multiplication calculations we use the two greatest lower bounds to obtain the greatest lower bound of the result. And we use the two least upper bounds to obtain the least upper bound of the result.

Example 10

Mr Robinson's garden is a rectangle with sides measured as 3.2 m and 5.9 m, both correct to 1 d.p. Calculate the greatest lower bound and the least upper bound for the area of the rectangle.

The largest area is found from the longest possible sides:

$$3.25 \times 5.95 = 19.3375 \text{ m}$$

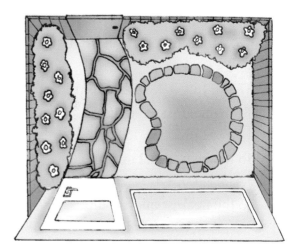

The smallest area is found from the shortest possible sides:

$$3.15 \times 5.85 = 18.4275 \text{ m}$$

(Note: the question is not about the least upper bounds and greatest lower bounds for the area (18.56 m ± 0.5) as if it were a measurement itself. It is the sides that have been measured. We are concerned with the possible error in the measurements and the effects that these can have on the resulting calculations.)

Exercise 23E

Using the least upper and greatest lower bounds for these measurements, find the least upper and greatest lower bounds for these quantities. The degree of accuracy of each measurement is given.

1 The area of a rectangle with sides 6 cm and 8 cm both measured to the nearest cm.

2 The perimeter of a rectangle with sides 6 cm and 8 cm both measured to the nearest cm.

3 The area of a square with side 7.5 cm measured to 2 s.f.

4 The perimeter of a square with side 7.5 cm measured to 2 s.f.

5 The area of a circle with radius 3 cm measured to 1 s.f.

6 The circumference of a circle with radius 0.34 m measured to 2 d.p.

7 The area of a circle with radius 41 mm measured to 2 s.f.

8 The area of a triangle with base length 8 cm and perpendicular height 6 cm, both measured to the nearest cm.

9 The area of a triangle with base length 8.0 cm and perpendicular height 6.0 cm both measured to 2 s.f.

10 The area of a triangle with base length 18 cm and perpendicular height 16 cm, both measured to the nearest 2 cm.

11 The area of a rectangle with sides 3.75 cm and 1.5 cm, both measured to the nearest 0.25 cm.

12 The circumference of a circle with radius 3.5 cm measured to the nearest 0.5 cm.

13 The volume of a sphere with radius:

(a) 15 cm measured to the nearest 5 cm

(b) 15 cm measured to 2 s.f.

(c) 15.0 cm measured to 1 d.p.

23.7 Calculations involving upper and lower bounds – subtraction and division

Care needs to be taken here as the least upper bound will result from a calculation using one least upper bound and one greatest lower bound.

Example 11

A piece of metal 32 cm long is cut from a 100 cm length. Both the measurements are correct to the nearest cm. Find the least upper and greatest lower bounds of the length of the remaining piece.

> 100 cm lies between 99.5 cm and 100.5 cm
>
> 32 cm lies between 31.5 cm and 32.5 cm

The least upper bound for the remaining piece = 100.5 – 31.5 = 69 cm.

The greater lower bound for the remaining piece = 99.5 – 32.5 = 67 cm.

If in doubt, do these four sums:

$$100.5 - 31.5 = 69$$
$$100.5 - 32.5 = 68$$
$$99.5 - 31.5 = 68$$
$$99.5 - 32.5 = 67$$
$$\text{l. u. b.} = 69$$
$$\text{g. l. b.} = 67$$

Example 12

The maximum and minimum temperatures, to the nearest °C, at Atlanta one day in 1993 were 27 °C and 8 °C respectively. Calculate the range of temperatures.

> 27 °C represents a temperature between 26.5 °C and 27.5 °C.
>
> 8 °C represents a temperature between 7.5 °C and 8.5 °C.

Least upper bound of the range   = 27.5 – 7.5 = 20 °C.

Greatest lower bound of the range   = 26.5 – 8.5 = 18 °C.

So the range of temperatures could be any value between 18 °C and 20 °C.

Careful thought is needed to combine the correct pair of initial bounds to find the least upper and greatest lower bounds of the result.

(Note: if you are in any doubt, then it is only necessary to do the sum with the four different pairs of least upper and greatest lower bounds. Then choose the largest and the smallest values for the bounds.)

Exercise 23F

1　A bar measures 44.8 cm long to 3 s.f. It is heated and expands to 45.7 cm. Calculate the least upper bound and greatest lower bound of the expansion.

2　Two recordings of a piece of music are measured, correct to the nearest minute, to be 34 minutes and 37 minutes. Calculate the least upper bound and greatest lower bound for the difference in the times of the two recordings.

3　Calculate the least upper bound and greatest lower bound for the difference in length of two pieces of paper. One measures 110 cm and the other 95 cm, both correct to 2 s.f.

4　Jeremy's time for a 100 m race is 14.7 seconds to the nearest 0.1 second. Calculate the least upper bound and greatest lower bound of Jeremy's average speed in metres per second for the race:

(a) assuming that the distance was accurately measured

(b) assuming that the distance was measured correct to the nearest metre.

5　Jennifer starts the day with a full tank of petrol. At the end of the day she has travelled 165 miles (correct to the nearest 5 miles). She fills the tank with 18 litres of petrol (correct to the nearest litre). Calculate the least upper bound and greatest lower bounds of the petrol consumption of Jennifer's car in:

(a) miles per litre

(b) miles per gallon (4.54 litres = 1 gallon).

6　In a triangle ABC angles A and B have been measured to be 35° and 72° respectively. Length BC is 15 cm. All the measurements are correct to 2 s.f. Use the sine rule to calculate the least upper bound and the greatest lower bound of the length AC.

7 A ball has mass 130 g and volume 40 cm³, both measurements correct to the nearest 10 units. Calculate the least upper bound and the greatest lower bound of the density of the ball (density = mass/volume).

8 Triangle *DEF* has sides of length 6 cm, 7 cm and 8 cm, all correct to the nearest centimetre. Use the cosine rule to calculate the least upper bound and the greatest lower bound of the smallest angle in the triangle.

23.8 Errors

 **The absolute error is the difference between the measured value and the notional value of a quantity.**

Example 13

A 100 metre running track is measured to be 98.7 m.

The absolute error is $100 - 98.7 = 1.3$ m.

It does not matter whether this is $+1.3$ m or -1.3 m, the absolute error will still be 1.3 m.

Example 14

A bag of potatoes is labelled as having 2.5 kg but is found to have a weight of 2.62 kg.

The absolute error is $2.62 - 2.5 = 0.12$ kg.

 **The percentage error is found by converting the fraction**

$$\dfrac{\textbf{absolute error}}{\textbf{notional value}} \textbf{ into a percentage.}$$

Example 15

Using the data in Example 13:

The percentage error for the 100 metre track is:

$$\tfrac{1.3}{100} \times 100\% = 1.3\%$$

Using the data in Example 14:

The percentage error for the bag of potatoes is:

$$\tfrac{0.12}{2.5} \times 100\% = 4.8\%$$

Example 16

Find the maximum possible absolute error and the maximum percentage error of a measurement of 16 cm to the nearest cm.

The least upper bound and the greatest lower bound are 15.5 cm and 16.5 cm.

The absolute error is $16.5 - 16 = 0.5$ cm

The percentage error is $\frac{0.5}{16} \times 100\% = 3.125\%$

Exercise 23G

1   A packet of rice is labelled 1 kg. The actual weight is found to be 984 g. Calculate the absolute error and the percentage error of the weight.

2   A pot of jam is labelled 454 g. The actual weight is found to be 467 g. Calculate the absolute error and the percentage error of the weight.

3   To estimate the calculation $2 \times \pi \times 74$, John used the approximation $2 \times 3 \times 70 = 6 \times 70 = 420$. Calculate the absolute error and the percentage error of this estimate.

4   Calculate the absolute error and the percentage error in using 3.14 as an approximation for π. (Use the value on your calculator as the more correct value.)

5   Yung measured and cut a piece of wood 4.5 m long, correct to 2 s.f. Calculate the maximum possible absolute error and the maximum percentage error of the length of the piece of wood.

6   Sam measures a length of 350 cm plastic tubing, correct to 3 s.f. Calculate the maximum possible absolute error and the maximum percentage error of the length of the tubing.

7   The side of a square is measured to be 7 cm long, correct to 1 s.f. Calculate the maximum absolute error and the maximum percentage error of:

   (a) a length of the side of the square

   (b) the area of the square.

8   The side the square in question 7 was measured more accurately and found to be 7.1 cm, correct to 1 d.p. Calculate the maximum absolute error and the maximum percentage error of:

   (a) a length of the side of the square

   (b) the area of the square.

9   The diameter of a circle is measured and found to be 18 cm, correct to 2 s.f. Calculate the maximum absolute error and the maximum percentage error for:

   (a) the diameter of the circle

   (b) the area of the circle.

Exercise 23H (Mixed questions)

1 Find a fraction equivalent to the decimal numbers:
   (a) 13.747474...   (b) 6.75321321321...

2 Write the least upper bound and greatest lower bound for:
   (a) 250 000 miles (correct to 2 s.f.)
   (b) 6×10^6 cm (correct to 1 s.f.)

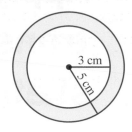

3 Explain the difference between the three lengths:
   4 cm, 4.0 cm and 4.00 cm.

4 Find the least upper bound and the greatest lower bound of the shaded area. The two circles have radii of 5 cm and 3 cm each correct to 1 s.f.

5 A cube has a volume 55 mm³ correct to the nearest mm³. Calculate the least upper bound and the greatest lower bound of the length of a side of the cube.

6 The diameter of a metal ball is measured and found to be 42 cm, correct to 2 s.f. Calculate the maximum absolute error and the maximum percentage error for:
   (a) the diameter of the ball
   (b) the surface area of the ball
   (c) the volume of the ball.

Summary of key points

1 A fraction is equivalent to a terminating decimal if the denominator is a factor of 10^n for some integer n.

2 A **recurring decimal** occurs when the denominator divides the numerator giving a repeating pattern.

3 A number written exactly using square roots is called a surd. For example $\sqrt{3}$ and $2 - \sqrt{5}$ are in surd form.

4 $\sqrt{a \times b} = \sqrt{a} \times \sqrt{b}$

5 $\sqrt{\dfrac{a}{b}} = \dfrac{\sqrt{a}}{\sqrt{b}}$

6 The conjugate of a number can be found by changing the sign. The conjugate of $a + \sqrt{b}$ is $a - \sqrt{b}$.

7 Simplified surds should never have a square root in the denominator.

8 The **absolute error** is the difference between the measured value and the actual value of a quantity.

9 The **percentage error** is found by converting the fraction:
   $$\frac{\text{absolute error}}{\text{notional value}}$$ into a percentage.

24 Applying transformations to sketch graphs

The previous units (7 and 18) involving graphical work have shown you how to draw up a table of given values of x, how to plot the points and join them using a smooth curve. The given range of values for x is chosen so as to include the main features of the graph. You will now consider examples where this is not the case. For example, the graph shows five plotted points for the equation
$y = x^2 - 9x + 10$ for $-2 \leqslant x \leqslant 2$.

Without plotting any more points, consider what the graph looks like for values of x up to 10.

This unit shows you how to use your knowledge of the graphs of some basic functions, apply suitable transformations (translations, reflections and stretches) to sketch the main features of more complicated graphs without the need to find and plot lots of points.

In previous units you have been using graphs with equations:

$$y = x^2, \ y = x^3, \ y = \tfrac{1}{x}$$

You will be using the graphs of these three functions along with the straight line $y = x$ in many examples during this unit. For reference we will call these four equations 'the basic functions'.

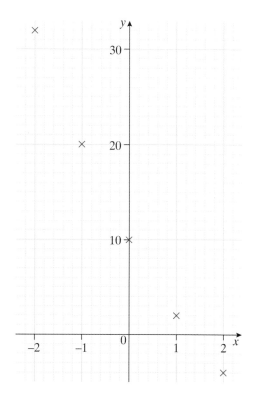

You will need tracing paper for many of the exercises in this unit.

24.1 Function notation

■ **A function is a rule which changes one number into another number.**

We can think of a function, f, as a machine which changes an input x into an output.

input x — f — output $f(x)$

■ **$f(x)$ which denotes the output is read as 'f of x'.**

If the function f is 'square' then, for an input x, the output is x^2 so $f(x) = x^2$.

In this case, for an input of 3 the output is 9.
We write it as $f(3) = 3^2 = 9$.

Similarly $f(-4) = 16$; and $f(k + 1) = (k + 1)^2 = k^2 + 2k + 1$.

A function f can also be equivalent to a sequence of operations (rules).

For example, $f(x) = 2x^2 + 1$ is the output from this flow diagram:

input x square   × 2   + 1 $2x^2 + 1$

Example 1

When $f(x) = 2x^2 + 1$ find:

(a) $f(3)$ (b) $f(-4)$

(c) $f(0)$ (d) $f(-k)$

(a) replacing x by 3 $f(3) = 2(3)^2 + 1 = 19$
(b) replacing x by -4 $f(-4) = 2(-4)^2 + 1 = 33$
(c) replacing x by 0 $f(0) = 2(0)^2 + 1 = 1$
(d) replacing x by $-k$ $f(-k) = 2(-k)^2 + 1 = 2k^2 + 1$

Example 2

When $f(x) = x^2$ find:

(a) $f(x+1)$ (b) $f(2x)$

(c) $1-f(-x)$ (d) $f(-\frac{1}{2}x) + 4$

(a) replacing x by $x + 1$ $f(x + 1) = (x + 1)^2 = x^2 + 2x + 1$
(b) replacing x by $2x$ $f(2x) = (2x)^2 = 4x^2$
(c) replacing x by $-x$ $1-f(-x) = 1 - (-x)^2 = 1 - x^2$
(d) replacing x by $(-\frac{1}{2}x)$ $f(-\frac{1}{2}x) + 4 = (-\frac{1}{2}x)^2 + 4 = \frac{x^2}{4} + 4$

Example 3

Given that $f(x) = x^2 + 2$, find the values of x when $f(x) = 11$:

When $f(x) = 11$, $11 = x^2 + 2$
Subtract 2: $9 = x^2$
Take the square root: $\pm 3 = x$

On page 482, the plotted points shown for the graph of $y = x^2 - 9x + 10$ come from the table:

$f(-2) = (-2)^2 - 9(-2) + 10$
$\quad\quad = 4 + 18 + 10$
$\quad\quad = 32$

x	-2	-1	0	1	2
y	32	20	10	2	-4

Using function notation, this is the same as the graph $y = f(x)$, where $f(x) = x^2 - 9x + 10$
$f(-2) = 32$, $f(-1) = 20$ and so on.

Example 4

Part of the graph $y = f(x)$, where $f(x) = x^2 + 1$ is drawn.

From the graph find:

(a) the value of (i) f(0) (ii) f(2)

(b) the values of x when $f(x) = 2$

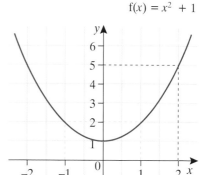

$f(x) = x^2 + 1$

Answer

(a) (i) f(0) is the output (y), when $x = 0$; from the graph, f(0) = 1
(ii) from the graph, f(2) = 5

(b) f(x) = 2; we need x when $y = 2$.
From the graph, when $y = 2$, $x = -1$ or $x = 1$
So when $f(x) = 2$, $x = \pm 1$

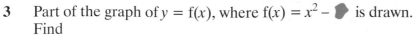

Exercise 24A

1 $f(x) = x^2 - 2$, find:

(a) f(2)   (b) f(4)   (c) f(0)
(d) f(-1)   (e) f(-3)   (f) $f(\frac{1}{4})$

2 $f(x) = x^2$, find:

(a) f(-x)   (b) f(3x)   (c) f(x + 2)
(d) f(x + 1) + 3   (e) $f(-\frac{x}{2} + 1) - 4$ (f) 5 - f(2x)
(g) f(kx + a) + b

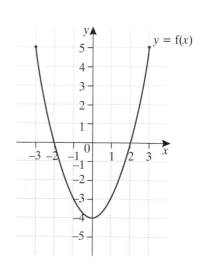

$y = f(x)$

Paul accidentally spilled some ink on question 3 in his text book. His teacher says that he should have been able to answer the questions without using the equation for f(x). Explain why Paul did not need the equation for f(x) then answer the question.

3 Part of the graph of $y = f(x)$, where $f(x) = x^2 -$ ⬤ is drawn.
Find
(a) the value of (i) f(3) (ii) f(1) (iii) f(0) (iv) f(-1)
(b) the values of x when f(x) = 0.

24.2 Applying vertical translations to graphs

Parabolas

The graphs of the parabolas $y = x^2$, $y = x^2 + 8$ and $y = x^2 - 15$ are shown below. Another parabola is also shown.

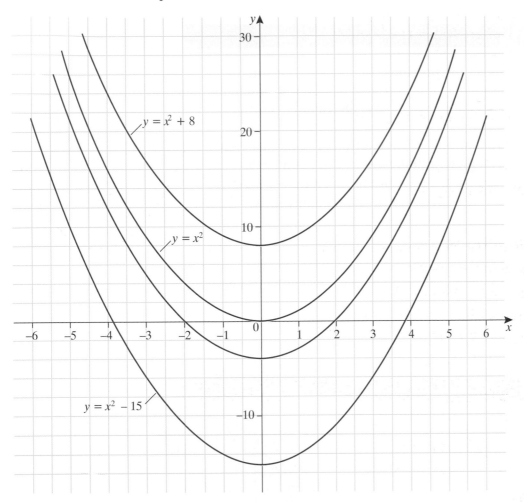

Step 1 Trace the graph of $y = x^2$.

Step 2 Slide your tracing paper vertically upwards along the y-axis so that your trace of $y = x^2$ coincides with the parabola $y = x^2 + 8$.

The tracing paper has moved *up* 8 units.

The graph of $y = x^2 + 8$ is the graph of $y = x^2$ **translated 8 units vertically** in the positive y-direction.

Step 3 Now put your tracing paper back on the parabola $y = x^2$.

This time slide your tracing paper vertically downwards so that it coincides with the parabola $y = x^2 - 15$.

The tracing paper has moved *down* 15 units.

The graph of $y = x^2 - 15$ is the graph of $y = x^2$ **translated 15 units vertically** in the negative y-direction.

■ **The graph of $y = x^2 + a$ is the graph of $y = x^2$ translated a units vertically in the** *positive y-direction if $a > 0$*
negative y-direction if $a < 0$.

Exercise 24B

1   Using the graphs on the previous page, place your tracing paper on the parabola $y = x^2$ and slide it so that it coincides with the unmarked parabola. What is the equation of this unmarked parabola?

2   Write down the equations of three different parabolas that always lie between the parabolas $y = x^2$ and $y = x^2 + 8$.

3   The parabola $y = x^2 + k$ always lies between the parabolas $y = x^2$ and $y = x^2 - 15$. Write down an inequality for k.

Other basic functions

The graphs below show how curves which are related to the basic functions $y = x$, $y = x^3$ and $y = \frac{1}{x}$ can be drawn by applying a vertical translation.

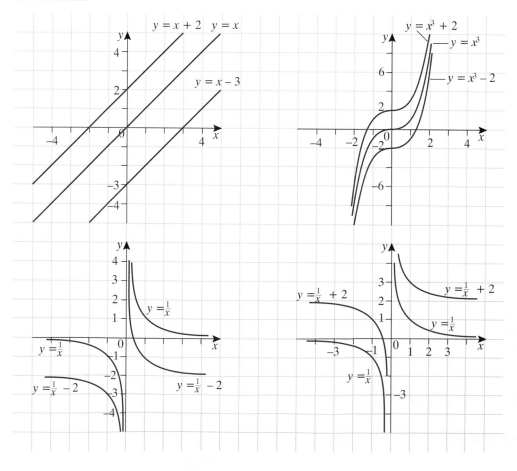

Example 5

Copy and complete:

(a) The graph of $y = x + 2$ is the graph of $y = x$ translated _____

(b) The graph of $y = x^3 - 2$ is the graph of $y = x^3$ translated _____

(c) The graph of $y = \frac{1}{x} + 2$ is the graph of $y = \frac{1}{x}$ _____

(d) The graph of $y = x^3 + 2$ is the graph of $y = x^3$ _____

(e) The graph of $y = \frac{1}{x} - 2$ is the graph of $y = \frac{1}{x}$ _____

(a) The graph of $y = x + 2$ is the graph of $y = x$ translated 2 units vertically in the positive y-direction

(b) The graph of $y = x^3 - 2$ is the graph of $y = x^3$ translated 2 units vertically in the negative y-direction.

(c) The graph of $y = \frac{1}{x} + 2$ is the graph of $y = \frac{1}{x}$ translated 2 units vertically in the positive y-direction.

(d) The graph of $y = x^3 + 2$ is the graph of $y = x^3$ translated 2 units vertically in the positive y-direction.

(e) The graph of $y = \frac{1}{x} - 2$ is the graph of $y = \frac{1}{x}$ translated 2 units vertically in the negative y-direction.

■ **The graph of $y = x + a$ is the graph of $y = x$ translated a units vertically in the *positive y*-direction if $a > 0$**
　　　　　　　　　　　　　negative y-direction if $a < 0$.

■ **The graph of $y = x^3 + a$ is the graph of $y = x^3$ translated a units vertically in the *positive y*-direction if $a > 0$**
　　　　　　　　　　　　　negative y-direction if $a < 0$.

■ **The graph of $y = \frac{1}{x} + a$ is the graph of $y = \frac{1}{x}$ translated a units vertically in the *positive y*-direction if $a > 0$**
　　　　　　　　　　　　　negative y-direction if $a < 0$.

Using function notation we can generalize all the results obtained so far as:

■ **For any function f, the graph of $y = f(x) + a$ is the graph of $y = f(x)$ translated a units vertically in the *positive y*-direction if $a > 0$**
$negative$ y-direction if $a < 0$.

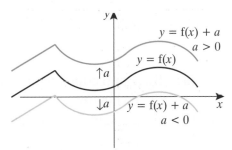

Example 6

The graph of $y = f(x)$, where $f(x) = x^3 - x$ is sketched here.

(a) Sketch, on the same axes, the graph of $y = x^3 - x + 10$.

(b) Describe the transformation which gives $y = x^3 - x + 10$ from $y = x^3 - x$.

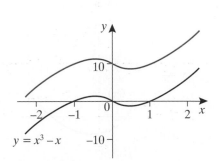

Answer

(a) $x^3 - x + 10 = f(x) + 10$. The graph of $y = x^3 - x + 10$ is shown in red.

(b) A vertical translation of 10 units in the positive y-direction.

Exercise 24C

1 In diagrams **(a)** to **(e)** the graphs of 'basic functions' are in red. Write down the translations which have to be applied to these basic functions in each of the questions in order to obtain the graphs labelled A, B, C.

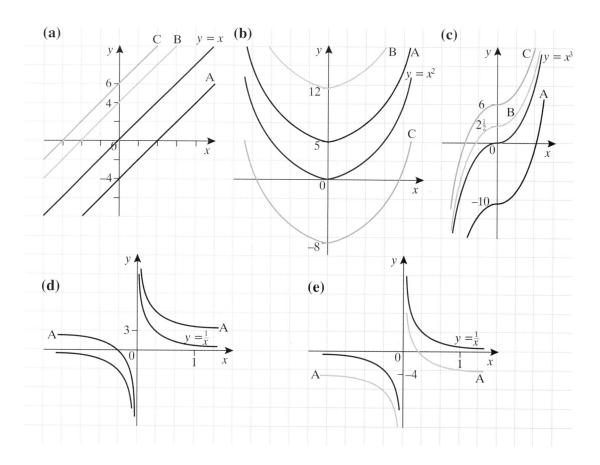

2 **(a)** Trace and copy the axes and sketch of the graph of $y = f(x)$ below.

   (b) On the same axes, draw the graph of $y = f(x) + 2$.

   (c) Describe the transformation which gives $y = f(x) + 2$ from $y = f(x)$.

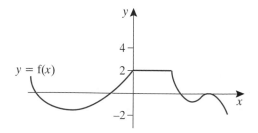

24.3 Applying horizontal translations to graphs

Parabolas

The graphs of the parabolas $y = x^2$, $y = (x + 2)^2$ and $y = (x - 4)^2$ are shown below. One other parabola is shown.

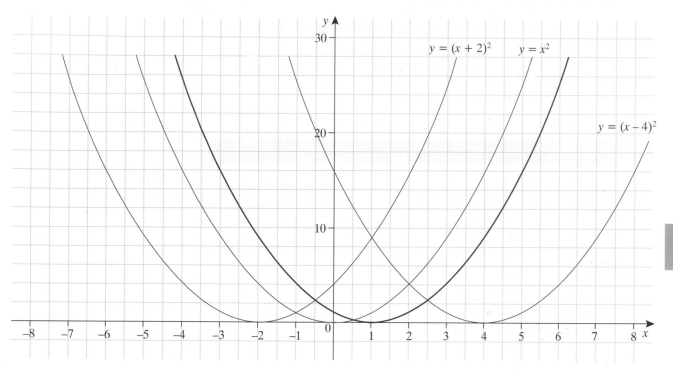

Get your tracing of the graph of $y = x^2$ (or trace out a new one).

Put your tracing paper on the parabola $y = x^2$.

Slide your tracing paper horizontally along the x-axis so that your trace of $y = x^2$ coincides with the parabola $y = (x + 2)^2$.

You should have slid the tracing paper along 2 units in the *negative* x-direction.

The graph of $y = (x + 2)^2$ is the graph of $y = x^2$ translated 2 units horizontally in the ***negative x-direction.***

Similarly, the graph of $y = (x - 4)^2$ is the graph of $y = x^2$ translated 4 units horizontally in the *positive x-direction*.

■ **The graph of $y = (x + a)^2$ is the graph of $y = x^2$ translated a units horizontally in the *negative x-direction* if $a > 0$**
$\qquad\qquad\qquad$ ***positive x-direction* if $a < 0$.**

■ **The vertex of the parabola $y = (x + a)^2$ is at the point $(-a, 0)$.**

Example 7

Sketch the graph of $y = x^2 + 6x + 9$. Mark the coordinates of the points where the graph crosses the axes.

Factorizing $x^2 + 6x + 9$ gives $(x + 3)^2$
so you need to sketch the graph of $y = (x + 3)^2$.

The graph of $y = (x + 3)^2$ is the graph of $y = x^2$ translated 3 units horizontally in the *negative* x-direction.

The graph of $y = (x + 3)^2$ is shown as a parabola with vertex $(-3,0)$. To find where this parabola crosses the y-axis, put $x = 0$ in $y = (x + 3)^2$ which gives you $y = 9$. This parabola crosses the y-axis at the point $(0,9)$ and touches the x-axis at the point $(-3,0)$.

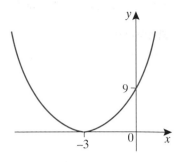

Exercise 24D

1   Using the graphs on the previous page, place your tracing paper on the parabola $y = x^2$ and slide it so that it coincides with the unmarked parabola. What is the equation of this parabola?

2   Sketch the graph of **(a)** $y = (x + 1)^2$ **(b)** $y = (x + 5)^2$.
In each case mark the coordinates of the points where the graph crosses the axes.

3   Sketch the graph of $y = x^2 - 4x + 4$. Mark the coordinates of the points where the graph crosses the axes.

Lines, cubes and reciprocals

Look back at the sketches of the lines on page 486.

The graph of the straight line $y = x - 3$ is the graph of $y = x$ translated *either* 3 units vertically down *or* 3 units horizontally in the positive x-direction.

The cubics and reciprocals below have undergone transformations. You should examine these sketches carefully to understand the **horizontal** translations that have been applied. Use tracing paper if necessary to help you.

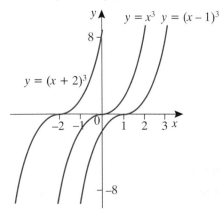

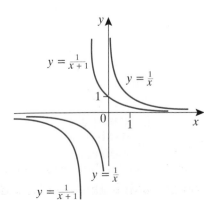

Example 8

Copy and complete:

(a) The graph of $y = \frac{1}{(x+1)}$ is the graph of $y = \frac{1}{x}$ _____

(b) The graph of $y = (x+2)^3$ is the graph of $y = x^3$ _____

Answer

(a) The graph of $y = \frac{1}{(x+1)}$ is the graph of

$y = \frac{1}{x}$ translated 1 unit horizontally in the negative x-direction.

(b) The graph of $y = (x+2)^3$ is the graph of $y = x^3$ translated 2 units horizontally in the negative x-direction.

> **In general, for any function f: the graph of $y = f(x + a)$ is the graph of $y = f(x)$ translated a units horizontally in the**
> ***negative x-direction if $a > 0$***
> ***positive x-direction if $a < 0$.***

Example 9

The graph of $y = f(x)$, where $f(x) = x(x + 2)$ is sketched below (in black).

(a) On the same axes sketch the graph of $y = x^2 + 4x + 3$.

(b) Describe the transformation which gives $y = x^2 + 4x + 3$ from $y = x(x + 2)$.

Answer

(a) Factorizing $x^2 + 4x + 3$ gives $(x + 1)(x + 3)$ so we need to sketch the graph of $y = (x + 1)(x + 3)$.

Comparing $x(x + 2)$ with $(x + 1)(x + 3)$, we see that the x in the first expression has been replaced by $(x + 1)$ to get the second.

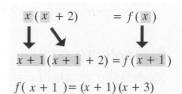

But $f(x) = x(x + 2)$

so $f(x + 1) = (x + 1)(x + 1 + 2) = (x + 1)(x + 3)$.

So sketching the graph $y = (x + 1)(x + 3)$ is the same as sketching $y = f(x + 1)$.

The graph of $y = f(x + 1)$ is the graph of $y = f(x)$ translated 1 unit in the negative x-direction.

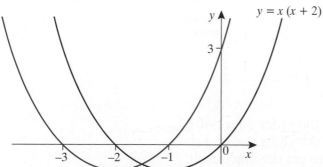

The graph of $y = x^2 + 4x + 3$ is sketched above (shown in red).

(b) A horizontal translation of 1 unit in the negative x-direction.

Exercise 24E

1   The graph of $y = x^3$ is transformed to $y = (x - 1)^3$. Write down the translation which has taken place.

2   The graph of $y = x(x + 2)$ is transformed to $y = (x - 1)(x + 1)$. Write down the transformation which has taken place.

3   The graph of $y = \dfrac{1}{x + 3} - 3$ is transformed to $y = \dfrac{1}{(x + 3)} + 4$.
    Write down the transformation which has taken place.

4   Graph A becomes $y = x^3 - 1$ following a vertical translation 3 units downwards. Write the equation of Graph A.

5   The graph of $y = x(x - 1)(x + 5)$ is translated by 4 units horizontally in the positive x-direction. Write down its new equation.

24.4 Applying double translations to graphs

This section shows you how to apply two translations and to sketch the graphs of various functions.

Example 10

Sketch the graph of $y = (x - 2)^3 + 1$.

This graph may be built up from the basic curve $y = x^3$ in the following way:

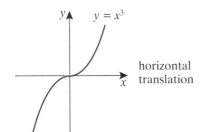
horizontal translation

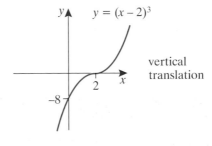
vertical translation

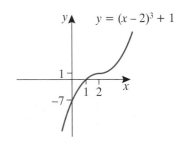

Note: in the case of **double translations**, the order does not matter. You should convince yourself by sketching $y = x^3 + 1$ then applying the horizontal translation to get the final result.

However, when combining **a translation with another transformation** the order of the operations becomes critical.

To sketch a quadratic function whose coefficient of x^2 is 1

Example 11

Sketch the graph of $y = x^2 + 4x - 1$. State the coordinates of the vertex and the points at which the curve crosses the x-axis.

$$y = x^2 + 4x - 1$$
$$= x^2 + 4x \quad - 1 \quad \text{complete the square}$$
$$= (x + 2)^2 - 4 - 1$$
$$= (x + 2)^2 - 5$$

check:
$(x + 2)^2 - 5$
$= x^2 + 4x + 4 - 5$
$= x^2 + 4x - 1$

$$y = (x + 2)^2 - 5$$

This graph may be built up from the basic curve $y = x^2$:

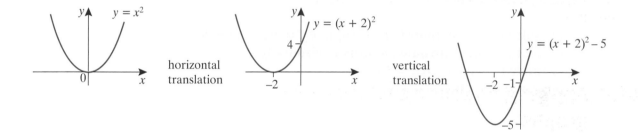

The coordinates of the vertex are $(-2, -5)$.

To find where the graph crosses the x-axis solve:
$$(x + 2)^2 - 5 = 0$$
$$(x + 2)^2 = 5$$
$$(x + 2) = \pm\sqrt{5}$$
$$x = -2 - \sqrt{5} \text{ or } -2 + \sqrt{5}$$

■ **To sketch the graph of $y = x^2 + bx + c$, complete the square and apply a double translation to the parabola $y = x^2$.**

Exercise 24F

1   Sketch the graphs of the following quadratic functions. In each case give the coordinates of the vertex.

(a) $y = x^2 + 2x + 4$

(b) $y = x^2 - 6x + 4$

(c) $y = x^2 + 3x - 1$

(d) $y = x^2 - 4x - \frac{1}{2}$

2   Describe the transformations which transform the graph of
    $y = (x + 1)^2$ to the graphs:
    (a) $y = (x + 7)^2$ **(b)** $y = (x - 3)^2 - 4$

3   Sketch the graphs of the quadratic functions:
    (a) $y = x^2 - 6x + 2$ **(b)** $y = x^2 + 2x - 4$
    Describe the transformations which give the graph in **(b)** from
    the graph in **(a)**.

4   What is the equation of the graph which is obtained by
    applying these transformations to the graph of $y = x^2 - 16x$:
    a horizontal translation of 1 unit in the positive x-direction
    followed by a vertical translation of 3 units down?

5   Sketch the graph of $y = 3 + \frac{1}{x + 2}$.
    Give the coordinates of the points where the graph and the
    asymptotes cross the axes.

■   **General result for all functions f: The graph of $y = f(x + a) + b$
    is the graph of $y = f(x)$ translated a units horizontally
    (in the negative x-direction if $a > 0$)
    (in the positive x-direction if $a < 0$)
    followed by a translation of b units vertically (upwards if $b > 0$)
    (downwards if $b < 0$).**

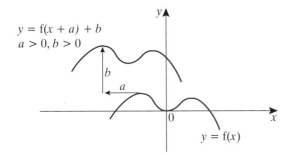

24.5 Applying reflections to graphs

In the x-axis

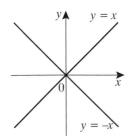 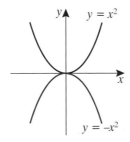 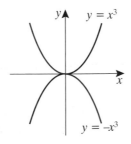 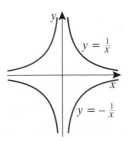

Sketches of $y = f(x)$ (in black) and $y = -f(x)$ (in red) are shown
above for each of our four basic functions.

The graph of $y = -f(x)$ is obtained by changing the sign of y in $y = f(x)$.

Each red graph is obtained by reflecting each black graph in the x-axis.

■ **For any function f, the graph of $y = -f(x)$ is obtained by reflecting $y = f(x)$ in the x-axis.**

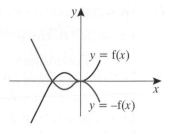

Tracing paper can help you reflect graphs in the x-axis:

● trace the function and the x-axis and mark the origin

● turn over the tracing paper and match up the origin and x-axis. The trace now shows the reflection of the original function in the x-axis.

Example 12

A sketch of the parabola $y = f(x)$, where $f(x) = -x^2 + 4x$ is shown in black. Its vertex has coordinates $(2, 4)$.

(a) Find the equation of $y = -f(x)$.

(b) Sketch the graph of $y = -f(x)$.

(c) Describe the transformation which is applied to $y = f(x)$ to obtain $y = -f(x)$.

(d) Write the coordinates of the vertex of $y = -f(x)$.

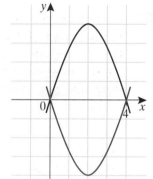

Answer

(a) $f(x) = -x^2 + 4x$
$y = -f(x) = -(-x^2 + 4x)$
so $y = x^2 - 4x$ is the equation of $y = -f(x)$.

(b) The graph of $y = -f(x)$ is shown in red.

(c) Reflection in the x-axis. **(d)** Vertex is $(2, -4)$.

In the y-axis

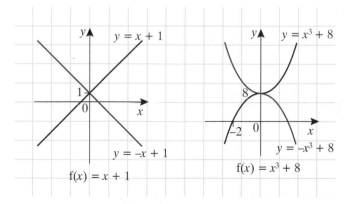

Sketches of $y = f(x)$ (in black) and $y = f(-x)$ (in red) are shown above for $f(x) = x + 1$ and $f(x) = x^3 + 8$.

The graph of $y = f(-x)$ is obtained by changing the sign of x in $y = f(x)$.

Each red graph is obtained by reflecting each black graph in the y-axis.

■ **For any function f, the graph of $y = f(-x)$ is obtained by reflecting $y = f(x)$ in the y-axis.**

Example 13

This is a sketch of the parabola $y = f(x)$, where $f(x) = -x^2 + 6x$.

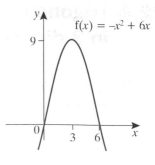

Its vertex has coordinates (3,9).

(a) Find the equation of $y = f(-x)$.

(b) Sketch the graph of $y = f(-x)$.

(c) Describe the transformation which is applied to $y = f(x)$ to obtain $y = f(-x)$.

(d) Write the coordinates of the vertex of $y = f(-x)$.

Answer

(a) $f(x) = -x^2 + 6x$

$y = f(-x) = -[(-x)^2] + 6(-x) = -[x^2] - 6x$

so $y = -x^2 - 6x$ is the equation of $y = f(-x)$.

(b)

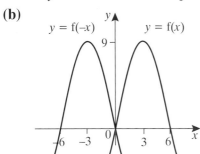

(c) Reflection in the y-axis.

(d) Vertex is $(-3, 9)$.

Even functions

For some functions the graph of $y = f(x)$ is symmetrical about the y-axis, for example $f(x) = x^2$.

If reflected in the y-axis, such graphs would remain unchanged so

$$f(-x) = f(x)$$

■ **Functions f which satisfy $f(-x) = f(x)$ are called even functions.**

Examples of even functions include $f(x) = 6$; $\quad f(x) = x^2$; $\quad f(x) = x^4$

$$f(x) = x^4 + x^2 + 6; \quad f(x) = \tfrac{1}{x^2}$$

Odd functions

For some functions the graph of $y = f(x)$, when reflected in the y-axis, gives the same graph as if being reflected in the x-axis so $f(-x) = -f(x)$.

■ **Functions f which satisfy $f(-x) = -f(x)$ are called odd functions.**

Examples of odd functions include $f(x) = x$; $\;f(x) = x^3$; $\;f(x) = \tfrac{1}{x}$.

A function does not have to fall into one of these two categories. For example, $f(x) = x + 2$ is neither even nor odd.

Exercise 24G

1 This is a sketch of $y = f(x)$, where $f(x) = x^2 - 2x$. The vertex of this parabola is $(1, -1)$.

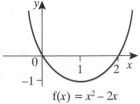

$f(x) = x^2 - 2x$

(a) Find the equation of $f(-x)$.

(b) Sketch the graph of $y = f(-x)$.

(c) Sketch the graph of $y = -x^2 + 2x$.

(d) Write the coordinates of the vertex of $y = -x^2 + 2x$.

2 A reflection is applied to the graph of $y = f(x)$, where $f(x) = x^3 + 2x^2$.

Find the equation of the new graph if:

(a) the reflection is in the x-axis

(b) the reflection is in the y-axis.

3 (a) Copy this sketch of the graph of $y = f(x)$.

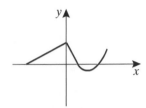

(b) On the same axes, sketch the graph of $y = -f(x)$.

(c) Describe fully the single transformation which when applied to the graph of $y = f(x)$ gives the graph of $y = -f(x)$.

4 (a) Copy this sketch of the graph of $y = f(x)$.

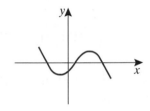

(b) On the same axes, sketch the graph of $y = f(-x)$.

(c) Describe fully the single transformation which when applied to the graph of $y = f(x)$ gives the graph of $y = f(-x)$.

5 Part of the graph of $y = f(x)$ is drawn for some positive values of x. Complete the graph of $y = f(x)$ for the corresponding negative values of x, given that $f(x)$ is an **even function**, in each of the following:

(a)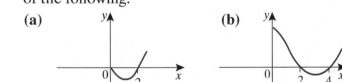

(b)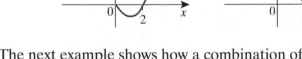

The next example shows how a combination of transformations can be used to sketch a function. It uses much of the work considered so far in this unit.

Example 14

Explain how to sketch the graph of $y = -x^2 + 6x + 3$ by applying transformations to the graph of $y = x^2$. Illustrate your answer by sketches and state the coordinates of the vertex of $y = -x^2 + 6x + 3$.

$$-x^2 + 6x + 3 = -(x^2 - 6x) + 3$$
$$= -[(x-3)^2 - 9] + 3 \quad \text{(complete the square)}$$
$$= -(x-3)^2 + 12$$

Check:
$-(x-3)^2 + 12$
$= -(x^2 - 6x + 9) + 12$
$= -x^2 + 6x - 9 + 12$
$= -x^2 + 6x + 3$

Let $f(x) = x^2$ so $y = x^2$ becomes $y = f(x)$
$\qquad\qquad\qquad\quad y = (x - 3)^2$ becomes $y = f(x - 3)$
$\qquad\qquad\qquad\quad y = -(x - 3)^2$ becomes $y = -f(x - 3)$
$\qquad\qquad\qquad\quad y = -(x - 3)^2 + 12$ becomes $y = -f(x - 3) + 12$

To transform the graph of $y = x^2$ to the graph of $y = -x^2 + 6x + 3$ apply the following transformations in this order:

1   horizontal translation of 3 units in the positive x-direction
2   reflection in the x-axis
3   vertical translation of 12 units in the positive y-direction.

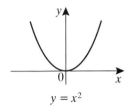 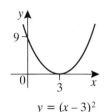 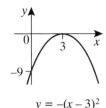 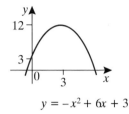

$\qquad y = x^2 \qquad\qquad\quad y = (x - 3)^2 \qquad\quad y = -(x - 3)^2 \qquad\quad y = -x^2 + 6x + 3$

The coordinates of the vertex of $y = -x^2 + 6x + 3$ are (3, 12).

24.6 Applying stretches to graphs

Stretch in the y-direction

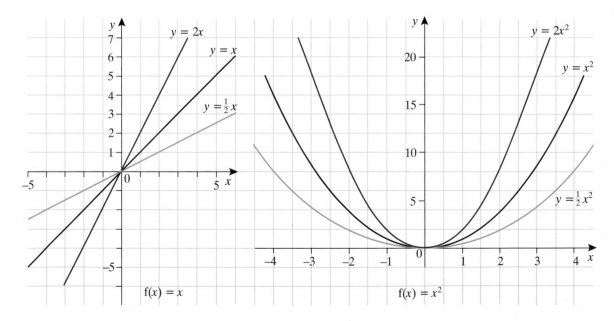

$\qquad f(x) = x \qquad\qquad\qquad\qquad\qquad\qquad f(x) = x^2$

The above graphs are of the type:

$y = f(x)$ (in black) $y = 2f(x)$ (in red) $y = \frac{1}{2}f(x)$ (in green)

for the two cases $f(x) = x$ and $f(x) = x^2$.

To obtain the graph of $y = 2f(x)$ from the graph $y = f(x)$, multiply the y-coordinates of each point on the graph of $y = f(x)$ by 2 but leave the x-coordinates unaltered. For example, the point (1,1) becomes (1,2) and so on.

This has the effect of stretching the curve out in the direction of the y-axis by a doubling factor. Points of the graph on the x-axis remain fixed.

> A stretch is similar to an enlargement but in one direction.

Similarly, by comparing the curves $y = x^2$ and $y = \frac{1}{2}x^2$ we can see that the stretch required to get $y = \frac{1}{2}f(x)$ from $y = f(x)$ is again in the direction of the y-axis but by a halving factor.

■ **For any function f, the graph of $y = af(x)$, where a is a positive constant, is obtained from $y = f(x)$ by applying a stretch of scale factor a parallel to the y-axis.**

Example 15

The graph of $y = f(x)$, where $f(x) = -x^2 + 6x - 5$ is shown below in black and labelled.

(a) Sketch the graph of:

    (i) $y = -2x^2 + 12x - 10$

    (ii) $y = -\frac{1}{2}x^2 + 3x - \frac{5}{2}$ on the same axes.

(b) Describe the transformation which is applied to $y = -\frac{1}{2}x^2 + 3x - \frac{5}{2}$ to get $y = -2x^2 + 12x - 10$.

Answer

(a)   (i) $y = -2x^2 + 12x - 10 = 2(-x^2 + 6x - 5) = 2f(x)$

Apply a stretch of scale factor 2 parallel to the y-axis.

[Note: $(0,-5) \rightarrow (0,-10)$; $(3,4) \rightarrow (3,8)$; $(5,0) \rightarrow (5,0)$]

The graph of $y = -2x^2 + 12x - 10$ is shown in red.

(ii) $y = -\frac{1}{2}x^2 + 3x - \frac{5}{2} = \frac{1}{2}(-x^2 + 6x - 5) = \frac{1}{2}f(x)$

Apply a stretch of scale factor $\frac{1}{2}$ parallel to the y-axis.

[Note: $(0,-5) \rightarrow (0,-\frac{5}{2})$; $(3,4) \rightarrow (3,2)$; $(5,0) \rightarrow (5,0)$]

The graph of $y = -\frac{1}{2}x^2 + 3x - \frac{5}{2}$ is shown in green.

(b) The transformation applied to $y = \frac{1}{2}f(x)$ to get $y = 2f(x)$ is a stretch of scale factor 4 parallel to the y-axis.

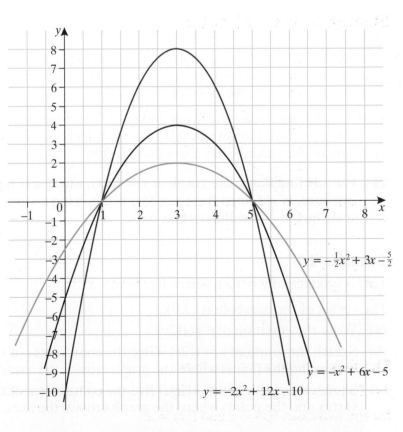

$y = -\frac{1}{2}x^2 + 3x - \frac{5}{2}$

$y = -x^2 + 6x - 5$

$y = -2x^2 + 12x - 10$

Stretch in the x-direction

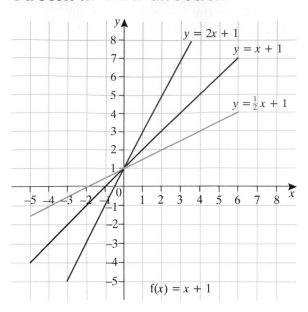

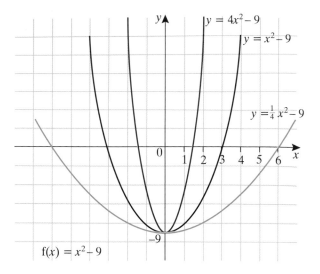

The graphs above are of the type:
$$y = f(x) \qquad y = f(2x) \qquad y = f(\tfrac{1}{2}x)$$
for the two cases $f(x) = x + 1$ and $f(x) = x^2 - 9$.

To get the graph of $y = f(2x)$ shown in red from the graph $y = f(x)$, you multiply the x-coordinates of each point on the graph of $y = f(x)$ by $\frac{1}{2}$ but leave the y-coordinates unaltered. For example, for the parabolas, $(4,7) \rightarrow (2,7)$; $(2,-5) \rightarrow (1,-5)$ and so on.

This has the effect of stretching the curve out (in fact it is 'squashed') in the direction of the x-axis by a halving factor. Points on the graph on the y-axis remain fixed.

Similarly, by comparing the black and green curves we can see that the stretch required to get $y = f(\tfrac{1}{2}x)$ from $y = f(x)$ is again in the direction of the x-axis but by a doubling factor.

■ **For any function f, the graph of $y = f(ax)$, where a is a positive constant, is obtained from $y = f(x)$ by applying a stretch of scale factor $\frac{1}{a}$ parallel to the x-axis.**

Example 16

This is a sketch of $y = f(x)$, where $f(x) = 6x - x^2$.

The vertex of the parabola is $(3,9)$.

(a) Write the equation of $y = f(2x)$.

(b) Sketch the graph of $y = f(2x)$. Give the coordinates of the vertex and the points where the graph crosses the axes.

(c) Sketch the graph of $y = 4f(x)$. Give the coordinates of the vertex and the points where the graph crosses the axes.

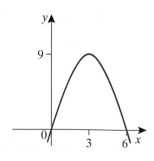

Answer

(a) $\quad$ f$(x) = 6x - x^2$

$\qquad y = f(2x) = 6(2x) - (2x)^2 = 12x - 4x^2 \quad$ so $\quad y = 12x - 4x^2$.

(b)

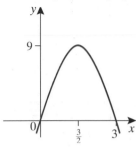

The vertex is at $(\frac{3}{2}, 9)$.
The graph crosses the
axes at $(0,0)$ and $(3,0)$.

(c)

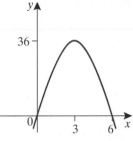

The vertex is at $(3,36)$.
The graph crosses the
axes at $(0,0)$ and $(6,0)$.

Example 17

Explain how to sketch the graph of $y = -4x^2 + 8x + 3$ by applying
transformations to the graph of $y = x^2$. Illustrate your answer by
sketches and state the coordinates of the vertex of $y = -4x^2 + 8x + 3$.

$$-4x^2 + 8x + 3 = -4(x^2 - 2x) + 3$$
$$= -4[(x-1)^2 - 1] + 3 \quad \text{(complete the square)}$$
$$= -4(x-1)^2 + 7$$

Check:
$-4(x-1)^2 + 7$
$= -4(x^2 - 2x + 1) + 7$
$= -4x^2 + 8x - 4 + 7$
$= -4x^2 + 8x + 3$

Let f$(x) = x^2 \quad$ so $\quad$
$\begin{aligned} y &= x^2 & \text{becomes } y &= \text{f}(x) \\ y &= (x-1)^2 & \text{becomes } y &= \text{f}(x-1) \\ y &= 4(x-1)^2 & \text{becomes } y &= 4\text{f}(x-1) \\ y &= -4(x-1)^2 & \text{becomes } y &= -4\text{f}(x-1) \\ y &= -4(x-1)^2 + 7 & \text{becomes } y &= -4\text{f}(x-1) + 7 \end{aligned}$

To transform the graph of $y = x^2$ to the graph of $y = -4x^2 + 8x + 3$
apply the following transformations in this order:

1 $\quad$ horizontal translation of 1 unit in the positive x-direction

2 $\quad$ stretch scale factor 4 parallel to y-axis

3 $\quad$ reflection in the x-axis

4 $\quad$ vertical translation of 7 units in the positive y-direction.

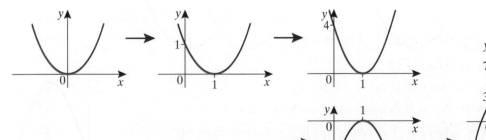

The vertex of $y = -4x^2 + 8x + 3$ is $(1,7)$.

Exercise 24H

1 List the transformations, in the correct order, which when applied to $y = x^2$ give the graphs of the following:

(a) $y = -x^2 + 6$

(b) $y = (x - 4)^2 + 2$

(c) $y = 2x^2 + 4x - 1$

(d) $y = 3x^2 - 12x + 2$

(e) $y = x^2 - 4x + 2$

(f) $y = -x^2 + 4x - 3$

Give a sketch of each graph.

2 List the transformations, in the correct order, which when applied to $y = x^3$ give the graphs of the following:

(a) $y = 4(x + 2)^3$

(b) $y = 8 - x^3$

(c) $y = 5 + (x - 1)^3$

(d) $y = (2x)^3 - 8$

Give a sketch of each graph.

3 Explain how you would use your graph to question 1(e) to sketch the graphs of these:

(a) $y = (x + 1)^2 - 4(x + 1) + 2$ (b) $y = x^2 + 4x + 2$

(c) $y = -x^2 + 4x$

24.7 Transformations applied to trigonometric functions

These are the graphs of $y = \sin x$ and $y = \cos x$ for $-360° \le x \le 360°$. You have met these before in Unit 13 (pages 283 and 284).

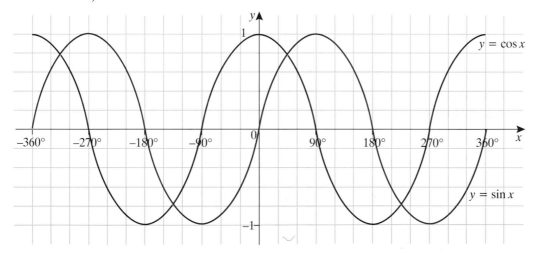

Example 18

Starting with $y = f(x)$, where $f(x) = \cos x$, apply a translation of 90° horizontally in the positive x-direction.

This leads to a trigonometrical identity. What is the identity?

From the result on page 489:

$y = f(x)$ translated 90 units in the positive x-direction becomes

$y = f(x - 90°)$.

Applying this translation to $f(x) = \cos x$ you obtain $y = \cos (x - 90°)$.

But from the graphs, when you apply a translation of 90° horizontally in the positive x-direction to $y = \cos x$, it coincides with the graph of $y = \sin x$.

Since $y = \cos (x - 90°)$ is the same graph as $y = \sin x$, we can deduce the trigonometrical identity:

$$\cos (x - 90°) = \sin x$$

Exercise 24I

1   Starting with $y = f(x)$, where $f(x) = \sin x$, apply a translation of 90° horizontally in the negative x-direction. This leads to a trigonometrical identity. What is the identity?

2   **(a)**   What is the transformation which needs to be applied to $y = f(x)$, where $f(x) = \cos x$ to obtain $y = \cos 2x$?
   (b)   Sketch the graph of $y = \cos 2x$ for $-360° \leqslant x \leqslant 360°$.
   (c)   How many solutions of the equation $\cos 2x = \frac{1}{2}$ lie within the range $-360° \leqslant x \leqslant 360°$?

3   **(a)**   What is the transformation which needs to be applied to $y = f(x)$, where $f(x) = \sin x$ to obtain $y = 2\sin x$?
   (b)   Sketch the graph of $y = 2\sin x$ for $-360° \leqslant x \leqslant 360°$.
   (c)   How many solutions of the equation $2\sin x = 1$ lie within the range $-360° \leqslant x \leqslant 360°$?
   (d)   Without any further sketches, state how many solutions of the equation $2\sin 2x = 1$ lie within the range $-360° \leqslant x \leqslant 360°$. Explain your answer.

4   **(a)**   For $f(x) = \sin x$, sketch the graph of (i) $y = f(-x)$; (ii) $y = -f(x)$
   (b)   Is the function $\sin x$ even, odd or neither? Explain your answer with reference to the two graphs you have drawn in **(a)**.

5   Explain why $\cos x$ is an even function.

6   The greatest value which $\cos x$ can have is $+1$. The least value that $\cos x$ can have is -1. Write down the greatest and least values for:
   (a) $5\cos x$ **(b)** $\cos 3x$ **(c)** $\cos (x + 60°)$
   (d) $-2 \cos x$ **(e)** $6 + 3\cos 4x$

Summary of key points

1 A function is a rule which changes one number into another number.

2 $f(x)$ which denotes the output is read as 'f of x'.

3 The graph of $y = x^2 + a$ is the graph of $y = x^2$ translated a units vertically in the *positive y-direction if $a > 0$*
negative y-direction if $a < 0$.

4 The graph of $y = x + a$ is the graph of $y = x$ translated a units vertically in the *positive y-direction if $a > 0$*
negative y-direction if $a < 0$.

5 The graph of $y = x^3 + a$ is the graph of $y = x^3$ translated a units vertically in the *positive y-direction if $a > 0$*
negative y-direction if $a < 0$.

6 The graph of $y = \frac{1}{x} + a$ is the graph of $y = \frac{1}{x}$ translated a units vertically in the *positive y-direction if $a > 0$*
negative y-direction if $a < 0$.

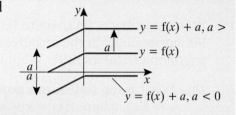

7 For any function f, the graph of $y = f(x) + a$ is the graph of $y = f(x)$ translated a units vertically in the *positive y-direction if $a > 0$*
negative y-direction if $a < 0$.

8 The graph of $y = (x + a)^2$ is the graph of $y = x^2$ translated a units horizontally in the
negative x-direction if $a > 0$
positive x-direction if $a < 0$.

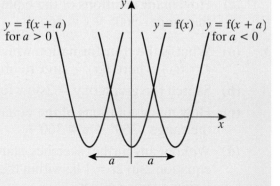

9 The vertex of the parabola $y = (x + a)^2$ is at the point $(-a, 0)$.

10 In general, for any function f: the graph of $y = f(x + a)$ is the graph of $y = f(x)$ translated a units horizontally in the
negative x-direction if $a > 0$
positive x-direction if $a < 0$.

11 To sketch the graph of $y = x^2 + bx + c$, complete the square and apply a double translation to the parabola $y = x^2$.

12 The graph of $y = f(x + a) + b$ is the graph of $y = f(x)$ translated a units horizontally
(in the negative x-direction if $a > 0$)
(in the positive x-direction if $a < 0$)
followed by a translation of
b units vertically (upwards if $b > 0$)
(downwards if $b < 0$).

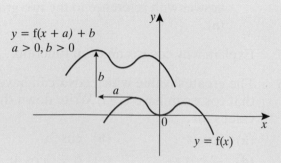

13 For any function f, the graph of $y = -f(x)$ is obtained by reflecting $y = f(x)$ in the x-axis.

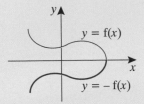

14 For any function f, the graph of $y = f(-x)$ is obtained by reflecting $y = f(x)$ in the y-axis.

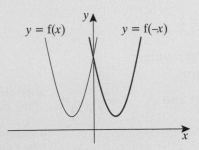

15 Functions f which satisfy $f(-x) = f(x)$ are called **even functions**.

16 Functions f which satisfy $f(-x) = -f(x)$ are called **odd functions**.

17 For any function f, the graph of $y = af(x)$, where a is a positive constant, is obtained from $y = f(x)$ by applying a stretch of scale factor a parallel to the y-axis.

18 For any function f, the graph of $y = f(ax)$, where a is a positive constant, is obtained from $y = f(x)$ by applying a stretch of scale factor $\frac{1}{a}$ parallel to the x-axis.

25 Vectors

25.1 Translations

A **translation** is a movement of a shape such that the shape is neither turned nor rotated. Examples of translations, T_1 and T_2, are shown here.

■ **Translations can be described using column vectors: $\binom{x}{y}$.**

The translation T_1 is given by the column vector $\binom{6}{1}$, meaning 6 units in the positive x direction and 1 unit in the positive y direction.

T_2 is given by $\binom{-1}{-3}$ meaning 1 unit in the negative x direction and 3 units in the negative y direction.

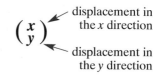

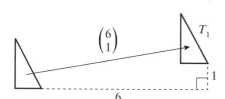 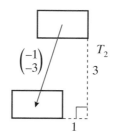

■ **An alternative notation which can be used to describe the translation is $\vec{AB}$, where A is the starting point and B is the finishing point.**

The translation that maps B to A can be described as $\vec{BA}$. The relation between $\vec{AB}$ and $\vec{BA}$ is:

$$\vec{AB} = -\vec{BA}$$

In terms of column vectors:

$$\vec{AB} = \binom{x}{y} \quad \vec{BA} = \binom{-x}{-y} \quad \text{and} \quad \binom{x}{y} = -\binom{-x}{-y}$$

The values of x and y in a column vector are called **components**.

Example 1

A is the point $(2,-3)$ and B is the point $(5,2)$. $\vec{AC} = \binom{-1}{3}$

(a) Write the column vector $\vec{AB}$.

(b) Write the column vector $\vec{BA}$.

(c) Write the coordinates of C.

(a) $\vec{AB} = \left(\begin{smallmatrix} 5-2 \\ 2--3 \end{smallmatrix}\right) = \left(\begin{smallmatrix} 3 \\ 5 \end{smallmatrix}\right)$

(b) $\vec{BA} = \left(\begin{smallmatrix} -3 \\ -5 \end{smallmatrix}\right)$

(c) $C = (2 + -1, -3 + 3) = (1,0)$.

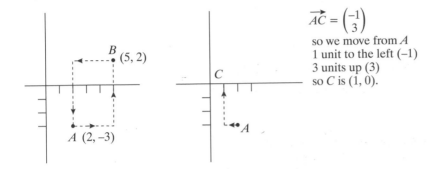

$\vec{AC} = \left(\begin{smallmatrix} -1 \\ 3 \end{smallmatrix}\right)$

so we move from A
1 unit to the left (–1)
3 units up (3)
so C is (1, 0).

Exercise 25A

1 A is the point (4,6), B is the point (2,5) and C is the point (–1,0).

Write the column vectors

(a) $\vec{AB}$ **(b)** $\vec{BC}$ **(c)** $\vec{CA}$ **(d)** $\vec{AC}$.

2 A is the point (2,4), B is the point (3,5) and C is the point (4,2). The vector $\vec{CD}$ is parallel to the vector $\vec{BA}$. The point D lies on the y-axis. Find the coordinates of D.

3 The point T_1 (1,2) is reflected in the line $x = 3$ to give the point T_2. The point T_2 is reflected in the line $x = 6$ to give T_3. Describe the single transformation that maps T_1 to T_3.

Generalize your result in the case where there are successive reflections in the lines $x = a$ and $x = b$, with $a < b$.

4 $\vec{AB}$ is the column vector $\left(\begin{smallmatrix} 4 \\ -3 \end{smallmatrix}\right)$. $\vec{BC}$ is the vector $\left(\begin{smallmatrix} 2 \\ 4 \end{smallmatrix}\right)$.

Find the column vector $\vec{AC}$. Draw a diagram to show your answer.

25.2 Vectors

A third way to describe a translation is to use bold type single letters such as **a**, **b**. When handwritten, they are underlined. Translations described in this way are simply referred to as **vectors**. The vectors **a** and **b** are shown here. The lines with arrows are called **directed line segments** and show a unique **length** and **direction** for each of vectors **a** and **b**.

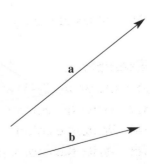

■ **A vector defined as 'a' has a unique length and direction.**

Addition of vectors

Look at the diagram below.

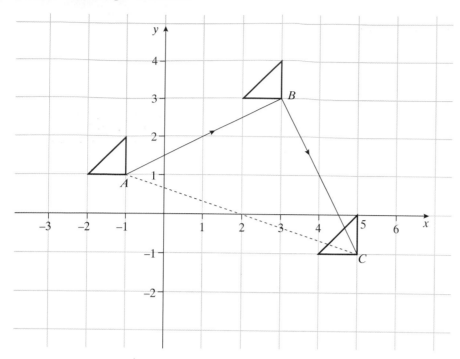

The vectors $\vec{AB}$ and $\vec{BC}$ can be written $\binom{4}{2}$ and $\binom{2}{-4}$.

$\vec{AB} + \vec{BC}$ can be interpreted as the result of two successive translations which are equivalent to $\vec{AC}$.

$$\binom{4}{2} + \binom{2}{-4} = \binom{4+2}{2-4} = \binom{6}{-2}$$
$$\vec{AB} + \vec{BC} = \qquad\qquad \vec{AC}$$

Here vectors $\vec{AB}$ and $\vec{BC}$ have been added to give vector $\vec{AC}$.

For any two vectors **a** and **b** it is possible to add them by placing them 'nose to tail'.

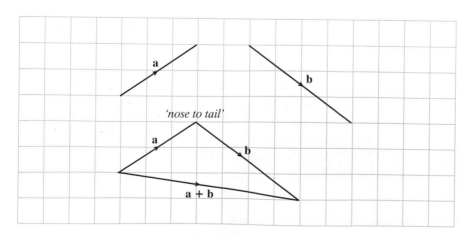

Note that
a + **b** = **b** + **a**
b + **a** gives a vector of the same length and direction.

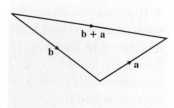

■ $\mathbf{a} + \mathbf{b} = \mathbf{b} + \mathbf{a}$

$$\binom{3}{2} + \binom{4}{-3} = \binom{7}{-1}$$

The diagram on the previous page shows the **triangle law of addition**.

Subtraction of vectors

Vectors can also be subtracted.

■ **p – q can be interpreted as p + (–q)**

Using column vectors, this means:

$$\binom{4}{5} - \binom{3}{2} = \binom{4}{5} + \binom{-3}{-2} = \binom{4-3}{5-2} = \binom{1}{3}$$

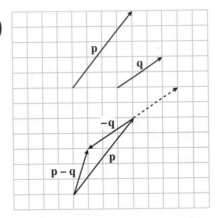

The parallelogram method

Another way to approach the **addition** and **subtraction** of two vectors is by using the **parallelogram** method.

a and **b** are two vectors as shown in the diagram.

The long diagonal in the direction shown is $\mathbf{a} + \mathbf{b}$.

The shorter diagonal in the direction shown is $\mathbf{a} - \mathbf{b}$.

If we put the short diagonal as **x** then $\mathbf{b} + \mathbf{x} = \mathbf{a}$.

So $\mathbf{x} = \mathbf{a} - \mathbf{b}$.

Complete the parallelogram:

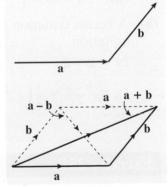

Multiplication of vectors by scalars

It is possible to multiply a vector by a scalar. Geometrically, the expression $2\overrightarrow{AB}$ means the directed line segment parallel to $\overrightarrow{AB}$, but twice the length. In other words when vector $\overrightarrow{AB}$ is multiplied by two the result is $2\overrightarrow{AB}$.

■ **In vector terms if a vector a is multiplied by a scalar *k* then the vector *k*a is parallel to a and equal to *k* times a.**

For the column vector $\binom{p}{q}$, $k \times \binom{p}{q} = \binom{kp}{kq}$

For example, $3\binom{2}{1} = \binom{3 \times 2}{3 \times 1} = \binom{6}{3}$

25.3 Vector algebra

The rules of algebra that you already know can be applied to vectors, providing you do not multiply or divide one vector by another.

■ $a + (b + c) = (a + b) + c$
$k (a + b) = ka + kb$
$(p + q) a = pa + qa$
for any vectors a and b and any scalars p, q and k.

Two column vectors are equal if they represent the same translation.

So, if $\begin{pmatrix} a \\ b \end{pmatrix} = \begin{pmatrix} c \\ d \end{pmatrix}$

then $a = c$ and $b = d$.

Example 2

$a = \begin{pmatrix} 3 \\ -2 \end{pmatrix}$ $b = \begin{pmatrix} -1 \\ -3 \end{pmatrix}$

(a) Write as column vectors
   (i) $a - b$ (ii) $3a$ (iii) $3a - 2b$

(b) Find the vector x such that
   $a + 2x = b$

Answer

(a)   (i) $a - b = a + (-b) = \begin{pmatrix} 3 \\ -2 \end{pmatrix} + -\begin{pmatrix} -1 \\ -3 \end{pmatrix} = \begin{pmatrix} 3--1 \\ -2--3 \end{pmatrix} = \begin{pmatrix} 4 \\ 1 \end{pmatrix}$

   (ii) $3a = 3\begin{pmatrix} 3 \\ -2 \end{pmatrix} = \begin{pmatrix} 9 \\ -6 \end{pmatrix}$

Remember:
$a - b = a + (-b)$.

   (iii) $3a - 2b = \begin{pmatrix} 9 \\ -6 \end{pmatrix} - 2\begin{pmatrix} -1 \\ -3 \end{pmatrix} = \begin{pmatrix} 9 \\ -6 \end{pmatrix} - \begin{pmatrix} -2 \\ -6 \end{pmatrix} = \begin{pmatrix} 9 \\ -6 \end{pmatrix} + \begin{pmatrix} 2 \\ 6 \end{pmatrix} = \begin{pmatrix} 11 \\ 0 \end{pmatrix}$

(b)   A vector equation can be solved in the same way as an ordinary equation.

   $2x = b - a$

   $x = \frac{b - a}{2} = \frac{1}{2}\begin{pmatrix} -1 -3 \\ -3 - -2 \end{pmatrix} = \frac{1}{2}\begin{pmatrix} -4 \\ -1 \end{pmatrix} = \begin{pmatrix} -2 \\ -\frac{1}{2} \end{pmatrix}$

Exercise 25B

1   $a = \begin{pmatrix} 2 \\ 1 \end{pmatrix}$ $b = \begin{pmatrix} -4 \\ 3 \end{pmatrix}$
   Calculate (i) $a + b$ (ii) $2a$ (iii) $2a - 3b$ (iv) $2 (a - b)$.
   Find a vector c such that $a + c$ is parallel to $\begin{pmatrix} 4 \\ 4 \end{pmatrix}$.

2   $a = \begin{pmatrix} 3 \\ 1 \end{pmatrix}$ $b = \begin{pmatrix} 2 \\ 2 \end{pmatrix}$
   Draw diagrams to show that (i) $2(a + b) = 2a + 2b$
                             (ii) $(2 + 3) a = 2a + 3a$

3   P is the point $(1,3)$, Q is the point $(2,4)$ and R is the point $(5,4)$.
   S is the point such that $\vec{PQ} = \vec{SR}$. Find the coordinates of S.

25.4 Finding the magnitude of a vector

■ **The magnitude of a vector is the length of the directed line segment representing it.**

If the vector is expressed in column form, Pythagoras' Theorem can be used to find the magnitude.

Example 3

Find the magnitude of the vector $\overrightarrow{AB} = \begin{pmatrix} 6 \\ -8 \end{pmatrix}$

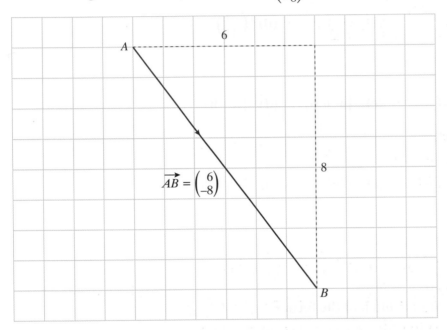

Draw $\overrightarrow{AB}$ as the hypotenuse of a right-angled triangle.

By Pythagoras the length AB is given by

$$AB^2 = 6^2 + 8^2$$
$$AB^2 = 36 + 64$$
$$AB^2 = 100$$
$$AB = 10$$

The magnitude of $\overrightarrow{AB}$ is 10 units.

■ **The term modulus can be used to describe the magnitude of a vector.**

You do not have to worry about using the term modulus in your GCSE exam. It is included here to introduce you to an important concept that will be developed in your A Level mathematics course.

The notation for modulus is a pair of bars written on each side of the vector: $|\overrightarrow{AB}|$ or $|\mathbf{a}|$.

For example with reference to $\vec{AB}$ in the last example

$$|\vec{AB}| = 10$$

■ **In general, the magnitude of the vector $\begin{pmatrix} x \\ y \end{pmatrix}$ is $\sqrt{x^2 + y^2}$**

Exercise 25C

1 Find the magnitude of each of these vectors:

(a) $\begin{pmatrix} -4 \\ 3 \end{pmatrix}$ (b) $\begin{pmatrix} 7 \\ 24 \end{pmatrix}$ (c) $\begin{pmatrix} -6 \\ -8 \end{pmatrix}$ (d) $\begin{pmatrix} 7 \\ 11 \end{pmatrix}$

2 $\mathbf{a} = \begin{pmatrix} 5 \\ 9 \end{pmatrix}$, $\mathbf{b} = \begin{pmatrix} 3 \\ 3 \end{pmatrix}$

Work out:

(a) $|\mathbf{a}|$ (b) $|\mathbf{b}|$ (c) $|\mathbf{a} + \mathbf{b}|$ (d) $|\mathbf{a} - \mathbf{b}|$

3 $\mathbf{c} = \begin{pmatrix} 2 \\ 3 \end{pmatrix}$, $\mathbf{d} = \begin{pmatrix} -3 \\ 4 \end{pmatrix}$

Work out:

(a) $|\mathbf{c}|$ (b) $2|\mathbf{c}|$ (c) $|2\mathbf{c}|$ (d) $|\mathbf{c}|\,|\mathbf{d}|$

25.5 Linear combinations

One useful and important application of vectors is to produce
special combinations of two given vectors.

■ **Combinations of the vectors a and b of the form**

$$p\mathbf{a} + q\mathbf{b}$$

**where p and q are scalars, are called linear combinations of the
vectors a and b.**

Generally, any two non-parallel vectors can be combined to give a
single vector in a different direction.

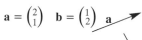

$$\mathbf{a} = \begin{pmatrix} 2 \\ 1 \end{pmatrix} \quad \mathbf{b} = \begin{pmatrix} 1 \\ 2 \end{pmatrix}$$

Example 4

$$\mathbf{a} = \begin{pmatrix} 2 \\ 1 \end{pmatrix} \quad \text{and} \quad \mathbf{b} = \begin{pmatrix} 1 \\ -2 \end{pmatrix}$$

Find a combination of **a** and **b** which is
parallel to the x-axis.

Answer

Draw **a** and **b** end to end.

If **b** is to end at the x-axis we need to start with $2\mathbf{a} = \begin{pmatrix} 4 \\ 2 \end{pmatrix}$.

The linear combination is $2\mathbf{a} + \mathbf{b} = 2\begin{pmatrix} 2 \\ 1 \end{pmatrix} + \begin{pmatrix} 1 \\ -2 \end{pmatrix} = \begin{pmatrix} 5 \\ 0 \end{pmatrix}$ which is
parallel to the x-axis.

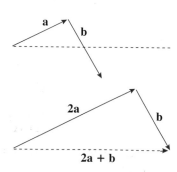

Example 5

$\mathbf{a} = \begin{pmatrix} 3 \\ 2 \end{pmatrix}, \quad \mathbf{b} = \begin{pmatrix} 1 \\ 3 \end{pmatrix}$

(a) Find scalars p and q such that $p\mathbf{a} + q\mathbf{b}$ is parallel to the x-axis.

(b) Find scalars r and s such that $r\mathbf{a} + s\mathbf{b}$ is parallel to the y-axis.

(a) $p\mathbf{a} + q\mathbf{b} = p \begin{pmatrix} 3 \\ 2 \end{pmatrix} + q \begin{pmatrix} 1 \\ 3 \end{pmatrix}$

The y component of vector $p\mathbf{a} + q\mathbf{b}$ must be zero because the vector lies in the direction of the x-axis.

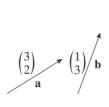 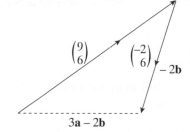

So $2p + 3q = 0$

Two possible values of p and q which satisfy the equation are:
$$p = 3, \quad q = -2$$
$3\mathbf{a} - 2\mathbf{b}$ is parallel to the x-axis.

(b) $r\mathbf{a} + s\,\mathbf{b} = r \begin{pmatrix} 3 \\ 2 \end{pmatrix} + s \begin{pmatrix} 1 \\ 3 \end{pmatrix}$

The x component must be zero because the vector is in the direction of the y-axis.

So $3r + s = 0$

Possible values of r and s are $r = 1, s = -3$.

$\mathbf{a} - 3\mathbf{b}$ is parallel to the y-axis.

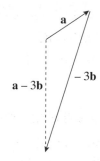

It is possible to find a linear combination of two vectors which is parallel to any required direction.

Example 6

Find a linear combination of vectors $\mathbf{a} = \begin{pmatrix} 3 \\ 2 \end{pmatrix}$ and $\mathbf{b} = \begin{pmatrix} 1 \\ 3 \end{pmatrix}$ which is equal to vector $\mathbf{c} = \begin{pmatrix} 9 \\ 13 \end{pmatrix}$.

There are two possible methods for finding the required linear combination.

Method 1 using simultaneous equations:
$$p\mathbf{a} + q\mathbf{b} = \mathbf{c}$$
$$p \begin{pmatrix} 3 \\ 2 \end{pmatrix} + q \begin{pmatrix} 1 \\ 3 \end{pmatrix} = \begin{pmatrix} 9 \\ 13 \end{pmatrix}$$
$$\begin{pmatrix} 3p \\ 2p \end{pmatrix} + \begin{pmatrix} q \\ 3q \end{pmatrix} = \begin{pmatrix} 9 \\ 13 \end{pmatrix}$$
$$3p + q = 9 \qquad (1)$$
$$2p + 3q = 13 \qquad (2)$$

Multiply equation (1) by 3:

$$9p + 3q = 27 \qquad (3)$$
$$2p + 3q = 13$$

Subtract (2) from (3):

$$7p = 14$$
$$p = 2$$

Substitute $p = 2$ in (1):

$$6 + q = 9$$
$$q = 3$$

The linear combination is $\qquad$ $2\mathbf{a} + 3\mathbf{b}$

Method 2 is a more complicated approach but is given here as an alternative to using simultaneous equations.

Using the results of Example 5:

$$3\mathbf{a} - 2\mathbf{b} = \begin{pmatrix} 7 \\ 0 \end{pmatrix} \qquad \mathbf{a} - 3\mathbf{b} = \begin{pmatrix} 0 \\ -7 \end{pmatrix}$$

$$\begin{pmatrix} 1 \\ 0 \end{pmatrix} = \tfrac{1}{7} \begin{pmatrix} 7 \\ 0 \end{pmatrix} = \tfrac{3}{7}\mathbf{a} - \tfrac{2}{7}\mathbf{b}$$

$$\begin{pmatrix} 0 \\ 1 \end{pmatrix} = -\tfrac{1}{7} \begin{pmatrix} 0 \\ -7 \end{pmatrix} = \tfrac{3}{7}\mathbf{b} - \tfrac{1}{7}\mathbf{a}$$

$$\begin{pmatrix} 9 \\ 13 \end{pmatrix} = \begin{pmatrix} 9 \\ 0 \end{pmatrix} + \begin{pmatrix} 0 \\ 13 \end{pmatrix} = 9\begin{pmatrix} 1 \\ 0 \end{pmatrix} + 13\begin{pmatrix} 0 \\ 1 \end{pmatrix}$$

$$= 9\left(\tfrac{3}{7}\mathbf{a} - \tfrac{2}{7}\mathbf{b}\right) + 13\left(\tfrac{3}{7}\mathbf{b} - \tfrac{1}{7}\mathbf{a}\right)$$

$$= 2\mathbf{a} + 3\mathbf{b}$$

$$= \mathbf{c}$$

Exercise 25D

1 $\mathbf{a} = \begin{pmatrix} -1 \\ 2 \end{pmatrix}$, $\mathbf{b} = \begin{pmatrix} -3 \\ 4 \end{pmatrix}$, calculate $\mathbf{x}$, given that $\mathbf{a} + \mathbf{x} = \mathbf{b}$.

2 $\mathbf{c} = \begin{pmatrix} 2 \\ -1 \end{pmatrix}$, $\mathbf{d} = \begin{pmatrix} 4 \\ -3 \end{pmatrix}$, calculate $\mathbf{x}$ given that $2\mathbf{x} + \mathbf{c} = \mathbf{d}$.

3 $\mathbf{e} = \begin{pmatrix} 4 \\ 1 \end{pmatrix}$, $\mathbf{f} = \begin{pmatrix} -2 \\ 3 \end{pmatrix}$, calculate $\mathbf{x}$ given that $2\mathbf{e} - \mathbf{x} = \mathbf{f}$.

4 $\mathbf{a} = \begin{pmatrix} 2 \\ -1 \end{pmatrix}$, $\mathbf{b} = \begin{pmatrix} 1 \\ 1 \end{pmatrix}$ Given that $2\mathbf{a} + p\mathbf{b}$ is parallel to the x-axis, find the value of p.

5 $\mathbf{a} = \begin{pmatrix} 2 \\ -1 \end{pmatrix}$, $\mathbf{b} = \begin{pmatrix} 1 \\ 1 \end{pmatrix}$ Given that $\mathbf{a} + q\mathbf{b}$ is parallel to the y-axis, find the value of q.

6 $\mathbf{a} = \begin{pmatrix} -3 \\ 4 \end{pmatrix}$, $\mathbf{b} = \begin{pmatrix} 4 \\ -3 \end{pmatrix}$.
Calculate (i) $|\mathbf{a}| + |\mathbf{b}|$ (ii) $|\mathbf{a} + \mathbf{b}|$
Given that for two vectors $\mathbf{x}$ and $\mathbf{y}$, $|\mathbf{x} + \mathbf{y}| = |\mathbf{x}| + |\mathbf{y}|$, what can you say about the vectors $\mathbf{x}$ and $\mathbf{y}$? Justify your answer.

7 Find the values of x and y, given that:
$$2\begin{pmatrix} 1 \\ y \end{pmatrix} + \begin{pmatrix} x \\ y \end{pmatrix} = \begin{pmatrix} 2 \\ -6 \end{pmatrix}$$

8 Find the values of the scalars p and q, given that:

$$p\binom{2}{1} + q\binom{-1}{2} = \binom{7}{-2}$$

9 P is a variable point which moves so that the vector $\vec{OP}$ is given by:

$$\vec{OP} = \binom{2}{6} + t\binom{1}{1}$$

Calculate the coordinates of P, for values of t from 0 to 5. Plot these coordinates. What is the path of P as t varies? Give the equation of the path in the form $y = mx + c$.

10 A is the point $(2,1)$. B is the point $(8,4)$ and C is the point $(6,6)$.

    (a) Calculate $\vec{AB}$.

    (b) Write an expression in terms of k for $\vec{AB} + k\,\vec{BC}$.
         The line BC is extended to a point D where AD is parallel to the y-axis.

    (c) Find the value of k and the coordinates of D.

25.6 Position vectors

The column vector $\binom{x}{y}$ denotes a translation.

There are an infinite number of points which are related by such a translation.

Look at the diagram, which shows several pairs of points linked by the same vector. The vector which translates O to P, $\vec{OP}$, is a special vector, the **position vector** of P.

It is called this because it fixes the position of point P relative to a fixed reference point which is the origin.

In this case $\vec{OP} = \binom{x}{y}$

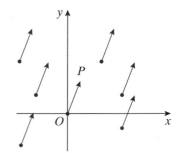

Example 7

P is the point $(2,3)$. $\vec{PQ} = \binom{-1}{2}$. Find the position vector of Q.

The position vector of Q is $\vec{OQ} = \vec{OP} + \vec{PQ}$

$$= \binom{2}{3} + \binom{-1}{2}$$

$$= \binom{1}{5}$$

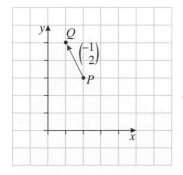

■ The position vector of a point P is $\overrightarrow{OP}$, where O is usually the origin.

Example 8

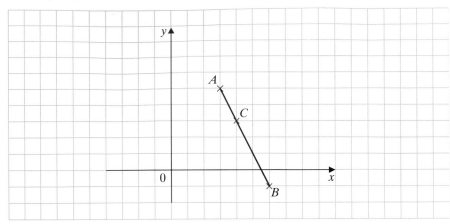

A is the point (3,5) and B is the point (6,–1). C is the point on AB such that $AC = \frac{1}{2} CB$. Find the position vector of the point C.

$$\overrightarrow{AB} = \begin{pmatrix} 3 \\ -6 \end{pmatrix}$$
$$\overrightarrow{AC} = \tfrac{1}{3} \overrightarrow{AB}$$
$$= \tfrac{1}{3} \begin{pmatrix} 3 \\ -6 \end{pmatrix}$$
$$= \begin{pmatrix} 1 \\ -2 \end{pmatrix}$$

The position vector of C is $\overrightarrow{OC}$

$$\overrightarrow{OC} = \overrightarrow{OA} + \overrightarrow{AC}$$
$$= \begin{pmatrix} 3 \\ 5 \end{pmatrix} + \begin{pmatrix} 1 \\ -2 \end{pmatrix}$$
$$= \begin{pmatrix} 4 \\ 3 \end{pmatrix}$$

You could also solve this problem using abstract vectors.

Example 9

If the position vector of A is $\mathbf{a}$ and the position vector of B is $\mathbf{b}$ find $\mathbf{c}$, the position vector of C, in terms of $\mathbf{a}$ and $\mathbf{b}$ so that $AC = \frac{1}{2} CB$:

$$\overrightarrow{AB} = \mathbf{b} - \mathbf{a}$$
$$\overrightarrow{AC} = \tfrac{1}{3} (\mathbf{b} - \mathbf{a})$$
$$\mathbf{c} = \overrightarrow{OC}$$
$$= \overrightarrow{OA} + \overrightarrow{AC}$$
$$= \mathbf{a} + \tfrac{1}{3} (\mathbf{b} - \mathbf{a})$$
$$= \tfrac{2}{3} \mathbf{a} + \tfrac{1}{3} \mathbf{b}$$

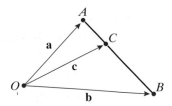

Check this result by using the answer to Example 8.

All the properties established for vectors in the earlier section hold for position vectors.

Here are two very useful results.

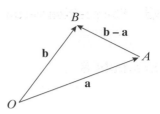

■ **If A and B have position vectors a and b respectively, then the vector:**

$\vec{AB}$ = b − a   (**Note the reversal of letters.**)

From the diagram $\vec{AB} = \vec{AO} + \vec{OB} = -\vec{OA} + \vec{OB} = -$a + b = b − a.

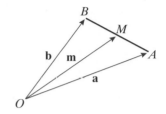

■ **If A and B have position vectors a and b, then the position vector of the midpoint, M, of the line joining A to B is**

$\vec{OM}$ = m = $\frac{1}{2}$ (a + b)

From the diagram

$$\vec{AM} = \tfrac{1}{2}\,\vec{AB}$$
$$\vec{OM} = \vec{OA} + \vec{AM}$$
$$= a + \tfrac{1}{2}(b - a)$$
$$= a + \tfrac{1}{2}b - \tfrac{1}{2}a$$
$$= \tfrac{1}{2}(a + b)$$

Exercise 25E

1 A is the point (1,2). B is the point (3,6). Write the position vector of the midpoint of AB.

2 C is the point (2,−2) and D is the point (8,4). E lies on the line CD, such that $CE = \frac{1}{2}ED$. Find the coordinates of E.

3 A is the point (2,3) and B is the point (4,7). C lies on the extension of the line AB, such that $BC = \frac{1}{2}AB$. Find the coordinates of C.

4 $PQRS$ is a parallelogram, with $P = (1,1)$, $Q = (5,3)$ and $R = (7,7)$. Find the position vector of S and of the mid-point of PR.

25.7 Proving geometrical results

Example 10

A and B are the mid-points of the sides OX and OY of a triangle. Prove that the line XY is parallel to the line AB and equal to twice the length of AB.

Look at the diagram.

Use letters **a** and **b** for the position vectors of A and B respectively.

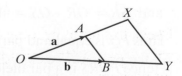

$$\vec{OX} = 2\mathbf{a}$$
$$\vec{OY} = 2\mathbf{b}$$
$$\vec{XY} = \vec{OY} - \vec{OX}$$
$$= 2\mathbf{b} - 2\mathbf{a}$$
$$= 2(\mathbf{b} - \mathbf{a})$$
but $\quad \vec{AB} = \mathbf{b} - \mathbf{a}$
$$\vec{XY} = 2\,\vec{AB}$$

So XY is twice the length of AB and XY is parallel to AB.

Example 11

Show that the diagonals of a parallelogram bisect one another.

In the parallelogram $OACB$, $\vec{OA} = \mathbf{a}$ and $\vec{OB} = \mathbf{b}$.

Opposite sides of a parallelogram are parallel and equal so $\vec{AC} = \mathbf{b}$.

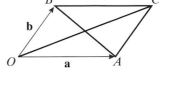

$$\vec{OC} = \vec{OA} + \vec{AC}$$
$$= \mathbf{a} + \mathbf{b}$$

If D is the mid-point of OC:

$$\vec{OD} = \tfrac{1}{2}\vec{OC}$$
$$= \tfrac{1}{2}(\mathbf{a} + \mathbf{b})$$

but $\tfrac{1}{2}(\mathbf{a} + \mathbf{b})$ is the position vector of the mid-point of AB (see page 517) which means that OC bisects AB.

Example 12

$OABC$ is a quadrilateral, with $\vec{OA} = \mathbf{a}$, $\vec{OB} = \mathbf{b}$ and $\vec{OC} = \mathbf{c}$.

P, Q, R and S are the midpoints of sides OA, AB, BC and CO.

Show that $PQRS$ is a parallelogram.

From the diagram,

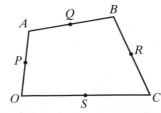

Remember: the result for two vectors $\mathbf{a}$ and $\mathbf{b}$:

$$\mathbf{m} = \tfrac{1}{2}(\mathbf{a} + \mathbf{b})$$

$\vec{OP} = \tfrac{1}{2}\mathbf{a}$ and $\vec{OS} = \tfrac{1}{2}\mathbf{c}$.

$\vec{OQ} = \tfrac{1}{2}(\mathbf{a} + \mathbf{b})$ and $\vec{OR} = \tfrac{1}{2}(\mathbf{b} + \mathbf{c})$.

Now $\vec{PQ} = \vec{OQ} - \vec{OP} = \tfrac{1}{2}\mathbf{b}$

and $\vec{SR} = \vec{OR} - \vec{OS} = \tfrac{1}{2}\mathbf{b}$

Thus PQ is equal and parallel to SR.

Hence $PQRS$ is a parallelogram.

Exercise 25F

1 $\vec{OX} = \mathbf{x}$ and $\vec{OY} = \mathbf{y}$. X and Y are $\frac{2}{3}$ of the way along OA and OB respectively. Write the vector $\vec{XY}$ in terms of $\mathbf{x}$ and $\mathbf{y}$.

Write the geometrical relationship between the line XY and the line AB.

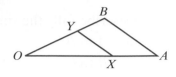

2 Express $\vec{AB}$ in terms of $\mathbf{a}$ and $\mathbf{b}$, where $\vec{OA} = \mathbf{a}$ and $\vec{OB} = \mathbf{b}$.

Given that $AC = 2OA$, $BD = 2OB$, OAC and OBD are straight lines, express $\vec{CD}$ in terms of $\mathbf{a}$ and $\mathbf{b}$. What are the geometrical relations between AB and CD?

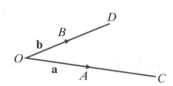

3 X is the mid-point of AB and Y is the mid-point of OA.
G is the point which is $\frac{2}{3}$ of the way along the line OX.
Find the position vector of G.
A point H is chosen to be $\frac{2}{3}$ of the way along the line BY.
Find the position vector of H.
What geometrical fact can you conclude from your answers?
The line AG, when extended, cuts the line OB at D.
Find the position vector of D.

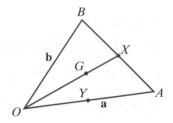

4 $OCDB$ is a trapezium, with OC parallel to BD. A is the mid-point of OC and E is the mid-point of BD. X is the point of intersection of AE and BC. $\vec{BD} = 4\mathbf{a}$.

(a) Show that $\vec{OX}$ can be written in the form $(1 + k)\mathbf{a} + k\mathbf{b}$ where k is a scalar.

(b) Show that $\vec{OX}$ can be written in the form $(2-2m)\mathbf{a} + m\mathbf{b}$, where m is a scalar.

(c) Find the values of k and m and hence the position vector of X.

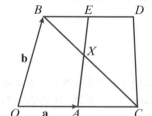

5 $OABC$ is a parallelogram. Points P, Q, R and S are taken $\frac{1}{3}$ of the way along each of the sides OA, AB, BC and CO respectively.

Find whether or not $PQRS$ is a parallelogram. Justify your answer by using position vectors.

Are the conclusions you have drawn still valid when $OABC$ is a quadrilateral, but not a parallelogram?

6 Find, in terms of $\mathbf{a}$ and $\mathbf{b}$:

(a) $\vec{AB}$ (b) $\vec{AF}$ (c) $\vec{AE}$

The midpoints of each of the sides of the regular hexagon $ABCDEF$ are joined to give a second hexagon. Find the position vectors of each of the corners of this second hexagon. This process is repeated to produce a third hexagon. What is the relationship between the first and third hexagons? Justify your answer.

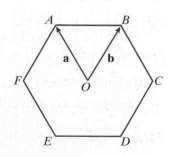

Exercise 25G (Mixed questions)

1   Describe fully the single translation which maps:
  (a) A to B **(b)** B to A
  (c) B to C **(d)** C to B

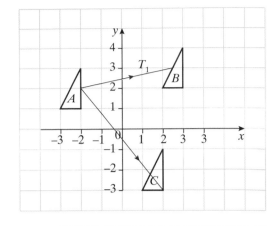

2   Write these as column vectors:
  (a) $\vec{OA}$ **(b)** $\vec{OB}$
  (c) $\vec{OA} + \vec{OB}$ **(d)** $\vec{OA} - \vec{OB}$
  (e) $2\vec{OA}$ **(f)** $\vec{OB} + \vec{AO}$
  Draw diagrams to show the operations in
  questions **(c)**, **(d)** and **(e)**.

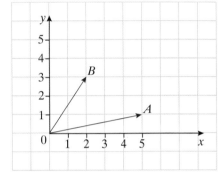

3   B is the point such that $OABC$ is a parallelogram.
  Copy the diagram and plot the point B.
  Write the vector $\vec{OB}$.
  Write the vector $\vec{OM}$ where M is the mid-point of OB.

4   $\vec{OA} = \binom{2}{1}$, $\vec{OB} = \binom{5}{6}$ Write down $\vec{AB}$.

5   $\vec{XY} = \binom{2}{3}$, $\vec{YO} = \binom{-1}{4}$ Write down $\vec{OX}$.

6   **(a)** $\binom{2}{1} + \binom{1}{-1}$ **(b)** $\binom{2}{1} - \binom{1}{-1}$

7   $3\binom{2}{1} - 2\binom{-2}{1}$

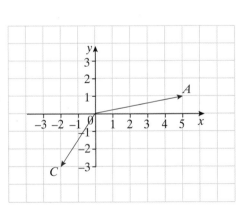

8   Given that $\binom{2}{x} + \binom{x}{y} = \binom{5}{6}$, find the values of x and y.

9   Solve the vector equation $\binom{a}{2} - \binom{-3}{b} = \binom{4}{b}$

10   $\mathbf{a} = \binom{3}{0}$, $\mathbf{b} = \binom{0}{2}$, $\mathbf{c} = \binom{3}{1}$
  Find the values of p and q such that $p\mathbf{a} + q\mathbf{b} = \mathbf{c}$.

11   $\vec{AB} = \binom{1}{3}$ and $\vec{BC} = \binom{-3}{1}$
  (a) Calculate $|\vec{AB}|$ and $|\vec{BC}|$.
  (b) Find $\vec{AC}$ as a column vector.
  (c) Calculate $|\vec{AC}|$.
  (d) What do you conclude about the vectors $\vec{AB}$ and $\vec{BC}$?

12 *ABCD* is a parallelogram.

Express in terms of **a** and **b**:

(a) $\vec{AC}$

(b) $\vec{DA}$

(c) $\vec{DB}$

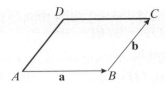

13 Given that $\vec{OC} = \frac{2}{5}\vec{OB}$,
express, in terms of **a** and **b**:

(a) $\vec{CB}$ (b) $\vec{BA}$

D is the point on *BA* so that *BD:DA* = 3:2.

Express, in terms of **a** and **b**:

(c) $\vec{BD}$ (d) $\vec{CD}$

What can you conclude about *CD* and *OA*?

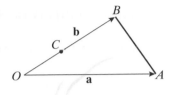

14 *OADEFC* is a regular hexagon and *B* is the point of
intersection of the diagonals. $\vec{OA} = $ **a**, $\vec{OB} = $ **b** and $\vec{OC} = $ **c**.
Express, in terms of **a**, **b** or **c** where appropriate, the vector $\vec{OD}$.
Find two different expressions for the vector $\vec{OE}$. Write down
an equality involving **a**, **b** and **c**.

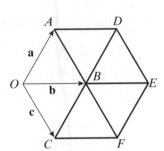

15 *OABC* is a parallelogram with $\vec{OA} = 4$**a** and $\vec{OB} = 4$**b**. The
diagonals intersect at *D*. *P*, *Q*, *R* and *S* are the mid-points of
OD, *AD*, *DB* and *DC* respectively.

Show, using vector algebra, that *PQRS* is a parallelogram.

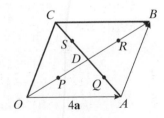

16 $\vec{OA} = $ **a**, $\vec{OB} = $ **b** and $\vec{OC} = $ **c**. *N* is the mid-point of *OB* and *M*
is the mid-point of *AC*.

Express:

(a) $\vec{AB}$ in terms of **a** and **b**

(b) $\vec{ON}$ in terms of **b**

(c) $\vec{AC}$ in terms of **a** and **c**

(d) $\vec{AM}$ in terms of **a** and **c**

(e) $\vec{OM}$ in terms of **a** and **c**

(f) $\vec{NM}$ in terms of **a**, **b** and **c**.

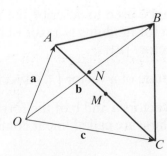

17 $\vec{OA} = \mathbf{a}$ and $\vec{OB} = \mathbf{b}$. X is the point such that $\vec{OX} = \frac{3}{2}\vec{OA}$ and Y is the point such that $\vec{OY} = 3\vec{OB}$.

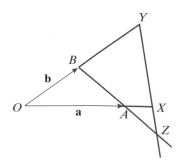

Express, in terms of $\mathbf{a}$ and $\mathbf{b}$:

(a) $\vec{OX}$ (b) $\vec{OY}$ (c) $\vec{YX}$

The lines YX and BA are extended to meet at Z.

Explain why the position vector of any point on YX extended can be written as $3\mathbf{b} + p\left(\frac{3}{2}\mathbf{a} - 3\mathbf{b}\right)$ and that the position vector of any point on BA extended can be written as $\mathbf{b} + q(\mathbf{a} - \mathbf{b})$.

Hence, or otherwise, find the position vector of Z.

18 $\vec{AB} = \begin{pmatrix} 1 \\ -4 \end{pmatrix}$, $A = (2, 3)$, $C = (-1, 15)$

Find the values of x and y if $x\vec{OA} + y\vec{OB} = \vec{OC}$.

Summary of key points

1 Translations can be described using column vectors $\begin{pmatrix} x \\ y \end{pmatrix}$.

An alternative notation which can be used is $\vec{AB}$, where A is the starting point and B is the finishing point.

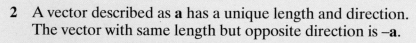

2 A vector described as $\mathbf{a}$ has a unique length and direction. The vector with same length but opposite direction is $-\mathbf{a}$.

3 For two vectors $\mathbf{a}$ and $\mathbf{b}$:

$$\mathbf{a} + \mathbf{b} = \mathbf{b} + \mathbf{a} \qquad \begin{pmatrix} 3 \\ 2 \end{pmatrix} + \begin{pmatrix} 11 \\ -3 \end{pmatrix} = \begin{pmatrix} 7 \\ -1 \end{pmatrix}$$

$$\mathbf{a} - \mathbf{b} = \mathbf{a} + (-\mathbf{b})$$

$$\mathbf{a} + (\mathbf{b} + \mathbf{c}) = (\mathbf{a} + \mathbf{b}) + \mathbf{c}$$

$$k\mathbf{a} = k \times \mathbf{a} \quad \text{(where } k \text{ is a scalar quantity)}$$

$$k(\mathbf{a} + \mathbf{b}) = k\mathbf{a} + k\mathbf{b}.$$

$$(p + q)\,\mathbf{a} = p\mathbf{a} + q\mathbf{a}$$

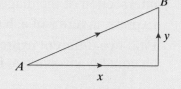
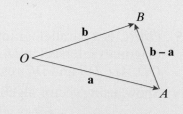

4 The magnitude of a vector is the length of the directed line segment representing it.

5 The term modulus can be used to describe the magnitude of a vector. The notation for modulus is a pair of bars written on each side of the vector: $|\mathbf{a}|$.

6 In general, the magnitude of a vector $\begin{pmatrix} x \\ y \end{pmatrix}$ is $\sqrt{x^2 + y^2}$.

7 Combinations of the vectors $\mathbf{a}$ and $\mathbf{b}$ of the form $p\mathbf{a} + q\mathbf{b}$, where p and q are scalars, are called linear combinations of the vectors $\mathbf{a}$ and $\mathbf{b}$.

8   The position vector of a point P is $\overrightarrow{OP}$, where O is usually the origin.

9   If A and B have position vectors **a** and **b** respectively, then the vector $\overrightarrow{AB} = \mathbf{b} - \mathbf{a}$

10 If A and B have position vectors **a** and **b** respectively, then the position vector of the midpoint, M, of the line joining A to B is $\overrightarrow{OM} = \mathbf{m} = \frac{1}{2}(\mathbf{a} + \mathbf{b})$.

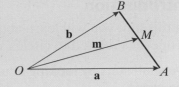

26 Circle theorems

Introduction

In this unit you will learn about calculating angles and using angle properties related to circles, and proof.

We shall start by looking at a collection of results known as **circle theorems**.

26.1 Chords and tangents

Theorem

■ **The perpendicular from the centre of a circle to a chord, bisects the chord.**

Proof

M is the point where the perpendicular from O meets the chord AB. In triangles OAM, OBM.

$OA = OB$ (radii)

OM is common to both

angle $O\hat{M}A$ = angle $O\hat{M}B$ = 90°

So triangles OAM and OBM are congruent (RHS).

So $AM = MB$.

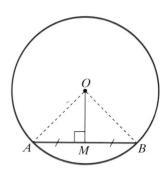

Example 1

Find the values of a and b.

$$\text{Angle } A\hat{O}B = 360° - 230° = 130°$$
$$a = \tfrac{1}{2} \text{ of } A\hat{O}B \text{ (symmetry)}$$

so
$$a = \tfrac{1}{2} \text{ of } 130°$$
$$a = 65°$$

In the triangle AMO:

$$b° + a° + 90° = 180°$$
$$b° = 180° - 90° - a°$$
$$b° = 90° - 65°$$
$$b° = 25°$$

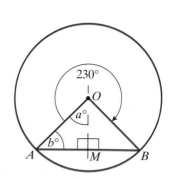

Example 2

The distance of the chord AB from the centre O is 5 cm. AB = 12 cm. Calculate the radius of the circle.

Using Pythagoras in triangle OAM, gives:

$OA^2 = 25 + 36$

$OA = \sqrt{61} = 7.81$ cm

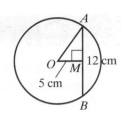

Theorem

■ **The angle between a tangent and a radius is 90°**

Draw lines parallel to AB. The radius OT intersects all of these chords at 90° even in the final, limiting case, where PT is a tangent.

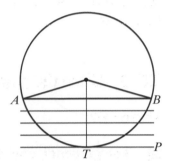

Example 3

Find angle $P\hat{T}Q$ ($x°$).

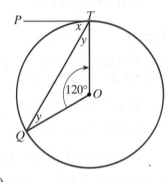

$$OT = OQ = \text{radius}$$

So triangle QOT is isosceles.

So   angle $O\hat{Q}T$ = angle $O\hat{T}Q$ (call each $y°$).

Then $y + y + 120° = 180°$ (angles in a triangle)

$2y + 120° = 180°$

$2y = 60°$

$y = 30°$

$x + y = 90°$ (PT is a tangent, OT a radius)

So $x = 90° - y$

$x = 90° - 30°$

$x = 60°$

Theorem

■ **The lengths of the two tangents from a point to a circle are equal.**

Proof

In triangles OPS, OPT:

$OS = OT$ (radii)

OP is common to both

angle $C\hat{S}P$ = angle $O\hat{T}P$ = 90° (tangent and radius)

So triangles OPS, OPT are congruent (RHS).

Therefore, $OS = PT$.

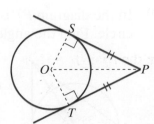

Exercise 26A

As you work through the exercises you may find it helpful to re-draw the given diagram and on your drawing fill in and label angles as you find them.

Approximate answers should be given correct to 3 s.f.

The centre of a circle is always O.

1 Find the value of x.

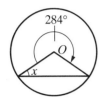

2 M is the midpoint of PQ.
Calculate angle $P\hat{O}M$ and angle $O\hat{P}M$.

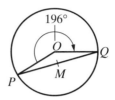

3 M is the midpoint of AB.
Find the lengths of AB and AN.

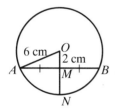

4 M is the midpoint of AC.
Find the lengths of AC and OM.

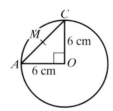

5 PT is a tangent. $OP = 8$ cm and $PT = 6$ cm.
Find the radius of the circle.

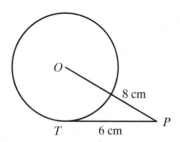

6 In the diagram, calculate SO and the area of $OSPT$.

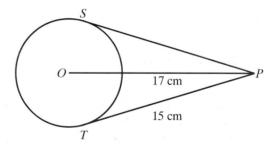

7 In the diagram PT is a tangent to the circle. Find the angles $O\hat{P}T$ and $O\hat{P}A$.

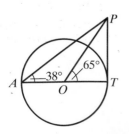

8 PT is a tangent. Find angle $P\hat{T}A$.

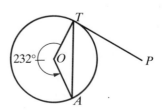

9 Calculate the length of *AB*.

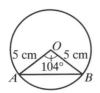

10 Calculate the length of *PQ*.

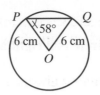

11 Calculate the size of angle $A\hat{O}B$.

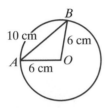

12 *M* is the midpoint of *AC*.
Calculate the sizes of angles $C\hat{O}B$ and $B\hat{O}M$.

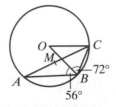

13 *TR* is a diameter. *PT* is a tangent.
Calculate the values of *a* and *b*.

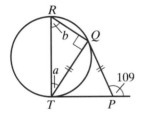

14 Calculate angle $O\hat{S}T$.

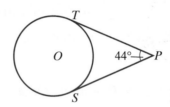

15 Calculate the length of *AB*.

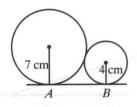

16 Calculate the size of angle $T\hat{X}S$.

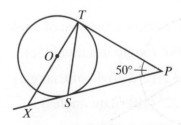

17 A bridge is built in the shape of the arc of a circle, centre *O*. The width, *AB*, is 12 m and the height is 4 m. Calculate the distance *AO*.

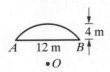

18 Draw accurately a triangle *ABC* which has all its angles acute. Draw the perpendicular bisectors of the sides *AC* and *AB*. These meet at the point *O*. Explain why *O* is the centre of the circle which passes through the 3 corners of the triangle.

26.2 Angle in a semicircle

Theorem

◼ **The angle in a semicircle is a right angle.**

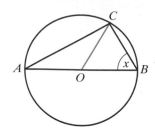

Proof

In triangle *BOC*:

angle $O\hat{C}B = x°$ (triangle *OCB* is isosceles with $OC = OB$)

angle $C\hat{O}B = 180° - 2x°$ (angle sum of triangle *OCB*)

In triangle *AOC*:

angle $A\hat{O}C = 2x°$ (angles on a straight line)

angle $A\hat{C}O = 90° - x°$ (triangle *AOC* is isosceles with $OC = OA$)

In triangle *ACB*:

angle $A\hat{C}B =$ angle $O\hat{C}B +$ angle $O\hat{C}A = 90°$

Example 4

In the diagram, calculate the size of angle $A\hat{O}X$.

Angle $B\hat{D}A = 90°$ (angle in semicircle)

Angle $B\hat{A}D = 32°$ (3rd angle of triangle)

Angle $A\hat{O}X = 32°$ (alternate angles)

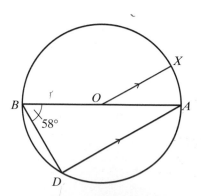

Exercise 26B

The centre of each circle is labelled O.

1 XY is a diameter.
Calculate the angles (i) $X\hat{O}Z$, (ii) $O\hat{Z}Y$
and (iii) $O\hat{Z}X$. Show that $X\hat{Z}Y = 90°$.

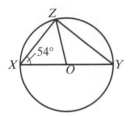

2 AB is a diameter.
Calculate the size of the angles
$C\hat{B}A$ and $C\hat{O}B$.

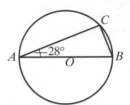

3 M is the midpoint of AC.
Calculate the size of angle $M\hat{O}C$.

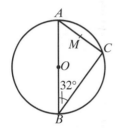

4 AB is a diameter.
Angle $A\hat{B}C = 54°$ and angle $A\hat{B}D = 32°$.
Calculate the sizes of angles $C\hat{X}D$ and
$C\hat{A}D$.

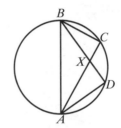

5 TQ is a diameter. PT is a tangent. Find, in
terms of x, (i) $Q\hat{P}T$, (ii) $R\hat{T}P$.

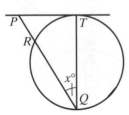

6 AB is a diameter.
Calculate the area of triangle ABD.

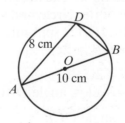

7 AB is a diameter.
Calculate the value of y.

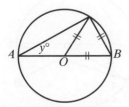

8 AB is a diameter.
Calculate the size of angles $A\hat{C}D$ and
$C\hat{O}D$.

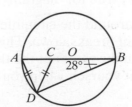

9 *AB* is a diameter. Calculate the size of angle $A\hat{B}C$.

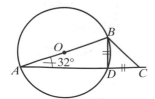

10 *PR* is a diameter. Calculate the area of *PQRS*.

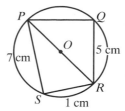

26.3 The angle at the centre

■ **The angle subtended at the centre is twice the angle at the circumference.**

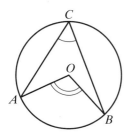

Proof

Angle $A\hat{O}B = 2 \times$ angle $A\hat{C}B$.

Look at the circle on the right. The arc *AB* of the circle subtends the angle $A\hat{O}B$ at the *centre* of the circle and subtends the angle $A\hat{C}B$ at the *circumference*. Angle $A\hat{O}B$ is split into two parts of sizes $2x$ and $2y$ respectively, so that $A\hat{O}B = 2x + 2y$.

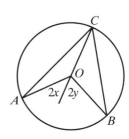

Now angle $A\hat{O}C = 180° - 2x$.
Since triangle *AOC* is isosceles, angle $O\hat{A}C =$ angle $O\hat{C}A = x$.

Similarly, angle $O\hat{B}C =$ angle $O\hat{C}B = y$.
So angle $A\hat{C}B = x + y$.
That is, angle $A\hat{O}B = 2 \times$ angle $A\hat{C}B$.

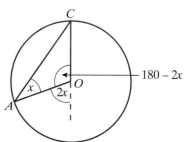

Since nothing has been assumed about the values of *x* and *y*, this result must be generally true for diagrams like the one above.

However, as *C* moves around the circumference, the diagram alters and the above argument breaks down. This happens when *C* gets closer to *B* as shown on the right but the theorem is still true. You can demonstrate it yourself when you do question 11 in Exercise 26C.

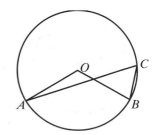

It also happens when C is on the *minor* arc AB as shown here.
In a similar way to the first demonstration, split the reflex angle
in parts of size $2x$ and $2y$.
In triangle $A\hat{O}C$, angle $A\hat{O}C = 180° - 2x$.
Since triangle $A\hat{O}C$ is isosceles,
angle $O\hat{A}C$ = angle $O\hat{C}A = x$.
Similarly, angle $O\hat{C}B = y$.
So angle $A\hat{C}B = x + y$ and thus
angle $A\hat{O}B = 2 \times$ angle $A\hat{C}B$.

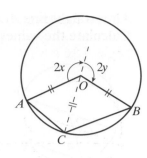

Thus, in all three cases, the theorem holds.

Example 5

Calculate the size of angles $A\hat{C}B$ and $A\hat{D}B$.

$$\text{Angle } A\hat{C}B = \tfrac{1}{2} \text{ angle } A\hat{O}B = 40°.$$
$$\text{Reflex angle } A\hat{O}B = 360° - 80° = 280°.$$
$$\text{Angle } A\hat{D}B = \tfrac{1}{2} \text{ reflex angle } A\hat{O}B = 140°.$$

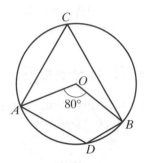

Exercise 26C

The centre of each circle is labelled O.

1 Angle $A\hat{C}B = 48°$.
Find angles $A\hat{O}B$ and $A\hat{B}O$.

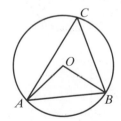

2 Angle $A\hat{O}B = 74°$.
Find angles $A\hat{C}B$ and $A\hat{D}B$.

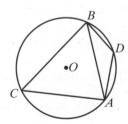

3 Angle $A\hat{O}B = 126°$. Angle $O\hat{A}C = 51°$.
Find angles $A\hat{C}B$ and $O\hat{B}C$.

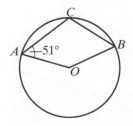

4 ABC is an isosceles triangle.
Angle $A\hat{O}B = 102°$.
Find the size of angle $O\hat{B}C$.

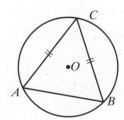

5   The straight line *AOB* is parallel to *DC*.
    Calculate the values of *a*, *b* and *c*.

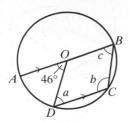

6   Calculate the sizes of angles $A\hat{D}B$ and $O\hat{D}A$.

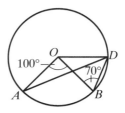

7   Calculate the sizes of angles $A\hat{B}C$ and $A\hat{D}C$.

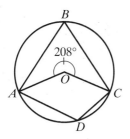

8   Calculate the size of angle $Q\hat{S}R$.

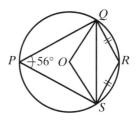

9   *OB* = *BC*.
    Calculate the size of angle $C\hat{A}B$.

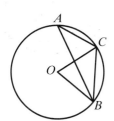

10 Calculate the size of angle $A\hat{D}C$.

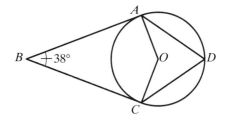

11 Show from first principles that angle
    $A\hat{C}B = 90° - x° - y°$.
    (This is a proof of the theorem in case 2.)

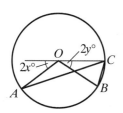

26.4 Angles in the same segment

■ **Angles in the same segment are equal.**

Proof

Angle $A\hat{C}B$ = angle $A\hat{D}B$

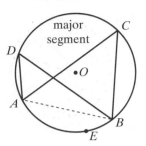

The work on loci, page 145, showed this result informally. Now we will look at it again formally and rigorously.

The diagram shows the arc AB making angles $A\hat{C}B$ and $A\hat{D}B$ at the circumference. Angles $A\hat{C}B$ and $A\hat{D}B$ are in the same (major) segment of the circle. The angle $A\hat{E}B$ stands in the opposite segment of the circle. For this section, just consider angles in the same segment. You can find out about angles in opposite segments in the section on cyclic quadrilaterals on page 534.

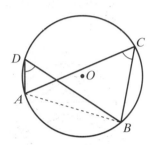

Using the result of **Section 28.3**:

$$\text{angle } A\hat{C}B = \tfrac{1}{2} \text{ angle } A\hat{O}B \text{ and}$$
$$\text{angle } A\hat{D}B = \tfrac{1}{2} \text{ angle } A\hat{O}B$$
$$\text{So angle } A\hat{C}B = \text{ angle } A\hat{D}B.$$

Example 6

Angle $A\hat{B}C$ = 102°, angle $C\hat{A}B$ = 31°.
Calculate the values of a and b.

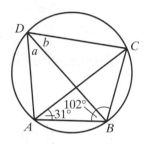

$$\text{Angle } A\hat{C}B = 47° \text{ (angle sum of a triangle)}$$
$$a = 47° \text{ (angles subtended by the arc } AB)$$
$$b = 31° \text{ (angles subtended by the arc } BC)$$

Exercise 26D

1 Find the sizes of angles $C\hat{B}D$ and $A\hat{D}B$.

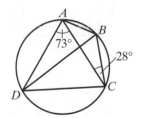

2 Find the sizes of angles $A\hat{C}B$ and $C\hat{B}D$.

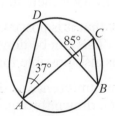

3 Find the sizes of angles $Q\hat{S}R$ and $Q\hat{P}R$.

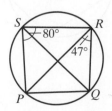

4 Find the sizes of angles $E\hat{A}C$ and $B\hat{E}A$.

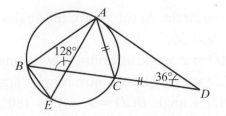

5   Find the sizes of the angles
    $A\hat{B}D, A\hat{C}D$ and $A\hat{E}D$.

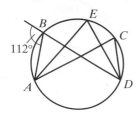

6   AB is a diameter. Find the sizes of the
    angles $A\hat{B}D, A\hat{C}D, A\hat{O}D$ and $A\hat{D}C$.

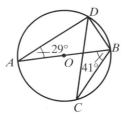

Questions 7–9 are based on the diagram below.

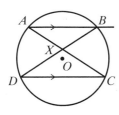

7   Explain why triangle ABX is isosceles.

8   Given that angle $B\hat{O}C = 78°$, find angle
    $A\hat{C}D$.

9   Given that angle $B\hat{O}C = 82°$ and that
    angle $X\hat{B}C = 68°$, find angle $A\hat{C}B$.

10 Draw a straight line AB of length 5 cm. Construct an isosceles
    triangle ABP, so that angle $A\hat{P}B = 60°$. Now find other
    positions of P such that angle $A\hat{P}B = 60°$. Describe the locus of
    P, as the position of P varies, subject to the angle $A\hat{P}B = 60°$.

    Investigate the locus of P for different values of angle $A\hat{P}B$.

2 minutes thought: can you
draw a circle through the
four corners of any
quadrilateral? The answer
to this can be seen from one
of the diagrams in this page.

26.5 Cyclic quadrilaterals

■   **Opposite angles of a cyclic quadrilateral are supplementary.**

Two angles are **supplementary** if their sum is 180°.

You have seen previously that it is possible to draw a circle through
each corner of a triangle. This can be done for any triangle. The
diagram shows that this is also possible for rectangles.

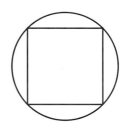

Points which lie on the circumference of the same circle are called
concyclic. If a circle can be drawn through all four corners of a
quadrilateral, then the quadrilateral is known as **cyclic**. You are now
going to work through the derivation of the condition for a
quadrilateral to be cyclic.

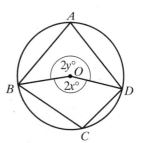

Look at the circle. At the centre the angles are $2x°$ and $2y°$.

   $2x + 2y = 360°$

angle $BAD = x°$ angle at centre is twice angle at circumference

angle $BCD = y°$ angle at centre is twice angle at circumference

angle BAD + angle $BCD = x° + y° = 180°$.

Since the sum of the angles of any quadrilateral is 360°,
then angle $C\hat{B}A$ + angle $C\hat{D}A = 180°$.

If the opposite angles of a quadrilateral are supplementary, a circle
can be drawn through all four of its corners.

Example 7

Find the sizes of angles $A\hat{B}C$ and $B\hat{A}C$.

Angle $A\hat{B}C = 146°$ (opposite angles of a cyclic quadrilateral)

Angle $B\hat{A}C = 17°$ (triangle ABC is isosceles)

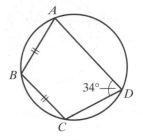

Exercise 26E

1   Calculate the sizes of angles $A\hat{D}C$
    and $C\hat{D}X$.

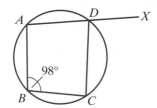

2   Calculate the sizes of angles $C\hat{D}E$
    and $A\hat{C}D$.

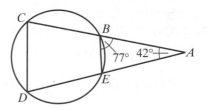

3   Calculate the sizes of angles $T\hat{P}R$
    and $T\hat{Q}R$.

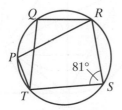

4   Calculate the sizes of angles $A\hat{C}B$
    and $C\hat{A}B$.

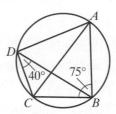

5   Find the value of angle $P\hat{Q}R$.

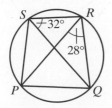

6   Calculate the values of angles
    $D\hat{O}B$, $D\hat{A}B$ and $O\hat{B}A$.

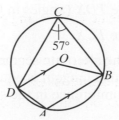

7 Find the values of angles $B\hat{D}C, A\hat{B}C$ and $D\hat{E}C$.

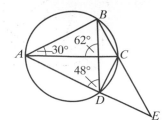

8 Find the sizes of the other three angles in the quadrilateral.

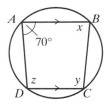

9 Find the values of the angles $D\hat{E}B$ and $A\hat{F}E$.

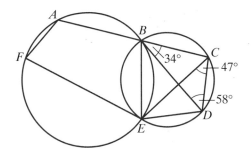

10 Angle $E\hat{C}D = 80°$. Explain why $AEDB$ is a cyclic quadrilateral. Calculate the size of angle $E\hat{D}A$.

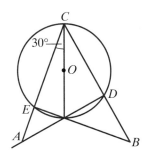

26.6 The alternate segment theorem

■ **The angle between a tangent and its chord is equal to the angle in the alternate segment.**

Proof

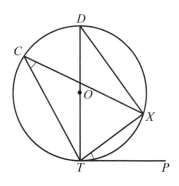

Look at the diagram.
TD is the diameter of the circle, so angle $T\hat{X}D = 90°$.

Angle $T\hat{D}X$ + angle $D\hat{T}X = 90°$ (angle sum of triangle is 180°).

PT is a tangent so angle $P\hat{T}D$ is 90°.

Thus angle $P\hat{T}X$ + angle $D\hat{T}X = 90°$.

Hence angle $P\hat{T}X$ = angle $T\hat{D}X$.

C is any other point on the circumference.

Now angle $T\hat{C}X$ = angle $T\hat{D}X$ (angles in the same segment).

Thus angle $P\hat{T}X$ = angle $T\hat{C}X$.

The theorem is called the alternate segment theorem because the angle a chord makes with a tangent is equal to the angle subtended by the chord at the circumference in the segment opposite to the tangent.

Example 8

Calculate the sizes of angles $T\hat{Q}X$ and $T\hat{X}Q$.

Angle $T\hat{Q}X = 36°$ (alternate segment)

Angle $T\hat{Q}X = 72°$ (angle at centre = twice angle at circumference)

Angle $T\hat{X}Q = 72°$ (TXQ is isosceles)

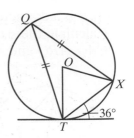

Exercise 26F

For diagrams **1–5**, calculate the value of the small letters

1

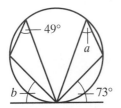

2

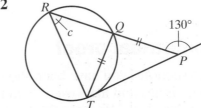

3

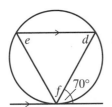

4

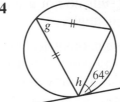

5

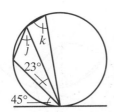

6 Calculate the sizes of angles $T\hat{B}A$ and $T\hat{C}A$.

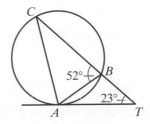

7 Given that AFD and FSG are tangents to both circles, and BSD and ASE are straight lines, calculate the sizes of angles $B\hat{A}S$, $S\hat{D}E$ and $S\hat{A}F$.

8 Calculate the sizes of angles $R\hat{T}P$ and $O\hat{X}T$.

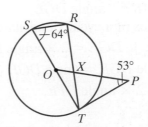

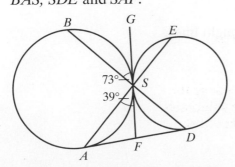

9 APB is a diameter of both circles. Calculate the sizes of the angles $C\hat{A}Q$, $A\hat{R}Q$ and $A\hat{B}C$.

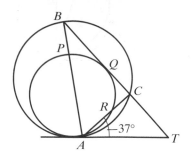

10 PT and PS are the tangents to a circle from the exterior point P. Given that $T\hat{P}S = 70°$, calculate the interior angles of the triangle QST.

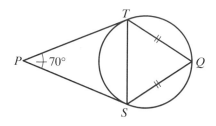

26.7 Geometrical proof

In the previous sections, you have been shown how to derive theorems about circles and have used them to work out values of angles in particular cases. An extension of this is to deal with more general cases where the results do not usually depend upon the values of a particular, given, angle.

It is important when writing geometrical proofs that you supply reasons for the steps that you take. This usually means stating the well-established theorems as justification for the steps in your proof.

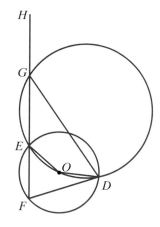

Example 9

O is the centre of the smaller circle. FEGH is a straight line.

Prove that:

(a) Angle $D\hat{G}H = 2 \times$ angle $D\hat{F}E$.

(b) $FG = DG$.

Answer

(a) In the small circle, angle $D\hat{O}E = 2 \times$ angle $D\hat{F}E$ (angle at centre = twice angle at circumference).

In the large circle,
angle $D\hat{O}E = 180° -$ angle $D\hat{G}E$
(opposite angles of a cyclic quadrilateral).
Thus $180° -$ angle $D\hat{G}E = 2 \times$ angle $D\hat{F}E$.
But $180° -$ angle $D\hat{G}E =$ angle $D\hat{G}H$ (FEGH is a straight line).
Angle $D\hat{G}H = 2 \times$ angle $D\hat{F}E$.

(b) Let angle $D\hat{F}E = x$.
Then from (a) angle $D\hat{G}E = 180° - 2x$.
Thus angle $F\hat{D}G = x$ (angle sum of triangle = 180°)
and $FG = DG$ (DFG is an isosceles triangle).

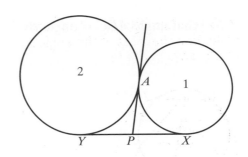

Example 10

Prove that:

(a) The tangent at *A* bisects the line *XY*.

(b) Angle $X\hat{A}Y = 90°$.

Answer

(a) $PX = PA$ (equal tangents to circle 1)

$PY = PA$ (equal tangents to circle 2)

P is the mid-point of *XY*.

(b) Triangle *XAY* consists of triangles *XPA* and *YPA*.

Let angle $P\hat{A}X = x$.

Then angle $P\hat{X}A = x$ (isosceles triangle *XPA*)
and angle $X\hat{P}A = 180° - 2x$.

Similarly, letting angle $P\hat{A}Y = y$, angle $A\hat{P}Y = 180° - 2y$.

Since *XPY* is a straight line, angle $A\hat{P}Y$ + angle $X\hat{P}A = 180°$.

Thus $2x + 2y = 180°$ and hence $x + y = 90°$.

Note, a quicker proof is to note that since $PX = PY = PA$, then
a circle can be drawn through *X*, *A* and *Y*, with *P* as centre.
Since *XY* is the diameter of the circle, then angle $X\hat{A}Y = 90°$
(angle in a semi-circle).

Exercise 26G

1 *O* is the centre of the circle. *AOC* is a
straight line. *TB* and *TC* are tangents.
Prove that the triangles *AOB* and *BTC*
are similar.

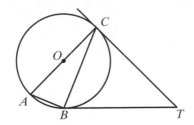

3 *AB* is parallel to *CD*. Prove that
(a) angle $A\hat{E}C = 2 \times$ angle $A\hat{B}C$
(b) $AE = BE$.

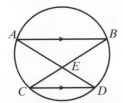

2 *PQ* and *RS* intersect at right angles. *PY* is
a tangent. *RXSY* is a straight line so that
$YX = XR$. *QSZ* is a straight line. Prove
that:
(a) angle $Y\hat{P}X =$ angle $X\hat{S}Q$
(b) angle $P\hat{Z}Q = 90°$

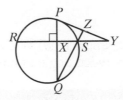

4 *ADX* and *BCX* are straight lines.
The straight line *XFE* bisects angle
$A\hat{X}B$. Prove that angle $D\hat{F}E =$ angle $F\hat{E}A$.

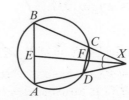

5   Prove that angle $A\hat{D}C$ = angle $B\hat{C}D$.

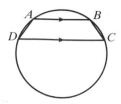

6   *ABCD* is a cyclic quadrilateral with *AB* = *AD*. Prove that the line *AC* bisects the angle $D\hat{C}B$.

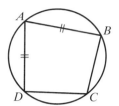

7   *ABCD* is a parallelogram. Prove that *AE* = *AD*.

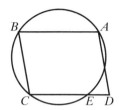

8   *ADE* is a straight line. *FD* bisects angle $E\hat{D}C$. Prove that *FB* bisects angle $A\hat{B}C$.

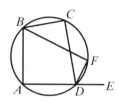

9   *AB* is parallel to the straight line *CDE*. *AF* bisects angle $D\hat{A}G$. Show that *AF* = *FE*.

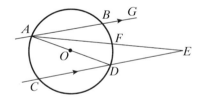

10   *C* is the point on the circumference such that *BT* bisects angle $C\hat{B}A$. Prove that angle $C\hat{O}A = 4 \times$ angle $A\hat{T}P$.

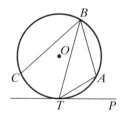

Exercise 26H (Mixed questions)

In questions **1–8**, calculate the values of *a* and *b*

1

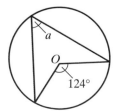

2

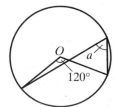

3

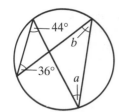

4

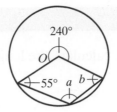

5

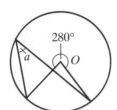

6

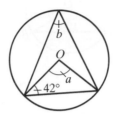

7

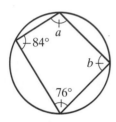

8

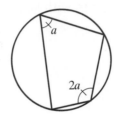

9 Calculate the lengths of:
(a) *AM*
(b) *AB*

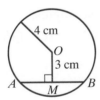

10 The radius of the circle is 25 cm.
Calculate the distance between *AB* and *CD*.

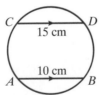

11 Calculate the radius of the circle.

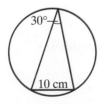

12 *AOB* is a diameter of the large circle. *OQ* is a tangent to the small circle. *OB* is a diameter of the small circle.

Prove that *P* is the mid-point of *BQ*.

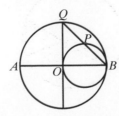

Exercise 26I

1   For each of the diagrams below, calculate the length of the
    chord *AB*

   (a)   **(b)**

   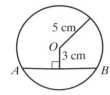   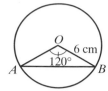

2   Angle $A\hat{B}C = 42°$.
    Name two other angles in the diagram with size 42°.

   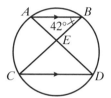

3   *QS* is a diameter. Triangle *PQS* is
    isosceles.
    Calculate the size of angle $P\hat{Q}R$.

   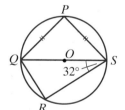

4   *PR* is a diameter of the semicircle. *PRY*
    is a straight line.
    Find the values of these angles:

   (a) $P\hat{Q}R$ **(b)** $Q\hat{R}P$ **(c)** $Q\hat{R}X$

   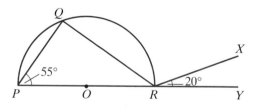

5   Calculate the sizes of these angles:
    (a) $C\hat{A}D$ **(b)** $A\hat{D}B$ **(c)** $A\hat{C}D$

   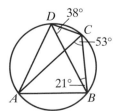

6   $A\hat{O}B$ is a diameter.
    Calculate the size of angle $A\hat{D}C$.

   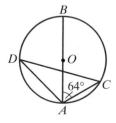

7 *AOB* is a diameter. Calculate the sizes of angles *OB̂C* and *OĈA*.

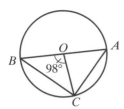

8 Calculate the radius of the circle and the length of the chord *PQ*.

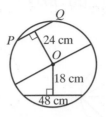

9 Triangle *ABC* is isosceles. Calculate the sizes of these angles:

   (a) *BÂD*   **(b)** *AĈB*   **(c)** *AD̂B*

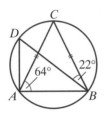

10 The straight lines *PSX* and *PTY* are tangents to the circle with centre *O*. Given that the radius of the circle is *r*, calculate the length of *PB* in terms of *r*.

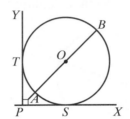

11 *PQRS* is a cyclic quadrilateral.
   PSB and *QRB* are straight lines. *PQA* and *SRA* are straight lines.
   Calculate the values of *x* and *y*.

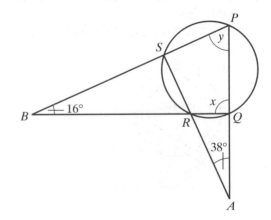

12 Triangle *AOB* is equilateral and *DA* is a tangent. Calculate the sizes of these angles:

   (a) *OÂB*   **(b)** *OÂD*
   (c) *BD̂A*   **(d)** *CÂB*
   (e) *AĈB*.

   Calculate the length of *AC*.

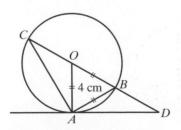

13 *TA* is a tangent to the circle, centre *O*. *DOB* is a diameter of the circle. Find the sizes of these angles:

    (a) $D\hat{B}A$ **(b)** $B\hat{D}A$ **(c)** $C\hat{B}D$

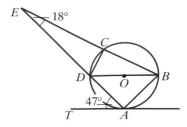

14 *AB* is a tangent to the circle. *DOE* is a diameter.
Find the values of these angles:

    (a) $D\hat{O}B$ **(b)** $B\hat{E}D$

Given that *AB* = 5.8 cm, calculate the radius.

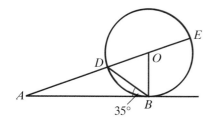

15 *TB* is a tangent to the circle, centre *O*. *AOC* is a diameter. Find the sizes of these angles:

    (a) $A\hat{B}T$ **(b)** $O\hat{A}B$ **(c)** $A\hat{E}D$

    (d) $A\hat{B}D$ **(e)** $C\hat{B}D$

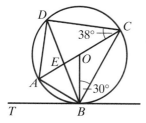

16 *SAT* is a tangent to the circle. *BC* is parallel to *ST*. Calculate the value of the angle $B\hat{A}C$.

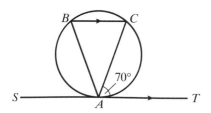

17 *TQ* is a tangent to the circle, centre *O*.
Calculate the size of angle *RQ̂T*.

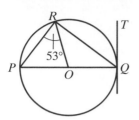

18 *BCE* and *ADE* are straight lines. *AB* and *DC* are parallel lines.
Calculate the sizes of these angles:

   (a) *AĈB*   **(b)** *DĈA*   **(c)** *CB̂A*

Show that triangle *EBA* is isosceles.

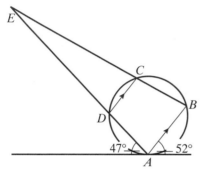

19 The circle, centre I, is the *inscribed* circle of triangle *ABC*.
Calculate the size of angle *ED̂F*.

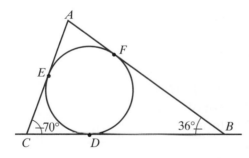

20 Calculate the sizes of angles *AĈB* and *CB̂A*. Give reasons for
your answers.

Prove that triangle *ABC* is isosceles.

Prove that triangles *PBR* and *ACR* are similar.

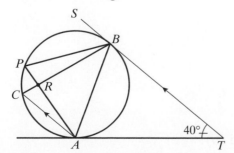

21 *SBT* is a tangent. Triangle *EAC* is isosceles. *BAE* and *CDE* are straight lines. Find the sizes of these angles:

(a) $C\hat{A}B$ (b) $A\hat{B}C$ (c) $E\hat{C}A$ (d) $E\hat{A}D$

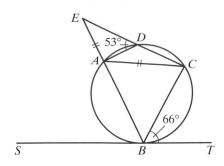

22 *C* is the centre of the large circle and *O* is the centre of the small circle. *PT* is a tangent to the small circle. *PCAO* is a straight line. The small circle has radius 1 cm. Calculate the radius of the large circle.

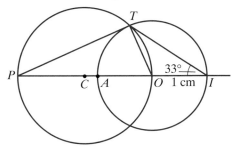

23 (a) Draw any acute-angled triangle. An **altitude** of a triangle is a straight line drawn from a corner to meet the opposite side at right angles. Draw the three altitudes of the triangle. Write what you notice.

(b) *AFB*, *AEC*, *BHE*, *AHD* and *FHC* are straight lines. *AOH* is a diameter. Find, in terms of *x* or *y* or both, the size of angle $B\hat{H}D$. Prove that *FBCE* is a cyclic quadrilateral. Hence, prove that *AH* is perpendicular to *BC*.

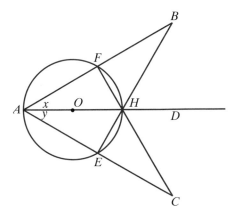

(c) What is the relationship between parts (a) and (b)?

Summary of key points

1 The perpendicular bisector of any chord passes through (bisects) the centre of the circle.

2 The angle between a tangent and a radius is 90°.

3 The tangents drawn to a circle from an exterior point are equal.

4 The angle in a semicircle is always a right angle.

5 The angle subtended at the centre is twice the angle at the circumference.

6 Angles in the same segment are equal.

7 Opposite angles of a cyclic quadrilateral are supplementary.

8 The angle between a tangent and its chord is equal to the angle in the alternate segment.

27 Histograms

27.1 Plotting data in a histogram

Equal-sized class intervals

This table shows how long an audience's applause lasted for 54 jokes told by a stand-up comedian:

Duration of applause in seconds	Frequency
0 to less than 5	8
5 to less than 10	12
10 to less than 15	15
15 to less than 20	10
20 to less than 25	9

The data can be displayed in a histogram like this:

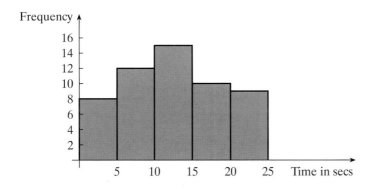

■ **In a histogram the areas of the rectangles are proportional to the frequencies they represent.**

In this case all the class intervals are the same (5 seconds) so the rectangles all have the same width and the height of each rectangle is proportional to its area and to the frequency.

Unequal class intervals

When the class intervals are unequal it is misleading to use heights to represent frequencies. For example, if the stop-watch used to time the applause is not accurate for times of less than 5 seconds the data can be recorded using different class intervals like this:

Duration of applause in seconds	Frequency
0 to less than 10	20
10 to less than 15	15
15 to less than 20	10
20 to less than 25	9

Here the first class interval is wider than the others.

We now construct a column which we call frequency density. We do this by dividing the frequency by class width.

Duration of applause in seconds	Frequency density
0 to less than 10	$20 \div 10 = 2$
10 to less than 15	$15 \div 5 = 3$
15 to less than 20	$10 \div 5 = 2$
20 to less than 25	$9 \div 5 = 1.8$

This histogram displays the data:

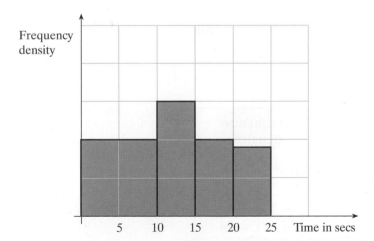

The areas of the rectangles are 20, 15, 10 and 9 respectively. They are in the correct proportion to the numbers of jokes (frequencies).

Notice that the vertical axis is labelled frequency density.

■ **Frequency density** $= \dfrac{\textbf{frequency}}{\textbf{class width}}$

550 Histograms

Standard class interval

Here is another example.

Juliet measured the weights of peas grown in her garden and grouped the weights into equal class intervals of 10 grams.

The first class interval (0 to less than 10) has a low frequency compared with the others and will not show up so clearly on a histogram with class intervals of equal width. It will show up more clearly if the first two class intervals are combined:

Weight in grams	Frequency
0 to less than 10	2
10 to less than 20	9
20 to less than 30	18
30 to less than 40	24
40 to less than 50	34
50 to less than 60	13

Weight in grams	Frequency
0 to less than 20	11
20 to less than 30	18
30 to less than 40	24
40 to less than 50	34
50 to less than 60	13

Here the first two class intervals from the previous table have been combined.

We could now simply calculate frequency densities by:

$$\text{frequency density} = \frac{\text{frequency}}{\text{class width}}$$

Note: if we take $\frac{11}{20}$ as the first frequency density it gives us 0.55.

However, there is an alternative to having decimals.

We define a **standard class interval**. This can be anything you wish, but in this case 10 grams is sensible. Then the weight 0 to less than 20 grams is 2×10 or 2 standard class intervals.

We now have:

Weight in grams	Weight in standard class intervals	Frequency	Frequency density
0 to less than 20	2	11	$11 \div 2 = 5.5$
20 to less than 30	1	18	$18 \div 1 = 18$
30 to less than 40	1	24	$24 \div 1 = 24$
40 to less than 50	1	34	$34 \div 1 = 34$
50 to less than 60	1	13	$13 \div 1 = 13$

Here the **frequency density** column has been calculated by:

$$\frac{\text{frequency}}{\text{class width in standard class intervals}}$$

The histogram below shows **frequency density** plotted against weight:

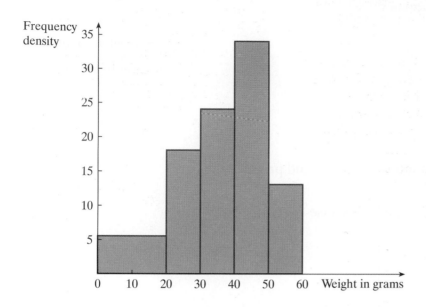

■ **An alternative for frequency density is:**

$$\frac{\textbf{frequency}}{\textbf{class width in standard class intervals}}$$

Here is an example of a set of data in which several class intervals have low frequencies, and one class interval has a frequency of 0:

Length in cm	Frequency
0 to less than 5	1
5 to less than 10	1
10 to less than 15	0
15 to less than 20	1
20 to less than 25	10
25 to less than 30	12
30 to less than 35	15
35 to less than 40	18
40 to less than 45	3
45 to less than 50	3

The class intervals at the top and bottom of the table have low frequencies. They can be regrouped like this:

Length in cm	Frequency
0 to less than 20	3
20 to less than 25	10
25 to less than 30	12
30 to less than 35	15
35 to less than 40	18
40 to less than 50	6

The results from 0 to 20 have been grouped together, as have the results from 40 to 50.

Here you set up a standard class interval of 5 cm and produce the table below.

0 to less than 20 becomes 0 to less than 4 because $20 \div 5 = 4$ and frequency density $= 0.75 = \frac{3}{4}$.

And 20 to less than 25 is

$$\frac{20}{5} = 4 \quad \text{and} \quad \frac{25}{5} = 5$$

So 20 to less than 25 becomes 4 to less than 5 and frequency density is:
$\frac{10}{1} = 10$

Length in cm standard class intervals	Class width in standard class intervals	Frequency	Frequency density
0 to less than 4	4	3	0.75
4 to less than 5	1	10	10
5 to less than 6	1	12	12
6 to less than 7	1	15	15
7 to less than 8	1	18	18
8 to less than 10	2	6	3

You calculate the frequency density thus:

$$\frac{\text{frequency}}{\text{class width in standard class intervals}}$$

(or 3rd column ÷ 2nd column).

The histogram of this data looks like this:

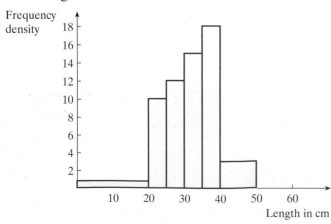

In a GCSE mathematics examination you could use either:

■ $$\text{frequency density} = \frac{\text{frequency}}{\text{class width}}$$

or

■ $$\text{frequency density} = \frac{\text{frequency}}{\text{class width in standard class intervals}}$$

unless you were told otherwise.

Handling continuous data

Sometimes data which is continuous is recorded as if it were discrete. You may wish to refer back to Unit 15, page 305.

Example 1

Agnes drives a van making deliveries for mail order catalogues. She records the distance travelled for each delivery correct to the nearest mile:

3	12	18	13	16	18	19	15	17	16
15	16	20	22	15	18	14	19	17	6
14	17	21	25	18	14	12	17	18	19
13	25	15	17	19	16	16	12	11	7
19	12	26	18	19	10	11	26	7	16
33	13	8	22	16	18	11	38	31	21

(a) Produce a frequency table using intervals 0–9, 10–14, 15–19, 20–24 and 25–39.

(b) Use your completed frequency table to draw a histogram.

As the distances are measured correct to the nearest mile the class widths are respectively 9.5, 5, 5, 5 and 15.

The 0 to 9 class is actually 0 to 9.5, width 9.5. The 10 to 14 class is actually 9.5 to 14.5, width 5 and so on.

(a) There is no advantage here in defining a standard class interval, so we will use:

$$\text{frequency density} = \frac{\text{frequency}}{\text{class width}}$$

This gives table:

Distance travelled in miles	Tally	Frequency	Frequency density
0 to 9	JHT	5	$5 \div 9.5 = 0.53$
10 to 14	JHT JHT IIII	14	$14 \div 5 = 2.8$
15 to 19	JHT JHT JHT JHT JHT IIII	29	$29 \div 5 = 5.8$
20 to 24	JHT	5	$5 \div 5 = 1$
25 to 39	JHT II	7	$7 \div 15 = 0.5$

(b) Using the frequency density, we can draw the histogram.

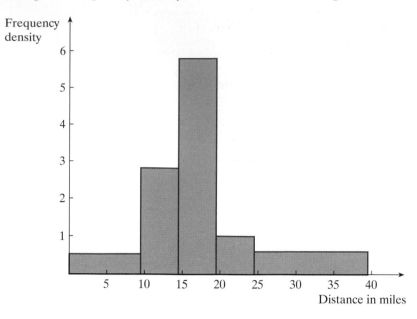

27.2 Interpreting histograms

The histogram below shows the distribution of the lifetimes of bees in a small colony. The table equivalent to the histogram is shown partially completed.

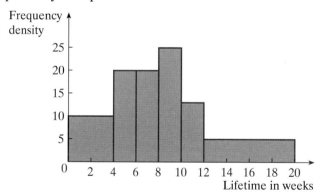

Lifetime (weeks)	Frequency density	Frequency
0 to less than 4	10	
4 to less than 6	20	
6 to less than 8	20	
8 to less than 10	25	
10 to less than 12	13	
12 to less than 20	5	

The frequency densities can be read straight from the histogram.

The frequencies are obtained by multiplying the frequency density by the class width. For the 0 to 4 class interval:

$$\text{frequency} = \text{frequency density} \times 4$$
$$= 10 \times 4 = 40 \text{ bees}$$

For the middle class intervals:

$$\text{frequency} = \text{frequency density} \times 2$$

For the last class interval:

$$\text{frequency} = \text{frequency density} \times 8$$
$$= 5 \times 8 = 40 \text{ bees}$$

Here is the completed table of frequencies:

Lifetime in (weeks)	Frequency density		Frequency
0 to less than 4	10	× 4 =	40
4 to less than 6	20	× 2 =	40
6 to less than 8	20	× 2 =	40
8 to less than 10	25	× 2 =	50
10 to less than 12	13	× 2 =	26
12 to less than 20	5	× 8 =	40

Example 2

Use this histogram to draw up a frequency table.

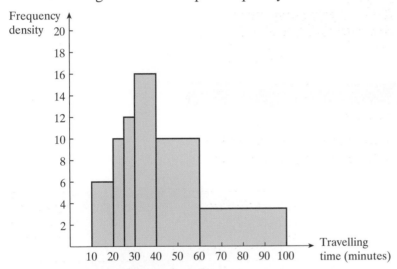

The standard class interval is 5 units, so the interval 10 to less than 20 is 10 units or 2 standard class intervals.

Here is the completed table:

Class interval	Frequency density	Frequency
10 to less than 20	6	6 × 2 = 12
20 to less than 25	10	10 × 1 = 10
25 to less than 30	12	12 × 1 = 12
30 to less than 40	16	16 × 2 = 32
40 to less than 60	10	10 × 4 = 40
60 to less than 100	3.5	3.5 × 8 = 28

A simpler method for giving frequencies is to draw a key to show what a standard area represents. In this histogram the key shows the size of the area which represents four people. This method can be used with all histograms.

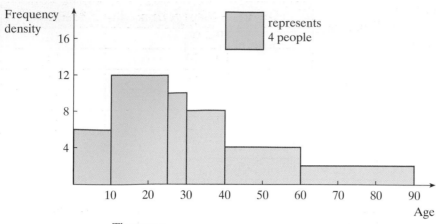

There are:
6 people aged between 0 and 10
18 people aged between 10 and 25
5 people aged between 25 and 30
and so on.

Exercise 27A

1   In a class survey of hand-lengths in centimetres these results were obtained:

20.1 13.8 17.9 14.6 20.3 16.3

21.6 17.9 18.3 17.7 21.8 18.4

22.1 20.3 18.7 18.7 23.1 19.1

21.7 21.0 21.8 18.4 19.7 18.7

21.5 22.3 20.9 17.1 18.8 19.6

20.2 23.1 15.8 18.2 22.0 20.8

21.5 22.1 19.9 20.7 21.6 22.0

 (a) Produce a frequency table with class intervals of 13–18, 18–20, 20–21, 21–22 and 22–25.

 (b) Draw a histogram to show this data.

2   This table shows the distribution of times required for subjects to memorize a list of 10 words:

Draw a histogram to display this data.

Time to memorize in secs	Frequency
0 to less than 30	6
30 to less than 60	8
60 to less than 80	12
80 to less than 90	14
90 to less than 100	10
100 to less than 150	6

3 Members of a group of people were asked to throw a tennis ball. The distances thrown were recorded in a table:

Draw a histogram to show this data, using a key of one square represents 2 throws.

Distance thrown in m	Frequency
0 – 9	18
10 – 14	14
15 – 19	16
20 – 24	15
25 – 34	17
35 – 50	10

4 Use this histogram to draw up and complete the frequency table beside it:

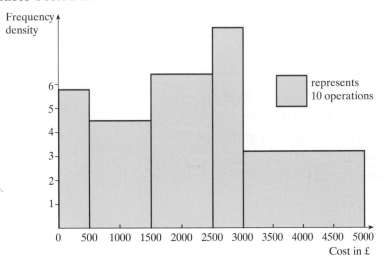

Class interval
0 – 500
501 – 1500
1501 – 2500
2501 – 3000
3001 – 5000

5 This histogram represents the number of spectators at professional football matches in England on one Saturday in 1993.

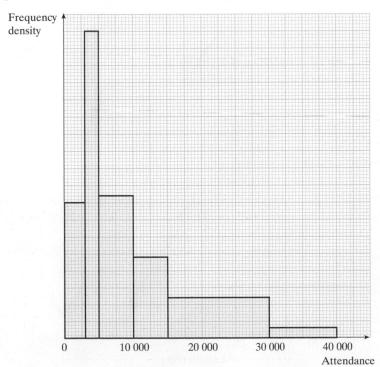

No match had more than 40 000 spectators.

At four matches the number of spectators was greater than or equal to 10 000 and less than 15 000.

(a) Use the information in the histogram to complete this frequency table:

Number of spectators (n)	Frequency
$0 \leqslant n < 3000$	
$3000 \leqslant n < 5000$	
$5000 \leqslant n < 10\,000$	
$10\,000 \leqslant n < 15\,000$	4
$15\,000 \leqslant n < 30\,000$	
$30\,000 \leqslant n < 40\,000$	

(b) Calculate the total number of professional football matches played in England on that Saturday.

6 This table shows the distribution of weights, in grams, of 30 portions of rice from a Take-Away:

Weight W (g)	Frequency
$0 \leqslant W < 30$	3
$30 \leqslant W < 50$	7
$50 \leqslant W < 60$	10
$60 \leqslant W < 70$	6
$70 \leqslant W < 90$	4

Draw a histogram to represent this information.

7 A student is studying the distribution of lengths of worms in a sample of topsoil from a field. The distribution of lengths is shown in this histogram:

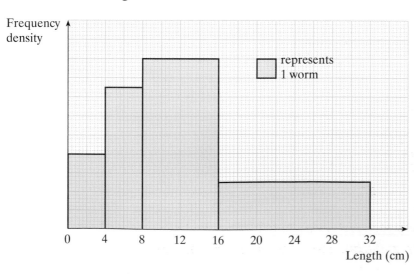

(a) Write down the number of worms with lengths less than 4 centimetres.

(b) Write down the number of worms with lengths greater than or equal to 8 centimetres and less than 16 centimetres.

(c) Calculate the total number of worms in the sample.

The area of the part of a field from which the student took her sample of topsoil was 10 m^2. The area of the whole field was 160 m^2.

(d) Estimate the number of worms in the topsoil of the whole field.

8 The unfinished histogram and table show information about the salaries, in pounds, of the teachers at Mathstown High School.

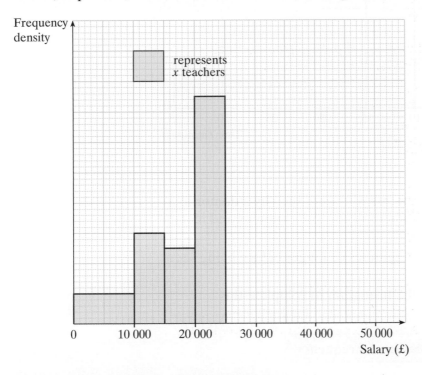

Salary (s) in pounds	Frequency
$0 \leqslant s < 10\,000$	4
$10\,000 \leqslant s < 15\,000$	6
$15\,000 \leqslant s < 20\,000$	5
$20\,000 \leqslant s < 25\,000$	...
$25\,000 \leqslant s < 30\,000$	8
$30\,000 \leqslant s < 50\,000$	4

(a) Calculate the value of x.

(b) Use the information in the histogram to complete the table.

(c) Use the information in the table to complete the histogram.

9   The waiting times for patients to be seen by a doctor after
     arriving at the accident department of a hospital during a
     weekend period were recorded. This histogram shows the
     results:

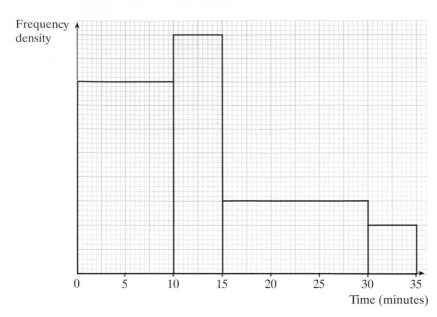

There were **exactly** 20 patients who were seen by a doctor in a
time which was greater than or equal to 10 minutes and less
than 15 minutes.

No patient had to wait 35 minutes or longer before being seen
by a doctor.

Use the information in the histogram to complete this
frequency table:

Waiting time in minutes (t)	Frequency
$0 \leqslant t < 10$	
$10 \leqslant t < 15$	20
$15 \leqslant t < 30$	
$30 \leqslant t < 35$	
$35 \leqslant t$	0

Summary of key points

1   In a histogram the areas of the rectangles are proportional
    to the frequencies they represent.

28 Introducing modelling

Simplifying a 'real-life' problem into a model which can be solved mathematically is called **mathematical modelling**. Estimating the height of a large waterfall using trigonometry is an example of mathematical modelling.

This unit shows how exponential functions and trigonometric functions may be used in such models. It also shows how experimental data can be analysed to obtain relationships between two variables.

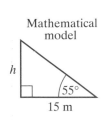

Mathematical model

h

$55°$

15 m

28.1 Modelling using exponential functions

2^x, $(0.5)^y$, 3^{2t} are all examples of exponential functions.

■ **The function a^x, where a is a positive constant and x is a variable, is called an exponential function.**

Example 1

Ben invests £2000 at Town Bank. Town Bank pays 9% per annum, added yearly. Ben does not intend to withdraw any interest.

(a) Find a formula which models this problem to give the value of the investment after t years.

(b) Use your formula to find when the investment is first worth more than £6000.

Answer

(a) Interest for the first year is $0.09 \times £2000$

Value of the investment after 1 year $= £2000 + 0.09 \times £2000$

$= (1 + 0.09) \times £2000$

$= 1.09 \times £2000$

$9\% = 0.09$

Interest for the second year $= 0.09 \times (1.09 \times £2000)$

Value of the investment after 2 years:

$= (1.09 \times £2000) + 0.09 \times (1.09 \times £2000)$

$= (1 + 0.09)(1.09 \times £2000)$

$= 1.09 \times (1.09 \times £2000)$

$= (1.09)^2 \times £2000$

Similarly, the value of the investment after 3 years:

$= (1.09)^3 \times £2000$

and the value of the investment after 4 years:

$= (1.09)^4 \times £2000$

This is a sequence which gives the value £V, of the investment after t years as the formula:

$$V = (1.09)^t \times £2000$$

(b) We need the least integer value of t so that $(1.09)^t \times £2000 > £6000$

$$\text{so } (1.09)^t > 3$$

Using the method of trial and improvement, try:

$t = 12$ $(1.09)^{12} = 2.8126...$ too small

$t = 13$ $(1.09)^{13} = 3.0658...$

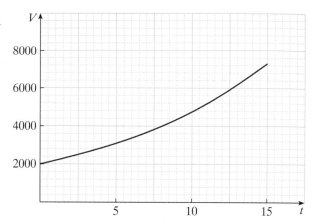

so the value of the investment is over £6000 for the first time after 13 years.

The graph shows the value of the investment £V after t years for Example 1.
This is an example of **exponential growth**.

The value of the investment is said to **grow exponentially** with a **multiplier** of 1.09.

Example 2

A student published an article in which she reported that the number of birds of prey nesting in an area had been decreasing by 8% per year since 1990. Assuming this same rate of decrease, find a formula for the number, N of these birds of prey that will nest in the same area t years after 1990, given that there were 600 in 1990.

Answer

Decrease of birds during 1991 $= 0.08 \times 600$

Number of birds in 1991 $= 600 - 0.08 \times 600$

$$= 0.92 \times 600$$

Decrease of birds during 1992 $= 0.08 \times (0.92 \times 600)$

Number of birds in 1992 $= (0.92 \times 600) - 0.08 \times (0.92 \times 600)$

$$= (1 - 0.08)(0.92 \times 600)$$

$$= (0.92)^2 \times 600$$

Similarly, the number of birds 3 years after 1990 $= (0.92)^3 \times 600$

the number of birds 4 years after 1990 $= (0.92)^4 \times 600$

so the number of birds t years after 1990 $= (0.92)^t \times 600$

$$N = (0.92)^t \times 600$$

The graph shows the number of nesting birds of prey N, t years after 1990.
This is an example of **exponential decay**.

The number of birds is said to **decay exponentially** with a **multiplier** of 0.92.

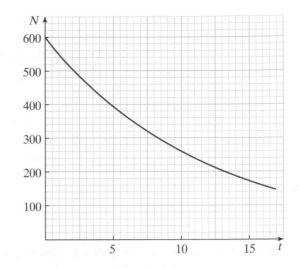

Exercise 28A

1 Rina invests £1500 at Shirebank. Shirebank pays 6% per annum, added yearly. Rina does not intend to withdraw any interest.

   (a) Find a formula which models this problem to give the value of the investment after t years.

   (b) Use your formula to find when the investment is first worth at least £3750.

2 A ball, dropped from a height x metres, rebounds to a height $0.6x$ metres. If the ball is dropped from a height of 8 metres, find:

   (a) a formula which models this problem to give the height of rebound after the nth bounce

   (b) the total distance the ball has travelled just before its third bounce.

3 The time taken for a mass of radioactive uranium to halve is 24 days. How long does it take for 40 mg to reduce to:

   (a) 10 mg

   (b) 2.5 mg.

4 Winston's debt of £40 000 is decreasing by 20% per year.

   (a) What is his debt after (i) 1 year (ii) 4 years (iii) n years?

   (b) After how many years will the debt first fall below £100?

28.2 Modelling using trigonometric functions

For problems where the motion repeats itself after a certain time, the functions sine and cosine are the most appropriate.

Unit 13 shows the graphs of these functions.

Example 3

A travelling fair comes to a Cheshire town every May.
The fair includes a big wheel which is constructed and tested.
The diameter of the wheel is 20 m.
Its centre is 12 m above the ground.
The wheel is first tested using one chair.
The wheel rotates anticlockwise and is timed for each revolution
once it reaches its working speed. The timing starts with the
chair moving upwards through the point level with the centre of
the wheel. The wheel rotates once every 36 seconds.

(a) Find the constants p and q so that:
$$y = p + q \sin (10t)°$$
is a suitable model for the height of the chair, y metres, above
the ground t seconds after timing starts.

(b) Find the times during the first minute when the chair is 17 m
above the ground.

(c) Sketch the graph of y against t for $0 \leqslant t \leqslant 54$.

Answer

(a) The diagram represents the big wheel with A, the position of
the chair when the timing starts and B, the highest point
reached by the chair shown.

$$y = p + q \ \sin (10t)°$$
When $t = 0, y = 12$ giving:
$$12 = p + q \sin 0°$$
$$12 = p + q \ (0)$$
$$12 = p$$

The chair reaches B when the wheel completes a $\frac{1}{4}$ of a
revolution.
The time taken is $\frac{1}{4} \times 36 = 9$ seconds.

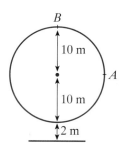

$$y = p + q \ \sin (10t)°$$
When $t = 9, y = 22$ giving:
$$22 = p + q \sin 90°$$
$$22 = p + q \ (1)$$
But $p = 12$, so:
$$22 = 12 + q$$
$$10 = q$$

(b) Using: $y = 12 + 10 \sin (10t)°$
When $y = 17, t = ?$ $17 = 12 + 10 \sin (10t)°$
$$5 = 10 \sin (10t)°$$
$$\sin (10t)° = 0.5$$

so $(10t)° = 30°, 180° - 30°, 360° + 30°, 540° - 30°, 720° + 30°$
$$t = 3, 15, 39, 51, 75$$

The required times are 3 secs, 15 secs, 39 secs and 51 secs.

Check the formula for y:
when $t = 18, y = 12$
when $t = 27, y = 2$
when $t = 36, y = 12$.

(c)

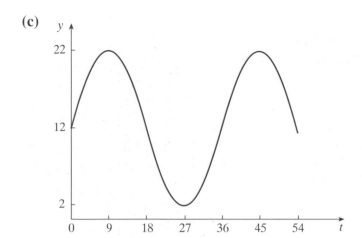

Another trigonometric modelling

In coastal areas the depth of the water depends on the time of the day.
The depth of the water is greatest at high tide.
The depth of the water is least at low tide.
The times when these two extremes occur are required by sailors.

This situation can be modelled using trigonometric functions.

TUGS WAIT FOR HIGH TIDE TO TRY TO REFLOAT FERRY

Example 4

Martin believes that the depth of water, d metres, at the end of a jetty, t hours after low tide can be modelled by a formula of the form:

$$d = a + b \cos (kt)°$$

where a, b and k are constants.

He measures the depth of water at low tide as 2 metres.

(a) Assuming that low tides occur every 12 hours, show that $k = 30$.

Martin also measures the depth of water at high tide as 6 metres.

(b) Calculate the values of a and b.

Martin needs at least 3 metres of water at the end of the jetty to sail his boat.

(c) Given that low tide on a particular day was at 09:00, find the earliest time that Martin could sail his boat.

Answer

(a) Low tide when t $= 0$ and $t = 12$

$$d = a + b \cos (kt)°$$

When $t = 0, d = 2$ $2 = a + b \cos 0°$ so $2 = a + b$

When $t = 12, d = 2$ $2 = a + b \cos (12k)°$ so $2 = a + b \cos (12k)°$

This gives: $\cos (12k)° = 1$

$$12k = 360$$

$$k = 30$$

$k \neq 0$ because the depth of water is not constant.

(b) High tide when $t = 6$

$$d = a + b \cos (30t)°$$

When $t = 6$, $d = 6$ $6 = a + b \cos 180°$ so $6 = a - b$.

but $2 = a + b$

so $a = 4$ and $b = -2$

(c) $d = 3, t = ?$

$$d = 4 - 2 \cos (30t)°$$

When $d = 3$ $3 = 4 - 2 \cos (30t)°$

$$2 \cos (30t)° = 1$$

$$\cos (30t)° = 0.5$$

$$30t = 60 \quad \text{so } t = 2$$

The **earliest** time that Martin can sail his boat is 11:00.

Exercise 28B

1   The depth, d metres, of water at the end of a jetty t hours after noon is modelled by the formula:

$$d = 4 + 2.5 \cos (30t)°$$

(a) Find the depth of water at:
(i) noon (ii) 2 pm (iii) 3 pm (iv) 6 pm (v) midnight.

(b) Find the first time, correct to the nearest minute, when the depth of water was 6 metres.

(c) Sketch the graph of d against t for $0 \leq t \leq 12$.

2   One end of a spring is fixed to a wall at point P. A mass M, which lies on a table, is attached to the other end. PM is horizontal.
Damian pushes the mass towards P and releases it.
He models the distance, y cm, of the mass from P at time t seconds after releasing it by the formula:

$$y = 15 - 5 \cos (45t)°$$

(a) Find the distance of the mass from the wall when:
(i) $t = 2$ (ii) $t = 4$ (iii) the mass is released.

(b) Sketch the graph of y against t for $0 \leq t \leq 8$.

3 t hours after midnight, the depth of water, d metres, at the entrance of a harbour is modelled by the formula:

$$d = 6 + 3 \sin (30t)°$$

   (a) What is the depth of water at (i) 1 am (ii) noon?

   (b) What is the depth of water at low tide?

   (c) Find the times of high tide during a complete day.

   (d) Sketch the graph of d against t for $0 \leqslant t \leqslant 24$.

4   The diameter of a big wheel is 16 m.
   Its centre is 9 m above thc ground. The wheel rotates clockwise.
   Mandy rides on the big wheel and starts to time it when her chair reaches its highest point.
   The wheel rotates once every 20 seconds.

   (a) Find the constants p and q so that:

$$y = p + q \cos (18t)°$$

   is a suitable model for the height of the chair, y metres, above the ground t seconds after timing starts.

   (b) Find the times during the first half minute when the chair is 13 m above the ground.

   (c) Find the times during the first half minute when the chair is 5 m above the ground.

   (d) Sketch the graph of y against t for $0 \leqslant t \leqslant 30$.

28.3 Using a line of best fit to obtain a relationship

Scientists frequently collect data from scientific experiments involving two quantities. These experiments are sometimes very costly so the scientists try to use the data to obtain a relationship between the two quantities.

They can then apply this relationship to other values of one of the variables and obtain results without having to carry out further experiments.

Experimental data is subject to errors in the measuring instruments so points are unlikely to lie exactly on a straight line when plotted.

A **line of best fit** (see Unit 4, page 89) is used.

Example 5

This scatter diagram shows the results of a scientific experiment involving two variables x and y.

The scientist has drawn in the line of best fit.

(a) Find the equation of the line of best fit.

(b) Assuming that this line is valid for larger values of x, find the value of y when $x = 52$.

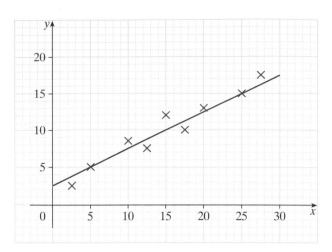

Answer

(a) (25,15) and (5,5) are two points on the line of best fit.

$$\text{Gradient } m = \frac{15-5}{25-5} = \frac{10}{20} = 0.5$$

Equation of the line takes the form $y = 0.5x + c$

$\quad$ (5,5) lies on the line so $5 = 0.5(5) + c \quad c = 2.5$

Equation of the line of best fit is $y = 0.5x + 2.5$

(b) $\qquad\qquad y = 0.5x + 2.5$

When $x = 52$, $\quad y = 0.5(52) + 2.5 = 26 + 2.5$

$\qquad\qquad$ so $y = 28.5 \quad$ when $x = 52$

You could use any two points on the line.

c can be read off from the graph as the y-intercept. In this example reading off between squares may not be very accurate.

Exercise 28C

These scatter diagrams show the results of experiments involving two variables x and y. The line of best fit is drawn on each.

1 (a) Find the equation of the line of best fit.

$\quad$ (b) Assuming that this line is valid for larger values of x, find the value of y when $x = 48$.

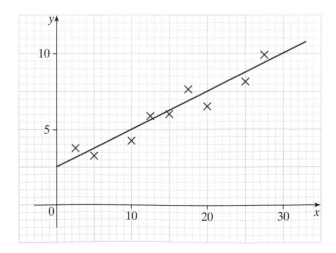

2   (a) Find the equation of the line of best fit.

     (b) Assuming that this line is valid for larger values of x, find the value of y when $x = 18$.

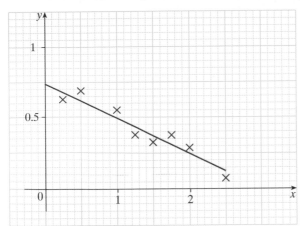

3   (a) Find the equation of the line of best fit.

     (b) Assuming that this line is valid for larger values of x, find the value of y when $x = 75$.

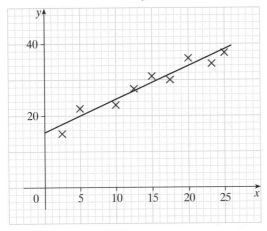

28.4 Reducing equations to linear form

Not all scientific data, when plotted as two variables, will lead to a straight line.

This section shows how to write some relationships in linear form. We shall use the general equation of the straight line as $Y = mX + c$, where Y and X are the two variables, m is the gradient of the line and c is the intercept on the y-axis.

We shall also use letters p and q to denote constants.

(Note: $Y = mX + c$ has three terms, two of which contain variables, that is Y and mX, and a third term, c, which does not contain a variable.)

Equations of the form $y = px^2 + q$

$$y = px^2 + q$$

has three terms, two with variables, one without.

variables   constant

$$\left. \begin{array}{l} y = px^2 + q \\ \text{Compare with } Y = mX + c \end{array} \right\} \quad \begin{array}{l} Y = y \ ; \\ m = p \ ; \end{array} \quad \begin{array}{l} X = x^2 \\ c = q \end{array}$$

If $Y = mX + c$ passes through the origin then $c = 0$.

◼ **Plotting y on the vertical axis and x^2 on the horizontal axis should lead to an approximate straight line if $y = px^2 + q$ is the relationship between y and x.**

The gradient of the line gives the value of p and the intercept on the y-axis gives the value of q.

Example 6

In an experiment these values of the variables V and R were obtained:

V	5	10	15	20	25
R	140	166	212	280	365

The variables V and R are thought to satisfy a relationship of the form $R = pV^2 + q$.

(a) Draw a graph to test this.

(b) Use your graph to estimate the values of the constants p and q.

(c) Use your relationship to find R when $V = 30$.

Answer

(a)

variables   constant

$$\left. \begin{array}{l} R = pV^2 + q \\ \text{Compare with } Y = mX + c \end{array} \right\} \quad \begin{array}{l} Y = R \ ; \\ m = p \ ; \end{array} \quad \begin{array}{l} X = V^2 \\ c = q \end{array}$$

Plotting R on the vertical axis and V^2 on the horizontal axis, should lead to an approximate straight line if $R = pV^2 + q$ is the correct relationship. The gradient of the line gives p and the intercept on the R-axis gives q.

V^2	25	100	225	400	625
R	140	166	212	280	365

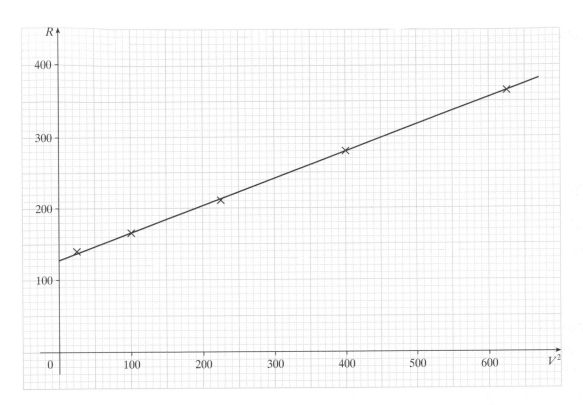

(b) Taking points **on the line** as (100, 166) and (400, 280):

the gradient, $m = p = \dfrac{280 - 166}{400 - 100} = \dfrac{114}{300} = 0.38$

and the y intercept, $q = 130$

so: $\quad R = 0.38V^2 + 130$

(c) When $V = 30$, $\quad R = 0.38(900) + 130$

$\qquad\qquad\qquad R = 472$

*You could take any two points **on the line**.*

From the equation of the line we see that $V^2 = 100$, $R = 168$. This discrepancy is due to the fact that 130 is an estimate for q.

Equations of the form $y = px^2 + qx$

$$y = px^2 + qx$$

All 3 terms contain variables so we cannot compare with $Y = mX + c$ directly.

Dividing each term of $\quad y = px^2 + qx \quad$ by x gives:

$$\left.\begin{array}{l} \dfrac{y}{x} = px + q \\[2mm] \text{Compare with:} \quad Y = mX + c \end{array}\right\} \quad \begin{array}{l} Y = \dfrac{y}{x} \; ; \quad X = x \\[2mm] m = p \; ; \quad c = q \end{array}$$

$y = px^2 + qx$

$\dfrac{y}{x} = \dfrac{px^2}{x} + \dfrac{qx}{x}$

$\dfrac{y}{x} = px + q$

■ Plotting $\frac{y}{x}$ on the vertical axis and x on the horizontal axis should lead to an approximate straight line if $y = px^2 + qx$ is the relationship between y and x.

The gradient of the line gives p and the intercept on the $\frac{y}{x}$-axis gives q.

Example 7

An object is fired vertically upwards and its height h metres, above the firing point, is recorded t seconds later. The table shows the results.

t	1	2	3	4	5	6
h	27	44	48	44	27	0

(a) Plot $\frac{h}{t}$ against t.

(b) Explain why your graph verifies that $h = pt^2 + qt$ and use your graph to estimate the values of the constants p and q.

(c) Use your relationship to find h when $t = 5.5$.

Answer

(a)

t	1	2	3	4	5	6
$\frac{h}{t}$	27	22	16	11	5.4	0

(b) The graph of $\frac{h}{t}$ against t gives a straight line so $\frac{h}{t} = pt + q$, where p is the gradient and q is the intercept on the vertical axis. Multiplying both sides of the equation by t leads to $h = pt^2 + qt$.

$$p = \text{gradient of the line} = \frac{-11}{2} = -5.5$$

$$q = \text{intercept on the vertical axis} = 33$$

(c) $$h = -5.5t^2 + 33t$$

When $t = 5.5$, $\quad h = -5.5(5.5)^2 + 33(5.5)$
$$= 15.125$$

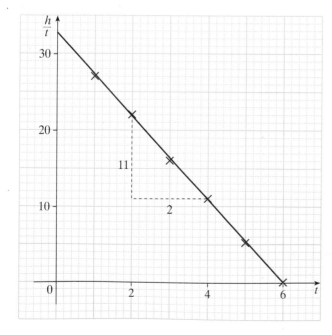

Exercise 28D

1

x	11	20	27	36	45
y	8	10.2	12.4	15	17

The table shows corresponding values of the variables x and y.
Peter believes that they satisfy a relationship of the form $y = ax + b$.

(a) Draw a graph to test if Peter is correct.

(b) Use your graph to estimate the values of the constants a and b.

2

x	1	2	3	4	5
y	3.5	9.3	19	33	52

y is approximately equal to $ax^2 + b$.

Plot y against x^2 and use the graph to estimate a and b.

3 Water is squirted horizontally from a hosepipe.
The height of the water is y metres at a distance x metres from
the hosepipe. Measurements of x and y are:

x	0	1	2	3	5	6
y	7.2	7.0	6.4	5.5	2.2	0.1

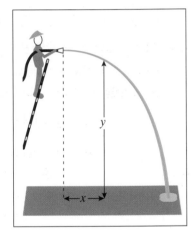

It is thought that the relationship between x and y is of the
form $y = ax^2 + b$.

Plot y against x^2 and use the graph to estimate a and b.

4

x	1	2	3	4	5
y	2.2	8.4	18.5	32.8	51.1

y is approximately equal to $ax^2 + bx$.

Plot $\frac{y}{x}$ against x and use the graph to estimate a and b.

5

x	1	2	3	4	5
y	2.5	4.1	4.6	4.0	2.5

y is approximately equal to $ax^2 + bx$.

Plot $\frac{y}{x}$ against x and use the graph to estimate a and b.

6

x	1	2	3	4	5
y	5.52	4.03	1.53	−1.96	−6.48

y is approximately equal to $ax^2 + b$.

Use a graph to estimate a and b.

28.5 Using points on a graph to find the constants in a given relationship where one of the variables appears as an index

To draw the graph of $y = 3^x$, we complete a table of values of x by finding the corresponding values of y.

x	0	1	2	3
y	1	3	9	27

We then plot these points and join them with a smooth curve.

The reverse process is also valid.

■ **If a point lies on a curve then the coordinates of the point satisfy the equation of the curve.**

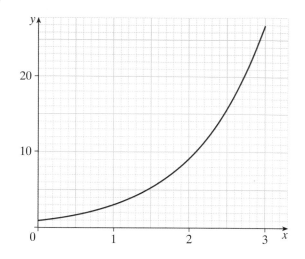

Example 8

This sketch shows part of the graph of $y = pq^x$.

It is known that the points $(0, 5)$, $(2, k)$ and $(3, 40)$ lie on this curve.

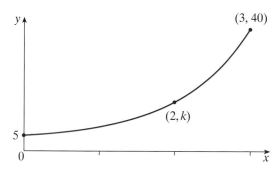

Use the sketch to find the values of p, q and k.

Point $(0, 5)$ lies on the curve $y = pq^x$ so $5 = pq^0$

$$5 = p(1) \quad \text{so} \quad p = 5$$

Point $(3, 40)$ lies on the curve $y = pq^x$ so $40 = pq^3$

$$40 = 5\,q^3 \quad \text{so} \quad q^3 = 8$$
$$q = 2$$

$$y = 5 \times 2^x$$

Point $(2, k)$ lies on the curve $y = 5 \times 2^x$ so $k = 5 \times 2^2$

$$k = 20$$

Exercise 28E

1 This sketch shows part of the graph of $y = pq^x$.
   Use the sketch to find the values of p, q and k.

2 This sketch shows part of the graph of $y = pq^x$.
   Use the sketch to find the values of p, q and k.

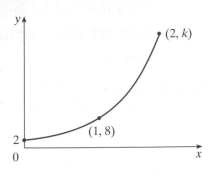

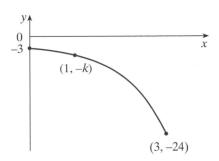

3 This sketch shows part of the graph of $y = a + b^x$.
   Use the sketch to find the values of a, b and k.

4 The point $(1, 3)$ lies on the curve $y = a^{-x}$.
   Calculate the value of a.

5 The point $(2, 2\frac{1}{4})$ lies on the curve $y = a^{-x}$.
   Calculate the two possible values of a.

6 The three points $(0, 5)$ $(1, 4\frac{1}{3})$ $(2, k)$ lie on the
   curve $y = a + b^{-x}$.
   Calculate the values of a, b and k.

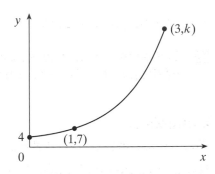

7 The point $(1, 8)$ lies on the curve $y = a^{-3x}$.
   Calculate the value of a.

8 **Challenge**
   The points $(2, 32)$ and $(5, 2048)$ both lie on the curve $y = pq^x$.

   (a) Find the values of the constants p and q.

   (b) Given also that the point $(k, 128)$ lies on this curve, find the
        value of k.

Summary of key points

1 The function a^x, where a is a positive constant and x is a
   variable, is called an exponential function.

2 **(a)** If $a > 1$ then ax is an example of exponential growth
        with a multiplier of a.

   (b) If $0 < a < 1$ then ax is an example of exponential decay
        with a multiplier of a.

3   A point lies on a curve if the coordinates of the point satisfy the equation of the curve.

4   To determine if the experimental results satisfy a given formula, reduce the formula to the form $Y = mX + c$.

Plot Y against X and if the points lie approximately on a straight line the formula is confirmed.

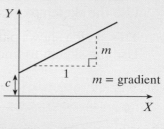

5   To test the formula $y = px + q$, plot y against x. If the points lie approximately on a straight line then p is the gradient of the line of best fit and q is the intercept on the vertical axis.

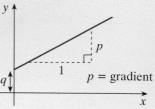

6   To test the formula $y = px^2 + q$, plot y against x^2. If the points lie approximately on a straight line then p is the gradient of the line of best fit and q is the intercept on the vertical axis.

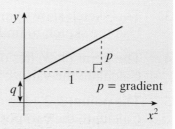

7   To test the formula $y = px^2 + qx$, plot $\frac{y}{x}$ against x.

If the points lie approximately on a straight line then p is the gradient of the line of best fit and q is the intercept on the vertical axis.

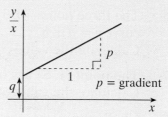

29 Calculators and computers

This unit shows you some ways of using scientific calculators, graphical calculators and computers to help you solve mathematical problems.

The examples will work on DAL calculators and most computers.
Your teacher will tell you if you need to change any of the instructions.

29.1 How well do you know your scientific calculator?

You can use your calculator to help you with fractions, decimals, percentages and money problems. Try these examples:

Example 1   fractions and decimals

Work out $\frac{2}{5} + 1\frac{1}{4}$.

Press the keys   **2** **ab/c** **5** **+** **1** **ab/c** **1** **ab/c** **4** **=**

Answer: $1\frac{13}{20}$.

Example 2   fractions and money

Work out $\frac{3}{8}$ of £6.

Press the keys   **3** **ab/c** **8** **×** **6** **=**

Press **ab/c** again to get the answer: £2.25.

Example 3   percentages

Work out 20% of 6 metres.

Press **6** **×** **2** **0** **SHIFT** **=**   Answer: 1.2 metres.

Example 4   percentage increase or decrease

Reduce £18 by 10%.

Press **1** **8** **×** **1** **0** **SHIFT** **=** **−**   Answer: £16.20.

You can make your calculator show pounds and pence correctly by using *Fix 2*.

This will always display two decimal places. Don't forget to return your calculator to *normal* mode after doing this.

> **Converting to fractions or decimals**
>   d/c
> Press **SHIFT** **ab/c** to convert your answer to an improper (top-heavy) fraction.
> Press **ab/c** again to convert the answer to a decimal.

Example 5　constant calculations

Work out:　　　　　£2.80 × 1.2

£8.90 × 1.2

£7.65 × 1.2

Press　　　**1** **.** **2** **×** **×** **2** **.** **8** **0** **=**

8 **.** **9** **0** **=**

7 **.** **6** **5** **=**

> Find out what happens when you press any of the keys **+** **−** **×** or **÷** *twice*.

Answers: £3.36,　　£10.68,　　£9.18.

Example 6　compound interest

A student borrows £600 to buy a computer. Interest is added to the loan at the rate of 5% per annum. How much does the student owe after 3 years if no repayments are made?

Each year the loan is increased by 5% to 105%, which is 1.05 times the previous value.

Press　　　**1** **.** **0** **5** **×** **×** **6** **0** **0** **=**

Now press　　　　　　**=** **=**　　Answer: £694.58.

> Pressing **=** multiplies the last answer by 1.05 each time.

Example 7　Squares and powers

(a)　Work out 47^2.

(b)　Work out 7^6.

(a)　Press　**4** **7** **x²**

The answer is 2209.

(b)　Press　**7** **xʸ** **6** **=**

The answer is 117 649.

> Try experimenting with different calculator keys.
>
> What do the **1/x** , **x^(1/y)** and **x!** keys do?

Example 8　brackets

Work out $\dfrac{7 + 12 + 5}{3}$

Press　**(** **7** **+** **1** **2** **+** **5** **)** **÷** **3** **=** Answer: 8.

Example 9　memory calculations

Find the total cost of the following items, and calculate the change from £10.

2 at £1.15,　　3 at 98p　　and 1 at £1.79

Press　**2** **×** **1** **.** **1** **5** **=** **Min**

3 **×** **0** **.** **9** **8** **=** **M+**

1 **.** **7** **9** **=** **M+**

MR　　The total is £7.03

1 **0** **−** **MR** **=**　　The change is £2.97

Example 10   trigonometry

Find angle $C\hat{A}B$ in degrees and the length of the hypotenuse AC in triangle ABC.

Press SHIFT TAN 5 ab/c 6 =   Answer: 39.8°.

Press √ (5 x² + 6 x²) =   Answer: 7.8102497

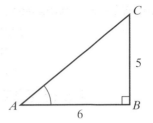

29.2 Using memories to represent formulae on a graphical calculator

Look at this problem:

This cone has a base radius R of 7 cm and a perpendicular height H of 24 cm. Find the area of the base, the slant height, the curved surface area, the total surface area, and the volume of the cone using the calculator memories.

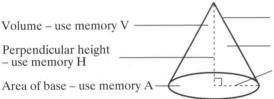

Volume – use memory V

Perpendicular height – use memory H

Area of base – use memory A

Use the L memory to represent the slant height

Curved surface area – use memory S

Radius – use memory R

Total surface area – use memory T

Example 11

It is useful, but not essential, to store each answer in memories R, H, A, L, S, T and V like this:

To get π you might have to press SHIFT first. To get R^2 you might have to press ALPHA R SHIFT x²

(a) store the number 7 in memory R

type 7 → R EXE

(b) store the number 24 in memory H

type 2 4 → H EXE

(c) multiply π by R^2 and store the answer in memory A

type π R x² → A EXE

(d) calculate $\sqrt{(R^2 + H^2)}$ and store the answer in memory L

type √ (R x² + H x²) → L EXE

(e) calculate $\pi \times R \times L$ and store the answer in memory S

type π R L → S EXE

(f) calculate $A + S$ and store the answer in memory T

type A + S → T EXE

(g) calculate $A \times H \div 3$ and store the answer in memory V

type A H ÷ 3 → V EXE

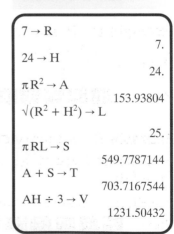

$7 \to R$

      7.

$24 \to H$

      24.

$\pi R^2 \to A$

    153.93804

$\sqrt{(R^2 + H^2)} \to L$

      25.

$\pi R L \to S$

   549.7787144

$A + S \to T$

   703.7167544

$A H \div 3 \to V$

   1231.50432

Example 12

(a) To find the area of a rectangle whose length L is 58 cm and whose width W is 27 cm, type:

`5` `8` `→` `L` `EXE`

`2` `7` `→` `W` `EXE`

Now type:

(i) `L` `W` `EXE` (ii) `W` `L` `EXE`

(iii) `L` `×` `W` `EXE` (iv) `W` `×` `L` `EXE`

(b) To find the perimeter of the rectangle type:

(i) `2` `L` `+` `2` `W` `EXE`

(ii) `2` `×` `L` `+` `2` `×` `W` `EXE`

(iii) `2` `(` `L` `+` `W` `)` `EXE`

(iv) `2` `×` `(` `L` `+` `W` `)` `EXE`

(a) All four answers are the same.

(b) All four answers are the same.

Entering formulae:
Formulae are typed exactly as you would write them. Multiplication signs are not required between letters or before brackets.

Exercise 29A

1 Calculate the area and perimeter of a rectangle 73 cm by 49 cm using any of the methods described above. Write down the sequence of keys as you press them.

2 Calculate the area and circumference of a circle whose radius is 12 cm. You should store the value 12 in memory R, then use the formulae:

$$\pi R^2 \to A \text{ and } 2 \pi R \to C$$

3 Calculate the area of a triangle whose base is 15 cm and whose height is 9 cm, using the formula:

$$(BH) \div 2 \to A$$

4 Calculate the volume of a rectangular box 120 cm by 60 cm by 20 cm using the formula:

$$L W H \to V$$

5 Calculate the volume of a cylinder whose base radius is 5 cm and whose height is 12 cm using the formula:

$$\pi R^2 H \to V$$

Write down the sequence of keys as you press them.

29.3 Saving formulae so you can use them again

If you want to use the same formula several times, you should save it as a **program**. Then you can use it again with different numbers.

Select the WRT mode on the calculator, and choose a program number, say Prog 3. Then type:

$\pi R^2 \rightarrow A$

Select the RUN mode on the calculator and then press:

0 **→** **R** **EXE** **Prog** **3** **EXE**

1 **→** **R** **EXE** **Prog** **3** **EXE**

2 **→** **R** **EXE** **Prog** **3** **EXE**

You can improve this program by adding more lines which tell you what to do:

'RADIUS'

$? \rightarrow R$

$\pi R^2 \rightarrow A$

To run this new program, type **Prog** **3** **EXE**.

When the ? appears type:

0 **EXE** **EXE**

1 **EXE** **EXE**

2 **EXE**

You can do the same thing by writing a computer program in BASIC:

```
10 PRINT 'RADIUS'
20 INPUT R
30 LET A = PI * R ^ 2
40 PRINT A
50 END
```

Find out how to write, run and clear programs on your graphical calculator.

```
10 → R
                    10.
Prog 3
             314.1592654
11 → R
                    11.
Prog 3
             380.1327111
```

? is used to input a number which is then stored in memory **R**.

Notice that **EXE** is pressed twice.

The first **EXE** enters the radius and **Prog** **3** produces the area.

The second **EXE** runs the program again.

```
RUN
RADIUS
?
10
314.1592654
RUN
```

Exercise 29B

1 Write a program for your graphical calculator or computer which will calculate the volume V of a cube whose sides are of length L.

Use it to find the volumes of cubes whose sides are 5 cm, 6 cm and 7 cm.

2 Write a program for your graphical calculator or computer which will convert a temperature in degrees Fahrenheit (°F) to degrees Celsius (°C) using the formula

$$C = (F - 32) \times \tfrac{5}{9}$$

Use it to convert 86 °F, 50 °F, 68 °F to Celsius.

3 Write a program for your graphical calculator or computer which will calculate the area A of a triangle using the formula:

$$A = (B \times H) \div 2$$

Use it to find the area of a triangle whose base is 15 cm and whose perpendicular height is 7 cm.

29.4 Solving problems and equations by trial and improvement

Look at this problem:

Find the radius of a circle whose area is 100 cm².

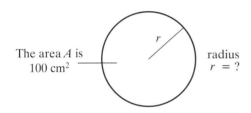

The area A is 100 cm²

radius $r = ?$

You can find the radius by guessing it and trying it in the formula:
$$A = \pi r^2$$

If at first you don't succeed, try, try and try again!

Sometimes you can solve problems like this by guessing a solution and trying it out.

You can improve your guess and try again until you get a good enough answer.

This method of solving problems is called trial and improvement.

You can use your calculator or computer to help you do this.

Example 13

Use a trial and improvement method to find the radius r of a circle whose area is 100 cm².

Your answer should be accurate to 2 d.p.

Method:

Guess a value for r.
Calculate πr^2.
Compare your answer with 100.

Decide whether your chosen value for r was too big or too small.
Make a better choice for r and repeat the process if you have not yet
achieved the accuracy you need.

Solution:

Type $\qquad ? \to r : \pi r^2$ **EXE**

You MUST include the colon
: as shown, or else
replace it by **SHIFT** **EXE** .

When the ? appears, enter your first guess. Suppose this is
$r = 10$ cm.

Type $\quad$ **1** **0** **EXE** **EXE**

Notice that **EXE** is pressed
twice.

The first **EXE** enters the
radius and the area is
calculated.

The second **EXE** repeats
the process and another ?
appears.

The answer, 314.1592654 is too big, so try a smaller value for r.

Type **5**	**EXE** **EXE**	giving	78.53981634	too small
Type **7**	**EXE** **EXE**	giving	153.93804	too big
Type **6**	**EXE** **EXE**	giving	113.0973355	too big
Type **5** **.** **5**	**EXE** **EXE**	giving	95.03317777	too small
Type **5** **.** **6** **5**	**EXE** **EXE**	giving	103.8689071	too big
Type **5** **.** **6** **5**	**EXE** **EXE**	giving	100.2874915	too big
Type **5** **.** **6** **4**	**EXE** **EXE**	giving	99.93280567	too small
Type **5** **.** **6** **4** **2**	**EXE** **EXE**	giving	100.0036926	too big

You can now see that the radius r lies between 5.64 cm and 5.642
cm. Therefore $r = 5.64$ cm to 2 d.p.

You can do the same thing by writing a computer program in
BASIC:

```
10 INPUT R
20 PRINT PI * R ^ 2
30 GOTO 10
40 END
```

RUN
?
5
78.53981634
?
7
153.93804
?

Try it using the numbers 5, 7, 6, 5.5, 5.75 and so on until you decide
when to stop.

Exercise 29C

1 Use a trial and improvement method to find the length L of the
side of a cube whose volume V is 100 cm^3.

The formula is $V = L^3$

Your answer should be accurate to 2 d.p.

Find out how to enter L^3
(L raised to the power 3)
on your graphical
calculator.

2 Use a trial and improvement method to find the length L of the side of a cube whose total surface area A is 50 cm^2.

The formula is $A = 6L^2$.

Your answer should be accurate to 2 d.p.

3 Use $? \rightarrow F : (F - 32) \times 5 \div 9$ to convert a temperature of 22 °C to Fahrenheit (°F) by trial and improvement.
Stop at 1 d.p. accuracy.

4 Use a trial and improvement method to solve the equation $x^3 + x = 19$, giving your answer correct to 1 d.p.

$? \rightarrow X : X^3 + X$

29.5 Using the **Ans** and **EXE** keys to produce number sequences

You can use these keys together to generate number sequences. For example:

Press **1** **EXE**

then **Ans** **+** **1** **EXE** **EXE** **EXE** ... keep pressing.

The calculator appears to be 'counting'. Each time **EXE** is pressed, 'Ans + 1' is calculated, where Ans is the *last displayed answer*.

```
1
                          1.
Ans + 1
                          2.
                          3.
                          4.
                          5.
```

Example 14

(a) Use the **Ans** and **EXE** keys to produce the even numbers, starting with 2.

(b) Show how the **Ans** and **EXE** keys can be used to produce the sequence 2, 6, 18, 54, ...

(a) **2** **EXE** **Ans** **+** **2** **EXE** **EXE** **EXE** ...

(b) **2** **EXE** **Ans** **×** **3** **EXE** **EXE** **EXE** ...

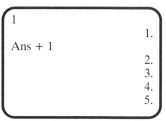

Ans recalls the most recent answer.

EXE performs (or repeats) the most recent calculation(s).

Exercise 29D

Write the key presses, including **Ans** and **EXE** , which will generate the following sequences:

1 5, 10, 15, 20, 25, ...

2 2, 4, 8, 16, 32, ...

3 200, 20, 2, 0.2, 0.02, ...

4 3, 9, 81, 6561, ...

5 256, 16, 4, 2, ...

6 3, 7, 3, 7, 3, ...

7 2, 0.5, 2, 0.5, ...

Example 15   compound interest (see also Example 6)

A student borrows £600 to buy a computer. Interest is added to the loan at the rate of 5% per annum. How much does the student owe after 3 years if no repayments are made?

Type   **6** **0** **0** **EXE**

Now type   **Ans** **+** **Ans** **×** **5** **÷** **1** **0** **0** **EXE** **EXE** **EXE**

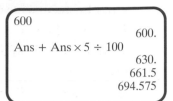

29.6 Drawing graphs on your calculator

Mcl

Choose the *default ranges* for x and y by pressing **Range** **SHIFT** **DEL** **Range** (or **Range** **INIT** **Range** **Range**)

Clear the graphics screen by pressing **SHIFT** **Cls** **EXE** .

Have a look at the graphics screen by pressing **G↔T**. Both axes are marked off in intervals of 1.

Find out how to set the ranges of x and y to your own chosen values.

Example 16

Draw the graphs of $y = x + 1$ and $y = x - 1$.

Press the **Graph** key, then type **X** **+** **1** **EXE** .

The first graph is drawn. To see the second graph, press the **Graph** key again and type **X** **+** **1** **EXE** .

Notice that the second graph is *superimposed*. It is drawn on the same axes as the first without erasing the first. This will continue to happen with further graphs until you *clear the graphics screen*. Clear the screen by pressing **SHIFT** **Cls** **EXE**.

If your calculator has a key marked **x,o,T** press it to get **X** .

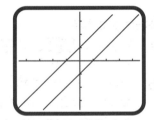

Example 17

Draw the graphs of $y = mx + c$ for several different values of m and c.

In this example $m = 2$ and c has values between -4 and 4.

Type 'M' ? → M 'C' ? → C : **Graph**   MX + C   **EXE** .

When the first ? appears, enter your value for M, then **EXE**.

When the second ? appears, enter your value for C, then **EXE**.

Wait until the graph is drawn.

Now press **EXE** again to draw another graph on the same axes.

Type:

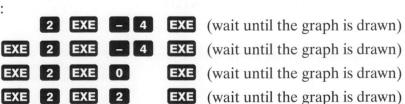

         2 **EXE** **–** **4** **EXE** (wait until the graph is drawn)

EXE **2** **EXE** **–** **4** **EXE** (wait until the graph is drawn)

EXE **2** **EXE** **0**   **EXE** (wait until the graph is drawn)

EXE **2** **EXE** **2**   **EXE** (wait until the graph is drawn)

EXE **2** **EXE** **4**   **EXE** (wait until the graph is drawn)

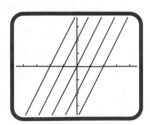

Clear the screen to finish. Try other values for m and c.

Exercise 29E

In each question draw all the graphs on the same axes and write down what you see.

Clear the screen before starting the next question.

1   $y = ax^2 + c$ for different values of a and c.
2   $y = ax^3 - c$ for different values of a and c.
3   $y = ax^3 + bx$ for different values of a and b.
4   $y = x + \frac{a}{x}$ for different values of a.

29.7 Solving equations by drawing graphs on your calculator

Example 18

You can solve the equation $x^2 - 4 = 2x - 3$ by drawing the graphs of $y = x^2 - 4$ and $y = 2x - 3$ on suitable axes and finding their point(s) of intersection.

Choose the *default ranges* for x and y and clear the graphics screen.

Press   **Graph** **X** **x²** **−** **4** **:** **Graph** **2** **X** **−** **3** **EXE**

Both graphs will be drawn. If they do not quite fit on the screen, *Zoom* out by a factor of 2.

Now use *Trace*. A flashing cross or dot is placed on the last line drawn and its x and y coordinates are displayed on the screen.

Press the right-arrow (and left-arrow) keys several times to make this dot move along the line. Stop when you reach a point where the two graphs cross and write down the x coordinate of the point of intersection.

The answers are 2.4 and −0.4 (correct to 1 d.p.).

Find out how to *Zoom* and *Trace* on your graphical calculator.

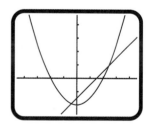

29.8 Investigating number sequences with a spreadsheet

You can generate many number sequences on the same spreadsheet and compare them.

Find out how to enter numbers and formulae in the cells of your spreadsheet.

Find out how to copy a formula from one cell to other cells.

Example 19

Generate the whole numbers up to 10 in column A.

Put the number 1 in cell A1.
Put the formula A1 + 1 in cell A2.
Copy the formula in A2 down column A as far as A10.

Example 20

Generate the odd numbers in column B.

Put the number 1 in cell B1.
Put the formula B1 + 2 in cell B2.
Copy the formula in B2 down column B as far as B10.

Example 21

Generate the triangular numbers in column C.

Put the number 1 in cell C1.
Put the formula C1 + A2 in cell C2.
Copy the formula in C2 down column C as far as C10.

Formula A2 = A1+1
Formula B2 = B1+2
Formula C2 = C1+A2
Formula D2 = B1+B2
Formula E2 = C1+C2

	A	B	C	D	E	F
1	1	1	1			
2	2	3	3	4	4	
3	3	5	6	8	9	
4	4	7	10	12	16	
5	5	9	15	16	25	
6	6	11	21	20	36	
7	7	13	28	24	49	
8	8	15	36	28	64	
9	9	17	45	32	81	
10	10	19	55	36	100	
11						

Example 22

Add consecutive odd numbers and put the answers in column D.

Put the formula B1 + B2 in cell D2.
Copy the formula in D2 down column D as far as D10.

You should now be able to see that the numbers in column D are multiples of 4.

Example 23

Add consecutive triangular numbers and put the answers in column E.

Put the formula C1 + C2 in cell E2.
Copy the formula in E2 down column E as far as E10.

What is the name of the sequence of numbers in column E?

29.9 Problem-solving with a spreadsheet

Look at this problem:

A farmer has 200 metres of fencing. He wants to use all the fencing to enclose a rectangular area of his field for his animals to graze. Find the length and width of the rectangle which gives his animals the maximum grazing area.

You can solve problems like this on a computer by using a spreadsheet.

Example 24

Think of all the rectangles you can draw whose perimeters are 200 metres. For example, some could be long and thin; others short and wide. Use a spreadsheet to find the length L and width W of the rectangle which has the maximum area.

There is more about this in **Section 31.2**.

The perimeter of each rectangle is 200 metres, so the equation is

$$2L + 2W = 200$$

You can divide this equation by 2 to make it simpler:

$$L + W = 100$$

Use your spreadsheet to try lots of values for L from $L = 0$ to $L = 100$ metres.

Increase L by 10 metres each time.

For each value of L calculate a value for W using:

$$W = 100 - L$$

Then multiply the values of L and W to get the area of each rectangle.

Make sure your spreadsheet has at least 12 rows and 3 columns. Use column A for the length, column B for the width and column C for the area of the rectangle.

You can now find the maximum value for the area in column C quite easily.

Drawing graphs of the data in each column of the spreadsheet can give you a better understanding of how the area changes as the lengths and widths of the rectangles change.

		(W=100-L)	(A=L×W)
	Formula	Formula	Formula
	A3=A2+10	B2=100-A2	C2=A2×B2

	A	B	C	D
1	L	W	LW	
2	0	100	0	
3	10	90	900	
4	20	80	1600	
5	30	70	2100	
6	40	60	2400	
7	50	50	2500	
8	60	40	2400	
9	70	30	2100	
10	80	20	1600	
11	90	10	900	
12	100	0	0	

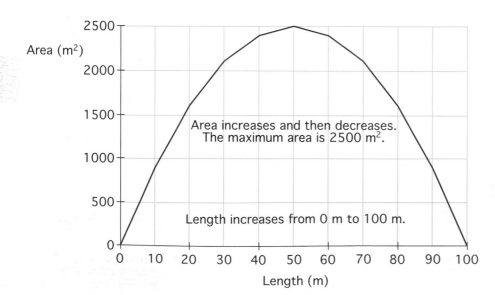

Area increases and then decreases. The maximum area is 2500 m².

Length increases from 0 m to 100 m.

Find out how your spreadsheet produces graphs.

Rectangles can be long and thin or short and wide, but the rectangle with the biggest area is **square**!

29.10 Trial and improvement on a spreadsheet

Example 25

The width of a rectangle is 2 cm less than the length. Use a trial and improvement method to find the length when the area of the rectangle is 30 cm². Your answer should be accurate to 2 d.p.

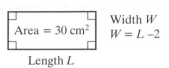

Width W
$W = L - 2$

Area = 30 cm²

Length L

Use L for the length and W for the width of the rectangle. Put the length in column A, the width in column B and the area of the rectangle in column C.

Use your spreadsheet to try lots of values for L.

For each value of L calculate a value for W using:
$$W = L - 2$$

Then multiply each value of L by the corresponding value of W to find the area of each rectangle.

Now look for those areas which are nearest to 30 cm².

You can use these answers to try new values for L and repeat the process until you achieve the accuracy you need.

Make sure your spreadsheet has at least 12 rows and 3 columns.

Values of L
from 0 to 10
in steps of 1

	A	B	C
	L	W	LW
1			
2	0	-2	0
3	1	-1	-1
4	2	0	0
5	3	1	3
6	4	2	8
7	5	3	15
8	6	4	24	too small
9	7	5	35	too big
10	8	6	48
11	9	7	63
12	10	8	80

Values:
A2 = 0
Formulae:
A3 = A2 + 1
B2 = A2 – 2
C2 = A2 × B2

Values of L
from 6 to 7
in steps of 0.1

	A	B	C
	L	W	LW
1			
2	6	4	24
3	6.1	4.1	25.01
4	6.2	4.2	26.04
5	6.3	4.3	27.09
6	6.4	4.4	28.16
7	6.5	4.5	29.25	too small
8	6.6	4.6	30.36	too big
9	6.7	4.7	31.49
10	6.8	4.8	32.64
11	6.9	4.9	33.81
12	7	5	35

Values:
A2 = 6
Formulae:
A3 = A2 + 0.1
B2 = A2 – 2
C2 = A2 × B2

Values of L
from 6.5 to 6.6
in steps of 0.01

	A	B	C
	L	W	LW
1			
2	6.5	4.5	29.25
3	6.51	4.51	29.360
4	6.52	4.52	29.470
5	6.53	4.53	29.580
6	6.54	4.54	29.691
7	6.55	4.55	29.802
8	6.56	4.56	29.913	too small
9	6.57	4.57	30.024	too big
10	6.58	4.58	30.136
11	6.59	4.59	30.248
12	6.6	4.6	30.36

Values:
A2 = 6.5
Formulae:
A3 = A2 + 0.01
B2 = A2 – 2
C2 = A2 × B2

Values of L
from 6.56 to 6.57
in steps of 0.001

	A	B	C
	L	W	LW
1			
2	6.56	4.56	29.913
3	6.561	4.561	29.924
4	6.562	4.562	29.935
5	6.563	4.563	29.946
6	6.564	4.564	29.958
7	6.565	4.565	29.969
8	6.566	4.566	29.980
9	6.567	4.567	29.991	too small
10	6.568	4.568	30.002	too big
11	6.569	4.569	30.013
12	6.57	4.57	30.024

Values:
A2 = 6.56
Formulae:
A3 = A2 + 0.001
B2 = A2 – 2
C2 = A2 × B2

You can now see that the length L lies between 6.567 cm and 6.568 cm.

So $L = 6.57$ cm correct to 2 d.p.

This is much easier than it looks!

Remember that the computer is doing all the really hard work.

Examination practice paper (non-calculator)

1   Two fair spinners can each score 1, 2, 3 or 4. They are spun at the same time.
    What is the probability that their sum will be 6? (3 marks)

2   (a) Solve the inequality $3(x + 2) \leq 7$ (3 marks)

    (b) Write the value of the greatest integer which satisfies this inequality. (1 mark)

3   Work out an estimate for the value of $\dfrac{23.2 \times 57.6}{0.43}$.

4   The diagram shows a rectangle with length $x + 2$ and width $2x - 7$.
    All measurements are given in centimetres.
    The perimeter of the rectangle is 17 cm.
    Find the value of x. (3 marks)

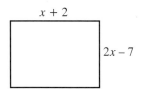

5   (a) Work out $\frac{5}{8} \div \frac{2}{3}$. Give your answer in its simplest form. (2 marks)

    (b) Work out $3\frac{3}{4} \times 2\frac{2}{5}$. Give your answer in its simplest form. (3 marks)

6

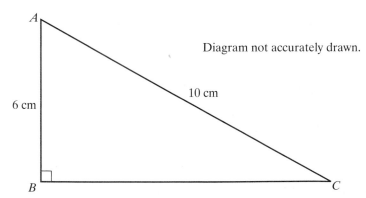

Diagram not accurately drawn.

A prism is 20 cm long. Its cross-section is a right-angled triangle ABC, in which $AB = 6$ cm and $AC = 10$ cm.

(a) Calculate the total surface area of the prism (6 marks)

A second, mathematically similar prism is 15 cm long.

(b) Calculate the total surface area of the second prism. (3 marks)

7   Here are the first five numbers of a sequence.

$$4, \quad 9, \quad 14, \quad 19, \quad 24.$$

Write down, in terms of n, an expression for the *n*th term of the
sequence.   (2 marks)

8
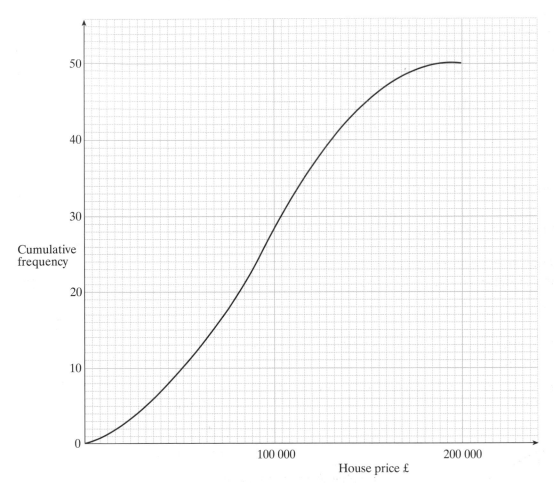

The cumulative frequency graph gives information about the
prices of 50 houses.

(a) Find the number of houses priced **below** £80 000. (1 mark)

(b) Find the median house price.
     Show clearly how you found your answer.   (1 mark)

(c) Find the interquartile range of the house prices.   (2 marks)

9   A computer performs 400 million calculations per second.

(a) Write 400 million in standard form.   (1 mark)

(b) Work out the number of calculations the computer
     performs in 10 minutes.
     Give your answer in standard form.   (2 marks)

10 **(a)** Expand and simplify $(4x + 3y)(4x - 3y)$. (3 marks)

(b) Factorise $10t - 15$. (1 mark)

(c) Factorise completely $9pq - 12q^2$. (2 marks)

11

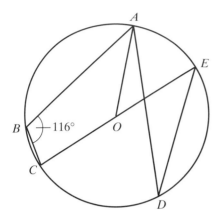

Diagram not
accurately drawn.

The diagram shows five points A, B, C, D and E on the
circumference of a circle.
The centre of the circle is at O.
CE is a diameter.
Angle $ABC = 116°$.

(a) State the size of angle CBE. (1 mark)

(b) Calculate the size of angle ADE. (2 marks)

(c) Calculate the size of angle AOE. (2 marks)

12 **(a)** Copy and complete the table of values for $y = x^3 - 2x^2 - 5x$.

x	-2	-1	0	1	2	3	4
y		2					

(2 marks)

(b) On graph paper plot the graph of $y = x^3 - 2x^2 - 5x$
for $-2 \leqslant x \leqslant 4$. (2 marks)

(c) Use your graph to solve the equation $x^3 - 2x^2 - 5x = 0$.
Where necessary, give your answers correct to 1 decimal
place. (2 marks)

(d) The x coordinates of the points of intersection of the curve
and a certain straight line give the solutions to the
equation $x^3 - 2x^2 - 6x + 1 = 0$. Find the equation of the
straight line. (2 marks)

13 **(a)** Evaluate (i) 7^0, (ii) 4^{-3}, (iii) $64^{\frac{1}{3}}$. (3 marks)

(b) Express $\dfrac{6}{\sqrt{2}}$ in the form $a\sqrt{b}$, where a and b are
integers. (2 marks)

20

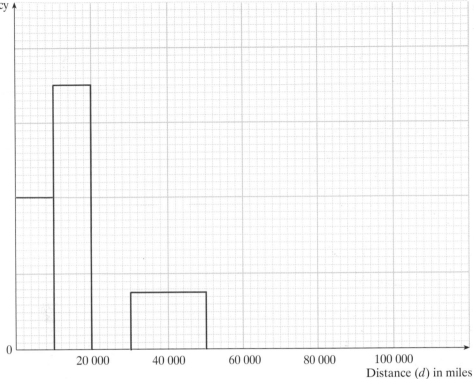

A hundred car owners were asked the distances, in miles, their cars had travelled.

The unfinished histogram and table show this information.

Distance (d) in miles	Frequency
$0 \leqslant d < 10\,000$	20
$10\,000 \leqslant d < 20\,000$	...
$20\,000 \leqslant d < 30\,000$	25
$30\,000 \leqslant d < 50\,000$	...
$50\,000 \leqslant d < 100\,000$	5

(a) Use the information in the histogram to complete the table. (2 marks)

(b) Use the information in the table to complete the histogram. (2 marks)

15 y is inversely proportional to the square of x.
$y = 4$ when $x = 5$.

(a) Find a formula for y in terms of x. (3 marks)

(b) Find the value of x when $y = 400$. (3 marks)

16 A is the point with coordinates $(2, 8)$ and B is the point with coordinates $(14, 4)$.

A straight line L is perpendicular to the line AB and passes through A.

Find the equation of the straight line L. **(4 marks)**

17 Prove algebraically that the sum of two consecutive square numbers is always an odd number. **(3 marks)**

18

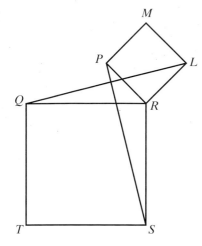

Diagram NOT accurately drawn

The diagram shows two squares $QRST$ and $PRLM$.
PS and LQ are straight lines.
Prove that triangle LQR and triangle PSR are congruent. **(3 marks)**

19 Rearrange $\dfrac{a}{x + b} = \dfrac{b}{x - a}$ to make x the subject. **(5 marks)**

20

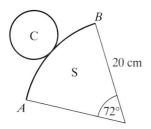

Diagram NOT accurately drawn

The diagram shows the net of a cone.
The net is made from a circle, **C**, and a sector **S**, of a circle centre O.
Angle $AOB = 72°$.

(a) Find the length of the arc AB.
Give your answer as a multiple of π. **(3 marks)**

The sector is folded to make a cone.

(b) Find the total surface area of the cone.
Give your answer as a multiple of π. **(5 marks)**

21   A bag contains only red beads and blue beads.
There are 4 red beads and n blue beads.
Sophie takes a bead at random from the bag.
She records its colour and then replaces it in the bag.
She then takes another bead at random from the bag and
records its colour.
The probability that she will take 2 blue beads is $\frac{4}{9}$.

   (a) Show that $5n^2 - 32n - 64 = 0$.   (4 marks)

   (b) **(i)** Solve $5n^2 - 32n - 64 = 0$.

   (ii) State the number of blue beads that are in
the bag.   (4 marks)

Examination practice paper (calculator)

1 Work out the value of

$$\frac{\sqrt{8.57^2 - 3.29^2}}{1.68 \times 3.41}$$

(3 marks)

2 The diagram shows a circle with its centre at C.

The tangent from A touches the circle at B.
Angle $C\hat{A}B = 21°$.

(a) Calculate the size of angle $A\hat{C}B$. (2 marks)

The length of AB is 7 cm.
BC is the radius of the circle.

(b) Calculate the length of BC.
Give your answer correct to 3 significant figures. (3 marks)

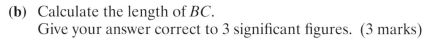

Diagram not
accurately drawn.

3 The table and the scatter graph show the number of units of
electricity used in heating a house on ten different days and the
average temperature for each day.

Average temperature (°)	6	2	0	6	3	5	10	8	9	12
Units of electricity used	30	39	41	34	33	31	22	25	23	22

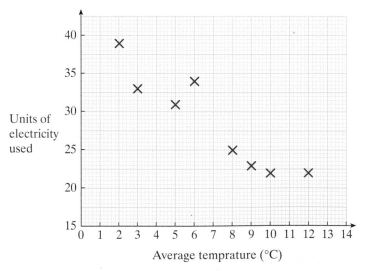

(a) Draw a line of best fit on your scatter graph. (1 mark)

Use your line of best fit to estimate
the average temperature if 35 units of electricity are
(b) used

its of electricity used if the average
re is 7°C. (1 mark)

4 **(a)** Show that the equation

$$x^3 - 7 = 81$$

has a solution in the range $4 < x < 5$ (2 marks)

    (b) Use a method of trial and improvement to obtain this solution correct to two places of decimals. (4 marks)

5 A year ago, Mrs Ford bought a new car. Its value has fallen by 12% since then. Its value now is £8624.

    (a) Calculate the value of the car when it was new. (2 marks)

The value of the car will continue to fall by 12% each year.

    (b) Calculate the value of the car when it is five years old. (3 marks)

6

Weekly rainfall (r mm)	Number of weeks
$0 < r \leqslant 10$	32
$10 < r \leqslant 20$	13
$20 < r \leqslant 30$	4
$30 < r \leqslant 40$	1
$40 < r \leqslant 50$	1
$50 < r \leqslant 60$	0
$60 < r \leqslant 70$	1

The grouped frequency table shows the distribution of weekly rainfall at Heathrow Airport in 1999.

Calculate an estimate for the mean weekly rainfall in 1989.

Give your answer correct to the nearest millimetre. (4 marks)

7 A factory employs 100 people.
76 of these people are men.
The other 24 are women.

The mean weekly wage of all 100 people is £407.
The mean weekly wage of the men is £395.

Work out the mean weekly wage of the 24 women. (5 marks)

8 Prove that the sum of the interior angles of any triangle is 180°. (4 marks)

9 (a) Solve each of the equations

(i) $\dfrac{4}{y} - 3 = 7$

(ii) $\dfrac{3x - 1}{5 - x} = -\frac{1}{2}$ (5 marks)

(b) Express

$$\frac{2}{3x - 1} + \frac{1}{x + 6}$$

as a single algebraic fraction.   (3 marks)

10 The diagram represents the cross-section of a church door.

The cross-section consists of a rectangle with a semi-circular top.

The door has a uniform thickness of 6 cm.
The door is made of metal of density 7.2 gm per cm^3

Work out the mass of the door.   (8 marks)

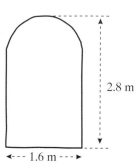

2.8 m

1.6 m

11

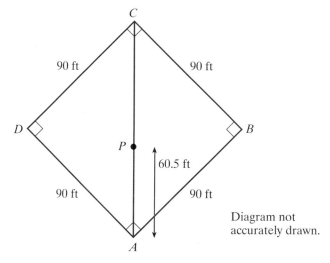

Diagram not accurately drawn.

The diagram shows some of the markings on a baseball field.
ABCD is a square with sides 90 ft long.
P is 60.5 ft from *A* along *AC*.

(a) Calculate the length of *AC*.
Give your answer in feet correct
to 1 decimal place.   (2 marks)

(b) Calculate the size of angle $C\hat{P}B$.
Give your answer correct to the nearest degree.   (4 marks)

12 **(a)** Factorize $x^2 - 5x - 14$ (2 marks)

(b) Solve the simultaneous equations
$$3x - 5y = 11$$
$$2x + 3y = 1$$ (4 marks)

13

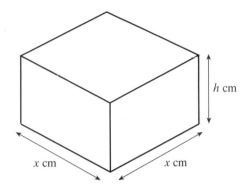

The diagram shows a square-based rectangular box. Each side of its base is x cm long. The box has no lid. Its height is h cm.

(a) Show that the surface area, A cm^2, of the box is given by the formula:
$$A = x^2 + 4hx$$ (2 marks)

(b) Make h the subject of the formula.   (2 marks)

(c) A square-based box 2 cm high has a surface area of 48 cm^2. Find the length of the sides of its base.   (2 marks)

(d) $A = 100$ and $h = 3$. Find the value of x. Give your answer correct to 3 significant figures.   (2 marks)

14 **(a)** On the axes below, sketch the graphs of
  (i) $y = \cos x$ (ii) $y = \sin 3x$

(4 marks)

(b) Solve the equation $\sin 3x = \dfrac{-2}{3}$
for all values of x in the range $0° < x < 180°$ (3 marks)

15 The diagram represents a side view of a cylindrical tin with 3 tennis balls in it. The diameter of each tennis ball is 6.5 cm.

Calculate, correct to 2 significant figures:

(a) the volume of the tin   (2 marks)

(b) the curved surface area of the tin   (2 marks)

(c) the total surface area of the three balls.   (2 marks)

Diagram not accurately drawn

16 A rectangle is 68 cm long and 8.4 cm wide. Each measurement is correct to 2 significant figures.

Calculate the greatest lower bound for the area of the rectangle, **giving with an explanation** your answer to an appropriate degree of accuracy.   (5 marks)

17 The points P, Q, R and S are at the four vertices of a square. The centre of the square is the point C.

The position vector of C is $\binom{2}{1}$,

The vector $\overrightarrow{CP} = \binom{2}{3}$

Work out the position vectors of P, Q, R and S.   (6 marks)

18 The equation of a straight line is:

$$y = ax + b, \quad \text{where } a \text{ and } b \text{ are constants.}$$

The equation of a circle is:

$$x^2 + y^2 = 64$$

The straight line is a tangent to the circle.

Prove that

$$a^2 + 1 = \frac{b^2}{64}$$ (6 marks)

Formulae sheet: Higher tier

Volume of prism = area of cross-section × length

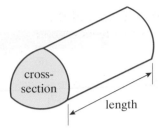

The quadratic equation

The solutions of $ax^2 + bx + c = 0$ where $a \neq 0$, are given by:

$$x = \frac{-b + \sqrt{(b^2 - 4ac)}}{2a}$$

Volume of sphere $= \frac{4}{3}\pi r3$

Surface area of sphere $= 4\pi r^2$

Volume of cone $= \frac{1}{3}\pi r^2 h$

Curved surface area of cone $= \pi r l$

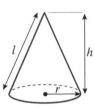

In any triangle ABC

Sine rule $\quad \dfrac{a}{\sin A} = \dfrac{b}{\sin B} = \dfrac{c}{\sin C}$

Cosine rule $\quad a^2 = b^2 + c^2 - 2bc \cos A$

Area of triangle $= \frac{1}{2} ab \sin C$

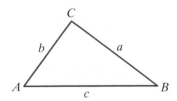

Answers

Unit 1 Exploring numbers

Exercise 1A

1 (a) 1, 2, 3, 4, 6, 8, 12, 24
   (b) 1, 2, 3, 4, 6, 9, 12, 18, 36
   (c) 1, 2, 4, 7, 11, 14, 22, 28, 44, 77, 154, 308
   (d) 1, 7, 11, 13, 77, 91, 143, 1001
   (e) 1, 2, 4, 5, 7, 8, 10, 14, 20, 25, 28, 35, 40, 50, 56, 70, 100, 140,
        175, 200, 280, 350, 700, 1400
   (f) 1, 53
2 2, 3, 5, 7, 11, 13, 17, 19, 23, 29, 31, 37, 41, 43, 47, 53, 59, 61, 67, 71,
   73, 79, 83, 89, 97
3 Yes $2 + 3 = 5, 2 + 5 = 7, 2 + 11 = 13$ etc.
   One of the primes *must* be 2.

Exercise 1B

1 (a) 12   (b) 36   (c) 28
2 (a) No HCF = 3   (b) No HCF = 3   (c) Yes HCF = 1

Exercise 1C

1 (a) 3   (b) 2   (c) 4
3 (a) 60   (b) 124   (c) 217   (d) 511   (e) 63
4 $p + 1$
6 $1 + p + p^2 + \ldots + p^n = \left(\dfrac{p^{n+1} - 1}{p - 1}\right)$

Exercise 1D

1 (a) 60   (b) 144   (c) 850   2   Every 120 secs (2 mins)
3 (a) pq (b) nm

Exercise 1E

1 (b) $\dfrac{n(n+1)}{2}$
2 (a) $T_3 + T_4 = 16 = S_4$ (b) $T_n + T_{n+1} = (n+1)^2 = S_{n+1}$
3 (a) 1, 4, 9, 16, 25, 36, 49, 64, 81, 100   (b) e.g. $9 + 16 = 25$
4 (a) Yes   (b) No   (c) No
5 $(n+1)^2 - n^2 = 2n + 1$
6 (d) Difference between $(n + p)$th square number and nth square
         number is a multiple of p.
   (e) Any of the type $(n^2)^3$

Exercise 1F

1 (a) 64   (b) 100 000   (c) 625   (d) 343
   (e) 20 736   (f) 0.59049
2 (a) 10^3 (b) 5^3 (c) 2^9 (d) 7^4
   (e) 5^4 (f) 3^9
3 $4^3 = (2^2)^3 = 2^6$ 4 2^5, by 7

Exercise 1G

1 (a) $5^6 = 15\,625$ (b) $5^8 = 390\,625$ (c) $10^2 = 100$
   (d) $12^3 = 1728$ (e) $3^7 = 2187$ (f) 1
   (g) 1   (h) $5^{-2} = 0.04$ (i) 6
   (j) 150   (k) 14   (l) $4^6 = 2^{12} = 4096$
   (m) $5^9 = 1\,953\,125$ (n) $2^{10} = 1024$ (o) $3^{-4} = \frac{1}{81}$
   (p) $4^5 = 1024$ (q) $6^{-1} = \frac{1}{6}$ (r) $5^5 = 3125$
   (s) $2^2 = 4$ (t) 4   (u) $5^{-6} = \frac{1}{15\,625}$
   (v) $3^{-4} = \frac{1}{81}$ (w) $2^{-4} = \frac{1}{16}$ (x) $3^2 = 9$
2 $2^4 = 4^2$

Exercise 1H

1 (a) 2   (b) $\frac{1}{2}$ (c) 8   (d) $\frac{1}{5}$ (e) 2   (f) $\frac{1}{2}$
   (g) $\frac{1}{8}$ (h) $\frac{1}{5}$ (i) $\frac{1}{10}$ (j) 10   (k) 125   (l) $\frac{1}{4}$
2 (a) $2^{10} = 1024$ (b) $3^2 = 9$ (c) 1
   (d) $5^0 = 1$ (e) 40   (f) $4^{-6} = \frac{1}{4096}$
   (g) $10^{-2} = \frac{1}{100}$ (h) $5^4 = 625$
3 (a) 3   (b) 5   (c) $\frac{1}{7}$ (d) 4   (e) 8   (f) $\frac{1}{100}$

Exercise 1I

1 (a) 8   (b) $\frac{1}{4}$ (c) 0.0001
   (d) 10 000 000 000   (e) 800 000
2 1000   3

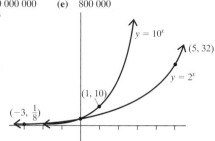

4 (a) $\frac{4}{5}$ (b) 1   (c) -4 5 $2^{n+1} - 1$

Exercise 1J

1 (b) $n + 1$ (c) $\tau(n \times m) = \tau(n) \times \tau(m)$ if n, m co-prime
2 nth term = $(n - 1)$th term + $(n - 3)$rd term
3 $1^3 + 2^3 + 3^3 + \cdots + n^3 = (1 + 2 + \cdots + n)^2$
4 (a) LD $(n \times m)$ = LD [LD$(n) \times$ LD(m)]
   (b) LD (square number) = 0, 1, 4, 5, 6, 9 only
   (c) Cannot because $10n + 7$ has LD = 7
5 (a) $\frac{12}{5}$ (b) $-\frac{4}{3}$ (c) $-\frac{5}{2}$ 6 $\dfrac{5^{n+1} - 1}{4}$

Exercise 1K

1 (a) $4\frac{13}{20}$ (b) $\frac{17}{4}$ or $4\frac{1}{4}$ (c) $1\frac{3}{4}$ (d) $\frac{380}{189}$ or $2\frac{2}{189}$
   (e) $\frac{68}{5}$ or $2\frac{18}{25}$ (f) $\frac{14}{9}$ or $1\frac{5}{9}$ (g) $\frac{342}{5}$ or $68\frac{2}{5}$ (h) $\frac{23}{280}$
   (i) $\frac{287}{69}$ or $4\frac{11}{69}$

Unit 2 Solving equations and inequalities

Exercise 2A

1 5	2 4	3 $2\frac{1}{2}$	4 $2\frac{3}{5}$	5 $\frac{3}{7}$
6 0	7 $\frac{5}{6}$	8 $\frac{7}{9}$	9 -3	10 -2
11 $-2\frac{1}{2}$	12 $-2\frac{2}{5}$	13 $-\frac{5}{8}$	14 $-\frac{1}{2}$	15 $-\frac{2}{5}$

Exercise 2B

1 4	2 3	3 $\frac{1}{2}$	4 $\frac{4}{5}$	5 -2
6 -1	7 $1\frac{1}{2}$	8 $3\frac{2}{3}$	9 $-1\frac{1}{3}$	10 $-2\frac{1}{5}$

Exercise 2C

1 1	2 -2	3 2	4 -4	5 -6
6 $-2\frac{3}{5}$	7 $1\frac{1}{2}$	8 $\frac{1}{6}$	9 $-\frac{1}{4}$	10 $-1\frac{4}{5}$

Exercise 2D

1	18	**2**	−6	**3**	24	**4**	$9\frac{1}{3}$	**5**	$4\frac{1}{2}$
6	14	**7**	−6	**8**	−12	**9**	6	**10**	$-6\frac{2}{3}$

Exercise 2E

1	1	**2**	$\frac{1}{2}$	**3**	$-1\frac{1}{2}$	**4**	−3	**5**	$\frac{3}{5}$
6	$-\frac{3}{4}$	**7**	$1\frac{1}{5}$	**8**	1	**9**	0	**10**	$-2\frac{1}{3}$

Exercise 2F

1	2	**2**	$-\frac{4}{5}$	**3**	$-\frac{1}{4}$	**4**	$1\frac{1}{2}$	**5**	1
6	8	**7**	$\frac{3}{4}$	**8**	$-1\frac{1}{2}$	**9**	$2\frac{1}{3}$	**10**	0
11	20	**12**	−12	**13**	$\frac{1}{5}$	**14**	$2\frac{2}{3}$	**15**	$2\frac{1}{4}$
16	$2\frac{1}{2}$	**17**	$2\frac{1}{5}$	**18**	−2	**19**	$-\frac{3}{4}$	**20**	$\frac{2}{3}$

Exercise 2G

1	4	**2**	62°	**3**	100, 140, 120	**4**	15, 19, 11	**5**	2
6	12	**7**	11	**8**	7	**9**	23	**10**	47
11	54	**12**	1.30 pm	**13**	10	**14**	both 42°	**15**	27

Exercise 2H

1 $q - p$ **2** $t - 7$ **3** $a + b$ **4** $\frac{w}{5}$

5 $\frac{p}{n}$ **6** $3f$ **7** md **8** $\frac{bc}{a}$

9 $\frac{p+q}{n}$ **10** $\frac{5-b}{a}$

11 $\frac{c-h}{d}$ **12** $\frac{p-q}{5}$

13 $\frac{b-a}{3}$ **14** $mp - mn$ or $m(p - n)$

15 $\frac{6-x}{2}$ or $3 - \frac{1}{2}x$ **16** $\frac{12-3x}{4}$ or $3 - \frac{3}{4}x$

17 $\frac{10-2x}{5}$ or $2 - \frac{2}{5}x$ **18** $\frac{x-8}{2}$ or $\frac{1}{2}x - 4$

19 $\frac{5x-20}{4}$ or $1\frac{1}{4}x - 5$ **20** $\frac{3x-24}{8}$ or $\frac{3}{8}x - 3$

Exercise 2I

1 $x > 1$ **2** $x \leqslant -2$ **3** $-3 \leqslant x < 2$ **4** $0 \leqslant x \leqslant 3$

5

6

7

8

9

10

Exercise 2J

1	$x \leqslant 3$	**2**	$x > \frac{2}{3}$	**3**	$x \leqslant 1\frac{2}{3}$	**4**	$x < -3$
5	$a \geqslant 1$	**6**	$b < -1$	**7**	$c > -4$	**8**	$d \leqslant 1\frac{1}{2}$
9	$e \leqslant -4$	**10**	$f < -28$	**11**	$g \leqslant 2$	**12**	$h \geqslant \frac{1}{2}$
13	$j > -1$	**14**	$k \leqslant 1\frac{2}{3}$	**15**	$m \geqslant -3\frac{1}{2}$	**16**	$n > -1\frac{1}{2}$

Exercise 2K

1 $1 < p \leqslant 6$ **2** $5 < q < 7\frac{1}{2}$ **3** $-1 \leqslant r \leqslant 2$ **4** $-6 < s < 4$

Exercise 2L

1	−2, −1, 0, 1	**2**	1, 2, 3, 4	**3**	−3, −2, −1
4	0, 1, 2, 3	**5**	e.g. $2 \leqslant x \leqslant 5$	**6**	e.g. $0 \leqslant x \leqslant 3$
7	e.g. $-3 \leqslant x \leqslant -1$	**8**	e.g. $-2 \leqslant x \leqslant 2$	**9**	2
10	1	**11**	−1	**12**	−3
13	−1	**14**	2	**15**	0
16	2	**17**	2, 3	**18**	2, 3
19	−2, −1, 0, 1, 2	**20**	0, 1		

Exercise 2M

1	$\frac{3}{5}$	**2**	$1\frac{2}{5}$	**3**	−1	**4**	4	**5**	−6
6	$\frac{2}{3}$	**7**	−2	**8**	$-2\frac{1}{2}$	**9**	$3\frac{3}{5}$	**10**	8
11	−8	**12**	6	**13**	2	**14**	$2\frac{1}{2}$	**15**	7
16	20p								

17 (a) $\frac{q+4}{p}$ (b) $\frac{n}{6}$ (c) $\frac{m-5}{2}$

18 (a)   (b)   (c)   (d)

19 (a) $x < 2$ (b) $x \leqslant -2$ (c) $x < 1\frac{1}{3}$

20 (a) (i) 3   (ii) 3   (b) (i) 3   (ii) 0

Unit 3 Shapes

Exercise 3A

1 (a) $a = 48°$ (b) $b = 106°$, $c = 74°$, $d = 74°$
  (c) $e = 111°$ (d) $f = 60°$, $g = 120°$ (e) $h = 51\frac{3}{7}°$

2 (a) $a = 35°$, $b = 50°$, $c = 50°$, $d = 95°$
  (b) $e = 68°$, $f = 68°$ (c) $g = 149°$, $h = 125°$ $i = 55°$

3 (d) 45°

Exercise 3B

1 (a) $a = 90°$, right-angled triangle
  (b) $b = c = 60°$, equilateral triangle
  (c) $d = 96°$, scalene triangle
  (d) $e = 83°$, scalene triangle
  (e) $f = 81°$, isosceles triangle

2 (a) $a = 95°$ (b) $b = 64°$, $c = 64°$, $d = 52°$
  (c) $e = 62°$, $f = 68°$, $g = 50°$

3 (c) EF = 7.8 cm   (d) XZ = 7.3 cm   (e) PR = 6.7 cm

4 9

Exercise 3C

1 (a) square   (b) kite   (c) trapezium
  (d) square, rhombus, kite

2 (a) (i) $A\hat{E}D = 52°$ (ii) $A\hat{E}F = 38°$ (iii) $A\hat{E}B = 83°$
  (b) (i) $A\hat{B}X = 41°$ (ii) $B\hat{A}X = 49°$
  (c) (i) $M\hat{N}P = 29°$ (ii) $P\hat{M}N = 122°$
  (d) (i) $P\hat{C}B = 30°$ (ii) $C\hat{P}B = 75°$ (iii) $A\hat{P}D = 75°$
  (iv) $A\hat{P}B = 150°$

3 (d) QS = 10.8 cm

Exercise 3D

1 (a) cylinder   (b) triangular-based pyramid
  (c) cuboid   (d) hexagonal prism

2 (a)

Solid	V	F	E
cube	8	6	12
cuboid	8	6	12
triangular prism	6	5	9
hexagonal prism	12	8	18
octagonal prism	16	10	24
triangular-based pyramid	4	4	6
square-based pyramid	5	5	8

  (b) V + F = E + 2   or   V + F − 2 = E   or   V − E + F = 2

3 (a)

Solid	V	F	E
tetrahedron	4	4	6
cube	8	6	12
octahedron	6	8	12
dodecahedron	20	12	30
icosahedron	12	20	30

(b)

Solid	Number of triangular faces meeting at a vertex
tetrahedron	3
octahedron	4
icosahedron	5

(c) Six regular triangular faces meeting at a vertex would have a total of 360°. This creates a surface *not* a vertex.

(d) Regular polyhedra require 3 or more faces to create a vertex.
3 interior angles of regular hexagons = 3 × 120° = 360°

4 (a) plan   front   side   **(b)** plan   front   side

5

6   **7**

8   **9**

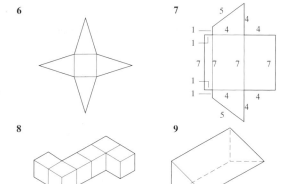

Exercise 3E

1 **(a)** (iii)   **(b)** (iii)   **(c)** (iii)
2 **(a)** ABC, YZX (SAS)   **(b)** ABC, EFD (AAS)
    (c) XYZ, QRP (RHS)
3 A and F   C and H   B and D   E and G
4 E and G   J and N   L and M   A and C
    B and I   D and F   H and K
5 AB = CD opposite sides of a parallelogram.
    AD = BC opposite sides of a parallelogram.
    BD = BD common
    Triangles ABD , CDB are congruent (SSS)
6 BC = BD − CD
    DE = CE − CD
    But BD = CE
    ∴ BC = DE
    CA = AD given
    ∠BCA = 180° − ∠ACD
    ∠ADE = 180° − ∠ADC
    But ∠ACD = ∠ADC base angles of an isosceles △.
    ∴ ∠BCA = ∠ADE
    Triangles ABC, ADE are congruent (SAS)
7 ∠ABE = 180 − 2 × ∠AEB 3rd angle in an isosceles △
    ∠DBC = 180 − 2 × ∠BDC
    But ∠AEB = ∠BDC
    ∴ ∠ABE = ∠DBC
    ∠ABD = ∠ABE + ∠EBD
    ∠EBC = ∠DBC + ∠EBD
    Therefore ∠ABD = ∠EBC
            AB = EB given
            BD = BC given
    Triangles ABD, EBC are congruent (SAS)

8 ∠ABC = ∠ADE   corresponding angles.
    ∠ACB = ∠AED   corresponding angles.
    But ∠ABC = ∠ACB   isosceles triangle
    ∴ ∠ADE = ∠AED
    and ∠ADE is isosceles.
    AD = AE   sides of and isosceles triangle
    AC = AB   sides of an isosceles triangle
    ∠A = ∠A   common
    Triangles ADC, ABE are congruent (SAS)

Exercise 3F

1 **(a)** 4   **(b)** 2.34   **(c)** 19.8   **(d)** $d = 6.72$, $e = 4.8$
2 **(a)** $x = 1.5$, $y = 3$ **(b)** 12.8
    (c) $x = 3.75$, $y = 5.25$, $z = 6.75$
3 A and C   B and E   D and F
4 squares   circles   equilateral triangles   regular hexagons
5 cubes   spheres   tetrahedrons   octahedrons
6 **(a)** (ii)   **(b)** (iii)   **(c)** (iii)
7 **(a)** Two pairs of sides in same ratio, included angle equal.
        $x = 4.72$
    (b) All corresponding angles equal.
        $y = 1.8$
    (c) All corresponding angles equal.
        $a = 7.68$, $b = 9.6$
8 **(a)** All corresponding angles equal.
    (b) $x = 3.2$, $y = 2.625$
9 **(a)** ABC similar to AXY
    (b) All corresponding angles equal.
    (c) AB = 1.91 (1 s.f.)
    (d) AY = 1.79 (to 3 s.f.)

Exercise 3G

1
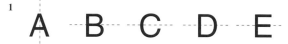

2
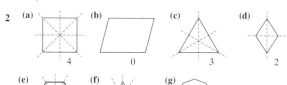

3 **(a)** 4   **(b)** 9   **(c)** infinite
4 **(a)** 4   **(b)** 3   **(c)** 4   **(d)** 7
5 **(a)** 4   **(b)** 2   **(c)** 1   **(d)** 5
    (e) 1   **(f)** 1   **(g)** 2
6 **(a)** (b) and (g) have point symmetry
7 For example: 8

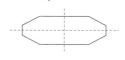
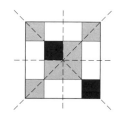

Unit 4 Collecting and presenting data

Exercise 4A

1 **(a)** Quantitative   **(b)** Qualitative   **(c)** Quantitative
    (d) Qualitative   **(e)** Qualitative
2 **(a)** Actually continuous but usually treated as discrete
    (b) Discrete   **(c)** Discrete   **(d)** Discrete
    (e) Continuous   **(f)** Continuous   **(g)** Continuous
    (h) Discrete

Exercise 4B

1 (a) Quantitative – usually treated as discrete
   (b) Quantitative – discrete   (c) Qualitative
   (d) Quantitative – continuous   (e) Qualitative
   (f) Usually considered quantitative especially in IQ tests –
        continuous
   (g) Qualitative

2 (a)

Goals	0	1	2	3	4	5
Frequency	4	5	6	3	0	2

   (b)

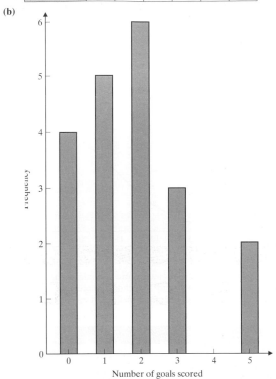

Number of goals scored

3

Number of dandelion plants per m²	0	1	2	3	4	5
Frequency	5	9	4	5	1	1

4 (a)

Hand span (cm)	Frequency
9.0 – 9.9	2
10.0 – 10.9	2
11.0 – 11.9	4
12.0 – 12.9	6
13.0 – 13.9	6
14.0 – 14.9	5
15.0 – 15.9	5

   (b)

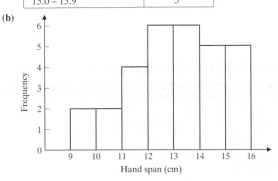

Hand span (cm)

5
7, 8, 9
3, 5, 6, 8, 9
1, 4, 6, 6, 7, 8
1, 3, 3, 3, 6, 7, 7, 7, 8
2, 3, 6, 7, 9
1, 3, 4, 5, 6, 7
2, 8
1, 2
1, 7

ian = 37

Exercise 4C

1 (a) Too vague – 'How many times have you been to the cinema in
        the last month?' *or* – 'How many films have you watched on
        video in the last month?'
   (b) Too complicated – 'Should people be allowed to smoke
        outdoors?'
   (c) Too leading – 'How much would you be willing to pay for dog
        food?'
   (d) Too invasive – could be OK if it is an anonymous
        questionnaire.
   (e) Too leading – 'What do you think is a suitable amount of food
        for a person?'

2 (a) Ignores different traffic flow during morning and evening rush
        hours.
   (b) Similar to (a). Also at school coming and going times the
        occupancy of cars will be different.
   (c) Too restricted to people of one age group and possibly similar
        abilities.
   (d) Ignores those who eat elsewhere. Probably overstates those
        who have school dinners.

3 (a) Select several short periods of time throughout the day.
   (b) As (a) or restrict your survey to a specific day, e.g. Saturday
        and say so in your report.
   (c) Ask a variety of ages and abilities and both male and female.
   (d) Select people at random from school lists.

4 Drop the ball several times from the same height and find the
   average rebound height. Repeat for other heights. Work out the
   ratio 'height:rebound height' to see if it is constant.

5 (a) His friends are likely to have similar tastes to him. Also 10 is
        probably not enough.
   (b) Ask a lot more students, selected at random.

6 Decide on the lengths for estimation.
   Ask a random sample of boys.
   Ask a random sample of girls.
   Look to see if there are any differences between the results.

Exercise 4D

1 (a) 25   (b) 8.5
   (c) She can't sample 0.5. She can double the sample size or
        decide, by tossing a coin, to sample 8 or 9.

2 (a) 800   (b) 1020   (c) 567

3 Every 50th person in the directory.

4 Use random numbers to select 60 from the school list.
   Select every 20th name from the school list.
   Select a suitable stratified sample.

Exercise 4E

1 (a)

Length (cm)		Frequency
1 up to but not including 2		0
2	3	2
3	4	6
4	5	8
5	6	8
6	7	4
7	8	2

(b)

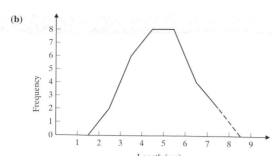

2 (a)

Time (s)		Frequency
10 up to but not including 20		0
20	30	3
30	40	3
40	50	7
50	60	4
60	70	3
70	80	2
80	90	1

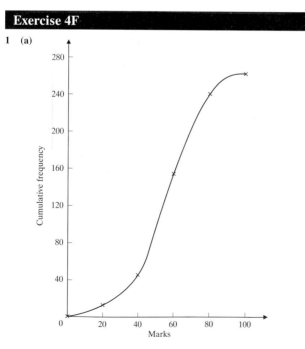

(b) The peak of the boys' frequency polygon is in the same interval as the girls. However, the boys' frequency polygon is extended towards longer times – especially if the two who gave up are included.

Exercise 4F

1 (a)

(b) 27%   **(c)** 51 marks

2 (a)

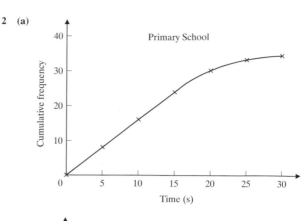

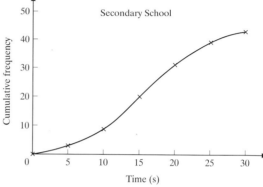

(b) There were proportionally more secondary school pupils lasting between 10 and 20 seconds. There were proportionally more primary school pupils lasting between 0 and 10 seconds.

Exercise 4G

1

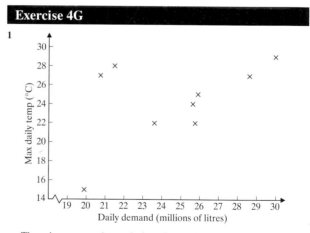

There is a greater demand when the maximum daily temperature is high.

2 (a),(b)

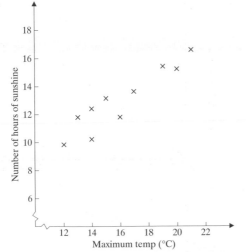

(c) As the number of hours of sunshine increases, the maximum temperature increases.

3 (a)

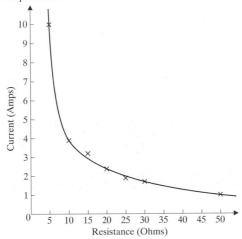

(b) 1.3 amps (c) 6.5 ohms

4

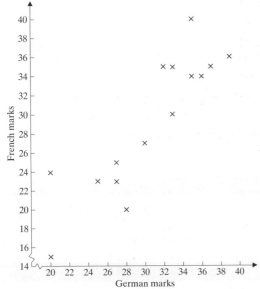

Exercise 4H

1 42, 110

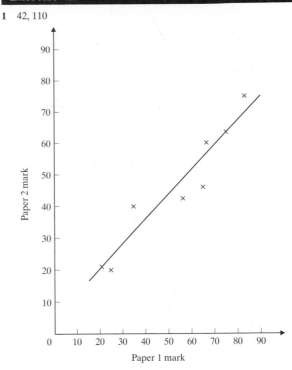

2 Temperature   260 litres

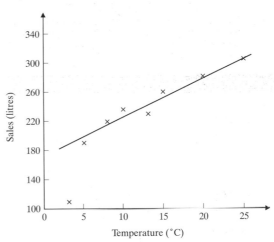

3 (a) The rate of decrease of population with area

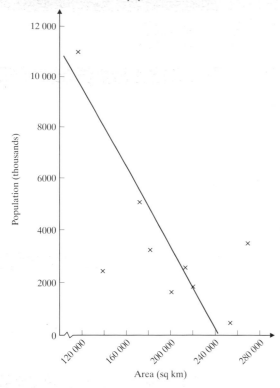

(b)

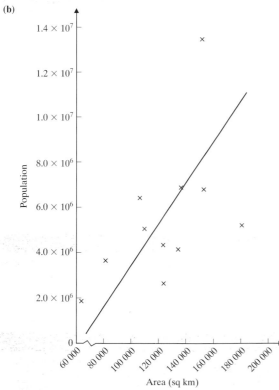

(c) In (a) population and area are negatively correlated.
In (b) population and area are positively correlated.

5 (b) (i) 14 to 15 hours   (ii) 15°C

Unit 5   Using basic number skills

Exercise 5A

1 (a) 112%   (b) 117%   (c) 130%   (d) 111%
  (e) 108.5%
2 (a) 1.24   (b) £2294
3 (a) 1.4   (b) £7.56
4 1.04
5 (a) £144   (b) 70 kg   (c) 2.784 m   (d) £1370.20
  (e) 128.52 cm

Exercise 5B

1 (a) 84%   (b) 89%   (c) 95%   (d) 78%
2 (a) 0.66   (b) £3.465
3 77.9 kg
4 (a) £96   (b) 42 kg   (c) 2.016 m   (d) £1109.80
  (e) 123.48 cm

Exercise 5C

1 (a) 1.06   (b) £106   (c)

Date	Amount
1/1/99	£100
1/1/00	£106
1/1/01	£112.36
1/1/02	£119.10
1/1/03	£126.25

2 (a) £12 320   (b)

Original cost	£14 000
value after 1 year	£12 320
value after 2 years	£10 102.40
value after 3 years	£9597.28
value after 4 years	£9117.42
value after 5 years	£8661.55

3 (a) 1.10   (b) 1.188

Exercise 5D

1 (a) $\frac{7}{9}$ (b) $\frac{1}{3}$ (c) 77.7%   (d) 33.3%
2 (a) 50%   (b) 60%   (c) 16.73%   (d) 20%
3 (a) 8 km/litre   (b) 10.6 km/litre   (c) 32.5%
4 (a) £54 600   (b) 58%
5 (a) 4.5 litres   (b) 75%

Exercise 5E

1 (a) £260   (b) 32.5%   (c) 20.6% to nearest 0.1%
2 (a) 33%   (b) 27%   (c) 12%   (d) 8%   (e) 55%
3 (a) 20%, 31.2%, 40%, 47.2%   (b) 12.8%
4 (a) 42.9%   (b) 33.3%   (c) 4.76% decrease

Exercise 5F

1 £428.40   **2** £54 538   **3** 13.04

Exercise 5G

1 (a) 90%   (b) presale price $\xrightarrow{\times 0.9}$ £540   (c) £600
2 (a) 117.5%   (b) ? $\xrightarrow{\times 1.175}$ £320   (c) £272.34
3 (a) 175   (b) 14
4 (a) £13 333.33   (b) £14 814.81

Exercise 5H

1 (a) £14.63   (b) 5.852%

2

Year	Amount at start of year	Interest	Total amount at year end
6	£334.56	£334.56 × 0.06	£354.63
7	£354.63	£354.63 × 0.06	£375.91
8	£375.91	£375.91 × 0.06	£398.46
9	£398.46	£398.46 × 0.06	£422.37
10	£422.37	£422.37 × 0.06	£447.71

3 (a) £47.71   **(b)** 19.1%   **(c)** 7.9%
4 (a) approx 12 years   **(b)** 16 years 8 months
5 (a) £1082   **(b)** £1370.59   **(c)** £1878.53
6 (a) £656   **(b)** £878.53 Difference £222.53
7 If an annual interest rate of 50% is paid, then after the first year the value of the investment increases to 150% of its original value, and during the second year the value of the investment increases by 50% of its value *at the end of the first year*, giving an increase of 125%.
8 (a) (i) 125%   **(ii)** 96%   **(b)** 41.4%
9 (a) £3840   **(b)** £153.60   **(c)** £332.80   **(d)** £404.90
10 (a)

Year	Value at start of year	Interest	Total at end of year
1	£100	£100 × 0.08	£108
2	£108 + 100	£208 × 0.08	£224.61
3	£224.64 + 100	£324.64 × 0.08	£350.61
4	£350.61 + 100	£450.61 × 0.08	£486.66
5	£486.66 + 100	£586.66 × 0.08	£633.59
6	£633.59 + 100	£733.59 × 0.08	£792.28

(b) £1564.55
11 (a) £10 800   **(b)** £9720   **(c)** £7085.88
(d) $6\frac{1}{2}$ years approx
12 (a) 1.602   **(b)** 2.027   **(c)** 3.247
13 approximately 6 years

Exercise 5I

1 (a) 87.17 km/h **(b)** 54 mph
2 25.71 mph
3 (a) 1.25 h   **(b)** 0.4 h   **(c)** 0.1 h   **(d)** 0.42 h
(e) 0.75 h   **(f)** 0.17 h
4 (a) 162 min   **(b)** 2.7 h
5 (a) Shannon by 4 minutes   **(b)** 1.63̇ h or 1 h 38 min
6 (a) 24 min   **(b)** 5 h 18 min   **(c)** 3 h 15 min
(d) 6 h 17.4 min   **(e)** 3 h 54 min   **(f)** 9 h 52.8 min
7 (a) 5 hours   **(b)** 5 km/h
8 (a) 456 km   **(b)** 6.85 h   **(c)** 66.57 km/h

Exercise 5J

1 14.5 km/l
2 (a) (i) 42 litres   **(ii)** 14 km/litre
(b) 84 km
3 (a) (i) regular 0.8125 p/h   longlife 0.78 p/h
(ii) regular 123 h/£   longlife 128 h/£
(b) Longlife cheaper per hour or more hours/£
4 (a) 1.424 marks/dollar
(b) 0.702 dollars/mark
5 (a) Cold 184 litres/min   Hot 115 litres/min
(b) 3 min 4.6 sec

Exercise 5K

1 100 hours
2 (a) (i) 0.00026 m³/sec   **(ii)** 15.6 litres/min
(b) 1 hr 4 min 6 sec
(c) 51 min 17 sec
3 (a) (i) 0.16 litres/km   **(ii)** 6.4 litres/hour
(b) (i) 10 km   **(ii)** 1.6 litres

Exercise 5L

1 (a) 32 000 cm³   **(b)** 3200 cm
2 (a) 6475 kg   **(b)** 0.463 m³
3 (a) 546 cm³   **(b)** 6.41 g/cm³

Exercise 5M

1 (a) (i) 2   **(ii)** 4   **(iii)** 6
(b) (i) $\frac{1}{6}$ **(ii)** $\frac{1}{3}$ **(iii)** $\frac{1}{2}$
2 (a) Sally $\frac{1}{5}$ Brian $\frac{3}{10}$ Mark $\frac{1}{2}$
(b) Sally £8   Brian £12   Mark £20
3 (a) 8 parts   **(b)** $\frac{3}{8}$ **(c)** £101.25   **(d)** $\frac{5}{8}$
(e) £168.75
4 (a) 5:2   **(b)** Danny $\frac{5}{7}$ (Melissa $\frac{2}{7}$)   **(c)** £19
(d) £60   **(e)** £10
5 (a) 4:1   **(b)** 18 marks
6 (a) $\frac{2}{7}$ **(b)** 68 boxes   **(c)** £32.64
7 (a) £120   **(b)** 2:1   **(c)** £282
8 (a) 540   **(b)** 180   **(c)** 324   **(d)** 216
9 (a) 86   **(b)** 86:5   **(c)** 86   **(d)** 80

Exercise 5N

1 (a) $\frac{5}{13}$ **(b)** $\frac{8}{13}$ **(c)** 60 **(d)** 195
2 9.37 km/litre
3 (a) 5:8   **(b)** No, 56 km/h   **(c)** 112 km/h
4 (a) 50   **(b)** 37.5   **(c)** 1:1.75
5 (a) 27 days   **(b)** 21 days   **(c)** 6 days   **(d)** 75 days
6 (a) 6:3:1   **(b)** 81 goals   **(c)** 135 goals
7 60 years, 30 years, 20 years, 12 years

Exercise 5O

1 (a) 10^2 **(b)** 10^3 **(c)** 10^6 **(d)** 10^{10}
2 (a) 6 million   **(b)** 3.4 million   **(c)** 7.8 million
(d) 5.5 million   **(e)** 2.65 million   **(f)** 7.642 million
3 (a) 3.1×10^6 **(b)** 4.3×10^6 **(c)** 0.5×10^6
(d) 2.4×10^6 **(e)** 7.8×10^6 **(f)** 8.6×10^6
(g) 4×10^6 **(h)** 9×10^6 **(i)** 0.4×10^6
4 (a) 10^3 **(b)** 10^4 **(c)** 10^1 **(d)** 10^6
(e) 10^5 **(f)** 10^3
5

Standard form
7.4×10^4
2.6×10^2
6.8×10^5
4.5×10^1
9.9×10^6
6.2×10^1
8×10^0

6 (a) 1.6×10 **(b)** 4.3×10^3
(c) 6.5×10^5 **(d)** 8.7×10^7
(e) 6.7×10^2 **(f)** 8.65×10^2
(g) 9.87×10^6 **(h)** 9.85×10^4
(i) 8.05×10^{11}

Exercise 5P

1 (a) 420   **(b)** 67 000   **(c)** 5500   **(d)** 7 500 000
(e) 620 000   **(f)** 73 000   **(g)** 24 000 000   **(h)** 11
(i) 7.25
2 (a) 9×10^2 **(b)** 9.6×10^3 **(c)** 4.05×10^4
(d) 3.91×10^5 (to 2 d.p.)

Exercise 5Q

1

Standard form
2.4×10^{-3}
2×10^{-1}
6×10^{-5}
1.5×10^{-1}
7×10^{-3}
4.5×10^{-4}
3.46×10^{-2}
1.25×10^{-3}

2 (a) 2×10^{-3} **(b)** 1.5×10^{-1}
(c) 4×10^{-4} **(d)** 5.4×10^{-2}
(e) 8×10^{-6} **(f)** 6.8×10^{-11}
(g) 3.46×10^{-1} **(h)** 9×10^{-2}
(i) 5.6×10^{-3}

3 (a) 0.35   (b) 0.060   (c) 0.000 72
   (d) 0.0022   (e) 0.000 013 5   (f) 0.000 005 33
   (g) 0.000 000 000 88   (h) 0.000 000 44   (i) 0.4999

Exercise 5R

1 (a) 9.6×10^{13} (b) 1.722×10^4 (c) 8.4×10^{-3}
   (d) 1.476×10^{-12} (e) 2.91×10^{-1} (f) 1.47×10^{-9}
   (g) 2.83×10^{30} (h) 1.0×10^3 (i) 7.84×10^6
   (j) 1.0648×10^{-5}
2 (a) 2.1×10^8 (b) 2.4×10^{-5} (c) 2.133×10^{-1}
   (d) 6.378×10^1 (e) 5.2   (f) 4.416×10^{-3}
   (g) 2.684×10^4 (h) 1.463×10^{-9}
3 (a) Earth   (b) 5.91×10^3 km (5910 km)
   (c) 1.949×10^4 (d) 6.543
4 Accept 400–600 secs or 6–10 minutes.

Exercise 5S

1 (a) £3150   (b) £7200
2 £224.97   **3** 37 mpg   **4** £1111.11
5 (a) 1.40×10^8 (b) 1.94×10^6
6 Accept £8 × 10^6 to £1.2 × 10^7.
7 Accept 3×10^8 to 4×10^8.

Unit 6 Transformations and loci

Exercise 6A

1
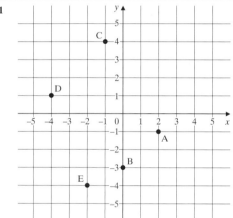

2 (a)
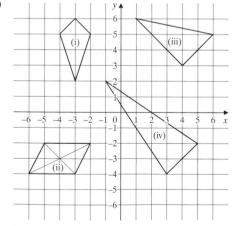

   (b) (i) $(-3, 5)$ (ii) $(-4, -3)$
3 (a) $(3, -4)$ (b) $(-5, -5)$ (c) $(5, 2)$ (d) $(-3, -1)$
4 $(-4, 5)$ $(-2, -3)$ $(6, -1)$

Exercise 6B

1 (a) (b) (c) (d) (e)
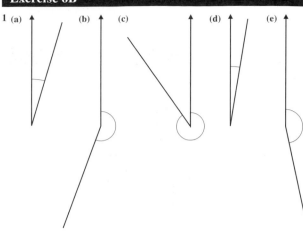

2 (a) 063°   (b) 170°   (c) 332°
3 (a) 235°   (b) 115°   (c) 295°
4 (a) 010°–020°, 560–580 m   (b) 001°, 520 m
   (c) 140°, 680 m   (d) 190°–195°

Exercise 6C

1 (a) $\begin{pmatrix} 2 \\ -2 \end{pmatrix}$ (b) $\begin{pmatrix} -2 \\ 2 \end{pmatrix}$ (c) $\begin{pmatrix} 5 \\ 0 \end{pmatrix}$ (d) $\begin{pmatrix} -7 \\ 2 \end{pmatrix}$ (e) $\begin{pmatrix} -6 \\ -3 \end{pmatrix}$
2 (a)
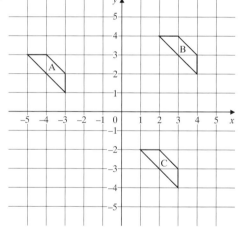

   (c) $\begin{pmatrix} 6 \\ -5 \end{pmatrix}$ (d) $\begin{pmatrix} -6 \\ 5 \end{pmatrix}$
3
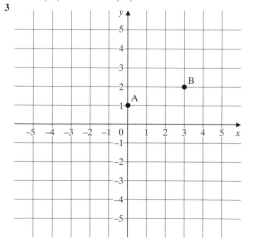

(a) A (3, 3) B (6, 4)
(b) A (6, 5) B (9, 6)
(c) $\begin{pmatrix} 6 \\ 4 \end{pmatrix}$
(d) A (1, 4) B (4, 5)
(e) A (1, 4) B (4, 5)
(f) They are the same.

Exercise 6D

1

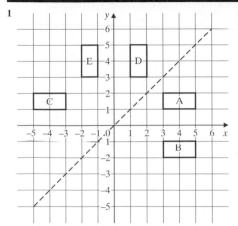

(a) (ii) (3, −1) (3, −2) (5, −2) (5, −1)
     (iii) x coordinate remained the same
            y coordinate changed sign
(b) (ii) (−3, 1) (−3, 2) (−5, 2) (−5, 1)
     (iii) x coordinate changed sign
            y coordinate remained the same
(e) Reflection in line y = −x

2

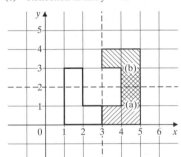

3

Reflection in:	Transforms
x-axis	point (a, b) to (a, −b)
y-axis	point (a, b) to (−a, b)
line y = x	point (a, b) to (b, a)
line y = −x	point (a, b) to (−b, −a)
line x = c	point (a, b) to (2c −a, b)
line y = d	point (a, b) to (a, 2d −b)

Exercise 6E

1

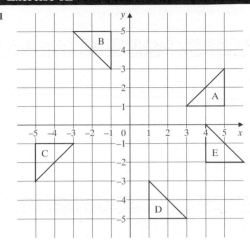

(a) (−1, 3) (−1, 5) (−3, 5)
(b) (−3, −1) (−5, −1) (−5, −3)
(c) Rotate 180° about the origin
(d) (1, −3) (1, −5) (3, −5)
(e) Rotate 180° about the origin
3 (a) (−3, −4)
  (b) (3, −1)
  (c) (−2, −1)

Exercise 6F

1 (a)

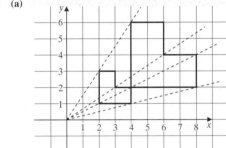

(b)

2 (a)

(b) 1:2

3

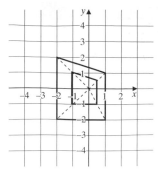

4

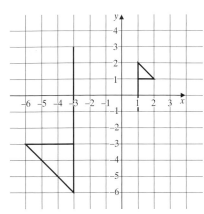

5 (a) (1, 1) scale factor 2 (b) (3, 3) scale factor 3
(c) (3, 3) scale factor $\frac{1}{3}$ (d) (2, 0) scale factor –1
6 Repeat 4 [FD 45 RT 90]

Exercise 6G

1

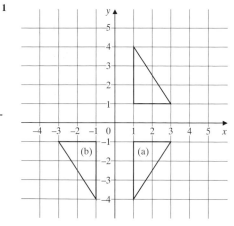

(c) Rotation 180° about origin.

2

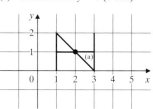

(c) Translation $\begin{bmatrix} 6 \\ 0 \end{bmatrix}$

3

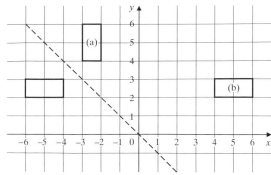

(c) Reflection in $x = 0$ (y-axis)

4

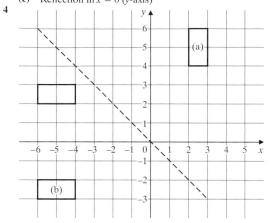

(c) Reflection in $y = 0$ (x-axis)

5

(b) Rotation 180° about point (2, 1)

6

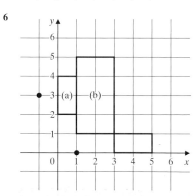

7

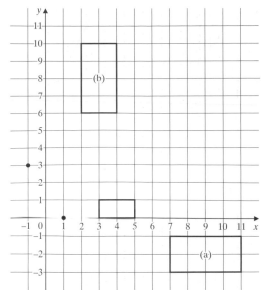

8 Rotation of 45° anti-clockwise and enlargement with scale factor √2.

Exercise 6H

1

Polygon	Number of sides	Interior angle	Tessellates?
Pentagon	5	108°	No
Hexagon	6	120°	Yes
Octagon	8	135°	No
Decagon	10	144°	No

Exercise 6I

1 (a) 7.6 m   (b) 4.3 m
2 (a) 80 m   (b) 1.5 m
3 (a) 13 km   (b) 30 km   (c) 2.7cm
4 (a) 1:200   (b) 3 cm by 2 cm   (c) 6 m by 4 m   (d) 2 m
5

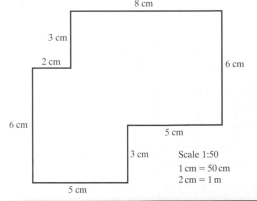

Exercise 6J

1 (a)

(b) 440 m to 3 s.f.

2 (a) 15 cm   (b)

30 cm

3

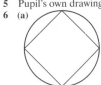

0.8
1.2   0.3

4 Daniel could attend Ayleton or Bankbury
5 Pupil's own drawing as answer
6 (a)   (b)   (c)

7 (a)   (c) 031°   (d) 56 km

30 km   L
75°
B
40 km
H

8 (a) ellipse   (b) parabola
9   **10**

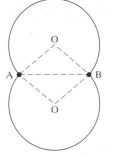 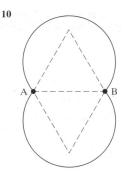

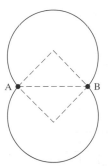

Exercise 6K

7 (a) $x = 120°$ (b) $y = 30°$
9 8.77 cm

Unit 7 Lines, simultaneous equations and regions

1 **(a)**

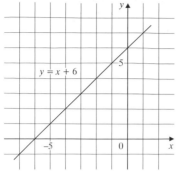

(b)

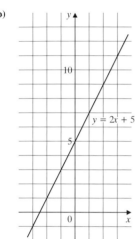

(c)

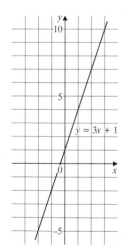

(d)

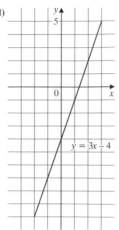

(e)

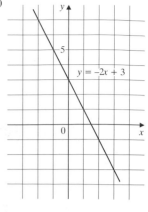

(f)

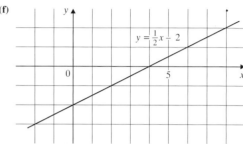

2 $y = 3x + 1$ and $y = 3x - 4$
3 **(a)** $y = \frac{1}{4}x + 1$ and $y = \frac{1}{4}x - 1$
 (b) Coefficients of x equal
4 **(a)** $(0, 5)$ **(b)** $(0, -2)$ **(c)** $(0, -5)$ **(d)** $(0, 4)$
5 $y = 4x + 3$ **6** $y = 5x - 3$
7 $y = 3x + 5$ **8** $y = mx + c$
9 $c = 4; (0, 4)$ **10** $c = -3; (0, -3)$
11 **(a)** $(0, 1)$ **(b)** 5
12 **(a)** $y = \frac{1}{3}x - 4$ **(b)** $(0, -4)$
13 **(a)**

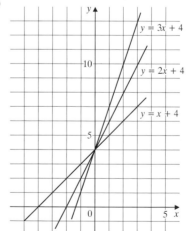

 (b) all $(0, 4)$
 (c) Lines with higher coefficients are steeper.
14 **(a)** 3 **(b)** 3 **15** **(a)** $\frac{1}{2}$ **(b)** $\frac{1}{2}$

1 **(i)** **(a)** 5 **(b)** $(0, 4)$ **(ii)** **(a)** 2 **(b)** $(0, -7)$
 (iii) **(a)** $\frac{1}{4}$ **(b)** $(0, 9)$ **(iv)** **(a)** -4 **(b)** $(0, -3)$
 (v) **(a)** 9 **(b)** $(0, 8)$ **(vi)** **(a)** -2 **(b)** $(0, 7)$

2 (i) (a) -2 (b) $(0, 3)$ (ii) (a) 3 (b) $(0, -5)$
(iii) (a) $\frac{1}{2}$ (b) $(0, -4)$ (iv) (a) $-\frac{1}{3}$ (b) $(0, 2)$
(v) (a) $\frac{2}{5}$ (b) $(0, -2)$ (vi) (a) $-\frac{3}{4}$ (b) $(0, 3)$
3 $y = 6x + 7$
4 (a) $y = 2x + 5$ (b) $y = \frac{1}{2}x - 1$
(c) $x + y = 8$ or equivalent (d) $5x + 2y + 10 = 0$
5 (a) 3 (b) $y = 3x + 2$
6 $y = 4x + 7$ **7** $y = -3x + 11$
8 (a) 4 (b) $y = 4x + 3$
9 $m = 10; c = 30$ **10** $a = 40; b = 20$

Exercise 7C

1 (a) $-\frac{1}{5}$, (b) $-\frac{1}{2}$ (c) -1 (d) $-\frac{4}{3}$ (e) $\frac{2}{5}$ (f) $\frac{7}{2}$
2 (b) **3** $y = -\frac{1}{3}x$ **4** $y = -x + 6$ **5** $y = \frac{5}{2}x - 9$
6 (c) **7** $y = -\frac{3}{2}x + 10$ **8** $y = \frac{3x}{4} - 5$

Exercise 7D

1 e.g. $(1, 9)$ **2** (b), (c), (d) **3** e.g. $(5, 2)$
4 (b), (c) **5** $x = 3, y = 1$
6 (a) $x = 2, y = 1$ (b) $x = 4, y = 2$ (c) $x = 4, y = 1$
(d) $x = -3, y = 4$ (e) $x = 3, y = -1$ (f) $x = 4, y = 3$

Exercise 7E

1 $x = 3, y = 2$ **2** $x = 4, y = 1$ **3** $x = 2, y = -1$
4 $x = -2, y = -3$ **5** $x = \frac{1}{2}, y = 1\frac{1}{2}$ **6** $x = 2\frac{1}{2}, y = -1$
7 $x = 7, y = -1\frac{1}{2}$ **8** $x = \frac{3}{4}, y = -3$ **9** $x = -1, y = -2$

Exercise 7F

1 $x = 4, y = 1$ **2** $x = 3, y = -2$ **3** $x = -3, y = -5$
4 $x = 3, y = 2$ **5** $x = \frac{1}{2}, y = -4$ **6** $x = 5, y = 3$
7 $x = -7, y = -4$ **8** $x = 3, y = -5$ **9** $x = 2\frac{1}{2}, y = 1$
10 $x = 1\frac{1}{3}, y = -1$

Exercise 7G

1 £5 per hour **2** $a = 25, b = 45$ **3** 29 and 37
4 £9.30 **5** 18 **6** $a = 5, b = -5$
7 £3 **8** 11p **9** 35p
10 (a) $a = 7, b = -2$ (b) (i) $3\frac{1}{2}$ (ii) $(0, -5\frac{1}{2})$

Exercise 7H

1

2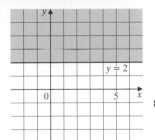

3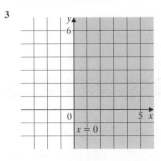

4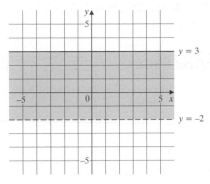

5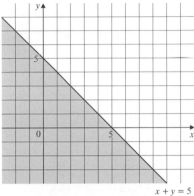

6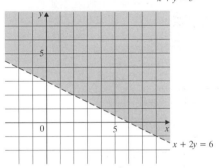

7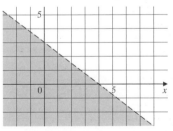

8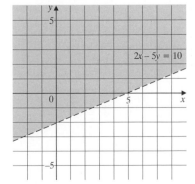

9

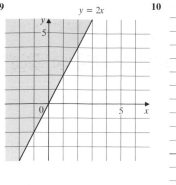

10

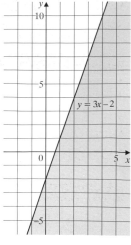

18

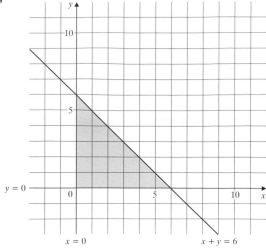

11

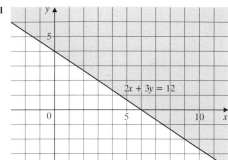

19

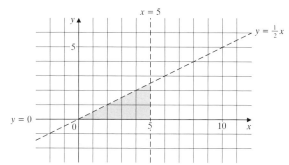

12

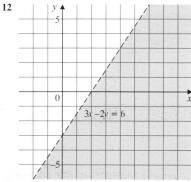

20

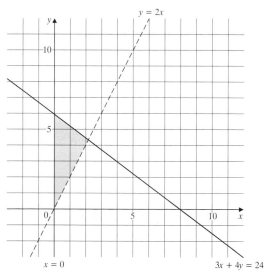

13 $x \leqslant 1$ **14** $x + y > 3$ **15** $y \geqslant x$ **16** $2x + y > 4$

17

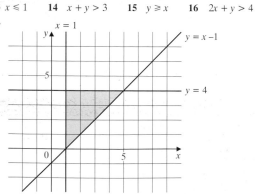

21

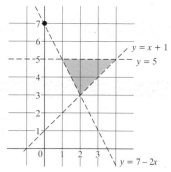

22

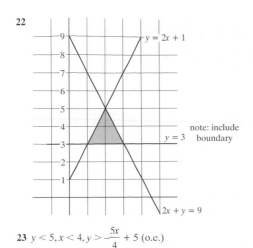

$y = 2x + 1$

$y = 3$ note: include boundary

$2x + y = 9$

23 $y < 5, x < 4, y > -\dfrac{5x}{4} + 5$ (o.c.)

Unit 8 Pythagoras' Theorem

Exercise 8B

1 (a) 5 cm   (b) 13 cm   (c) 13.04 cm
(d) 8.94 cm   (e) 8 cm   (f) 41 cm
2 26 cm   **3** 14.14 cm   **4** 9.8 m   **5** 129.7 km

Exercise 8C

1 (a) $x = 13$ cm $y = 19.85$ cm   (b) $x = 6$ cm $y = 12.04$ cm
(c) $x = 9$ cm $y = 50.8$ cm

Exercise 8D

1 (a) 5   (b) 4.47   (c) 11.31   (d) 14.87
(e) 5   (f) 12.08
2 5.83 units   **4** Any eight of: $(5, 0), (0, 5), (-5, 0), (0, -5)$
$(3, 4), (3, -4), (-3, 4), (-3, -4)$
$(4, 3), (4, -3), (-4, 3), (-4, -3)$

Exercise 8E

1 (a) $x^2 + y^2 = 16$ (b) $x^2 + y^2 = 49$
2 (a) $x^2 + y^2 = 25$ (b) $x^2 + y^2 = 144$ (c) $x^2 + y^2 = 225$
3 $x^2 + y^2 = 169$

Exercise 8G

1 (a) obtuse   (b) acute   (c) obtuse   (d) right
(e) obtuse   (f) obtuse
3 (a) Yes   (b) Yes   (c) Yes   (d) No
(e) Yes   (f) No

Exercise 8H

1 15.26 cm
3 (a) (i) 13 cm   (ii) 17.72 cm   (iii) 20.81 cm   (b) acute

Exercise 8I

1 12.37 cm   **2** 6.24 m   **3** 30.81 km
4 $x = 26$ cm $y = 14.97$ cm   **5** 13.6   **6** obtuse
7 (a) No   (b) Yes   (c) No   **8** 20.81 cm
9 (a) 13 cm   (b) 23.32 cm   (c) 23.85 cm   **10** $x = 5$

Unit 9 Probability

(Any correct fraction, decimal or percentage equivalents are acceptable unless otherwise stated.)

Exercise 9A

1 (a) $\frac{4}{9}$ (b) $\frac{2}{9}$
2 (a) $\frac{3}{10}$ (b) $\frac{8}{10}$
3 (a) $\frac{1}{6}$ (b) $\frac{3}{6}$ (c) $\frac{3}{6}$ (d) $\frac{5}{6}$
4 (a) $\frac{3}{8}$ (b) $\frac{2}{8}$ (c) $\frac{5}{8}$
5 (a) $\frac{6}{20}$ (b) $\frac{10}{20}$ (c) $\frac{16}{20}$ (d) $\frac{16}{20}$
   (e) Add the answers to (a) and (b)
   (f) Subtract p from 1

Exercise 9B

1 (a)

+	1	3	5	7
2	3	5	7	9
4	5	7	9	11
6	7	9	11	13
8	9	11	13	15

(b) $\frac{3}{16}$

2 (Only correct fraction equivalents are acceptable.)
(a) $\frac{1}{6}$ (b) $\frac{9}{36}$ (c) $\frac{6}{36}$ (d) $\frac{15}{36}$
3

H	H	H	H	T	H	H	H	T	T	T	H	T	T	T	T
H	H	H	T	H	H	T	T	H	H	T	T	H	T	T	T
H	H	T	H	H	T	H	T	H	T	H	T	T	H	T	T
H	T	H	H	H	T	T	H	T	H	H	T	T	T	H	T

(a) $\frac{1}{16}$ (b) $\frac{6}{16}$ (c) $\frac{15}{16}$
4

Red	Red	Red	Red	Red	Red	Red	Red	Red
Bro	Bro	Bro	Gre	Gre	Gre	Yel	Yel	Yel
Blu	Pin	Bla	Blu	Pin	Bla	Blu	Pin	Bla
Whi	Whi	Whi	Whi	Whi	Whi	Whi	Whi	Whi
Bro	Bro	Bro	Gre	Gre	Gre	Yel	Yel	Yel
Blu	Pin	Bla	Blu	Pin	Bla	Blu	Pin	Bla

$\frac{3}{18}$
5 (a) $\frac{15}{36}$ (b) $\frac{18}{36}$ (c) $\frac{27}{36}$ (d) $\frac{22}{36}$
6 (a) $\frac{1}{6}$ (b) $\frac{1}{8}$ (c) $\frac{4}{48}$ (d) $\frac{6}{48}$
(e) $\frac{6}{48}$ (f) $\frac{3}{48}$

Exercise 9D

1 (a) $\frac{1}{5}$ (b) (i) $\frac{1}{25}$ (ii) $\frac{1}{5}$
2 (a) $\frac{1}{12}$ (b) $\frac{1}{10}$ (c) $\frac{26}{120}$

Exercise 9E

1 (a) $\frac{1}{60}$ (b) $\frac{24}{60}$ (c) $\frac{9}{60}$ (d) $\frac{10}{60}$
2 (a) $\frac{2}{60}$ (b) $\frac{13}{60}$ (c) $\frac{45}{60}$

Unit 10 Brackets in algebra

Exercise 10A

1 25   **2** 5   **3** 20   **4** 18   **5** 2
6 5   **7** 25   **8** −1   **9** 1.8   **10** 0.9
11 −21   **12** 0   **13** $4 \times (6 - 1)$
14 $(7 \times 3) - 2$ **15** $9 - (5 - 4)$
16 $(7 - 2) \times (6 + 3)$ **17** $(6 \times 8) - (5 \times 7)$
18 $2 \times (8 + 2) \times 3$
19 (a) (i) 11   (ii) 11   (b) (i) 1   (ii) 7
   (c) (i) 30   (ii) 30   (d) (i) 1   (ii) 4
20 (a) (i) 40   (ii) 40   (b) (i) 28   (ii) 28

Exercise 10B

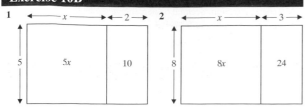

3	$7x + 21$	4	$8x - 16$	5	$5x - 25$	6	$6x + 24$
7	$12x - 30$	8	$27x + 18$	9	$35x - 5$	10	$18x + 14$
11	$x^2 + 4x$	12	$x^2 - 8x$	13	$2x^2 + 9x$	14	$5x^2 - 7x$
15	$-5x - 10$	16	$-8x + 24$	17	$-21x + 14$	18	$-6x - 42$
19	$45x^2 - 36x$	20	$28x^2 + 32x$	21	$-12x^2 + 18x$	22	$ax - 8a$
23	$x^2y - xy^2$	24	$15ax - 6a$	25	$24ax + 16a^2$	26	$20a^2x - 4a$
27	$3x + x^2$	28	$2x^2 - 9x$	29	$x^3 + 7x^2$	30	$4x^2 - 3x^3$

Exercise 10C

1	$5(x - 4)$	2	$8(x + 3)$	3	$6(2x + 3)$
4	$5(4x - 5)$	5	$8(x^2 - 3)$	6	$9(x^2 + 4)$
7	$5(2x^2 - 3)$	8	$x(x + 6)$	9	$x(x - 1)$
10	$p(x - 3)$	11	$q(x + q)$	12	$5x(x - 3)$
13	$3x(2x + 3)$	14	$5x(3x - 7)$	15	$4x(2 - 3x)$
16	$ax(x + 1)$	17	$2a(2x - 3)$	18	$3ax(2x + 5)$
19	$5n + 10 = 5(n + 2)$	20	$4n + 6 = 2(2n + 3)$		

Exercise 10D

1	$7x + 1$	2	$7x + 9$	3	$12x + 5$	4	$36x$
5	$2x - 8$	6	$x + 29$	7	$2x - 3$	8	$4x - 1$
9	x	10	21	11	$6x + 12$	12	$11x - 12$
13	$10x - 6$	14	$x - 13$	15	$7x$	16	$6x + 24$
17	$11x$	18	-17	19	$4x + 3$	20	$2 - x$
21	$10x + 15$	22	$9x - 29$	23	$-9x - 2$	24	$6x + 1$
25	39						

Exercise 10E

1	4	2	-3	3	$2\frac{1}{6}$	4	$3\frac{1}{3}$	5	$\frac{4}{5}$
6	$6\frac{1}{2}$	7	2	8	-2	9	2	10	4
11	$\frac{1}{2}$	12	15	13	-14	14	0	15	$\frac{1}{2}$

Exercise 10F

1	18	2	7	3	5	4	39	5	10 years
6	7	7	$2\frac{1}{2}$ hours	8	5	9	3 hours		

Exercise 10G

1	$x > 4$	2	$x > 11\frac{2}{3}$	3	$x \leqslant -3\frac{1}{2}$	4	$x \geqslant 2$
5	$x > -\frac{1}{2}$	6	$x \leqslant -1$	7	$x > 2\frac{1}{2}$	8	$x < -1$
9	$x \leqslant 3\frac{1}{5}$	10	$x < 0$	11	$x < -1\frac{1}{3}$	12	$x \geqslant -\frac{1}{2}$
13	$x < -8$	14	$x < -2$	15	$x \leqslant 3\frac{2}{5}$		

Exercise 10H

1	$\dfrac{ab + c}{a}$	2	$\dfrac{c}{a - b}$	3	$\dfrac{ab}{c - a}$	4	$\dfrac{c - a}{b - d}$
5	$\dfrac{c - ab}{a + d}$	6	$\dfrac{a - c}{b + d}$	7	$\dfrac{cd - ab}{a - c}$	8	$ab + ac$
9	$\dfrac{e - bd}{a + b}$	10	$\dfrac{bc}{1 + ab}$	11	$\dfrac{ac}{1 - ab}$	12	$\dfrac{abc}{1 - ab}$
13	$\dfrac{e - ab - cd}{a + c}$	14	$\dfrac{a + bc - e}{b + d}$	15	$\dfrac{de - bc}{a + b - d}$		

Exercise 10I

1	$x = 2, y = 7$	2	$x = 1, y = -1$	3	$x = -2, y = 3$
4	$x = -3, y = -6$	5	$x = \frac{1}{2}, y = 1$	6	$x = 1\frac{1}{3}, y = -3$
7	$z = 23t - 24$	8	$z = t + 24$	9	$z = 7t - 14$
10	$z = 3t + 14$				

Exercise 10J

1 (i) (a) $x^2 + 3x + 2$ **(b)**

(ii) (a) $x^2 + 10x + 16$ **(b)**

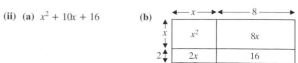

(iii) (a) $x^2 + 6x + 9$ **(b)**

(iv) (a) $x^2 + ax + bx + ab$ **(b)**

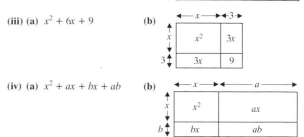

2	$x^2 + 5x + 4$	3	$x^2 - x - 20$	4	$x^2 - 7x - 18$
5	$x^2 - 12x + 35$	6	$x^2 + 5x - 24$	7	$x^2 - 7x + 6$
8	$x^2 + 6x - 16$	9	$10 + 3x - x^2$	10	$x^2 - 9x + 8$

Exercise 10K

1	$x^2 + 10x + 25$	2	$x^2 - 6x + 9$	3	$x^2 + 18x + 81$
4	$x^2 - 4x + 4$	5	$x^2 + 2ax + a^2$	6	$x^2 - 2ax + a^2$
7	$x^2 + 6x + 9$	8	$x^2 - 14x + 49$	9	$x^2 - 2ax + a^2$

10 (a) $(x + 4)^2 = x^2 + 8x + 16$ **(b)** $(x - 10)^2 = x^2 - 20x + 100$
     (c) $(x + 1)^2 = x^2 + 2x + 1$ **(d)** $(x - 12)^2 = x^2 - 24x + 144$

Exercise 10L

1	$x^2 - 25$	2	$x^2 - 1$	3	$x^2 - 100$	4	$x^2 - 81$
5	$x^2 - 49$	6	$x^2 - 9$	7	$x^2 - a^2$	8	$36 - x^2$
9	$9 - x^2$	10	$49 - x^2$				

Exercise 10M

1	$15x^2 + 11x + 2$	2	$14x^2 + 29x - 15$	3	$16x^2 + 18x - 9$
4	$12x^2 - 23x + 5$	5	$15x^2 + 23x + 4$	6	$12x^2 + 4x - 5$
7	$6x^2 - 23x + 21$	8	$36x^2 + 41x - 5$	9	$5x^2 - 20x - 60$
10	$4x^2 - 4$	11	$30x^2 + 4x - 2$	12	$36x^2 - 3x - 18$
13	$4x^2 + 28x + 49$	14	$16x^2 - 24x + 9$	15	$25x^2 - 10x + 1$
16	$9x^2 + 12x + 4$	17	$9x^2 - 1$	18	$1 - 16x^2$
19	$4x^2 - 25$	20	$49x^2 - 4$		

Unit 12 Estimation and approximation

Exercise 12A

1	**(a)** 270	**(b)** 130	**(c)** 70	**(d)** 90
	(e) 600	**(f)** 840	**(g)** 10	**(h)** 0
2	**(a)** 3	**(b)** 7	**(c)** 98	**(d)** 1
	(e) 0	**(f)** 80	**(g)** 4	**(h)** 6

3 (a) 4×7 (b) 4×4 (c) 9×3 (d) 12×3
(e) 5×5 (f) 0×6
4 (a) 28 (b) 16 (c) 27 (d) 36
(e) 25 (f) 0
5 (a) (i) 49 440 (ii) 49 400 (iii) 49 000

Exercise 12B

1 (a) 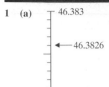 (b) 46.383

2 (a)

Name	Time
R Grey	1 min 45.48 sec
M Hobson	1 min 45.48 sec
T Knight	1 min 45.48 sec

(b) M. Hobson, T. Knight, R. Grey

3 (a) 137.3 (b) 0.6738 (c) 5.0 (d) 9.00
(e) 17.994 (f) 2.0097
4 (a) 95.047 (b) 9.760 (c) 10.91 (d) 176.0

Exercise 12C

1 (a) 3.17 (b) 965 (c) 16 (d) 10
(e) 55.764 (f) 55.90
2 (a) 196.161 (b) 196 (c) 196.16 (d) 200
(e) 196 (f) 200 (g) 200 (h) 196.1614
3 (a) 43 670 (b) 44 000 (c) 40 000
4 (a) 1940 (b) 4450 (c) 9800 (d) 86 000
(e) 1000 (f) 60 000
5 (a) 0.003 (b) 0.000 88 (c) 0.0090 (d) 0.0009
(e) 0.01 (f) 0.0004
6 (a) 0.007 (b) 0.066 67 (c) 0.000 0029
7 (a) 0.39 (b) 0.039 (c) 39.4 (d) 40 000

Exercise 12D

1 (*Note:* there are alternatives to each (i) and (ii).)
(a) (i) 9×9 (ii) 81 (iii) 82.7896
(b) (i) $4 \times (2 - 0)$ (ii) 8 (iii) 7.3875
(c) (i) 40×4 (ii) 160 (iii) 160.7296
(d) (i) $(2000 \times 300) + 400$ (ii) 600 400
(iii) 633 086
(e) (i) $\dfrac{200 \times 75}{50}$ (ii) 300 (iii) 294.831 3152
(f) (i) $\dfrac{0.5 \times 0.02}{0.05}$ (ii) 0.2 (iii) 0.262 550 588
2 (a) 8.5 (b) 40.488 (c) 4.6706 (d) 134.3632
(e) 76.704 (f) 8.0736
3 (a) 7 (b) 6 (c) 9 or 10 (d) 3
(e) 8 (f) 10 (g) 7 (h) 8
(i) 6 or 7 (j) 3 (k) 10 (l) 3 or 4 or 5
4 10 $(\sqrt{4^2 + 9^2} = \sqrt{97} \simeq \sqrt{100})$

Exercise 12E

1

School	Attendance to the nearest 100	Least possible attendance	Greatest possible attendance
Southpark	1400	1350	1450
Westown	400	350	450
Eastgrove	800	750	850
Northbury	1200	1150	1250
Southone	2100	2050	2150

2 (a) 3.85 cm–3.95 cm (b) 17.6195 m–17.6205 m
(c) 52.95 cm–53.05 cm (d) 3.0095 m–3.0105 m
(e) 28.5 m–29.5 m
3 4 min 7.555 sec to 4 min 7.565 sec
4 16.5 miles to 17.5 miles

Exercise 12F

1 (a) 29.5 cm (b) 18.45 cm, 18.35 cm
2 (a) 3.65 cm (b) 15.0 cm, 14.6 cm
(c) 2 s.f. (d) Yes, 2 s.f.
3 (a) 99.5 m
(b) (i) 14.75 seconds (ii) 14.85 seconds
(c) (i) 6.700 3367 m/s (ii) 6.813 5593 m/s
(d) (i) 7 m/s
(ii) Answers in (c) agree to 1 s.f.
4 (a) 5.65 cm, 5.55 cm
(b) 120.5 cm 119.5 cm
(c) 6042.32 cm^3 5781.94 cm^3
(d) 11.298 96 cm 11.1343 cm
5 (a) 8 450 000 km^2–8 550 000 km^2
(b) 17.1 people/km^2, 16.78 people/km^2
6 (a) (i) 815 (ii) 805
(b) (i) 2.935 (ii) 2.925
(c) (i) 278.632 478 6 (ii) 274.2759796 (d) 276.5 to 4 s.f.
(There are alternatives; the quoted answer is the mean of the upper and lower bounds.)

Unit 13 Basic trigonometry

Exercise 13A

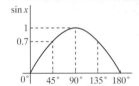
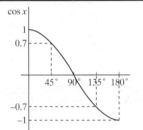

Answers to include:

Angle x	$\sin x$	$\cos x$
0	0	1
45	0.7	0.7
90	1	0
135	0.7	−0.7
180	0	−1

Exercise 13B

(All answers to 4 d.p.)
(a) 0.7314 (b) 0.5299 (c) 0.3839 (d) 8.1443 (e) 0.9063
(f) 0.8660 (g) −3.4874 (h) −0.0523 (i) 0.7431

Exercise 13C

(All answers to nearest 0.1°)
(a) 20.6° (b) 82.2° (c) 54.7° (d) 23.6° (e) 9.8°
(f) 62.3° (g) −45° (h) 135°

Exercise 13D

(All answers to nearest 0.1°)
(a) 44.4° (b) 38.7° (c) 48.2° (d) 21.8° (e) 24.6°
(f) 60° (g) 18.4° (h) 24.3° (i) 52.4°

Exercise 13E

(All answers to 2 d.p.)
(a) 7.66 (b) 7.71 (c) 12.99 (d) 5.34 (e) 7.52
(f) 6.75 (g) 6.36 (h) 12.12 (i) 7.16

Exercise 13F

1 (a) 12.37 m (b) 72.1° 2 (a) 8.15 m (b) 38.7°
3 7.71 m 4 021.6° 5 (a) 19.36 cm (b) 75.5°
6 14.9° 7 Sin 35° = 0.5736 not 0.7
8 (a) $h = 1247.47$ m (b) 5868.89 m
9 (a) 23.6° (b) 25.24 m (c) 33.7° (d) 53.08 m
10 (a) 26.4° (b) 10.6 m (c) 22.7 m (d) 125.9 m

Exercise 13H

1 (a)

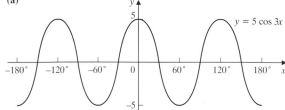

(b) 120° (c) Max of 5 at 0°, 120°, −120°
Min of −5 at 60°, −60°, 180°, −180°

2 (a)

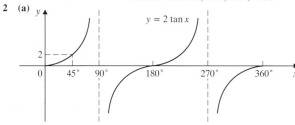

(b) 180°

3 (a)

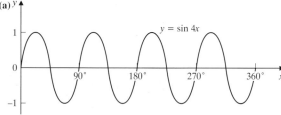

Period 90°

(b)

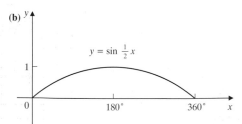

Period 720°

Exercise 13I

1 60°, −60°, 300°, −300°
2 (b) 23.6°, 156.4°, 383.6°, 516.4°
3 63.4°, −116.6°
4 (a)

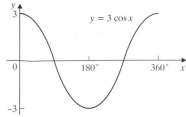

(b) 70.5°, 289.5°
5 (b) 17.71°, 42.29°, 137.71°, 162.29°, −102.29°, −77.71°

Exercise 13J

1 (a) 20.17 cm (b) 57.2°
2 (a)

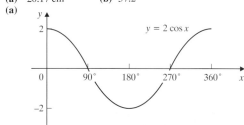

(b) 120°, 240°

Unit 14 Sequences and formulae

Exercise 14A

1 2, 4, 6, 8, 10 2 3, 4, 5, 6, 7
3 10, 13, 16, 19, 22 4 13, 11, 9, 7, 5
5 6, 9, 14, 21, 30 6 2, 11, 26, 47, 74
7 4, 9, 16, 25, 36 8 1, 8, 27, 64, 125
9 2, 4, 8, 16, 32 10 −2, 4, −8, 16, −32
11 $\frac{1}{2}, \frac{1}{4}, \frac{1}{8}, \frac{1}{16}, \frac{1}{32}$ 12 1, 3, 6, 10, 15
13 $1, \frac{1}{2}, \frac{1}{3}, \frac{1}{4}, \frac{1}{5}$ 14 $\frac{1}{2}, \frac{2}{3}, \frac{3}{4}, \frac{4}{5}, \frac{5}{6}$
15 17

Exercise 14B

1 8 2 12 3 17 4 7 5 12
6 9 7 21 8 8 9 7 10 20

Exercise 14C

1 $n + 2$ 2 $2n + 3$ 3 $4n + 3$ 4 $10 - n$
5 $14 - 2n$ 6 $10 - 3n$ 7 $\frac{1}{n + 1}$ 8 $\frac{n + 1}{n}$
9 n^2 10 $n^2 + 1$ 11 $2n^2$ 12 $(n + 2)^2$
13 $(n - 3)^2$ 14 5^n 15 10^n 16 $(-3)^n$
17 $(0.1)^n$ 18 $\frac{1}{2n - 1}$ 19 $\frac{n}{2n - 1}$ 20 $\frac{1}{(n + 1)^2}$

Exercise 14D

1 −11 2 $53\frac{1}{8}$ 3 30.2 4 26.3 5 5
6 47.8 7 223 8 −45 9 100.152 10 −43
11 47.4 12 195.5 13 −65 14 82 15 9.09
16 10.8 17 3.44 18 1.30 19 9.71 20 23

Exercise 14E

1	18	**2**	5.4	**3**	22	**4**	0.8	**5**	4
6	4.5	**7**	−8	**8**	$−1\frac{1}{4}$	**9**	6	**10**	$7\frac{1}{2}$
11	5	**12**	12.6	**13**	7 or −7	**14**	3.99	**15**	4.2
16	27.6	**17**	201	**18**	15.3	**19**	1.5	**20**	4.8

Exercise 14F

1 $I = \dfrac{V}{R}$ **2** $L = \dfrac{A}{B}$ **3** $d = \dfrac{C}{\pi}$ **4** $B = \dfrac{V}{LH}$

5 $u = v - at$ **6** $t = \dfrac{v - u}{a}$ **7** $h = \dfrac{C}{2\pi r}$ **8** $R = \dfrac{100I}{PT}$

9 $L = \dfrac{P - 2B}{2}$ **10** $h = \dfrac{A - 2\pi r^2}{2\pi r}$ **11** $x = \dfrac{y}{3}$

12 $T = \dfrac{PV}{k}$ **13** $x = \sqrt{\dfrac{y}{5}}$ **14** $x = \sqrt{\dfrac{E}{k}}$ **15** $r = \sqrt{\dfrac{A}{\pi}}$

16 $n = \dfrac{2S}{a + d}$ **17** $a = \dfrac{2S - nd}{n}$ **18** $x = \sqrt{\dfrac{y}{k}}$

19 $h = \dfrac{3V}{\pi r^2}$ **20** $r = \sqrt{\dfrac{3V}{\pi h}}$ **21** $\sqrt{c^2 - b^2}$

22 $b = \sqrt{\dfrac{a^2 + 8}{4}}$ **23** $s = \dfrac{gt^2}{2}$ **24** $L = \dfrac{gT^2}{4\pi^2}$

25 $y = \sqrt{1 - x^2}$ **26** $u = \sqrt{v^2 - 2as}$ **27** $x = \sqrt{\dfrac{a^2 w^2 - v^2}{w^2}}$

28 $x = \sqrt{\dfrac{9 - y^2}{9}}$ **29** $x = \dfrac{y - 5}{y + 1}$ **30** $u = \dfrac{vf}{v - f}$

Exercise 14G

1 $P = \dfrac{V^2}{R}$ **2** $s = \frac{1}{2}gt^2$ **3** $A = \dfrac{\pi d^2}{4}$ **4** $a = rw^2$

5 $y = 2x + 5$ **6** $A = \dfrac{p^2}{16}, p = 4\sqrt{A}$ **7** $x = \dfrac{y^2}{4a}$

8 $A = 2\pi r^2 + \dfrac{2V}{r}$ **9** $F = \dfrac{2A}{l} + l$ **10** $A = 2x^2 + \dfrac{4V}{x}$

Exercise 14H

1 **(a)** (i) $g = 10$ $t = 20$ (ii) 2000 **(b)** 2748.12
2 **(a)** 4.35 m, 4.25 m; 3.75 m, 3.65 m **(b)** 16.3 m², 15.5 m²
3 **(a)** $\pi = 3, r = 7, h = 10$ give 1470 cm³ **(b)** 1860 cm³
    (c) 1891.88 cm³
4 **(a)** $f = 4(v - 55)$ **(b)** $v = \dfrac{f}{4} + 55$ or $v = \dfrac{f + 220}{4}$

Unit 15 Averages and measures of spread

Exercise 15A

1 **(a)** (i) 26.4 (ii) 34 (iii) 28.5
    (b) Mean is lowest – the '4' depresses its value compared to the mode and median.
    (c) 23.4, 31, 25.5
2 **(a)** 32.3̇ **(b)** No – because the ages are so spread out
3 **(a)** 2.83̇ **(b)** $(1 + 2 + 2 + 3 + 4 + 5) \div 6$
    (c) Not an integer
    (d) 2 – because it appears twice (assuming each section has the same probability of landing)
4 It does not specify which average. If it is the mean, then its value may be affected greatly by a few very well paid people.
5 For the typical family size she could use the mode, as most families would be this size. For the amount of money, she could use both the mean and median – the mean because it takes into account all the data, and the median because it will give the 'middle' amount.

Exercise 15B

1

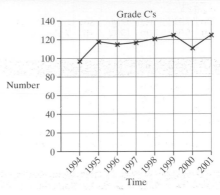

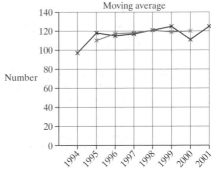

The data has been smoothed out. The trend is increasing. The number is increasing, but might be because the school is getting bigger.

2

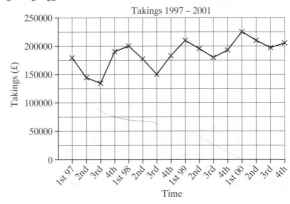

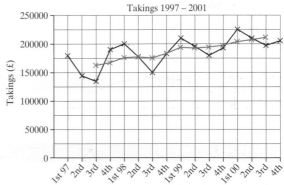

The data has been smoothed out. It shows a steady increase in takings.

3

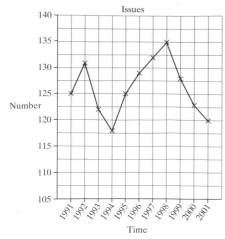

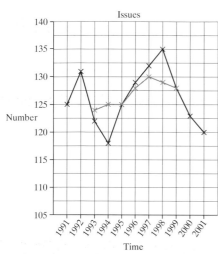

The original data varied considerably. There is some smoothing out of the data. The average increased but has dipped.

Exercise 15C

1 (a) 21.8, 23, no mode   (b) 121.8, 123, no mode
   (c) 8.8, 4, no mode   (d) $3x$, $3x$, no mode
2 (a) 8   (b) 3
3 (a) £15 333, median = mode = £12 000   (b) £15 717
4 (a) 20.8   (b) Lies between 17 and 21
5 (a) 3 cm   (b) (i) 12 cm   (ii) 11 cm^2
6 1.67, 2, 0
7 (a) 5   (b) 4.73, 5
8 (a) 2.56, 2, 2
   (b) Yes, because team mean = total goals ÷ number of teams.

Exercise 15D

1 41.3 s   **2** 62.7 g, 62.99 g (2 d.p.)
3 2.15 cm (2 d.p.)   **4** (a) (i) 6–10 (ii) 11–15   (b) 12.95

Exercise 15E

1 (a) 41.2 s   (b) 50 s   (c) 20 s
2 (a) 2.9 cm   (b) 1.8 cm
3 (a) 38.3 years   (b) 30%
4 (a) 139 000 km^2 (b) 141 000 km^2

5 (a)

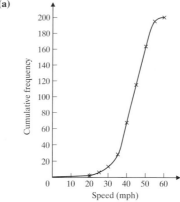

   (b) (i) 43 mph   (ii) 10.7 mph   (iii) 72%

6 (a)

Diameter (cm)	Cumulative frequency
Up to but not including 0.5	8
Up to but not including 1.0	20
Up to but not including 1.5	35
Up to but not including 2.0	52
Up to but not including 2.5	59
Up to but not including 3.0	64

   (b)

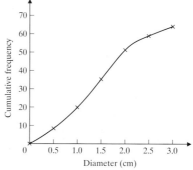

   (c) (i) 1.4 cm   (ii) 1 cm
7 (a) 45%   (b) 26, 45, 67, 82, 91, 96, 100
   (c)

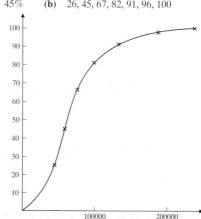

   (d) £82 700

Exercise 15F

1

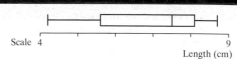

2 Minimum: 0 minutes
Maximum: 432 seconds
Median: 150 seconds
Lower quartile: 60 seconds
Upper quartile: 264 seconds.

3
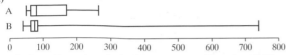

Length (cm)

4 Class B performed better than class A, as the median was higher. The range of performance of the middle 50% was narrower in class A than in class B.
5 Graph A.
6 (a)

A
B

0   100   200   300   400   500   600   700   800

(b) Assistant B may have had only one extreme time value of 740 seconds, the rest were low.

Exercise 15G

1 Not necessarily. Drapeway Ltd may have a few very high wage earners which will push up the value of the mean.
2 On average students did better on paper A (higher median). There was a bigger spread of marks on paper A, with the spread of the middle 50% of candidates being over $2\frac{1}{2}$ times that on paper B.
3 Club 2 has an older median but a smaller spread. Generally there are older people in Club 2 but because the spread in Club 1 is larger, it has some members who are as old as the Club 2 members.
4 (a) On average pupils scored lower on paper 1 than on paper 2. However, there was a lot more variability in their marks.
  (b) On average pupils did better on paper 2 than on paper 3. The spread in marks was similar on both papers.

Unit 16 Measure and mensuration

Exercise 16A

1 (a) 3.35 m, 3.45 m (b) 4.55 m, 4.65 m
  (c) 15.2425 m^2, 16.0425 m^2
2 (a) 194.4625 km (b) 190.5458$\dot{3}$ km

Exercise 16B

1 (a) $6x + 8$ cm (b) 7 (c) 136 cm^2
2 (a) 4 cm (b) 16 cm^2
3 62 500 m^2

Exercise 16C

1 (a) 30 cm^2 (b) 6 cm^2 (c) 45 cm^2
2 (a) $4x - 5$ (b) 8.5

Exercise 16D

1 (a) 40 cm^2 (b) 36 cm^2 (c) 30 cm^2
2 (a) 70 cm^2 (b) 8.75 cm
3 9.49

Exercise 16E

1 58.4 cm^2 **2** 2.4 cm

Exercise 16F

1 (a) 31.42 cm, 78.54 cm^2 (b) 50.27 cm, 201.06 cm^2
  (c) 37.70 cm, 113.10 cm^2

2 (a) 15.71 m (b) 19.63 m^2
3 $\pi r^2 = 2\pi r, r^2 = 2r, r = 2$ (or 0)
4 (a) 2.39 cm (b) 17.90 cm^2 **5** 37.85 cm
6 (a) Area = 39.27 cm^2
  Perimeter = 25.71 cm
  (b) Area = $\dfrac{25\pi}{2}$ cm^2
  Perimeter = $(10 + 5\pi)$ cm
7 (a) 181.70 cm (b) 1666.19 cm^2

Exercise 16G

1 (a) 125 cm^3 (b) 1728 cm^3 (c) 54.872 cm^3
2 (a) 10 cm (b) 600 cm^2
3 (a) 180 cm^3 (b) 480.24 cm^3
4 (a) 4 cm (b) 352 cm^2

Exercise 16H

1 (a) 504 cm^3 (b) 396 cm^2
2 240 cm^3 **3** 12 cm **4** 576 cm^3

Exercise 16I

1 $4\pi r^2$, dimension is 2
2 $n + m$ must equal 3
  so $n = 0, m = 3$ or $n = 1, m = 2$ or $n = 2, m = 1$ or $n = 3, m = 0$

Exercise 16J

1 86.625 m^3
2 (a) 317.08 m (b) 5963.50 m^2
3 50 cm^2 **4** 154.06 cm^2 **5** 960 cm^3
6 (a) $x^2h = 50, x = \sqrt{\dfrac{50}{h}}$ (b) 3.54 (c) 2
7 Volume since dimension is 3.

Unit 17 Proportion

Exercise 17A

1 (a) Yes (b) No (c) No
  (d) Yes (e) Yes (f) No
2 (a) 12 (b) 4
3 (a) 24 (b) $2\frac{1}{2}$
4 (a) 12 (b) 54

Exercise 17B

1 (a)

(10, 8)

  (b) $l = 0.8h$
  (c) $15.625 = 15.63$ (2 d.p.)
  (d) 12

2 (a)

A

(14, 42)

  (b) $A = 3l$
  (c) 97.2 cm^2
  (d) 8.2

3 (a)
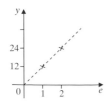

(b) Student's explanation is 2 → 24 so 1 → 12 etc.
(c) 84 ÷ 12 = 7

Exercise 17C

1 (a) $w = 1.05\,h$, $s = \frac{41}{24}\,h$, $p = \frac{39}{34}\,l$, $p = \frac{9}{4}\,a$
    (b) 12.6, 28.7, 31.7, 248.4
2 (a) 9.36 volts, 5.2 amps   (b) $V = \frac{9}{5}I$
    (c) 6.4 amps   (d) 4.428 volts
3 (b) 4.6 should be 5.6   (c) $V = 2.8\,I$
    (d) 9.8   (e) 7.321
4 (a) 1.67   (b) 16.53
5 (a) $C = 3.6\,A$ (b) 360 p
6 (a) $d = 38\,t$ (b) 190 miles   (c) 11 hours

Exercise 17D

1 $12\frac{1}{3}, 86\frac{1}{3}$ 2 $\frac{2}{3}, 4\frac{2}{3}$ 3 $z = k\,x$, 8.4   4 $z = k\,w$, 4725
5 (a) 15   (b) 39   (c) $3\frac{1}{3}$
6 (a) 225   (b) 1250   (c) 72
7 $y = 2.8$, $p = 1.43$ (2 d.p.)   8 5.71 cm^3 (2 d.p.)   9 40%
10 (a) $h = 7.5d$
     (b) 18.75 cm
     (c) $42\frac{2}{3}$ cm

Exercise 17E

1 5, 80
2 (a) 6.25   (b) 156.25   (c) 1.265
3 (a) 2   (b) 128   (c) 4
4 (a) $l = k\,m^3$ (b) (i) 25.6   (ii) 0.952
5 25 000 N
6 (a) 45   (b) 101.25
7 $p = 0.75$, $q = 3$ 8 0.002
9 (a) 1 m   (b) 1.5625 m
10 4   11 125%   12 4.14 m   13 $z = 2$, $w = 1.5$

Exercise 17F

1 (a) 72   (b) 7.2
2 (a) 4.8   (b) 9.6
3 (a) 512   (b) 128
4 (a) 2.5   (b) 40
5 (a) 233   (b) 291

6 (a) $l \propto \frac{1}{w}$, $l = \frac{k}{w}$ (b) area of the rectangle

7 4 units

8 $f = \dfrac{33024}{w}$ (b) 384 cycles/sec   (c) 96 cm

Exercise 17G

1 9.6 units   2 5.56 kg   3 28 800 km
4 (a) $h = \frac{35}{64}s^2$ (b) 315 m   5 18
6 (a) 31.25 m   (b) 45 m
7 1 hour 48 minutes   8 0.576 units
9 (a) 20 202.02   (b) (i) 0.3   (ii) 0.6

Unit 18 Graphs and higher order equations

Exercise 18A

1

x	−4	−3	−2	−1	0	1	2	3	4
y	21	14	9	6	5	6	9	14	21

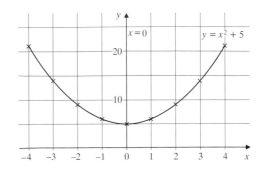

2

x	−4	−3	−2	−1	0	1	2	3	4
y	6	−1	−6	−9	−10	−9	−6	−1	6

$x = 0$

3

x	−4	−3	−2	−1	0	1	2	3	4
y	48	27	12	3	0	3	12	27	48

$x = 0$

4

x	−4	−3	−2	−1	0	1	2	3	4
y	8	4.5	2	0.5	0	0.5	2	4.5	8

$x = 0$

5

x	−4	−3	−2	−1	0	1	2	3	4
y	−16	−9	−4	−1	0	−1	−4	−9	−16

$x = 0$

6

x	−4	−3	−2	−1	0	1	2	3	4
y	−32	−18	−8	−2	0	−2	−8	−18	−32

$x = 0$

7

x	−4	−3	−2	−1	0	1	2	3	4
y	8	3	0	−1	0	3	8	15	24

$x = -1$

8

x	−4	−3	−2	−1	0	1	2	3	4
y	4	0	−2	−2	0	4	10	18	28

$x = -1\frac{1}{2}$

9

x	−4	−3	−2	−1	0	1	2	3	4
y	9	4	1	0	1	4	9	16	25

$x = -1$

10

x	−4	−3	−2	−1	0	1	2	3	4
y	36	25	16	9	4	1	0	1	4

$x = 2$

11

x	−4	−3	−2	−1	0	1	2	3	4
y	31	20	11	4	−1	−4	−5	−4	−1

$x = 2$

12

x	−4	−3	−2	−1	0	1	2	3	4
y	23	13	5	−1	−5	−7	−7	−5	−1

$x = 1\frac{1}{2}$

13

x	−4	−3	−2	−1	0	1	2	3	4
y	24	16	10	6	4	4	6	10	16

$x = \frac{1}{2}$

14

x	−4	−3	−2	−1	0	1	2	3	4
y	15	4	−3	−6	−5	0	9	22	39

$x = \frac{-3}{4}$

15

x	−4	−3	−2	−1	0	1	2	3	4
y	66	41	22	9	2	1	6	17	34

$x = \frac{2}{3}$

Exercise 18B

1

x	−3	−2	−1	0	1	2	3
y	−22	−3	4	5	6	13	32

$y = x^3 + 5$

2

x	−3	−2	−1	0	1	2	3
y	−37	−18	−11	−10	−9	−2	17

3

x	−3	−2	−1	0	1	2	3
y	−54	−16	−2	0	2	16	54

4

x	−3	−2	−1	0	1	2	3
y	−13.5	−4	−0.5	0	0.5	4	13.5

5

x	−3	−2	−1	0	1	2	3
y	27	8	1	0	−1	−8	−27

6

x	−3	−2	−1	0	1	2	3
y	54	16	2	0	−2	−16	−54

7

x	−3	−2	−1	0	1	2	3
y	−125	−64	−27	−8	−1	0	1

8

x	−3	−2	−1	0	1	2	3
y	−8	−1	0	1	8	27	64

9

x	−3	−2	−1	0	1	2	3
y	−45	−16	−3	0	−1	0	9

10

x	−3	−2	−1	0	1	2	3
y	−18	−4	0	0	2	12	36

11

x	−3	−2	−1	0	1	2	3
y	−42	−18	−6	0	6	18	42

12

x	−3	−2	−1	0	1	2	3
y	−12	2	4	0	−4	−2	12

13

x	−3	−2	−1	0	1	2	3
y	6	12	8	0	−6	−4	12

14

x	−3	−2	−1	0	1	2	3
y	−38	−16	−6	−2	2	12	34

15

x	−3	−2	−1	0	1	2	3
y	−43	−14	−1	2	1	2	11

Exercise 18C

1

x	−3	−2	−1	−0.5	−0.2	0.2	0.5	1	2	3
y	−0.7	−1	−2	−4	−10	10	4	2	1	0.7

$y = \frac{2}{x}$

2

x	−3	−2	−1	−0.5	−0.2	0.2	0.5	1	2	3
y	0.3	0.5	1	2	5	−5	−2	−1	−0.5	−0.3

3

x	−3	−2	−1	−0.5	−0.2	0.2	0.5	1	2	3
y	1	1.5	3	6	15	−15	−6	−3	−1.5	−1

4

x	−3	−2	−1	−0.5	−0.2	0.2	0.5	1	2	3
y	3.7	3.5	3	2	−1	9	6	5	4.5	4.3

5

x	−3	−2	−1	−0.5	−0.2	0.2	0.5	1	2	3
y	6.3	7	9	13	25	−15	−3	1	3	3.7

6

x	−3	−2	−1	−0.5	−0.2	0.2	0.5	1	2	3
y	−4.3	−5	−7	−11	−23	17	5	1	−1	−1.7

Exercise 18D

1

x	−1	0	1	2	3	4
y	6	1	−2	−3	−2	1

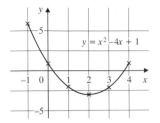

$y = x^2 - 4x + 1$

2

x	−3	−2	−1	0	1	2
y	−12	−3	−2	−3	0	13

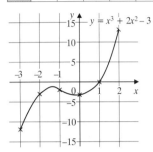

$y = x^3 + 2x^2 - 3$

3

x	−2	−1	0	1	2	3
y	−8	0	0	−2	0	12

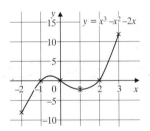

$y = x^3 - x^2 - 2x$

4

x	−3	−2	−1	0	1	2	3
y	−10	4	6	2	−2	0	14

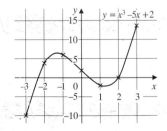

$y = x^3 - 5x + 2$

5

x	−2	−1	0	1	2	3	4
y	−7	−2	1	2	1	−2	−7

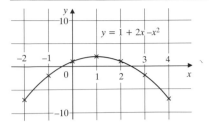

$y = 1 + 2x - x^2$

6

x	−3	−2	−1	−0.5	−0.2	0.2	0.5	1	2	3
y	−6.3	−4.5	−3	−3	−5.4	5.4	3	3	4.5	6.3

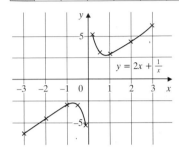

$y = 2x + \frac{1}{x}$

7

x	−3	−2	−1	−0.5	−0.2	0.2	0.5	1	2	3
y	9.3	4.5	2	2.3	5.0	−5.0	−1.8	0	3.5	8.7

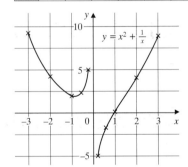

$y = x^2 + \frac{1}{x}$

8

x	0.1	0.2	0.5	1	2	3	4	5
y	22.1	12.2	6.5	5	5	5.7	6.5	7.4

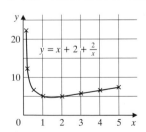

$y = x + 2 + \frac{2}{x}$

9

x	0.2	0.5	1	2	3
y	−5.0	−1.9	0	7.5	26.7

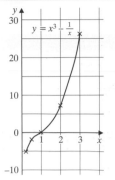

$y = x^3 - \frac{1}{x}$

10

x	−3	−2	−1	−0.5	−0.2	0.2	0.5	1	2	3
y	11.3	5	0	−3.3	−9.8	9.8	3.8	2	3	6.7

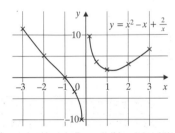

$y = x^2 - x + \frac{2}{x}$

Exercise 18E

1	5.6 or 1.4	**2**	2.3 or −1.3	**3**	2.8 or −1.3
4	1.3 or 0.3	**5**	−2.1, 0.3 or 1.9	**6**	−2.3, 0 or 1.3
7	2.2	**8**	0 or 2	**9**	2.2
10	0.7 or 2.9				

Exercise 18F

1	1.3	**2**	2.1	**3**	2.1	**4**	2.47	**5**	6.6
6	8.56	**7**	5.6	**8**	6.89	**9**	4.35	**10**	2.4

Exercise 18G

1 (a) Sharon.
(b) Tracey overtakes Sharon.
(c) Pass each other travelling in opposite directions.

2

3 (a) 11:00 (b) 12:30 (c) 2 hrs (d) 60 mph

4

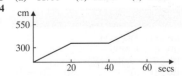

5 (a) 7 m/s² (b) steady speed of 36 m/s
(c) −9 m/s² (d) 274 m

6 (a) 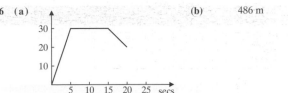 (b) 486 m

7 (a) E (b) A (c) B (d) C

8 (a) (b) (c) (d)

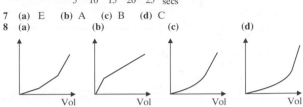

9 **10**

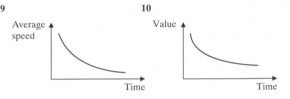

11 Starts slowly then speeds up. Has a break. Continues more slowly than before.
(a) 09:00 (b) Post Office closed
12 (a) B (b) D

Unit 19 Advanced mensuration

(All answers correct to 2 d.p. for Exercises 19A to 19D.)

Exercise 19A

1	(a) 9.77 cm	(b) 10.47 cm	(c) 7.38 cm
	(d) 39.10 cm	(e) 47.12 cm	(f) 60.21 cm
2	(a) 50.13°	(b) 80.21°	(c) 47.75°
	(d) 222.82°	(e) 73.52°	(f) 39.14°

Exercise 19B

1	(a) 22.34 cm²	(b) 61.09 cm²	(c) 30.54 cm²
	(d) 45.95 cm²	(e) 139.63 cm²	(f) 279.25 cm²
2	72.15°	**3** 10.06 cm	

Exercise 19C

1	(a) 9.06 cm²	(b) 10.98 cm²	(c) 0.51 cm²
	(d) 3.22 cm²	(e) 308.83 cm²	(f) 192.04 cm²

Exercise 19D

1 (a) 68.75° (b) 135 cm²
2 (a) 82.08 m (b) 83.78 m (c) 5026.55 m²
(d) 398.48 m²
3 (a) 7.16 m (b) 2.99 m² (c) 0.21 m³

Exercise 19E

1 314 cm³, 283 cm² **2** 156 cm³, 211 cm²
3 0.942 m³ **4** 11 972 cm³
5 (a) 8310 cm³ (b) 2354 cm²

Exercise 19F

1 960 cm³ **2** 140 cm³ **3** 6.18 cm
4 (a) 301.59 cm³ (b) 6.71
5 7.08

Exercise 19G

1 (a) $2144.7\,\text{cm}^3$, $804.25\,\text{cm}^2$ (b) $3591.4\,\text{cm}^3$, $1134.1\,\text{cm}^2$
2 $10.61\,\text{cm}$
3 9.67

Exercise 19H

1 $11.70\,\text{cm}$
2 $\left(\frac{12.5}{12}\right)^3 \approx 1.13$ i.e. 13% increase
3 (a) $4:9$ (b) $8:27$ 4 $\frac{1}{50^3} = \frac{1}{125000}$
5 (a) $12.59\,\text{cm}$ (b) $1:1.31$

Exercise 19I

1 $618.6\,\text{cm}^3$ 2 $112.72\,\text{cm}^3$
3 $496.37\,\text{cm}^3$ 4 $45389.33\,\text{cm}^3$

Exercise 19J

1 (a) $6.76\,\text{cm}$ (b) $6.98\,\text{cm}$ (c) $27.93\,\text{cm}^2$ (d) $3.41\,\text{cm}^2$
2 $9048\,\text{cm}^3$
3 $268.08\,\text{cm}^3$
4 $\frac{4}{3}\pi r^3 = r, r^2 = \frac{3}{4\pi}, r = \sqrt{\frac{3}{4\pi}}$
5 $1120\,\text{cm}^3$
6 $9366.42\,\text{cm}^3$
7 $26.64\,\text{cm}$
8 $2165.60\,\text{cm}^3$

Unit 20 Simplifying algebraic expressions

Exercise 20A

1 x^8 2 y^8 3 $6x^7$ 4 $5a^4$
5 $24a^6$ 6 $48y^5$ 7 $10x^5y^2$ 8 $24a^5b^5$
9 $a^2b^2c^2$ 10 $18a^2b^3c^5$ 11 $6x^2y^3$ 12 x^8
13 $9a^4$ 14 $8x^6$ 15 $2a^{20}$ 16 $2x^4y^6$
17 $63x^8$ 18 $14a^{13}b^6$ 19 $36x^2$ 20 $27x^6$

Exercise 20B

1 x^3 2 $3a^3$ 3 $6xy^3$ 4 x^{-2}
5 a^6 6 $\frac{3}{2}a^3x^{-1}$ 7 4 8 x^4

Exercise 20C

1 5 2 2 3 4 4 $\frac{1}{7}$
5 $\frac{1}{3}$ 6 6 7 $\frac{1}{5}$ 8 $\frac{1}{10}$
9 $\frac{1}{2}$ 10 0.4

Exercise 20D

1 125 2 32 3 32 4 $\frac{1}{4}$
5 32 6 25 7 $\frac{1}{4}$ 8 $\frac{1}{32}$

Exercise 20E

1 (a) 2 (b) $\frac{1}{100}$ (c) 4 (d) 32 (e) 81
2 (a) $\frac{1}{2}$ (b) 0 (c) -3 (d) $\frac{2}{3}$ (e) $\frac{7}{2}$
   (f) $\frac{9}{2}$ (g) 5 (h) 4 (i) 3 (j) $\frac{1}{2}$
   (k) $\frac{3}{2}$ (l) 3

3 (a) 9 (b) a^2 (c) $x^{\frac{3}{2}}y^{\frac{5}{2}}$ (d) $3x$
   (e) $\frac{2}{x}$ (f) $\frac{a^2}{5}$ (g) $729x^3$ (h) 16
   (i) $24y^2$

Exercise 20F

1 $\frac{x}{4}$ 2 $\frac{b}{d}$ 3 $4x$ 4 $\frac{4t}{3r}$
5 $3ab$ 6 2 7 $\frac{(x+2)}{2x}$ 8 $\frac{3}{(x-4)}$

Exercise 20G

1 (a) $\frac{3x}{5}$ (b) $\frac{y+x}{xy}$ (c) $\frac{13}{5x}$
   (d) $\frac{11x}{12}$ (e) $\frac{7a+2}{12}$ (f) $\frac{x+17}{10}$
   (g) $\frac{(x+3)}{(x+1)(x+2)}$ (h) $\frac{2}{x(2-x)}$
   (i) $\frac{(y-1)}{2(y-5)}$
2 (a) 18 (b) $x^2(x-1)$ (c) $12(x-1)(x+1)$
   (d) $x^3(1-x)(1+x)$
3 (a) $\frac{5a}{18}$ (b) $\frac{(x-2)}{x^2(x-1)}$
   (c) $\frac{(x+2)}{12(x-1)(x+1)}$ (d) $\frac{x^2+x+2}{x^3(1-x)(1+x)}$
4 $\frac{2(x+10)}{(x-2)(x+2)}$

Exercise 20H

1 $(p+q)(x+y)$ 2 $(x+y)(3-z)$ 3 $(x-y)(5-x)$
4 $2(a-b)(5+x)$ 5 $(y+2)(x+z)$ 6 $(a+1)(y+2b)$
7 $(x-b)(1-x)$ 8 $(x-y)(x-z)$ 9 $2(b+1)(a-5)$
10 $3(a-2c)(1-b)$

Exercise 20I

1 (a) $x(x+3)$ (b) $8(x+3)$ (c) $(x+3)(x+8)$
   (d) $x(x+6)$ (e) $2(x+6)$ (f) $(x+6)(x+2)$
   (g) $x(x-5)$ (h) $8(x-5)$ (i) $(x-5)(x+8)$
   (j) $(x+5)(x+3)$ (k) $(x+4)(x-8)$ (l) $(x-6)(x-2)$
   (m) $(x+4)(x-3)$ (n) $(x+3)(x+5)$ (o) $(x-7)(x-9)$
2 x^2-5x-6
3 (a) $-2, -3$ (b) $1, 6$ (c) $-2, -6$
   (d) $4, 15$ (e) $4, -6$ (f) $-2, 12$
   (g) $3, -16$ (h) $-4, 12$
4 (a) $(x-4)(x-1)$ (b) $(x+5)(x+2)$ (c) $(x-5)(x+3)$
   (d) $(x+3)(x+2)$ (e) $(x-10)(x-1)$ (f) $(y-4)(y+3)$
5 (a) $(2x-3)(x-1)$ (b) $6(x+2)(x+5)$ (c) $(2y-5)(y+2)$
   (d) $2(3a+2)(a+1)$ (e) $(4x+3)(x-2)$ (f) $5(5+4x)(3-x)$
   (g) $(5x-1)(x-3)$ (h) $(x^2+2)(x^2+3)$ (i) $2(x^2+4)(x^2+3)$
6 $(x^n+1)(x^n+8)$
7 $x(x^{2n}+4)(x^{2n}+2)$

Exercise 20J

1 (a) $(x-1)(x+1)$ (b) $(x-5)(x+5)$
   (c) $(2x-3)(2x+3)$ (d) $(2x+3)(4x-3)$
   (e) $2(p-7)(p+7)$ (f) $2(x-2y)(x+2y)$
   (g) $(x+2)^2$ (h) $(x-3)^2$ (i) $2(x+7)^2$
2 (a) $(x-2)(x+2)$ (b) $x(x-3)$ (c) $(x+1)(x-4)$
   (d) $4(x-3)(x+3)$ (e) $2x(2x-35)$ (f) $2(2x+1)(x-18)$
   (g) $(x+1)^2$ (h) $x(x+2)$ (i) $2(4x-7)(x+2)$

3 **(a)** $(x-4)(x-1)$ **(b)** $(x+12)(x-3)$ **(c)** $(x+8)(x-4)$
    (d) $(x-1)(2x-5)$ **(e)** $(4x-7)(x+1)$ **(f)** $(5x+3)(x-2)$
    (g) $(x-1)(x-3)$ **(h)** $(y+2)(y+5)$ **(i)** $(x+5)(x-3)$
    (j) $(x-7)(x+3)$ **(k)** $(y+4)^2$ **(l)** $(2x-5)^2$
    (m) $(5x-3)(x-2)$ **(n)** $(4x+5)(x-1)$ **(o)** $(5x-3)(x+2)$

6   4.7

7   **(a)** $\dfrac{84}{x}$ km/h   **(c)**   7

Exercise 20K

1 $\dfrac{x+2}{2}$ **2** $\dfrac{x-1}{x+4}$ **3** $\dfrac{(x+2)}{(x-2)^2}$ **4** $\dfrac{(x+9)}{(x+1)}$

5 $\dfrac{(x+2)}{x}$ **6** $\dfrac{4}{(x+1)(x+2)}$ **7**   2

Unit 21   Quadratics

Exercise 21A

1 **(a)** $-2, 1$ **(b)** $0, 3$ **(c)** $-\frac{1}{2}, 4$ **(d)** $0, 1\frac{1}{2}$
   (e) $\frac{1}{2}, 1\frac{1}{3}$ **(f)** $2, 5$

2 **(a)** $0, 3$ **(b)** ± 4 **(c)** $-1, 4$ **(d)** $-7, -4$
   (e) $-3, \frac{1}{2}$ **(f)** $-\frac{3}{4}, 1\frac{1}{3}$ **(g)** $\frac{3}{4}, 2$ **(h)** $0, 9$
   (i) $-\frac{1}{2}, 2\frac{1}{3}$

3 **(a)** ± 5 **(b)** $0, \frac{1}{3}$ **(c)** $-2, \frac{1}{3}$ **(d)** ± 2
   (e) $0, 2\frac{1}{2}$ **(f)** $-7, 5$ **(g)** $-3, 4$ **(h)** $1, 3$
   (i) $-2\frac{1}{3}, 2$ **(j)** $\frac{1}{2}, 1$

4   Both sides have been divided by $6y$ and $y = 0$ is a solution.
    $y = 0, y = 2$

Exercise 21B

1 **(a)** $(x+2)^2 - 4$ **(b)** $(x-7)^2 - 49$ **(c)** $(x+\frac{3}{2})^2 - \frac{9}{4}$
   (d) $(x+\frac{1}{2})^2 - \frac{1}{4}$

2 **(a)** $2(x+4)^2 - 32$ **(b)** $3(x-2)^2 - 12$ **(c)** $2(x+\frac{1}{4})^2 - \frac{1}{8}$
   (d) $5(x-\frac{3}{2})^2 - \frac{45}{4}$

Exercise 21C

1 **(a)** $-5 \pm \sqrt{22}$ **(b)** $4 \pm \sqrt{18}$ **(c)** $\dfrac{-9 \pm \sqrt{69}}{2}$
   (d) $1 \pm \sqrt{\dfrac{2}{3}}$ **(e)** $\dfrac{3 \pm \sqrt{41}}{4}$ **(f)** $1 \pm \sqrt{\dfrac{9}{2}}$

2 **(a)** $-0.65, 4.65$ **(b)** $0.35, 5.65$ **(c)** $-3.12, 1.12$
   (d) $-0.69, 2.19$ **(e)** $-0.15, 2.15$ **(f)** $-1.29, 1.54$

Exercise 21D

1 **(a)** 5   **(b)** 8   **(c)** 44   **(d)** 288
   (e) 41   **(f)** 28

2 **(a)** $-2.62, -0.38$ **(b)** $-0.41, 2.41$ **(c)** $-3.16, 0.16$
   (d) ± 1.06 **(e)** $-2.35, 0.85$ **(f)** $-0.82, 1.82$

3 **(a)** $0.27, 3.73$ **(b)** $0.21, 4.79$ **(c)** $-2.13, -0.12$
   (d) $-0.65, 1.15$ **(e)** $0.13, 7.87$ **(f)** $-5.65, -0.35$

4 $9 - \sqrt{60}$ and $9 + \sqrt{60}$
5 $3 + \sqrt{24}$ and $-3 + \sqrt{24}$

Exercise 21E

1 $-\frac{4}{7}, 3$ **2** $0.22, 2.28$ **3** $-\frac{2}{3}, 2$
4 $-0.11, 1.11$ **5** $-7, 4$ **6** $-1.30, 2.30$

Exercise 21F

1 $4, 5$ **2**   5   **3** $-8, 3$ **4**   247 cm

5   **(a)** £$\dfrac{400}{x}$ **(c)**   £50

Exercise 21G

1 **(a)** $x = 4, y = 16;\ x = -4, y = 16$
   (b) $x = 6, y = 36$
   (c) $x = 7, y = 49;\ x = -5, y = 25$
   (d) $x = 9, y = 81;\ x = -2, y = 4$
   (e) $x = \frac{3}{2}, y = \frac{9}{2};\ x = -1, y = 2$
   (f) $x = -\frac{2}{3}, y = \frac{4}{3};\ x = 3, y = 27$
   (g) $x = 2, y = -2;\ x = -3, y = -7$
   (h) $x = \frac{3}{4}, y = \frac{7}{4};\ x = \frac{-4}{3}, y = \frac{49}{3}$

2 **(a)** $(1, 1), (-3, 9)$ **(b)** $(2, 16), (-\frac{5}{4}, \frac{25}{4})$
   (c) $(5, 10), (-2, 3)$ **(d)** $(0, 7), (-\frac{1}{2}, 5\frac{1}{2})$

3 **(a)** none:
   (b) one: $(5, 25)$
   (c) two: $(-1, 2)$ and $(\frac{7}{3}, \frac{49}{9})$
   (d) none:
   (e) two: $(3, 45)$ and $(-\frac{1}{5}, \frac{1}{5})$
   (f) two: $(3, 36)$ and $(-\frac{1}{2}, 1)$
   (g) one: $(2, 2)$
   (h) two: $(\frac{5}{4}, \frac{25}{2})$ and $(-\frac{1}{3}, 3)$

Exercise 21H

1 **(a)** $x = 4, y = 3;\ x = -4, y = 3$ meet in 2 points
   (b) $x = 0, y = 5;\ x = -5, y = 0$ meet in 2 points
   (c) $x = -3, y = 4;\ x = -\frac{24}{5}, y = -\frac{7}{5}$ meet in 2 points
   (d) $x = 3, y = -4;\ x = -4, y = 3$ meet in 2 points
   (e) $x = 5, y = -5$ meet in 1 point, line is a tangent
   (f) $x = 1, y = -7;\ x = -\frac{49}{65}, y = \frac{457}{65}$ meet in 2 points
   (g) $x = 3, y = -1$ meet in 1 point, line is a tangent
   (h) $x = 2, y = 1$ meet in 1 point, line is a tangent
   (i) $x = -\frac{7}{5}, y = -\frac{1}{5}$ meet in 1 point, line is a tangent
   (j) $x = 0, y = 2;\ x = \frac{48}{25}, y = -\frac{14}{25}$

3 **(a)** $y = x + 4$ touches circle at $x = -2$ only
   (b) (i) $y = x - 4; (2, -2)$
       (ii) $4\sqrt{2}$ or $2\sqrt{8}$
   (c) $2r$

Exercise 21I

1 $-5 \le x \le 5$ **2** $x < -7, x > 7$ **3** $-1.5 < x < 1.5$
4 $-4 \le x \le 4$ **5** $x \le -3, x \ge 3$ **6** $x < -1, x > 1$
7 $x \le -3, x \ge 3$ **8** $-3.5 < x < 3.5$ **9** $x < -1.5, x > 1.5$
10 $-5 < x < 5$ **11** $x < -\frac{1}{3}, x > \frac{1}{3}$ **12** $-8 \le x \le 8$
13 $x \le -3, x \ge 3$ **14** $x \le -3, x > 3$ **15** $-3.5 \le x \le 3.5$
16 $-6 \le x \le 4$ **17** $x < -5, x > 9$ **18** $-6.5 < x < 5.5$

Exercise 21J

1 **(a)**   **(b)**

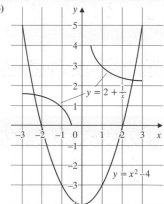

(c) (i) −1.4, 1.4   (ii) −0.8   (iii) −2.4, 2.5
2 **(a)** (i) $y = 1$ (ii) $y = -2$ (iii) $y = 2x + 1$
   (iv) $y = -2x + 4$
  (b) $x^3 - 2x = 4$ **(c)** $x^4 - 4x^2 = 1$
3 **(a)**

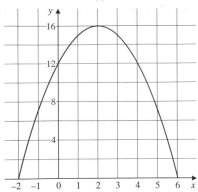

  (b) 1.5 m, 5.4 m
4 **(a)**

x	−3	−2	−1	0	1	2
y	−8	5	6	1	−4	−3

  (b) (c) (i)

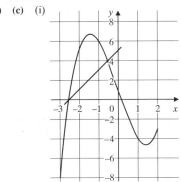

   (ii) −2.5, −0.5
  (d) $x^3 - 8x - 4 = 0$ **(e)** $y = 2x + 1$ **(f)** $0, \pm\sqrt{8}$

Unit 22 Advanced trigonometry

Exercise 22A

(All answers correct to 2 d.p.)
1 **(a)** 17.32 cm²   **(b)** 29.60 cm²   **(c)** 76.31 cm²
  (d) 32.14 cm²   **(e)** 40.26 cm²   **(f)** 111.54 cm²
2 492.62 m²   3 73.54 cm²   4 53.13°   5 123.56°

Exercise 22C

1 **(a)** 9.92 cm   **(b)** 11.43 cm   **(c)** 14.95 cm
  (d) 8.28 cm   **(e)** 13.32 cm   **(f)** 6.24 cm
2 **(a)** 33.27°   **(b)** 23.71°   **(c)** 13.73°
3 **(a)** 61.90 km   **(b)** 37.82 km

Exercise 22D

$P\hat{Q}R = 63.73°$
or $P\hat{Q}R = 20.27°$

Exercise 22E

1 **(a)** 6.22 cm   **(b)** 6.25 cm   **(c)** 11.53 cm   **(d)** 24.70 cm
2 **(a)** 36.34°   **(b)** 101.54°   **(c)** 100.95°   **(d)** 98.25°
3 146.46 m
4 145.26 km

Exercise 22F

1 **(a)** 15.76 cm²   **(b)** 6.30 cm   **(c)** 38.71°
2 **(a)** 183.97 km   **(b)** 320.64°   **(c)** 28.53 km
3 **(a)** 9.94 km   **(b)** 151.18°   **(c)** 3.19 km

Exercise 22G

1 **(a)** 13 cm   **(b)** 23.85 cm   **(c)** 23.32 cm
  (d) 22.62°   **(e)** 12.10°   **(f)** 17.75°
2 **(a)** 14.14 cm   **(b)** 14.35 cm   **(c)** 63.77°
  (d) 26.23°
3 **(a)** 17.44 cm   **(b)** 17.89 cm   **(c)** 54.50°

Unit 23 Exploring numbers

Exercise 23A

1 0.75, 0.6, 0.428571.., 0.375, 0.3, 0.2727..
  $\frac{3}{7}$ and $\frac{3}{11}$ are recurring decimals
2 $\frac{5}{8}, \frac{5}{16}$ are terminating decimals because the denominators divide exactly into powers of 10. Others are recurring.
3 0.142857.., 0.285714.., 0.428571.., 0.571428.., 0.714285.., 0.857142..
  Same digits in the same order.
4 16
5 $\frac{1}{13}, \frac{3}{13}, \frac{4}{13}, \frac{9}{13}, \frac{10}{13}, \frac{12}{13}$ use the same digits in the same order
  $\frac{2}{13}, \frac{5}{13}, \frac{6}{13}, \frac{7}{13}, \frac{8}{13}, \frac{11}{13}$ use the same digits in the same order and each fraction is double a value in the first sequence.
6 **(a)** Decimal recurs because 17 does not divide exactly into any power of 10.
  (b) When dividing by 17 there can only be 16 different remainders and 16 different subtraction sums.

Exercise 23B

1 **(a)** $\frac{2}{3}$ **(b)** $\frac{7}{9}$ **(c)** $\frac{34}{99}$ **(d)** $\frac{91}{99}$
  (e) $\frac{2}{11}$ **(f)** $\frac{125}{999}$ **(g)** $\frac{19}{37}$ **(h)** $\frac{91}{909}$
  (i) $\frac{1279}{9999}$ **(j)** $\frac{9}{101}$ **(k)** $1\frac{31}{75}$ **(l)** $4\frac{119}{165}$
2 1; There is no number between 0.9999... and 1.

Exercise 23C

1 $\pm\sqrt{30}$
2 $\sqrt{40} = 2\sqrt{10}$
3 **(a)** Perimeter = 12 units
  (b) Area = 4 (units)²
4 **(a)** $\dfrac{\sqrt{7}}{7}$ **(b)** $\dfrac{3\sqrt{5}}{5}$ **(c)** $\dfrac{\sqrt{17}}{17}$ **(d)** $\dfrac{\sqrt{11}}{11}$ **(e)** 4
5 **(a)** $x = 3 \pm \sqrt{7}$ **(b)** $-5 \pm \sqrt{11}$
7 $\sqrt{11}$ units
8 **(a)** $\frac{3}{2}$ (units)²   **(b)** $\sqrt{22}$ units
9 $\dfrac{\sqrt{6}}{2}$

Exercise 23D

1 3.5, 4.5; 13.5, 14.5; 103.5, 104.5; 9.5, 10.5; 99.5, 100.5
2 25, 35; 45, 55; 175, 185; 3015, 3025; 5, 15; 95, 105
3 4.65, 4.75; 2.85, 2.95; 13.55, 13.65; 0.25, 0.35; 157.45, 157.55; 9.95, 10.05; 99.95, 100.05
4 36.5, 37.5; 49.5, 50.5; 179.5, 180.5; 3.15, 3.25; 9.45, 9.55; 9350, 9450; 9.5, 10.5; 95 or 99.5, 105 or 100.5
5 4.25, 4.75; 7.25, 7.75; 16.25, 16.75; 2.75, 3.25; 15.25, 15.75; 9.75, 10.25; 99.75, 100.25
6 3.3, 3.5; 3.1, 3.3; 3.9, 4.1; 9.3, 9.5; 12.1, 12.3; 24.5, 24.7; 9.9, 10.1; 99.9, 100.1
7 3.625, 3.875; 3.375, 3.625; 4.125, 4.375; 5.875, 6.125; 15.375, 15.625; 9.875, 10.125; 99.875, 100.125
8 (a) 7.5 m, 6.5 m; 6.5 m, 5.5 m
    (b) 735 cm, 725 cm; 595 cm, 585 cm
    (c) 7.325 m, 7.315 m; 5.945 m, 5.935 m
    (d) Yes, the range of (c) contains the range of (b) which contains the range of (a).
9 (a) 74.365 s, 74.355 s
    (b) It is not necessarily the fastest lap time.

Exercise 23E

Answers are either exact or given to 4 s.f.
1   55.25 cm^2, 41.25 cm^2 2   30 cm, 26 cm
3   57.0025 cm^2, 55.5025 cm^2 4   30.2 cm, 29.8 cm
5   38.48 cm^2, 19.63 cm^2 6   2.168 m, 2.105 m
7   5411 mm^2, 5153 mm^2 8   27.625 cm^2, 20.625 cm^2
9   24.35 cm^2, 23.65 cm^2 10   161.5 cm^2, 127.5 cm^2
11 6.297 cm^2, 4.984 cm^2 12   23.56 cm, 20.42 cm
13 (a) 22 450 cm^3, 8181 cm^3 (b) 15 599 cm^3, 12 770 cm^3
    (c) 14 279 cm^3, 13 996 cm^3

Exercise 23F

Answers are either exact or given to 4 s.f.
1   1 cm, 0.8 cm   2   4 min, 2 min   3   20.5 cm, 9.5 cm
4   (a) 6.826 m/s, 6.780 m/s   (b) 6.860 m/s, 6.746 m/s
5   (a) 9.571 miles/l, 8.784 miles/l   (b) 43.45 mpg, 39.88 mpg
6   23.68 cm, 26.10 cm
7   3.857 g/cm^3, 2.778 g/cm^3
8   54.77°, 39.59°

Exercise 23G

Answers are either exact or given to 4 s.f.
1   16 g, 1.6%   2   13 g, 2.863%   3   44.96, 9.669%
4   0.001593, 0.05070%   5   0.05 m, 1.111%   6   0.5 cm, 0.143%
7   (a) 0.5 cm, 7.143%   (b) 7.25 cm^2, 14.80%
8   (a) 0.05 cm, 0.7042%   (b) 0.7125 cm^2, 1.413%
9   (a) 0.5 cm, 2.778%   (b) 57.33 cm^2, 5.633%
                                              (correct to 4 s.f.)

Exercise 23H

1   (a) $13\frac{74}{99}$ (b) $6\frac{25082}{33300}$
2   (a) 255 000, 245 000
    (b) 6.5×10^6, 5.5×10^6
3   4 cm is correct to the nearest cm
    4.0 cm is correct to 1 d.p.
    4.00 cm is correct to 2 d.p.
4   75.398 cm^2 25.133 cm^2
5   3.81 mm, 3.79 mm
6   (a) 0.5;   1.1905%
    (b) 132.7 cm^2;   2.395%
    (c) 1402.0 cm^3; 3.614%

Unit 24 Applying transformations to sketch graphs

Exercise 24A

1   (a) 2   (b) 14   (c) −2   (d) −1
    (e) 7   (f) $-1\frac{15}{16}$
2   (a) x^2 (b) $9x^2$ (c) $x^2 + 4x + 4$
    (d) $x^2 + 2x + 4$ (e) $\frac{1}{4}x^2 - x - 3$ (f) $5 - 4x^2$
    (g) $k^2x^2 + 2akx + a^2 + b$
3   (a) (i) 5   (ii) −3   (iii) −4   (iv) −3   (b) −2, 2

Exercise 24B

1   $y = x^2 - 4$
2   $y = x^2 + k$ for any three values of k between 0 and 8
3   $-15 < k < 0$

Exercise 24C

1   (a) A: 4 units vertically in the negative y-direction.
         B: 4 units vertically in the positive y-direction.
         C: 6 units vertically in the positive y-direction.
    (b) A: 5 units vertically in the positive y-direction.
         B: 12 units vertically in the positive y-direction
         C: 8 units vertically in the negative y-direction.
    (c) A: 10 units vertically in the negative y-direction.
         B: 2.5 units vertically in the positive y-direction.
         C: 5 units vertically in the positive y-direction.
    (d) A: 3 units vertically in the positive y-direction.
    (e) A: 4 units vertically in the negative y-direction.
2   (c) Translation 2 units vertically in the positive y-direction.

Exercise 24D

1   $y = (x - 1)^2$
2   (a)

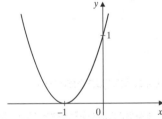

    (b)

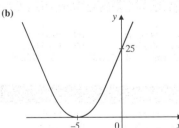

3

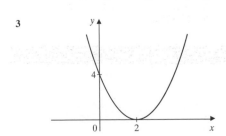

Exercise 24E

1 A horizontal translation of 1 unit in the positive *x*-direction.
2 A horizontal translation of 1 unit in the positive *x*-direction.
3 A vertical translation of 7 units in the positive *y*-direction.
4 $y = x^3 + 2$
5 $y = (x - 4)(x - 5)(x + 1)$

Exercise 24F

1 (a) 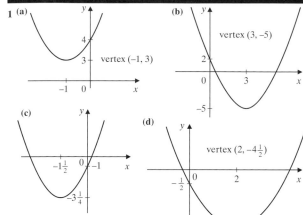 vertex (−1, 3)
   (b) vertex (3, −5)
   (c) vertex $(-1\frac{1}{2}, -3\frac{1}{4})$
   (d) vertex $(2, -4\frac{1}{2})$

2 (a) A horizontal translation of 6 units in the negative *x*-direction.
   (b) A horizontal translation of 4 units in the positive *x*-direction
        followed by a vertical translation of 4 units in the negative
        y-direction.

3 (a) 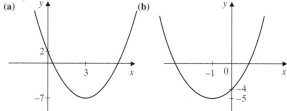 (b)

   A horizontal translation of 4 units in the negative *x*-direction
   followed by a vertical translation of 2 units in the positive
   y-direction.

4 $y = x^2 - 18x + 14$

5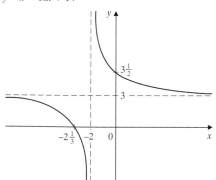

Exercise 24G

1 (a) $f(-x) = x^2 + 2x$
   (b) 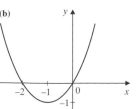   (c)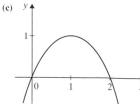
   (d) (1,1)

2 (a) $y = -x^3 - 2x^2$ (b) $y = -x^3 + 2x^2$

3 (b)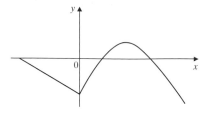
   (c) Reflection in the *x*-axis.

4 (b)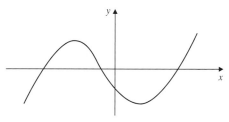
   (c) Reflection in the *y*-axis.

5 (a) 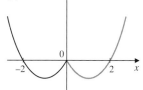   (b)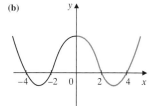

Exercise 24H

1 (a) Reflection in *x*-axis then a vertical translation of 6 units in the
        positive *y*-direction.

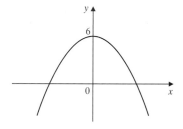

(b) Horizontal translation of 4 units in the positive x-direction then a vertical translation of 2 units in the positive y-direction.

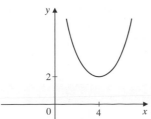

(c) Horizontal translation of 1 unit in the negative x-direction then a stretch scale factor 2 parallel to the y-axis then a vertical translation of 3 units in the negative y-direction.

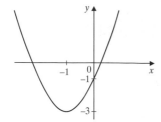

(d) Horizontal translation of 2 units in the positive x-direction then a stretch scale factor 3 parallel to the y-axis then a vertical translation of 10 units in the negative y-direction.

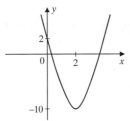

(e) Horizontal translation of 2 units in the positive x-direction then a vertical translation of 2 units in the negative y-direction

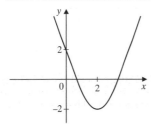

(f) Horizontal translation of 2 units in the positive x-direction then a reflection in x-axis then a vertical translation of 1 unit in the positive y-direction.

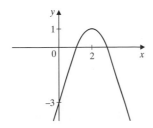

2 (a) Horizontal translation of 2 units in the negative x-direction then a stretch scale factor 4 parallel to the y-axis.

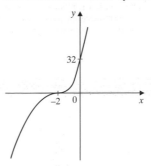

(b) Reflection in the x-axis then a vertical translation of 8 units in the positive y-direction.

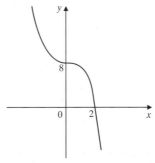

(c) Horizontal translation of 1 unit in the positive x-direction then a vertical translation of 5 units in the positive y-direction.

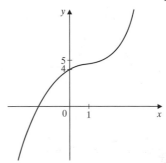

(d) Stretch scale factor 0.5 parallel to the x-axis then a vertical translation of 8 units in the negative y-direction.

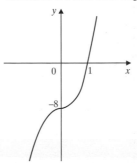

3 (a) Horizontal translation of 1 unit in the negative x-direction.
(b) Horizontal translation of 4 units in the negative x-direction, or a reflection in the y-axis.
(c) Reflection in the x-axis then a vertical translation of 2 units in the positive y-direction.

Exercise 24I

1 $\sin(x + 90°) = \cos x$
2 (a) Stretch scale factor 0.5 parallel to the x-axis.
   (b)

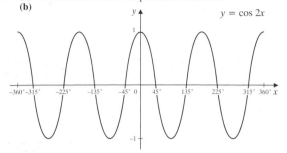

$y = \cos 2x$

   (c) 8
3 (a) Stretch scale factor 2 parallel to the y-axis.
   (b)

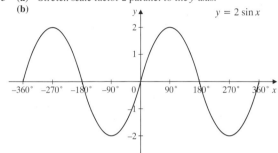

$y = 2\sin x$

   (c) 4   (d) 8
4 (a) (i)

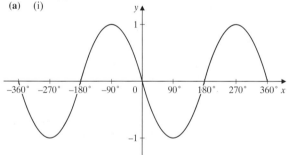

   (ii) Same graph for both $y = f(-x)$ and $y = -f(x)$.
   (b) odd
5 $f(x) = \cos x$
   $f(-x) = \cos(-x) = \cos x$
   $\therefore f(x) = f(-x)$
   Hence $\cos x$ is an even function.
6 (a) greatest: 5,   least: −5
   (b) greatest: 1,   least: −1
   (c) greatest: 1,   least: −1
   (d) greatest: 2,   least: −2
   (e) greatest: 9,   least: 3

Unit 25 Vectors

Exercise 25A

1 (a) $\begin{pmatrix} -2 \\ -1 \end{pmatrix}$ (b) $\begin{pmatrix} -3 \\ -5 \end{pmatrix}$ (c) $\begin{pmatrix} 5 \\ 6 \end{pmatrix}$ (d) $\begin{pmatrix} -5 \\ -6 \end{pmatrix}$
2 $(0, -2)$
3 Translation by $\begin{pmatrix} 6 \\ 0 \end{pmatrix}$; translation by $\begin{pmatrix} 2b-2a \\ 0 \end{pmatrix}$
4 $\begin{pmatrix} 6 \\ 1 \end{pmatrix}$

Exercise 25B

1 (i) $\begin{pmatrix} -2 \\ 4 \end{pmatrix}$ (ii) $\begin{pmatrix} 4 \\ 2 \end{pmatrix}$ (iii) $\begin{pmatrix} 16 \\ -7 \end{pmatrix}$ (iv) $\begin{pmatrix} 12 \\ -4 \end{pmatrix}$,

   for example $c = \begin{pmatrix} 2 \\ 3 \end{pmatrix}$ or $\begin{pmatrix} -1 \\ 0 \end{pmatrix}$
3 $(4, 3)$

Exercise 25C

1 (a) 5   (b) 25   (c) 10   (d) $\sqrt{170} = 13.0$
2 (a) $\sqrt{106} = 10.3$ (b) $\sqrt{18} = 4.24$
   (c) $\sqrt{208} = 14.4$ (d) $\sqrt{40} = 6.32$
3 (a) $\sqrt{13} = 3.61$ (b) $2\sqrt{13} = 7.21$
   (c) $\sqrt{52} = 7.21$ (d) $\sqrt{13} \times 5 = 18.0$

Exercise 25D

1 $\begin{pmatrix} -2 \\ 2 \end{pmatrix}$ 2 $\begin{pmatrix} 1 \\ -1 \end{pmatrix}$ 3 $\begin{pmatrix} 10 \\ -1 \end{pmatrix}$ 4 $p = 2$
5 $q = -2$
6 (i) 10   (ii) $\sqrt{2}$, x and y are parallel.
7 $x = 0, y = -2$
8 $p = 2.4, q = -2.2$
9 $(2, 6), (3, 7), (4, 8), (5, 9)$ $(6, 10)$ $(7, 11)$. A straight line, $m = 1$ and $c = 4$.
10 (a) $\begin{pmatrix} 6 \\ 3 \end{pmatrix}$ (b) $\begin{pmatrix} 6 \\ 3 \end{pmatrix} + k\begin{pmatrix} -2 \\ 2 \end{pmatrix} = \begin{pmatrix} 6-2k \\ 3+2k \end{pmatrix}$
   (c) $k = 3$, $D = (2, 10)$

Exercise 25E

1 $\begin{pmatrix} 2 \\ 4 \end{pmatrix}$ 2 $(4, 0)$ 3 $(5, 9)$
4 $\begin{pmatrix} 3 \\ 5 \end{pmatrix} \begin{pmatrix} 4 \\ 4 \end{pmatrix}$

Exercise 25F

1 $y - x$. XY is parallel to AB and $\frac{2}{3}$ of the length of AB.
2 $b - a$. $3b - 3a$. CD is parallel to AB and 3 times the length of AB.
3 $\frac{1}{3}(a + b), \frac{1}{3}(a + b)$. G and H are the same point. $\frac{1}{2}b$.
4 $k = \frac{1}{3}, m = \frac{1}{3}$, $\frac{4}{3}a + \frac{1}{3}b$
5 e.g. $\overrightarrow{OA} = 3a$, $\overrightarrow{OC} = 3b$. $\overrightarrow{PQ} = 2a + b$, $\overrightarrow{SR} = b + 2a$. PQRS is a parallelogram. No longer true.
6 (a) $b - a$ (b) $-b$ (c) $-b - a$
   Midpoints are $\frac{1}{2}(a + b), \frac{1}{2}(2a - b), \frac{1}{2}(a - 2b), -\frac{1}{2}(a + b), \frac{1}{2}(b - 2a)$, $\frac{1}{2}(2b - a)$
   The third hexagon is an enlargement, scale factor 0.75, centre 0, of the first hexagon.

Exercise 25G

1 (a) $\begin{pmatrix} 5 \\ 1 \end{pmatrix}$ (b) $\begin{pmatrix} -5 \\ -1 \end{pmatrix}$ (c) $\begin{pmatrix} -1 \\ -5 \end{pmatrix}$ (d) $\begin{pmatrix} 1 \\ 5 \end{pmatrix}$
2 (a) $\begin{pmatrix} 5 \\ 1 \end{pmatrix}$ (b) $\begin{pmatrix} 2 \\ 3 \end{pmatrix}$ (c) $\begin{pmatrix} 7 \\ 4 \end{pmatrix}$ (d) $\begin{pmatrix} 3 \\ -2 \end{pmatrix}$
   (e) $\begin{pmatrix} 10 \\ 2 \end{pmatrix}$ (f) $\begin{pmatrix} -3 \\ 2 \end{pmatrix}$
3 $\begin{pmatrix} 3 \\ -2 \end{pmatrix}, \begin{pmatrix} 1.5 \\ -1 \end{pmatrix}$
4 $\begin{pmatrix} 3 \\ 5 \end{pmatrix}$ 5 $\begin{pmatrix} -1 \\ -7 \end{pmatrix}$ 6 (a) $\begin{pmatrix} 3 \\ 0 \end{pmatrix}$ (b) $\begin{pmatrix} 1 \\ 2 \end{pmatrix}$
7 $\begin{pmatrix} 10 \\ 1 \end{pmatrix}$ 8 3, 3   9 $a = 1, b = 1$
10 $p = 1, q = \frac{1}{2}$

11   (a) $\sqrt{10}, \sqrt{10}$ (b) $\begin{pmatrix} -2 \\ 4 \end{pmatrix}$ (c) $\sqrt{20}$

    (d) Perpendicular and equal in length.

12   (a) $a + b$ (b) $-b$ (c) $a - b$

13   (a) $\frac{3}{5}b$ (b) $a - b$ (c) $\frac{3}{5}a - \frac{3}{5}b$

    (d) $\frac{3}{5}a$

    CD is parallel to OA and $\frac{3}{5}$ of the length of OA.

14 $a + b, a + b + c, 2b, a + c = b$

15   (a) b (b) $b + 2a$ (c) $3b$ (d) $3b - 2a$

    PQRS is a parallelogram

16   (a) $b - a$ (b) $\frac{1}{2}b$ (c) $c - a$

    (d) $\frac{1}{2}(c - a)$ (e) $\frac{1}{2}(a + c)$ (f) $\frac{1}{2}(a + c) - \frac{1}{2}b$

17   (a) $\frac{3}{2}a$ (b) $3b$ (c) $\frac{3}{2}a - 3b; 2a - b$

18 $x = 4, y = -3$

Unit 26 Circle theorems

Exercise 26A

1 $52°$ 2 $82°, 8°$ 3 11.3 cm, 6.93 cm

4 8.49 cm, 4.24 cm   5 5.29 cm   6 8 cm, $120\,\text{cm}^2$

7 $25°, 27°$ 8 $64°$ 9 7.88 cm

10 6.36 cm   11 112.9°   12 $36°, 16°$

13 $19°, 71°$ 14 $22°$ 15 10.6 cm

16 $40°$ 17 6.5 m

18 The sides are chords of the circle. So, the perpendicular bisectors pass through the centre. Hence, their point of intersection is the centre.

Exercise 26B

1   (i) $72°$ (ii) $36°$ (iii) $54°$

2 $62°, 56°$ 3 $32°$ 4 $112°, 22°$

5 $90° - x°, x°$ 6 $24\,\text{cm}^2$ 7 $30°$

8 $62°, 56°$ 9 $103°$ 10 $16\,\text{cm}^2$

Exercise 26C

1 $96°, 42°$ 2 $37°, 143°$ 3 $117°, 66°$

4 $25.5°$ 5 $46°, 113°, 67°$ 6 $50°, 20°$

7 $76°, 104°$ 8 $28°$ 9 $30°$

10 $71°$

Exercise 26D

1 $73°, 28°$ 2 $58°, 37°$ 3 $33°, 33°$

4 $56°, 72°$ 5 $68°, 68°, 68°$ 6 $61°, 61°, 122°, 41°$

7 Angle BAX = angle ACD (alternate angles)

    Angle ABX = angle ACD (angles in the same segment)

    $\therefore \triangle$ ABX is isosceles

8 $39°$

9 $30°$

10

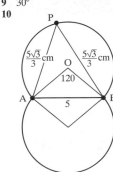

locus is given by the arc, subtended by AB, of the circle whose radius is $\frac{5\sqrt{3}}{3}$ cm, as shown in the diagram.

Exercise 26E

1 $82°, 98°$ 2 $77°, 61°$ 3 $99°, 99°$

4 $65°, 40°$ 5 $120°$ 6 $114°, 123°, 66°$

7 $30°, 102° 34'$ 8 $70°, 110°, 110°$ 9 $92°, 81°$

10 Angle AEB = angle ADB = 90°

    AEDB is a cyclic quadrilateral (angles in the same segment).

    30°

Exercise 26F

1 $73°, 49°$ 2 $50°$ 3 $70°, 70°, 40°$

4 $64°, 58°$ 5 $45°, 68°$ 6 $128°, 29°$

7 $73°, 39°, 39°$ 8 $64°, 117°$ 9 $26.5°, 116.5°, 37°$

10 $55°, 62.5°, 62.5°$

Exercise 26G

1 angle TCB = angle OAB (alternate segment)

    angle OBA = angle OAB (isosceles triangle)

    angle TBC = angle TCB (equal tangents)

    Since two angles are equal then the third must also be equal.

    Hence triangles AOB, BTC are similar.

2   (a) P lies on the perpendicular bisection of RY

        $\therefore$ PR = PY and triangle PRY is isosceles

        $\therefore$ angle YPX = angle XPR.

        But angle XPR = angle XSQ (angles in the same segment)

        $\therefore$ angle YPX = angle XSQ

    (b) angle XSZ + angle XSQ = 180° (angles on a straight line)

        angle XSZ + angle YPX = 180° (from part (a))

        then PZSX is a cyclic quadrilateral

        $\therefore$ angle PZS = 180° - angle PXS = 90°

3   (a) angle BAD = angle ADC (alternate angles)

        angle ABC = angle ADC (angles in the same segment)

        Hence, triangle ABE is isosceles.

        $\therefore$ angle AEB = 180° - 2 × angle ABE

        Also, angle AEB = 180° - angle AEC (angles on a straight line)

        $\therefore$ angle AEC = 2 × angle ABC

    (b) Since triangle ABE is isosceles, AE = BE

4 Let angle CXE = $x°$ = angle AXE and angle XCD = $y°$

    The angle DFE = 180° - $x°$ - $y°$ (angle sum of a triangle)

    angle XAB = $y°$ (cyclic quadrilateral)

    In triangle AXE, angle XEA = 180° - $x°$ - $y°$ (angle sum of a triangle)

    $\therefore$ angle DFE = angle FEA

5 angle ADC = 180° - angle DAB (interior angles of parallel lines)

    angle BCD = 180° - angle DAB (opposite angles of a cyclic quadrilateral)

    $\therefore$ angle ADC = angle BCD

6 Join D to B and A to C.

    Triangle ABD is isosceles. $\therefore$ angle ABD = angle ADB

    angle ABD = angle ACD (angles in the same segment)

    angle ADB = angle ACB (angles in the same segment)

    $\therefore$ angle ACD = angle ACB

    $\therefore$ AC bisects angle BCD.

7 Join A to E

    angle AEC + angle ABC = 180° (cyclic quadrilateral)

    angle AEC + angle AED = 180° (angles on a straight line)

    $\therefore$ angle ABC = angle AED.

    But angle ABC = angle ADC (opposite angles of a parallelogram)

    $\therefore$ angle ADC = angle AED

    $\therefore$ AE = AD (isosceles triangle)

8 Let angle CDF = angle FDE = $x°$

    angle CBF = angle CDF = $x°$ (angles in the same segment)

    angle CBA = 180° - angle CDA (cyclic quadrilateral)

    angle CDE = 180° - angle CDA (straight line)

    $\therefore$ angle CBA = $2x°$

    $\therefore$ FB bisects angle ABC

9 Join F to D. Let angle DAF = angle FAB = $x°$

    Then angle AED = angle EAB = $x°$ (alternate angles)

    angle AFD = 90° (angle in a semi-circle)

    Hence FD is a line symmetry of triangle AED $\therefore$ AF = FE

10 angle COA = 2 × angle CBA (angle at the centre)
            = 4 × angle TBA
    But angle TBA = angle ATP (alternate segment)
    ∴ angle COA = 4 × angle ATP

Exercise 26H

1 62°   2 60°   3 $a = 36°, b = 44°$
4 $a = 120°, b = 65°$ 5 40°
6 $a = 96°, b = 48°$ 7 $a = 104°, b = 96°$ 8 60°
9 (a) $\sqrt{7}$ cm   (b) $2\sqrt{7}$ cm
10 48.3 cm
11 10 cm
12 Angle OPB = 90°. Triangle QOB is isosceles.
    ∴ OP is a line of symmetry of OBQ.

Exercise 26I

1 (a) 8 cm   (b) 10.4 cm
2 A$\hat{D}$C, B$\hat{C}$D.
3 103°
4 (a) 90°   (b) 35°   (c) 125°
5 (a) 21°   (b) 53°   (c) 68°
6 26°
7 41°, 49°
8 30 cm, 36 cm
9 (a) 86°   (b) 52°   (c) 52°
10 $(1 + \sqrt{2})r$
11 101, 63
12 (a) 60° (b) 90°   (c) 30°   (d) 90°
    (e) 30°; 6.93 cm
13 (a) 47° (b) 43°   (c) 25°
14 (a) 70° (b) 35°   (c) radius = 2.11 cm
15 (a) 30° (b) 60° (c) 98°   (d) 38°   (e) 52°
16 40°
17 53°
18 (a) 52° (b) 47°   (c) 81°
19 53°
20 70°, 40° B$\hat{A}$C = B$\hat{C}$A = 70°
    In △ PBR and △ ACR
        P$\hat{B}$R = C$\hat{A}$R (angles in same segment)
        B$\hat{P}$R = R$\hat{C}$A (angles in same segment)
        P$\hat{R}$B = C$\hat{R}$A (virtually opposite)
        ∴ △ PBR is similar to △ ACR (AAA)
21 (a) 66° (b) 53°   (c) 33°   (d) 94°
22 1.23 cm
23 (a) The altitudes meet at a point.
    (b) Angle BHD = 90° – y. Join B to C so that BC cuts AD at K.
        BC subtends 90° at F and at E, so is the diameter of a circle
        through FBCE.
        angle EFC = y (angles in the same segment (EH))
        angle EBC = y (angles in the same segment (EC))
        From triangle BHK,
        angle BKH = 180° – angle BHD – angle HBK
                = 180° – (90° – y) – y
                = 90°

Unit 27 Histograms and dispersion

Exercise 27A

1 (a)

Hand length (cm)	Frequency	Frequency density
13 to less than 18	8	1.6
18 to less than 20	12	6
20 to less than 21	7	7
21 to less than 22	8	8
22 to less than 25	7	2.3̇

(b)

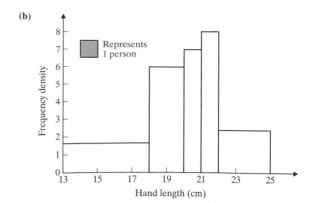

2

Class widths (s)	Frequency density
30	2
30	2.6̇
20	6
10	14
10	10
50	1.2

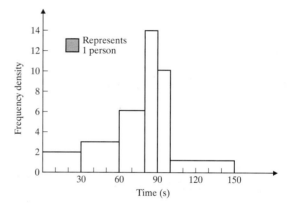

3

Distance (m)	Frequency density
0 – 9	9
10 – 14	14
15 – 19	16
20 – 24	15
25 – 34	8.5
35 – 50	3.3

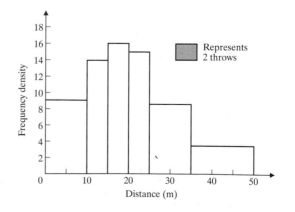

4

Class interval	Frequency density	Frequency
0 – 500	4.5	45
501 – 1500	3.5	70
1501 – 2500	5.0	100
2501 – 3000	6.5	65
3001 – 5000	2.5	100

5 (a) 4 6 7 4 6 1
(b) 28

6

Class interval (g)	Frequency density
30	1
20	3.5
10	10
10	6
20	2

7 (a) 8 (b) 36 (c) 79 (d) 1264
8 (a) $x = 2$ (b) 15
(c) Frequency densities are 2, 6, 5, 15, 8, 1
9

Waiting time (min)	Frequency density	Frequency
$0 \leq t < 10$	16	32
$10 \leq t < 15$	20	20
$15 \leq t < 30$	6	18
$30 \leq t < 35$	4	4
$35 \leq t$	0	0

Unit 28 Introduction to modelling

Exercise 28A

1 (a) £1500$(1.06)^t$ (b) 16 years
2 (a) $8(0.6)^n$ metres (b) 23.36 metres
3 (a) 48 days (b) 96 days
4 (a) (i) £32 000 (ii) £16 384 (iii) £40 000$(0.8)^n$
(b) 27 years

Exercise 28B

1 (a) (i) 6.5 m (ii) 5.25 m (iii) 4 m (iv) 1.5 m
(v) 6.5 m
(b) 1.14 pm
(c)

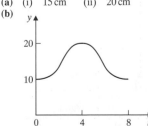

2 (a) (i) 15 cm (ii) 20 cm (iii) 10 cm
(b)

y graph

3 (a) (i) 7.5 m (ii) 6 m
(b) 3 m
(c) 3 am, 3 pm

(d)

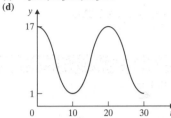

4 (a) $p = 9, q = 8$
(b) $3\frac{1}{3}$ sec, $16\frac{2}{3}$ sec, $23\frac{1}{3}$ sec
(c) $6\frac{2}{3}$ sec, $13\frac{1}{3}$ sec, $26\frac{2}{3}$ sec
(d)

y graph

Exercise 28C

1 (a) $y = 0.25x + 2.5$ (b) 14.5
2 (a) $y = -0.25x + 0.75$ (b) −3.75
3 (a) $y = \frac{14}{15}x + \frac{46}{3}$ (b) $85\frac{1}{3}$

Exercise 28D

(The following answers are estimates from a line of best fit. Your answers may therefore differ slightly and still be correct.)
1 $a = 0.28, b = 5$
2 $a = 2, b = 1$
3 $a = -0.2, b = 7.2$
4 $a = 2, b = 0.2$
5 $a = -0.5, b = 3$
6 $a = -0.5, b = 6$

Exercise 28E

1 $p = 2, q = 4, k = 32$
2 $p = -3, q = 2, k = 6$
3 $a = 3, b = 4, k = 67$
4 $\frac{1}{3}$
5 $\frac{2}{3}$ and $-\frac{2}{3}$
6 $a = 4, b = 3, k = \frac{37}{9}$
7 $\frac{1}{2}$
8 (a) $p = 2, q = 4$
(b) 3

Unit 29 Calculators and computers

Exercise 29A

1 7 3 → L EXE
4 9 → W EXE
L W → A EXE
2 L + 2 W → P EXE
Answers 3577 cm^2, 244 cm
2 1 2 → R EXE
SHIFT π R x^2 → A EXE
2 SHIFT π R → C EXE
Answers 452.389 cm^2, 75.398 cm

3

[1] [5] [→] [B] [EXE]

[9] [→] [H] [EXE]

[(] [B] [H] [)] [÷] [2] [→] [A] [EXE]

Answer 67. 5 cm²

4

[1] [2] [0] [→] [L] [EXE]

[6] [0] [→] [W] [EXE]

[2] [0] [→] [H] [EXE]

[L] [W] [H] [→] [V] [EXE]

Answer 144 000 cm³

5

[5] [→] [R] [EXE]

[1] [2] [→] [H] [EXE]

[SHIFT] [π] [R] [x²] [H] [→] [V] [EXE]

Answer 942.5 cm³

Exercise 29B

1 L x^y 3 → V EXE   Answers 125 cm³, 216 cm³, 343 cm³
2 (F – 32) × 5 ÷ 9 → C EXE   Answers 30°C, 10°C, 20°C
3 (B H) ÷ 2 → A EXE   Answer 52.5 cm²

Exercise 29C

1 ? → L : L x^y 3 EXE   Answer 4.64 cm
2 ? → L : 6 L x^y 2 EXE   Answer 2.89 cm
3 Answer 71.6 °C
4 Answer 2.5

Exercise 29D

1 [5] [EXE] [Ans] [+] [5] [EXE] [EXE] ...

2 [2] [EXE] [Ans] [×] [2] [EXE] [EXE] ...

3 [2] [0] [0] [EXE] [Ans] [÷] [1] [0] [EXE] [EXE] ...

4 [3] [EXE] [Ans] [x^y] [2] [EXE] [EXE] ...

5 [2] [5] [6] [EXE] [√] [Ans] [EXE] [EXE] ...

6 [3] [EXE] [1] [0] [–] [Ans] [EXE] [EXE] ...

7 [2] [EXE] [1] [÷] [Ans] [EXE] [EXE] ...

Examination practice paper (non-calculator)

1 $\frac{3}{16}$
2 (a) $x \leqslant \frac{1}{3}$ (b) 0
3 3000
4 $4\frac{1}{2}$
5 (a) $\frac{15}{16}$ (b) 9
6 (a) 528 cm²   (b) 297 cm²
7 $5n - 1$
8 (a) 19–20   (b) 90000–96000   (c) 58000–64000
9 (a) 4×10^8 (b) 2.4×10^{11}
10 (a) $16x^2 - 9y^2$ (b) $5(2t - 3)$ (c) $3q(3p - 4q)$
11 (a) 90°   (b) 26°   (c) 52°
12 (a)

x	−2	−1	0	1	2	3	4
y	−6	2	0	−6	−10	−6	12

(b)

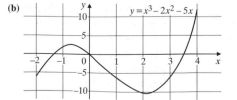

$y = x^3 - 2x^2 - 5x$

(c) 0, −1.3 to −1.6, 3.3 to 3.6
(d) $y = x - 1$
13 (a) (i) 1   (ii) $\frac{1}{64}$ (iii) 4   (b) $3\sqrt{2}$
14 (a) 35 and 15 are missing frequencies
(b)

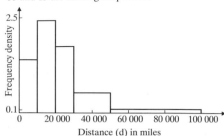

15 (a) $y = \dfrac{100}{x^2}$ (b) $\frac{1}{2}$ (or $-\frac{1}{2}$)
16 $y = 3x + 2$
17 $n^2 + (n + 1)^2 = n^2 + n^2 + 2n + 1 = 2(n^2 + n) + 1$
$= 2 \times$ integer $+ 1 \rightarrow$ odd.
18 $LR = PR$, $QR = SR$, angle $LRQ =$ angle PRS
$(=$ angle $PRQ + 90°$); 2 sides and the included angle.
19 $x = \dfrac{a^2 + b^2}{a - b}$ 20 (a) 8π cm   (b) 96π cm²
21 (b) (i) $n = 8$ or $n = -1\frac{3}{5}$ (ii) 8

Examination practice paper (calculator)

1 1.381   2 (a) 69°   (b) 2.69 cm
3 (b) (i) 3°–4°   (ii) 26–30
4 (a) $4^3 - 7 = 57, 5^3 - 7 = 118$ (b) 4.45
5 (a) £9800   (b) £5172
6 12 mm   7 £445
8 Construct a line, XY, through A which is parallel to BC

$X\hat{A}B = b$ (alternate angles, XY || BC)
$Y\hat{A}C = c$ (alternate angles, XY || BC)
$X\hat{A}B + B\hat{A}C + Y\hat{A}C = 180°$ (angles on a straight line)
Therefore $a + b + c = 180°$
i.e the interior angles of $\triangle ABC$ add up to 180°.

9 (a) (i) $\frac{2}{5}$ (ii) $-\frac{3}{5}$ (b) $\dfrac{5x + 11}{(3x - 1)(x + 6)}$
10 1817 kg
11 (a) 127.3 ft   (b) 87°
12 (a) $(x - 7)(x + 2)$ (b) $x = 2, y = -1$
13 (a) base x^2, sides $4hx$ (b) $\dfrac{h = A - x^2}{4h}$
(c) 4 cm   (d) 5.66 cm
14 (a) (i)

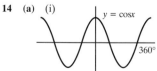

$y = \cos x$

(ii)

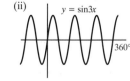

$y = \sin 3x$

(b) 73.9°, 106.1°
15 (a) 650 cm³   (b) 46 cm²   (c) 400 cm²
16 Greatest lower bound for length = 67.5 cm
Greatest lower bound for width = 8.35 cm
Greatest lower bound for area = 560 cm² (2 s.f.)
17 (4, 4), (−1, 3), (0, −2), (5, −1)
18 Line crosses circle when $x^2 + (ax + b)^2 = 64$
$$x^2(a^2 + 1) + 2abx + b^2 - 64 = 0$$
for equal roots $4a^2b^2 = 4(a^2 + 1)(b^2 - 64)$
$$a^2 + 1 = \frac{b^2}{64}$$

Index